The Complete Capuchin

Capuchin monkeys are one of the most widely distributed genera of primates in Central and South America. Capuchins captivate the imagination of scientists and the lay public alike with their creative and highly variable behavior, their grace and power in action, and their highly developed social character. In this, the first scholarly book devoted to the biology of the genus *Cebus* (Primates: Platyrrhine), we summarize the taxonomy, distribution, life history, ecology, anatomy, development, perception, cognition, motor skills, and social and sexual behavior of these monkeys. The book also describes how humans have viewed, used, and studied these monkeys from ancient times to the present. The authors' explicitly organismic and inclusive treatment provides a picture in unparalleled detail of the capuchin over its lifetime for all those with an interest in these fascinating animals.

DOROTHY M. FRAGASZY is Professor of Psychology and the Chair of the Neuroscience and Behavior Program at the University of Georgia, where she teaches development, developmental psychobiology, and comparative cognition. She has also co-edited *The Biology of Traditions* with Susan Perry (2003; 0 521 81597 5, hardback).

ELISABETTA VISALBERGHI is Research Director at the Istituto di Scienze e Tecnologie della Cognizione of the Consiglio Nazionale delle Ricerche in Rome. She studies comparative primate cognition and behavioral biology. Her research focuses on capuchin monkeys.

LINDA M. FEDIGAN is Professor of Anthropology and Canada Research Chair at the University of Calgary. She teaches primatology, physical anthropology, and gender studies. Her research interests include life histories of female monkeys, social relations between the sexes, and conservation of neotropical primates.

The Complete Capuchin

The Biology of the Genus *Cebus*

Dorothy M. Fragaszy
Elisabetta Visalberghi
Linda Marie Fedigan

CAMBRIDGE UNIVERSITY PRESS
Cambridge, New York, Melbourne, Madrid, Cape Town, Singapore, São Paulo, Delhi

Cambridge University Press
The Edinburgh Building, Cambridge CB2 8RU, UK

Published in the United States of America by Cambridge University Press, New York

www.cambridge.org
Information on this title: www.cambridge.org/9780521667685

First published 2004

A catalogue record for this publication is available from the British Library

ISBN 978-0-521-66116-4 hardback
ISBN 978-0-521-66768-5 paperback

Additional resources for this publication at www.cambridge.org/9780521667685

Transferred to digital printing 2009

For Rick, for everything – DF

For my parents Noela and Aldo, to Liz Bates an extraordinary friend, and to the capuchins Cammello, Toko, Carlotta, Roberta, Gal . . . They all taught me how to think. – EV

For John, helpmate and fellow naturalist – LF

Contents

Contents

Preface

Do you like jigsaw puzzles? We do. We like to look at the individual pieces, to appreciate the characteristics of each piece and the similarities and differences in contour and color among pieces; then to try different ways of aligning them with each other; ultimately to see emerging from our constructive efforts a coherent picture as we succeed in putting the puzzle together. This has been our approach in this book about capuchin monkeys. We have the opportunity to do this now because information about capuchins has accumulated rapidly in the past 20 years.

We believe that we are in a transition period, where capuchins are shifting from a little-known to a well-known genus, but the knowledge at present is scattered and therefore not readily accessible outside the small community of "capuchinologists." For many years, in fact, we have been hampered in our efforts to educate others about the genus with which we three work the most at present because there is no general reference volume about it. So, this book is for the many scientists and personnel working where these monkeys occur naturally, and for our students, who so often have asked "Is there a book about capuchins that would help me to learn something about them?". This innocent question has always been disconcerting for us. We became tired of replying, "Well, not really – but here is a collection of articles touching on some features of their biology that might interest you," and handing them a sheaf of papers the thickness of a substantial log. We finally responded to this pebble in our shoes – it seems to have grown to the size of a boulder – by taking on the task of writing the reference volume ourselves.

Our usual strategy in beginning to solve a puzzle is to begin with the easier parts, usually the boundaries or particularly distinctive sections. Once at least some of these are relatively defined, we begin to piece in more ambiguous elements. We have followed this general strategy in our exploration of the capuchin puzzle. We realize that we have described a puzzle that we can only partially complete at this time (despite the hubris of our title). There are still many pieces missing, and one can rearrange the pieces that we have provided to produce somewhat different pictures than we have. There are omissions, of course, most notably of physiological and disease-related topics, that reflect our limitations. Nevertheless, we hope that this volume will prove useful to students looking for a general picture of capuchins in historical, geographic, physical, social, ecological, experiential, and phylogenetic context. Our intent was to write a book that would be accessible to the person interested in learning about capuchin monkeys but who is not necessarily a specialist trained in biology, psychology, or anthropology. We also wanted the book to be useful to those with a professional interest in these monkeys – that is, our colleagues in research and education, those responsible for caring for capuchins in captivity, those who work to conserve and manage the monkeys and the forests where they occur in nature, and those who answer the public's questions about them. Thus we have covered as broad a spectrum of subjects as we felt knowledgeable to write about. We have, to the best of our ability, presented capuchin monkeys from an organismic perspective – that is, as whole individuals participating in groups, and groups in ecosystems, and as individuals with evolutionary, developmental, physical, and experiential facets to their existence. This perspective is the source of our fascination with these monkeys, and we hope to convey this fascination to our readers.

Unfortunately, deciding to write a book occurs on a vastly different time scale than accomplishing that task. In our case, we wanted the book to be a product of our joint efforts, not simply a collection of pieces written by three people. We wrote most of the book, literally, together, over the course of three furiously productive periods. We had the good fortune to spend these periods at a serene spot facing the sandy shores of Peconic Bay, on Long Island, in New York. The surroundings nurtured all manner of creative thought; long walks along shore and through wood allowed us to set these thoughts in order. For this good fortune we thank Russ and Carolyn McCall, who generously lent us the Red House, and the whole McCall family for their hospitality. We also visited with Stephen Nash, whose art work graces this book, and during these visits we were inspired by his incomparable collection of images of primates. We also thank Anthony Rylands for his invaluable collaboration on Chapter 1 (taxonomy), a subject which he knows deeply and we do not, and which we thought critical to cover thoroughly in this book.

Acknowledgements

A project of this magnitude reflects the collaborative efforts of many people, and this particular project is no exception. We have many individuals to thank for sharing the burdens of preparing this book. For providing photographs, we thank Jim Anderson, Sue Boinski, Brian Grafton, Kathy Jack, Charles Janson, Katie McKinnon, Russ Mittermeier, Briseida Resende, Giovanna Spinozzi and Peter Strick. For providing images of published figures, we thank Brendan McGonigle, Margaret Chalmers and Patrizia Potì. For the cover photo, we thank Margherita Stammati. For sharing unpublished data, including dissertations, we thank Mary Baker, Dave Bergeson, Sue Boinski, Sarah Brosnan, John Buckley, Sarah Cummins-Sebree, Mario Di Bitetti, Carolyn Hall, Patricia Izar, Katherine Jack, Katrina Landau, Katie MacKinnon, Lynne Miller, Betsy Mitchell, Carlos Nagle, Robert O'Malley, Melissa Panger, Susan Perry, Lisa Rose and Kym Snarr. For informing us about the collection of radiographs of capuchins followed longitudinally through development, we thank John Fleagle. For helping us to locate old prints, we thank Gustl Anzenberger. For commenting in conversation on various aspects of the work, we thank Gustl Anzenberger, Sue Boinski, Mario Di Bitetti, Andrea Hohman, Charles Janson, Susan Perry, John Robinson and Lisa Rose.

For reading and commenting on drafts of chapters, we thank Leah Adams-Curtis, John Addicott, Elsa Addessi, Jim Anderson, Pamela Asquith, Sue Boinski, Karly Branch, Sarah Carnegie, Monica Carosi, Colin Chapman, Marianne Christel, Sarah Cummins-Sebree, Thomas Defler, Maud Drapier, Erin Ehmke, Cam Fielding, Joan Fragaszy, Spartaco Gippoliti, Katherine Jack, Larry Jacobsen, Gram Jones (who merits a special thanks from DF for much help with the finer points of comparative morphology), Katie Leighty, Jessica Lynch, Katie MacKinnon, Carlotta Maggio, Rob O'Malley, Lauren McCall, Floriano Papi, Susan Perry, Patrizia Potì, Carrie Rosengart, Anthony Rylands, Giovanna Spinozzi, Karen Strier, Roger Thomas, Ari Vreemzaal, Anne Weaver and Toni Ziegler.

For helping with references, tables, figures, appendices and assisting with other editorial work, we thank Elsa Addessi, Gustl Anzenberger, Karly Branch, Sarah Carnegie, Sarah Cummins-Sebree, Amy Fuller, Alison Grand, Erica Hoy, Katie Leighty, Stefano Marta, Carlotta Maggio, John McCall, Deborah Nicol, Andy de Paoli, Carrie Rosengart, Louise Seagraves, Michelle Tremblay and Amy Venable.

The distribution maps in Chapter 1 were kindly drawn by Kimberly Meek, Publications Department, and the base maps were prepared by Mark Denil of the Conservation Mapping Program in the GIS and Mapping Laboratory, at the Center for Applied Biodiversity Science, Conservation International, Washington, DC.

We would like to acknowledge the use of the Primate Information Network (PIN), an information resource maintained jointly by the Washington and Wisconsin Regional Primate Research Centers and the Libraries of the University of Wisconsin, Madison, Wisconsin, USA. In particular, we made extensive use of the Primate Lit database maintained by the PIN to prepare our bibliography and to search the literature.

Many of the drawings in this book, including the color plates, were prepared by Stephen Nash, to whom we offer a very special, deep thanks for generous help and enduring enthusiasm.

DMF acknowledges financial support for research with capuchins from the National Science Foundation and the National Institutes of Health, USA, and the LSB Leakey Foundation. EV acknowledges financial support from the Consiglio Nazionale delle Ricerche, Italy. LMF acknowledges 20 years of continuous support for capuchin research from the Natural Sciences and Engineering Research Council of Canada (NSERC). She also thanks the Canada Research Chairs Program for financial support.

Finally, we thank our first editor at Cambridge University Press, Tracey Sanderson, for her moral support and good advice throughout this long adventure. She never lost faith that we would finish this project. Thanks also to Maria Murphy, who took over Tracey's role as we finished the book.

A colour version of the plates in this section are available for download from www.cambridge.org/9780521667685

Plate 1 The weepers. Engraved by Milne in William MacGillivray, *The Edinburgh Journal of Natural History and of the Physical Sciences*, (Volume 1, 1835–1839. Edinburgh). The species names listed are 1. *Cebus c. apella*, Common Weeper (bottom right); 2. *Cebus* var. *trepidus*, Fearful Weeper (top right); 3. *Cebus capucinus*, Capuchin (top left); 4. Var. *hypoleucus*, White throated (bottom right); 5. *Barbatus*, Bearded (center right); 6. *Fatuellus*, Horned (center left). The taxonomy of the genus *Cebus* has changed many times since MacGillivray's book was published (see Tables 1.1 and 1.2). Photo by Elisabetta Visalberghi.

Plate 2 Drawings by Stephen Nash of *Cebus capucinus* (top), *C. apella apella* (middle), and *C. xanthosternos* (bottom).

Cebus albifrons

Cebus olivaceus

Cebus kaapori

Plate 3 Drawings by Stephen Nash of *C. albifrons* (top), *C. olivaceus* (middle) and *C. kaapori* (bottom).

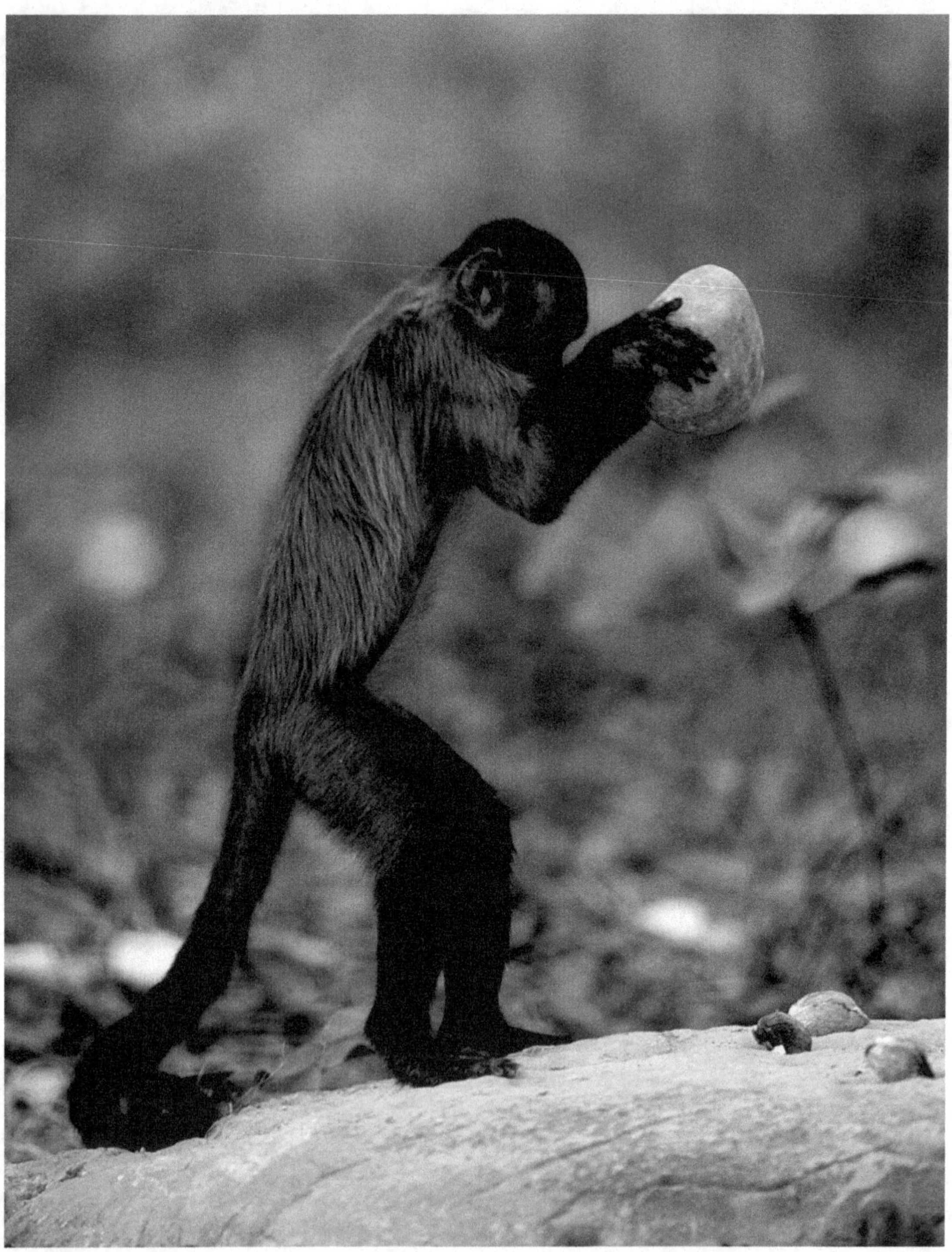

Color plate 4 A wild capuchin monkey uses a stone to crack open a very hard palm nut that it has placed on a stone anvil. This is the most complex form of tool use observed to date in wild capuchin monkeys. See Chapter 10. This monkey, according to the most recent taxonomy, is a *Cebus libidinosus* (see Table 1.1). Photo by Peter Oxford, and printed with permission.

Prologue
Capuchins in human worlds

CAPUCHINS IN HISTORY

Five hundred years ago two events occurred that, although unrelated at the time, have become linked in science, especially in primatology. First, the Europeans (re-)discovered the Americas and were astonished by the novel natural landscapes. Second, a group of friars belonging to the Franciscan Order wanted to return to a literal observance of the rule of St. Francis of Assisi and so split from the founding monastic order to become the autonomous branch named, to this day, the Order of Friars Minor Capuchin. These friars wore distinctive robes of rich brown, with generous cowls, or hoods, covering their heads. When European explorers encountered small monkeys whose crowns of brown hair resembled the cowls of capuchin monks, they named them capuchins! In the *Histoire Naturelles des Mammifères* (1824), the French Enlightenment scientists Geoffroy Saint-Hilaire and Cuvier described a specimen (possibly a *Cebus xanthosternos*) noticing that [translated from French] "his muzzle of a tanned color, . . . with the lighter color around his eyes that melts into the white at the front, his cheeks . . . , give him a look which involuntarily reminds us of the appearance that historically in our country represents ignorance, laziness, and sensuality" (i.e., the monks) (*Le saï a grosse tête mâle*, p. 1, 1824, Vol. 1). The conflicts between scientists and clerics at the time are clearly evident in this statement. Even today, there may be some initial confusion if one says "I study the capuchins in the forest"; people wonder why the friars are residing in the forest.

The chronicles of the Europeans traveling in the Americas in the fifteenth and sixteenth centuries often contain references to capuchin monkeys (Urbani 1999a). A few travelers (e.g., Galeotto Cey, citation in Urbani) reported that monkeys were actively traded to Europe and that some of them (possibly capuchins) threw stones and branches at humans. Several monkeys, most of them dead, reached scientists who started to describe them meticulously without knowing, in most cases, from where these animals originated. Others were relegated to the *ménageries* of noble, aristocratic people and became a status symbol (see Figure 0.1). For example, in Versailles the Marquise de Pompadour kept and successfully bred capuchins (Buffon 1770, Vol. 12).

When the first capuchin monkeys survived the trip to Europe and zoologists had the opportunity to observe them alive, they found them particularly attractive. Not only did these monkeys have four dexterous hands (!), an interesting semi-prehensile tail, unusually shaped genitals, a petulant and curious personality and bird-like vocalizations; they were also very friendly to humans. Geoffroy Saint-Hilaire and Cuvier (1824) write of a capuchin that (translated from French) "he was extremely sweet and so intelligent: his eyes were remarkably penetrating; he seemed to read in my eyes what was going on inside me . . ." (Vol. 1, *Le saï a gorge blanche mâle*, p. 2).

Capuchins are visitors' favorites in zoos and monkey parks (Vermeer 2000). What makes them so attractive? People who visit our laboratories (including those already familiar with other monkey species, such as macaques) are impressed by the deep involvement they feel in the presence of capuchins, and by how closely and intensely capuchins look at them. There are a number of good reasons for this. First, capuchins are generally friendly and tolerant of each other and toward humans. Second, unlike macaques, they do not perceive a stare (our wondering gaze) as a threat, because in affiliative contexts they often watch each other with a similar intensity. Third, during courtship they use some of the same behaviors we do (e.g., gaze into each other's eyes, tilt their heads, raise their forehead, and blink their eyes). So, what is friendly and touching for us is moving for them as well. In addition, because humans are easily prompted to imitate monkeys (but not vice versa),

Figure 0.1. Catharine of Aragon by Lucas Hombolte (1494–1544). After the rediscovery of the Americas by Europeans, young capuchins (as well as other New World primates, such as marmosets) were taken to Europe as pets for the noble class. In fact, several portraits show them as the status symbols of their fortunate owners, instead of more common pets, such as dogs.

visitors might try to copy the friendly behaviors of the capuchins and leave our laboratories with the feeling that a friendly exchange has occurred.

Thomas Belt was an English mining engineer and naturalist, whose book *The Naturalist in Nicaragua* (1874/1985) was considered by Charles Darwin as "the best of all natural history journals which have [sic] ever been published" (Darwin F. 1888, p. 188, letter of Charles Darwin to Sir J. D. Hooker). Belt was fascinated by the white-faced capuchin (*Cebus capucinus*, see Table 1.1 and color plate 2). He described its feeding habits in the forest while it "is incessantly on the look out for insects, examining the crevices in trees and withered leaves, seizing the largest beetles and munching them up with great relish" (p. 118), as well as its intelligent and mischievous behavior. This pet capuchin was an expert tool user who fished for ducklings using pieces of bread, retrieved things that were out of reach by means of sticks, and "could loosen any knot in a few minutes" (p. 118).

CAPUCHINS AS THEY ARE PERCEIVED IN THEIR NATIVE COUNTRIES

Ancient Egyptians considered the hamadryas baboon (*Papio hamadryas*) to be a god responsible for the art of writing, and often depicted them as dictating to a kneeling human scribe (Kummer 1995). Similarly, in their funerary ceramics ancient Mayans portrayed scribes that appear to have a mixture of characteristics belonging to humans and monkeys. In fact, the Mayans associated monkeys with writing, painting, music, and dance (Coe 1978). Given their current distribution, howlers (*Alouatta palliata* and *A. pigra*) and spider monkeys (*Ateles geoffroyi*) are thought to have been the monkeys represented in Mayan iconography. However, on the basis of morphological and behavioral comparisons among capuchin, howler, and spider monkeys, and on historic accounts, faunal analysis and linguistic evidence, Baker (1992) has argued that Mayans were inspired by capuchins in their depictions of monkey scribes.

The subsistence economy of traditional indigenous societies includes hunting. Especially for those people living so far from rivers that fishing is not an option, monkeys are an important source of protein (Hill and Padwe 2000, Sponsel 1997; see Figure 0.2). However, some indigenous groups, such as the Xavante community in Central Brazil studied by Leeuwenberg and Robinson (2000), do not hunt for primates. According to Vickers' (1984) study of eight indigenous and four mestizo communities in Amazonia, mammals account for most of the meat that is hunted (64–74% of the individuals hunted and 91–92% of the meat weight) and monkeys rank second (by individuals) and third (by weight). Larger monkeys (spider monkeys, woolly monkeys, howlers, and capuchins) are preferred to other primates (Peres 2000a). On the basis of 867 hunts by 13 Waorani hunters, Yost and Kelley (1983) report that among the nine species of monkeys hunted, *C. albifrons* was third in terms of number of individuals hunted and by weight. Similarly in Suriname, where most rural

Figure 0.2. Left, capuchins are killed for meat. An Amerindian hunter (of the indigenous group of the Tirio, at the border between Suriname and Brazil) holds a recently killed *Cebus apella*. (Photograph courtesy of Mark Plotkin.) Right, Guajá in the Carú Reserve in western Maranhão (Brazil) have pet monkeys. The capuchin in this picture belongs to the very rare subspecies *C. olivaceus kaapori*. (Photograph courtesy of Loretta Cormier.)

people eat monkeys, interviews as well as data on freshly killed animals show that *C. apella* and *C. olivaceus* are among the most popular (Mittermeier 1991). The capture of infants for pets is usually a by-product of hunting for meat, and 45% (38 out of 84) of the monkey pets are tufted capuchins (Mittermeier 1991).

A very interesting "ethnoprimatological" project was carried out by Cormier (2002, 2003) on the Guajá Amerindians in the Carú Reserve in western Maranhão (Brazil) to study the relations between humans and nonhuman primates. This reserve hosts eight species of primates: the Guajá (*Homo sapiens*), the red-handed howler monkey (*Alouatta belzebul*), the black-bearded saki (*Chiropotes satanas*), the brown capuchin (*C. apella*), the recently identified Ka'apor capuchin (*C. kaapori* [or *C. olivaceus kaapori*], see Table 1.2), the owl monkey (*Aotus infulatus*), the squirrel monkey (*Saimiri sciureus*) and the golden-handed tamarin (*Saguinus midas*). These monkeys serve as both food and pets for the Guajá Amerindians.

Guajá show a heavy seasonal reliance on monkeys; in the wet season, they account for 30% of the animals eaten. Orphaned monkeys whose mothers were hunted are brought back to the village and treated as human children by the women. The infant monkey is in constant physical contact with a Guajá woman, is breast-fed, played with, bathed, sung to, and even given previously masticated food to eat from the woman's mouth. The monkey is also given a personal name and a kinship term of address. In contrast, the Guajá are known to severely mistreat dogs, even puppies. Therefore a pet monkey acquires a nearly human status, and because in the Guajá culture a woman is sexually attractive especially when pregnant and/or lactating, it follows that the presence of a monkey (regardless of the species, and usually clinging to the woman's headdress) increases her "sex appeal" and her image of fertility. The presence in Cormier's study area of 90 monkey pets and 110 humans shows the importance of having one or more monkeys per woman. Cormier argues that the apparent incongruity of having monkeys as surrogate children and as preferred food should be considered in the light of the role of the symbolic endocannibalism (eating of kin) in the Guajá religion in which "Eating is not an act that merely satisfies hunger; it also has the transformational power to make another sacred" (Cormier 2002, p. 79). In the Guajá culture, monkeys, and especially howlers, are considered very close to humans and eating what is similar to you is a way of making it sacred. Therefore, eating is no less respectful to the monkeys than keeping them as pets (Figure 0.2).

Unfortunately, as they grow older, monkeys gradually become independent, sometimes aggressive, and certainly less suited to be "pet-babies". Juvenile monkeys (especially capuchins, given their destructive propensities, see Chapters 2 and 7) become a nuisance; consequently they are tied up and become more and more isolated from humans. This dramatic change in the way monkeys live has deleterious psychological consequences and they develop stereotyped and abnormal behaviors and many die. Hunting a monkey, having it as a pet, a child, a brother, or eating it, so that the monkey becomes sacred and lives a "paradisiacal existence" in the "sky-home", are all fundamental aspects of the life and culture of the Guajá Indians. For Westerners, the only way to understand the puzzling lifestyle of the Guajá is to take a multidisciplinary approach, as Cormier has done, and not to lose sight of any of its aspects.

At the present time, some capuchin groups live in parks of several South American cities and towns, and research has begun on their behavior in such urban environments (Amaral and De Yong 1999, Balestra and Bastos 1999). Monkeys are not hunted in urban parks and people generally have positive attitudes toward them. However, people do not know how to behave properly towards monkeys, interpret their behavior, or assess their needs. Instead they usually think of capuchins as animals that can be hand-fed any food to see what they will do with different items. For example, in Brasilia people go to swim or picnic in the *Parque Nacional de Brasilia*, a 430 km^2 ecological reserve established in 1969. In recent years, the capuchins in this park have come to rely more and more on human food and to be increasingly brazen in stealing food (Figure 0.3). Their population has grown tremendously. It is very likely that the capuchins will shortly be considered a nuisance or even a pest, and face a changed attitude on the part of the citizens of Brasilia. It is possible that this will lead to a decrease in provisioning of human food (preventing access to garbage cans by frequent emptying), an increase in social conflicts, higher mortality rates, and a decrease in the population size. We believe that urban monkeys such as these provide an excellent opportunity for studying, among other interesting questions, the effects of human disturbance on monkey populations and vice versa, and the ways in which both parties can learn to live together.

Many people believe that capuchins have surprising skills and succeed in doing things that other species cannot. In the Northeast, Central and Amazonian parts of Brazil, natives tell stories of capuchins raiding corn plantations and of how they collect and transport more than one ear at a time by tying them together around the neck (Carlos Tomaz, pers. comm.). Even the *castanheira* (*Bertholletia excelsa*, see Figure 2.4d), the tough husk (pyxidium) that encases 12–22 Brazil nuts, seems not to be a problem for the ingenious capuchin. Natives in Belem (Pará, Brazil) say that the monkey has a trick: it inserts a finger into the operculum (hole) and the husk . . . magically opens (Carlos Tomaz, pers. comm.). Note that this does not actually work! In Brazil, there is also another large husked fruit of the Lecythidaceae family in the forest that is called "the monkey cup" because local people report having seen capuchin monkeys drinking from it. In Caratinga (Minas Gerais, Brazil) the elderly report that when workers are making treacle sugar, a few capuchins threaten and distract them, while others steal the sugar; moreover, the young adults also remove the tiles from the roof of a granary and steal the corn (José Rimoli, pers. comm.). Therefore, it is not surprising that the capuchin is called the "*pivete da floresta*" (the rascal of the forest). In a beginners' Brazilian-Portuguese language class, we learned this idiom: *macaco velho não põe a mão em cumbuca* (an old capuchin does not insert its hand in a pumpkin). In other words, an experienced capuchin learns not to insert its hand through a small hole to retrieve something out of it, and by doing so prevents getting its fist stuck. Finally, in 1992, the *Skank*, a rock band from Belo Horizonte (Brazil) released a song with the title *Macaco prego*, the Brazilian name for

Figure 0.3. Left, capuchins living close to urbanized areas learn to exploit picnic leftovers. Here is a *Cebus libidinosus* in the Parque Agua Mineral Brasilia, Brazil. (Photograph by Elisabetta Visalberghi.) Right, capuchins are easy to train. This *C. albifrons* is an organ grinder monkey; the photograph was taken in Lima (Peru). (Photograph courtesy of Russell Mittermeier.)

C. apella (this name means "nail monkey" and alludes to the nail-shaped penis of the male). In this song a male capuchin advertises himself to a female with a swinging hair cut, and boasts of all his "merits" – including his determination, his Elvis Presley tufts, his free time and his lust for satisfaction.

Given their obvious skills at raiding crops and people's attitudes toward them, it is not surprising that as soon as capuchins and people compete for the same foods, capuchins become the enemy and are hunted in all possible ways. In Costa Rica, where *C. capucinus* are called *Monos muy bravos* (monkey very brash, obnoxious, tough, destructive; Fedigan, pers. obs.), Guaymi Amerindians are reported to prefer eating the meat of spider monkeys, but choose to shoot capuchins because they raid their crops (Gonzales-Kirchner and Sainz de la Maza 1998).

CAPUCHINS AS PETS

Capuchins were and still are commonly kept as pets in their native countries, as well as in the United States and Europe (Cormier 2002, 2003). In most cases Westerners own capuchins as a result of coming across them fortuitously in a pet shop or during a trip abroad where a young capuchin needed care (luckily, the international trade in primates has been curtailed significantly since the drafting of the Convention of Trade in Endangered Species of Wild Flora and Fauna (CITES) in 1973 and the imposition of export bans by producing countries; see Chapter 1). Barnard (1960, p. 1) at the beginning of a book on the adventurous life of a pet capuchin describes the typical emotional reaction of a young girl when she first saw the capuchin: "I have seen the saddest little monkey in that awful pet store where there are always dead birds in the cages and dead fish in the tanks. It must be a baby, because it's hardly got any teeth, but all it has to eat is an enormous dog-biscuit. And it's so cold! It's sitting there without even a blanket, shivering and trying to fold all its arms and legs inside each other and then wrap its tail round everything: Ma, what can we do? I'm sure we can do something." A typical rescue follows and the mother agrees to keep the tiny monkey at home. Unfortunately, monkeys grow and behave like monkeys and, sooner or later, despite all possible efforts keeping them at home becomes a nightmare. Some of them end up in a zoo. However, integrating human-reared monkeys into social groups is not always possible, and the longer the pet has been kept in a home the more difficult it is to "rehabilitate" it to live with other monkeys.

Until a few years ago, there was a sign at the Jersey Zoo by the exhibit of *Piccolo*, a capuchin, telling the story of the lonely *C. apella* who came from Brazil. "He was a pet of a sailor who sold him to a local restaurateur, who in turn gave him to the Zoo, way back in 1960. When he arrived here, he was already permanently crippled, having been kept in a small cage and fed on an inadequate diet. Despite expert veterinary care, his condition has not improved. He is kept on his own as he does not get on with other monkeys. He does like people, however, and has many friends who visit him regularly. Provided he is not in any pain, he will be able to live out his days here. Once he goes to the tropical rain forest in the sky, he will not be replaced". Visitors were also warned not to have a monkey as a pet. When *Piccolo* died at about 45 years of age, the Jersey Preservation Trust (now the Durrell Wildlife Conservation Trust) built a tomb in his memory to remind people of the sad lives of pet monkeys. Unfortunately, capuchin monkeys are still available for sale in many countries!

Sometimes curiosity and interest in the animal's behavior play an important role. As a young student in Vienna, Konrad Lorenz (who later won the Nobel prize for his ethological studies), could not resist the temptation of taking as a pet a capuchin that he named Gloria (Lorenz 1949). Gloria was left free to roam in his room while he was at home but was otherwise kept in a cage. One evening, on entering his room, Lorenz realized that something had happened: Gloria had escaped and left a trail of destruction. Among many other disasters, Gloria had dropped the lamp in the aquarium (which resulted in a short circuit and electrified its contents); then Gloria opened the latch of the book cabinet and dunked pages and pages of Lorenz's huge and heavy medical volumes into the aquarium. The sea anemones were all dead and paper was hanging on their tentacles!

Dressed in clothes and trained to perform tricks, play music, and hold out collection cups, capuchins often accompany organ-grinders on city streets in the New World (in the Old World, macaques were used for the same purposes) (Figure 0.3). Though not common nowadays, this can still be seen in both old and recent movies. In *The Cameraman* (1928, directed by Edward Sedgwick), Buster Keaton is helped by an organ-grinder monkey (who switches from asking for money to operating the camera) in filming the riots in the Chinese neighborhood. The duo survives all kinds of hilarious risks; unfortunately, at the very end of this adventure Keaton's character discovers that he forgot to load the camera with film, so that the whole event ends up being undocumented! Capuchins had roles in all kinds of movies from drama (e.g., *Citizen Kane*, 1941, directed by Orson Welles) to comedy, such as *Babes in Toyland* with Stanley Laurel and Oliver Hardy (1934, directed by Gus Meins).

Hollywood movie and television productions continue to portray capuchins as the "typical" monkey and sometimes place capuchins in countries where they do not belong, such as Egypt (*Raiders of the Lost Ark*, 1981 directed by Steven Spielberg). In *Outbreak* (1995, directed by Wolfgang Petersen), a capuchin (playing the part of an African monkey) is supposed to have spread a deadly Ebola-like virus from Africa to the United States. In *Monkey Trouble* (1995, directed by Franco Amurri), a capuchin is a good money-maker not only because it adorably holds out its hat but also because it is trained by a thief to be a pickpocket. And the capuchin Marcel (whose role is played by two females) was briefly a major attraction of the TV show *Friends*, a NBC Production that was first aired in the 1990s. A capuchin monkey participated as a cheerleader for a team that played in the 2002 World Series of baseball. Named Rally Monkey, it became so popular that fans in the stadium waved little stuffed toy monkeys at every opportunity to cheer on their team. The movie *Monkey Shines – An Experiment in Fear* (1988, directed by George A. Romero inspired by Stewart's book *Monkey Shines*, 1983) deals with a theme of interest to primatologists: the use of capuchin monkeys as aides for quadriplegic humans and the strong emotional relationship that develops between the monkey and the human.

CAPUCHINS AS AIDES FOR DISABLED HUMANS

In the 1970s, Willard, a behavioral psychologist in the United States, started the *Helping Hands* project which aimed at placing trained capuchin monkeys with quadriplegic humans to provide them with functional and psychological help (Willard *et al.* 1982, 1985). Their dexterous hands and high trainability make capuchins well suited to perform simple tasks that are impossible for a quadriplegic person. Though the procedure followed by *Helping Hands* has varied somewhat through time, it involves hand rearing by volunteer families. Then, when the capuchin is sufficiently mature, it is trained by positive reinforcement for about 1 year. Eventually, after careful scrutiny of the needs and dispositions of both the disabled person and the monkey, the latter is moved to its new home and trained to perform specific tasks to assist a particular individual. This initiative raised strong concerns among the public and scientific community, one of the reasons being that by preventing the capuchin from engaging in normal social interactions, its psychological well-being would be compromised. Nevertheless, this project was followed by similar ones in Israel (founded in 1984 with the assistance of the American project), France, and Belgium (*Programme d'Aide Simienne aux Personnes Tétraplégiques*, Deputte and Busnel 1997).

One aim of the latter project was to investigate the learning processes involved in training and in human–monkey communication, so as to evaluate the effectiveness of the training procedures (Deputte *et al.* 1995, Deshaies *et al.* 1996, Hervé and Deputte 1993, Lejeune *et al.* 1998). A second aim of the project, because there was a lack of information about the costs (e.g., time, money, number of monkeys, etc.), was to assess the success of the placement program both from the disabled person's point of view and in relation to the psychological consequences for the monkeys throughout their lives. At the moment, we know from a few reports that the presence of a capuchin has a positive influence on the social environment of the disabled (Hien and Deputte 1997; see also Willard *et al.* 1985) and that some of the matches between the disabled and the monkey succeed whereas other capuchins went back to zoos or similar facilities. In Israel the project was terminated for several reasons: (a) only a small number of disabled persons fit the criteria as candidates or were interested in having a monkey; (b) as the project progressed it was realized that standards for the quality of life of the capuchins would be difficult to meet in a home; and (c) the program operated on a voluntary basis, making its activities very difficult to sustain (Tamar Fredman, pers. comm.). After termination of the project, the monkeys were found a suitable home and some have since participated in experimental behavioral research (Custance *et al.* 1999).

CAPUCHINS IN BIOMEDICAL RESEARCH

In addition to habitat destruction and hunting, the demand for primates as laboratory animals has added further pressure on neotropical species (Mittermeier *et al.* 1994). Capuchins (especially *C. apella*) are among the neotropical primate species commonly used for biomedical research, although to a lesser extent than squirrel monkeys or owl monkeys.

Capuchins have been extensively used in research in immunology (de Palermo *et al.* 1988), in reproductive biology (Nagle *et al.* 1989, 1994), pharmacological and metabolic studies (Bergeron *et al.* 1992), parasitology (Garcez *et al.* 1997, Pereira *et al.* 1993), physiology (Terpstra *et al.* 1991), neuroscience (Bortoff and Strick 1993, Leichnetz and Gonzalo Ruiz 1996, Yamada *et al.* 1996), and neurotoxicity (Lifshitz *et al.* 1991, 1997, O'Keeffe and Lifshitz 1989). For example, evidence for a local mechanism controlling the alternating ovulatory performance of the ovaries in primates has come from capuchin monkeys (Nagle *et al.* 1994). They have also participated extensively in studies investigating the effects of antipsychotic medications, as they can develop some of the extrapyramidal side effects (e.g., tardive dyskinesia and dystonias) seen in schizophrenic patients (Casey 1984, Linn *et al.* 2001). Capuchins are also popular in behavioral pharmacology research as models for schizophrenia (Linn *et al.* 1999, Linn and Javitt 2001) or cognitive deficits associated with aging and Alzheimer's disease (Bartus *et al.* 1980, 1983, Bartus *et al.* 1982, Bartus and Dean 1988).

CAPUCHINS IN BEHAVIORAL RESEARCH

Capuchins were the object of scientific behavioral investigation up to the 1940s and again from the 1970s until now. It is unclear why there were 30 years of relative disinterest. Cruz Lima (1945) wrote that "*Cebus* are the most common and most generally known neotropical monkeys, although from a scientific point of view the opposite is the case" (p. 133–134). However, it is possible that the life history of a very influential psychologist of this century, Harry Harlow, had something to do with switching the interest of comparative psychologists toward another species. Few people know that Harlow's contributions to the field of primatology include a film, an unpublished manuscript, a one-page essay, and a few lines in various articles on primate learning that testify to his interest and experimental work on capuchin monkeys (Cooper and Harlow 1961, Harlow 1951, Harlow and Settlage 1934, Harlow unpubl.). In particular, Harlow carried out a few experiments on tool use in capuchin monkeys, including the combined use of a climbing pole and a box to reach a suspended reward. In this task, the monkey's behavior following the initial success was "short of perfection" (1951, p. 219; see Figure 10.3). The only detailed description Harlow gave is of a capuchin monkey spontaneously using a stick to strike a conspecific and, on another occasion, some rhesus monkeys (Cooper and Harlow 1961). This anecdote was reported in a scientific journal 25 years after observation because the authors were reluctant to report "this unusual observation until they had achieved established reputations" (Cooper and Harlow 1961, p. 418). Their reluctance was probably due to the general conviction at that time that capuchins are not as intelligent as Old World monkeys (Yerkes 1916; cf. Chapter 7 and 9).

Harlow gave up his research on capuchins following a tuberculosis epidemic in his colony, which killed all his specimens.[1] Harlow decided to concentrate all future efforts solely on the more convenient rhesus monkeys. Therefore, this practical consideration strongly selected for the use of macaques in comparative psychology and subsequently in biomedical research. Despite the fact that Harlow was aware that "the rhesus monkey lacks the gay abandon of *Cebus*" and that the difference between the two species "is all the difference between a Southern belle and a New England store keeper" (Harlow unpubl., p. 15), the rhesus became the classic laboratory monkey worldwide.

For many years capuchins seemed forgotten by behavioral scientists for no obvious reason. It is only in the last two decades that capuchins have been the object of renewed interest. Visalberghi (1997) documented the recent surge of studies on the behavior of capuchin monkeys. Through a computer search on the bibliographic database of the Primate Information Center of the Washington Regional Primate Research Center (USA) using *Cebus* and behavior and *Cebus* and cognition as subject entries, she found that the number of articles published in a recent 5-year block (1991–1995) had increased 10 times in comparison with an earlier 5-year block (1976–1980). A similar increase in the number of studies between these two time blocks was also found for the genus *Pan*, the chimpanzees and bonobos.

CONSERVATION STATUS AND THE WELL-BEING OF CAPTIVE CAPUCHINS

In the past, capuchins were extremely popular in zoos: their social attitudes and manipulative skills were much appreciated by visitors. However, their destructive behavior, their cleverness in escaping and the fact that they are generally not threatened in their natural habitats were all contributing factors to their decline in popularity in zoo exhibits. Recently, however, an international breeding program, coordinated by an International Recovery and Management Committee established by the Brazilian government, was set up for *C. xanthosternos* and *C. apella robustus* (recently re-classified as *C. nigritus robustus*, see Table 1.2). These two capuchins are now considered to be "Critically Endangered" and "Vulnerable", respectively (Hilton-Taylor 2002), within their habitat, Brazil's Atlantic Forest, now among the "hotspots" for conservation priorities (Myers *et al.* 2000; see Chapter 1 and Table 1.4 for further information).

Visalberghi and Anderson (1999) provide specific information on the husbandry, and physical and social environments suited for captive capuchins (see also the more general advice of the Committee on Well-Being of Nonhuman Primates 1998 and Appendix IV). In particular, they stress the capuchins' need for social companionship despite the risks involved in group formation and in the integration of individuals into groups (Cooper *et al.* 2001, Fragaszy *et al.* 1994a; see also Visalberghi and Anderson 1993). Recently, Dettmer and Fragaszy (2000) reported that for pair-living individuals, gaining access to the social companion took precedence over gaining access to food even when the monkeys were very hungry. The presence of objects may allow capuchins to compensate for reduced opportunities for social play; Visalberghi and Guidi (1998) showed that young capuchins lacking same-age peers to play with increase their play with objects.

Given their manipulative propensities, capuchins benefit from access to objects they can handle and play with (blocks of wood, tennis balls and sturdy toys are excellent for their destructive play) as well as devices that encourage foraging activity (see also Anderson and Visalberghi 1991). While captive capuchins respond to the introduction of novel objects, it is nevertheless important to maintain or renew their interest by alternating frequently the types of objects provided. The provision of litter (5–10 cm deep) on the floor, such as ground corn cobs, woodchips, wood wool or peat, encourages locomotion and foraging (Ludes and Anderson 1996). Wood wool (long shavings of wood) and peat were more effective as enrichment; wood wool promoted the most play, while peat elicited communal peat-bathing and enhanced social contacts (Ludes-Fraulob and Anderson 1999).

Boinski *et al.* (1999a) used plasma and fecal cortisol measures and behavioral data to assess the effectiveness of four forms of enrichments for singly housed *C. apella*. The study shows significant reductions in abnormal and undesirable behaviors and an increase in normal behavior as the level of enrichment increases (control condition, two plastic toys, foraging box, toys plus foraging box). Though they did not differ significantly across conditions, the plasma cortisol measures decreased as the proportion of normal behavioral measures increased. Boinski, Gross and Davis (1999b) also found that the frequency of alarm vocalizations (usually elicited by terrestrial predators in wild monkeys and common in other stressful situations in captivity) are a good indicator of the effectiveness of the enrichment. Alarm vocalizations varied significantly across conditions, being significantly less frequent in the enriched conditions than in the control condition. Moreover, alarm vocalizations were positively related (though not significantly) to the plasma cortisol values. Therefore, at least for singly housed monkeys, measuring the frequency of alarm vocalizations might be an easy way to evaluate monkeys' current psychological well-being.

Undoubtedly, capuchins are a very special and important primate genus. They are widely popular as pets and in exhibits, easy to keep in good health, active, highly trainable, interesting to zoologists in their own habitat, and kept by medical and behavioral scientists for reasons of expedience as well as for their special characteristics. In the following chapters the reader will find other reasons why they are so special for us and for scientific study.

ENDNOTE

1 Capuchins are less susceptible to tuberculosis than rhesus monkeys when kept in good conditions and provided with an adequate diet.

Part I
Capuchins in nature

1 • Taxonomy, distribution and conservation: where and what are they and how did they get there?

When I arrived in Venezuela to study wedge-capped capuchins, I had never seen one. However, I had read all that I could about them and I expected to see monkeys in shades of brown, with dark brown caps on the crown of their heads and individually distinctive faces. When I first saw them, high overhead in the dim light of early morning, they looked more like gray than brown monkeys, but I could see their distinctive tails, watch their distinctive patterns of movement, and hear their distinctive calls. These were definitely capuchin monkeys! Later, when the sun had fully risen, I was surprised to see that the main study group that I would be following for the next several months contained monkeys of several colors, including one so pale blonde that its cap, its darkest part, was the color of honey, and another so uniformly dark that I could not even notice a cap. The entire spectrum of coat colors was present in this one group. I wondered then at the specificity with which "the wedge-capped capuchin" was described so precisely in the books I had read. Some of these monkeys did not look at all like the published descriptions! At that moment I realized the difficulty facing the naturalist trying to identify species by something distinctive in their appearance. What should one do when appearance is so variable within the very same group, let alone over a geographic region? These monkeys could not belong to any other species of capuchins than *Cebus olivaceus*, given where I was and the accepted taxonomy of the genus, but it was not at all clear to me that I could identify any particular characteristic of all these monkeys that supported that classification. Only much later did I learn that taxonomists do indeed have more than the normal degree of difficulty in sorting out the varieties of capuchin monkeys.

Capuchin monkeys are medium-sized, robustly built monkeys that exhibit moderate sexual dimorphism, and have arms and legs of nearly equal length, large brain-to-body ratios and furry semi-prehensile tails. They weigh about 2.5–5 kg and are named for the distinctive caps on their crowns that appear in various colors and shapes in different species (see color plates 1–3; also well illustrated in Napier and Napier 1967). Arboreal quadrupeds that prefer the middle canopy, capuchin monkeys sometimes come to the ground to forage or cross gaps in the forest. Their tails lack the naked distal portion and precise grasping ability found in atelids (woolly monkeys, spider monkeys, howling monkeys and muriquis), but they are prehensile, and capuchins use them during foraging to brace their bodies and to facilitate suspension in an inverted posture. Any lack of precision in tail use is more than compensated for by their strong and highly dexterous hands. They also have robust mandibles, large canines, and thickly enameled molars that are well adapted to crushing seeds and tearing open hard fruits and tough substrates. The manipulative abilities and behavioral inclinations of capuchins have no doubt facilitated their wide distribution through Central and South America as probably the most omnivorous of the neotropical primates, renowned for their ability to adjust facultatively the proportions of plant and animal matter in their diet according to local and seasonal circumstances.

TAXONOMY AND DISTRIBUTION

As stated by Hill (1960, p. 405): "the classification of the numerous forms of the present genus [*Cebus*] is one of the most vexatious problems in Primate taxonomy and has been so since early times." Lönnberg (1939, pp. 1–2) summarized the complications: "The prevailing confusion has thus several different causes, firstly the often insufficient diagnoses, which have been misunderstood, secondly the great variability of the animals, thirdly the often scanty material on which the original descriptions have been based and fourthly the often prevailing uncertainty about the geographical origin of the material described." Even within the same geographical area, and within the same group, there is great variation among individuals; moreover, sex-specific developmental changes occur throughout their lifetime. Torres (1988) noted that numerous type specimens represent just one of various phenotypes found in the type locality (the geographical place of capture or collection). If the phenotypic characters used to identify them were valid for distinguishing subspecies, there would be two or three at each locality! To add further confusion, recognized forms have been given different names by different authors. Finally, although genetic investigations have been initiated, they have not yet provided a satisfactory understanding of the variability that underlies the phenotypic diversity in the genus. A lively historical account of the taxonomic changes, discoveries and scientific debates up through the 1950s was provided by Hill (1960).

Over the last decades, the genus *Cebus* has been considered to include four species. Following Elliot (1913), Hershkovitz (1949) divided them into two groups: the "tufted group" (for Hershkovitz consisting of just one species, *Cebus apella* (Linnaeus, 1766)); and the "untufted group" (including *Cebus capucinus* (Linnaeus, 1758)), *Cebus albifrons* (Humboldt, 1812), and *Cebus nigrivittatus* (Wagner, 1848)). Although Hershkovitz (1949), and many others through the 1970s, used the name *nigrivittatus* for the Guianan weeper capuchin, in 1941 Von Pusch combined the genera *Saimiri* and *Cebus*. Then *nigrivittatus* Wagner, 1848 had to be considered a homonym (two species of the same genus having the same name), due to the prior description in 1846 of the squirrel monkey *Chrysothrix nigrivittatus* Wagner. Although a temporary situation, this invalidated the use of the name *nigrivittatus*, and the correct one, therefore, became that given by Schomburgk in 1848: *Cebus olivaceus* (see Groves 2001, Rylands 1999). The untufted group lacks tufts on the top of the head; conversely, the tufted capuchins have them, though their size, shape and color can vary considerably with age, sex and geographic origin.

A revision by Groves (2001) proposed the elevation of three subspecies of *C. apella* to species status, and reduced a number of subspecies in the untufted group listed by Hershkovitz (1949) to synonyms (names given to forms believed to be an already named taxon) (see Table 1.2). His departure from the more traditional view of just one tufted capuchin species was supported by the work of Torres de Assumpção (1983, Torres 1988). She identified five phenotypic groupings but did not apply names to them. In 1992, Queiroz described a fifth species of untufted capuchin, *Cebus kaapori*, from the eastern Amazon. The Ka'apor capuchin was maintained as a valid species by Groves (2001) but listed as a subspecies of *C. olivaceus* by Rylands *et al.* (2000), following the genetic analysis of Harada and Ferrari (1996). Although the taxonomy of the genus still awaits exhaustive molecular analysis to better resolve the phenotypic variability, the taxonomic arrangement proposed by Groves (2001), while undoubtedly not definitive, represents an important starting point. In the untufted "*capucinus*" group he placed *C. capucinus*, *C. albifrons*, *C. olivaceus*, and *C. kaapori*, and in the tufted "*apella*" group, *C. apella*, *C. libidinosus* Spix, 1823, *C. nigritus* Goldfuss, 1809 and *C. xanthosternos* Wied, 1820. Table 1.1 presents the eight species proposed by Groves (2001), and a description of their appearances and geographical distributions, based on a variety of published sources, as well as our personal knowledge and personal communications from experts in the relevant countries.

Rylands *et al.* (2000) continued to follow Hershkovitz (1949) in listing 11 (poorly defined) subspecies of *Cebus albifrons*, but Groves (2001) reduced the number to just six. Rylands *et al.* (2000) did not list *C. albifrons unicolor* as a result of a recent study by Defler and Hernández-Camacho (2002), which concluded that it is a junior synonym (the later established of two names for the same taxon) of *C. albifrons albifrons*. Hershkovitz (1949) mapped the type localities, but not the distributions, of four subspecies of *C. olivaceus* (referred to by him as *C. nigrivittatus*, see above), but Groves (2001) placed them all as junior synonyms of *C. olivaceus*. Likewise, Groves (2001) did not recognize the three subspecies of *C. capucinus* given by Hershkovitz (1949) and listed by Rylands *et al.* (2000). Table 1.2 lists the subspecies according to the most updated information (Groves 2001, Rylands *et al.* 2001).

The most exhaustive and thorough revision of the capuchin monkeys to date, building on that of Torres (1988) but as yet unpublished, was carried out by Silva Jr. (2001). While not recognizing any subspecies, he separated the untufted and tufted species into subgenera: *Cebus* Erxleben, 1777, and *Sapajus* Kerr, 1792,

Table 1.1. *Description of eight species of* Cebus, *taxonomy follows Groves (2001)*[a]. *A key to the traditional four species of the genus,* C. apella, C. albifrons, C. olivaceus (=griseus) and C. capucinus, *is provided by Hill (1960).*

Species	Description
C. capucinus White-faced or white-throated capuchin	This is the only species with jet black fur on body, limbs and tail. The black coat extends up over the back of the head forming a black cap. White fur occurs frontally and laterally all around the black cap, as well as on the throat, shoulders, chest and upper arms. The facial skin is usually pink and the amount of white fur on the face is variable across age/sex classes.
C. albifrons White-fronted capuchin	A brown color replaces the black color of *C. capucinus* and the hands, feet and distal parts of the tail are lighter (sometimes red-yellow or reddish). A whitish-beige color replaces the clear white of *C. capucinus*. The brown cap extends forward, and is rounded or slightly pointed in front and well demarcated from the lighter forehead. A frontal superciliary brush (furry eyebrows) can be present.
C. olivaceus Wedged-capped or weeper capuchin	The general appearance of this species is light brown to brown, with gray or yellow tinges on different parts of the body. The back is uniformly light brown, the tail, hands, feet and limbs are darker brown, and the upper arms lighter than the lower arms. The face and forehead are light-gray brown and a dark brown wedge-shaped cap is present. The facial skin is whitish to grayish pink.
C. kaapori Ka'apor capuchin	The face, shoulders, mantle and tail tip are silvery grey, and the hands and feet are black or dark brown. The crown has a Y-shaped black cap extending toward the nose.
C. apella Tufted or brown capuchin[b]	The crown is black, contrasting with the body, which is gray-fawn to dark brown. The face is light gray-brown; the sideburns are thick and black. The lower limbs and tail are also black. The upper arms are fawn and the underside is yellowish or red. Variably well-developed dorsal stripe; tufts on crown.
C. libidinosus Black-striped or bearded capuchin.	Formerly considered a subspecies of *C. apella*, but elevated to species status by Groves (2001). It has a yellow to white head, with dark sideburns. The body is light. The underside and shoulders are yellowish or reddish. A marked black dorsal stripe is present, with limbs mainly dark to blackish. Upper arms not lighter than body. Tail is much longer than head and body.
C. xanthosternos Yellow-breasted, buff-headed, or golden-bellied capuchin	Formerly considered a subspecies of *C. apella*, but recognized as a species by Mittermeier *et al.* (1988), Coimbra-Filho *et al.* (1991/1992), Groves (2001), Silva Jr. (2001) and others. It has a distinctive yellow to golden red chest, belly and upper arms. Its facial skin is fawn, its cap is black or light brown and its tufts are not very evident and oriented towards the rear of the skull. A band of short hair around the upper part of the face, colored salt and pepper, contrasts with darker fur of the sideburns and beard. Its limbs and tail are black.
C. nigritus Black or black horned capuchin	Formerly considered a subspecies of *C. apella*, but elevated to species by Groves (2001) and Silva Jr. (2001). It has a very dark brown or gray, even blackish body with no (or very vague) dorsal stripe. Its limbs are darker than its body and its underside is deep reddish with black overlay. The face is white and contrasts with the rest of the body. The cap is dark and tufts are evident; tufts can be erected or directed sideways or ahead. Silva Jr. (2001) recognized *C. robustus*, given as a subspecies of *C. nigritus* by Groves (2001).

[a] Table 1.2 presents other taxonomic divisions of the genus *Cebus*.

[b] As a common name, tufted capuchin is loosely used for all members of the tufted capuchin group, but most frequently refers to Amazonian *C. apella*.

Table 1.2. *Species and subspecies of capuchin monkeys,* Cebus, *according to the three most recent taxonomic revisions – Rylands* et al. *(2000), Groves (2001) and Silva, Jr. (2001). The shaded entries indicate the species recognized by each of the authors.*

Rylands *et al.* (2000)	Groves (2001)	Silva, Jr. (2001)
C. capucinus capucinus (Linnaeus, 1758)	*C. capucinus* (Linnaeus, 1758)	*C. (Cebus) capucinus* (Linnaeus, 1758)
C. capucinus limitaneus Hollister, 1914	Synonym of *C. capucinus* (Linnaeus, 1758)	Synonym of *C. capucinus*
C. capucinus imitator Thomas, 1903	Synonym of *C. capucinus* (Linnaeus, 1758)	Synonym of *C. capucinus*
C. capucinus curtus Bangs, 1905	Synonym of *C. capucinus* (Linnaeus, 1758)	Synonym of *C. capucinus*
C. albifrons albifrons (Humboldt, 1812)	*C. albifrons albifrons* (Humboldt, 1812)	*C. (Cebus) albifrons* (Humboldt, 1812)
C. albifrons adustus Hershkovitz, 1949	Synonym of *C. albifrons versicolor*	Synonym of *C. albifrons*
C. albifrons aequatorialis Allen, 1914	*C. albifrons aequatorialis* Allen, 1914	Synonym of *C. albifrons*
C. albifrons cesarae Hershkovitz, 1949	Synonym of *C. albifrons versicolor*	Synonym of *C. albifrons*
C. albifrons cuscinus Thomas, 1901	*C. albifrons cuscinus* Thomas, 1901	Synonym of *C. albifrons*
C. albifrons malitiosus Elliot, 1909	Synonym of *C. albifrons versicolor*	Synonym of *C. albifrons*
C. albifrons trinitatis Von Pusch, 1941	*C. albifrons trinitatis* Von Pusch, 1941	Synonym of *C. albifrons*
C. albifrons versicolor Pucheran, 1845	*C. albifrons versicolor* Pucheran, 1845	Synonym of *C. albifrons*
C. albifrons yuracus Hershkovitz, 1949	Synonym of *C. albifrons cuscinus*	Synonym of *C. albifrons*
C. albifrons leucocephalus Gray, 1865	Synonym of *C. albifrons versicolor*	Synonym of *C. albifrons*
Synonym of *C. albifrons albifrons*	*C. albifrons unicolor* Spix, 1823	Synonym of *C. albifrons*
C. olivaceus olivaceus Schomburgk, 1848	*C. olivaceus* Schomburgk, 1848	*C. (Cebus) olivaceus* Schomburgk, 1848
C. olivaceus apiculatus Hershkovitz, 1949	Synonym of *C. olivaceus*	Synonym of *C. olivaceus*
C. olivaceus brunneus Allen, 1914	Synonym of *C. olivaceus*	Synonym of *C. olivaceus*
C. olivaceus castaneus I. Geoffroy, 1851	Synonym of *C. olivaceus*	Synonym of *C. olivaceus*
C. olivaceus kaapori Queiroz, 1992	*C. kaapori* Queiroz, 1992	*C. (Cebus) kaapori* Queiroz, 1992
C. apella apella (Linnaeus, 1758)	*C. apella apella* (Linnaeus, 1758)	*C. (Sapajus) apella* (Linnaeus, 1758)
C. apella fatuellus (Linnaeus, 1766)	*C. apella fatuellus* (Linnaeus, 1766)	Synonym of *C. apella*
C. apella macrocephalus Spix, 1823	*C. apella macrocephalus* Spix, 1823	*C. (Sapajus) macrocephalus* Spix, 1823
C. apella peruanus Thomas, 1901	*C. apella peruanus* Thomas, 1901	Synonym of *C. macrocephalus*
C. apella tocantinus Lönnberg, 1939	*C. apella tocantinus* Lönnberg, 1939	Synonym of *C. apella*
C. apella margaritae Hollister, 1914	*C. apella margaritae* Hollister, 1914	Synonym of *C. apella* (provisional)
C. libidinosus libidinosus Spix, 1823	*C. libidinosus libidinosus* Spix, 1823	*C. (Sapajus) libidinosus* Spix, 1823
C. libidinosus pallidus (Gray, 1866)	*C. libidinosus pallidus* (Gray, 1866)	Synonym of *C. cay* (provisional)
		C. (Sapajus) cay Illiger, 1815
C. libidinosus paraguayanus Fischer, 1829	*C. libidinosus paraguayanus* Fischer, 1829	Synonym of *C. cay*
C. libidinosus juruanus Lönnberg, 1939	*C. libidinosus juruanus* Lönnberg, 1939	Synonym of *C. macrocephalus*
C. nigritus nigritus (Goldfuss, 1809)	*C. nigritus nigritus* (Goldfuss, 1809)	*C. (Sapajus) nigritus* (Goldfuss, 1809)
C. nigritus robustus Kuhl, 1820	*C. nigritus robustus* Kuhl, 1820	*C. (Sapajus) robustus* Kuhl, 1820
C. nigritus cucullatus Spix, 1823	*C. nigritus cucullatus* Spix, 1823	Synonym of *C. nigritus*
C. xanthosternos Wied-Neuwied, 1826	*C. xanthosternos* Wied-Neuwied, 1826	*C. (Sapajus) xanthosternos* Wied, 1820

Figure 1.1. The distribution of the white-faced or white-throated capuchin monkey, *Cebus capucinus*, in extreme northwestern Ecuador, western Colombia, Panama, Costa Rica, Nicaragua, and Honduras. There are also populations on the Island of Gorgona, off the Pacific coast of Colombia (*C. c. curtus*), and the Islands of Coiba and adjacent Jicarón, Panama (*C. c. imitator*). Despite reports of its occurrence in Belize, its presence there is uncertain. Sources: Hernández-Camacho and Cooper (1976), Hershkovitz (1949), Marineros and Gallegos (1998), Reid (1997), Rodríguez-Luna *et al.* (1996). (Drawing by Kimberly Meek.)

respectively. His untufted species are the same as those listed by Groves (2001). For the tufted capuchins, he recognized seven species (Table 1.2). While Groves (2001) identified six taxa in Amazonia, Silva Jr. recognized only two, consistent with the findings of Torres (1988): *C. apella* of the Guianas and eastern and central Amazon, east of the Rio Madeira, and *C. macrocephalus* Spix, 1823, of the upper Amazon, west of the Rios Negro and Madeira. He recognized the form *robustus* Kuhl, 1820, as a full species (Groves listed it as a subspecies of neighboring *nigritus*), and the only nomenclatural divergence from Groves is in his use of the name *Cebus cay* Illiger, 1815 (*paraguayanus* Fischer, 1829 as a junior synonym) as the form in the Mato Grosso, Paraguay, southern central Bolivia, and northwest Argentina.

We must warn the reader that our descriptions of the physical appearances of the species in Table 1.1 are generalizations and far from being conclusive or adequately accounting for the marked variation within species, within sexes and across ages (see, for example, Silva Jr. 2001, Torres 1988, Torres de Assumpção 1983). Further, although we will provide verbal descriptions in the text of the hypothetical geographic distribution of Groves' (2001) proposed species, their exact boundaries have yet to be carefully documented, and Figures 1.1–1.6 present only our best hypothesis for each (Table 1.1); neither the taxonomy nor the ranges are definitive. Finally, almost all of the published literature on capuchins refers to the four traditional species (*C. apella*, *C. capucinus*, *C. albifrons* and *C. olivaceus* (*nigrivittatus*)) and thus, in order to be consistent with the original authors of the research, the subsequent chapters of our book will, with a few exceptions, refer only to these four species. Therefore *C. apella*

Figure 1.2. The geographic distribution of the white-fronted capuchin, *Cebus albifrons*, in Venezuela, Colombia, Ecuador, Amazonian Peru and the Tumbes region in northwest Peru, the upper Amazon basin of Brazil, northern Bolivia, and Trinidad. This includes the ranges of five subspecies recognized by Groves (2001): *C. a. albifrons* (Amazonia), *C. a. versicolor* (northern Colombia), *C. a. aequatorialis* (western Ecuador and extreme northwest Peru), *C. a. cuscinus* (southern Peru), and *C. a. trinitatis* (Island of Trinidad). Sources: Bodini and Pérez-Hernández (1987), Encarnación and Cook (1998), Hernández-Camacho and Cooper (1976), Hershkovitz (1949), Hill (1960), T. R. Defler (pers. comm.). (Drawing by Kimberly Meek.)

in this usage refers to the tufted capuchin indiscriminately.

Along with the howling monkeys (*Alouatta*), capuchins have the widest distribution of any New World primate genus, extending from Honduras in Central America (Marineros and Gallegos 1998), through Venezuela, Colombia, Ecuador, Peru, Bolivia, the Guianas and Brazil, to southern Paraguay (Stallings 1985) and northern Argentina (Brown 1986, 1989). This is in large part due to the widespread and phenotypically variable tufted capuchin group which is found across much of northern South America. In Amazonia, the tufted capuchins are sympatric with the South American untufted capuchins, *C. albifrons*, *C. olivaceus*, and *C. kaapori*.

Cebus capucinus (Linnaeus, 1758) White-faced or white-throated capuchin

C. capucinus is the only capuchin monkey in Central America, occurring from Honduras in the north, through Nicaragua, Costa Rica, and Panama, while

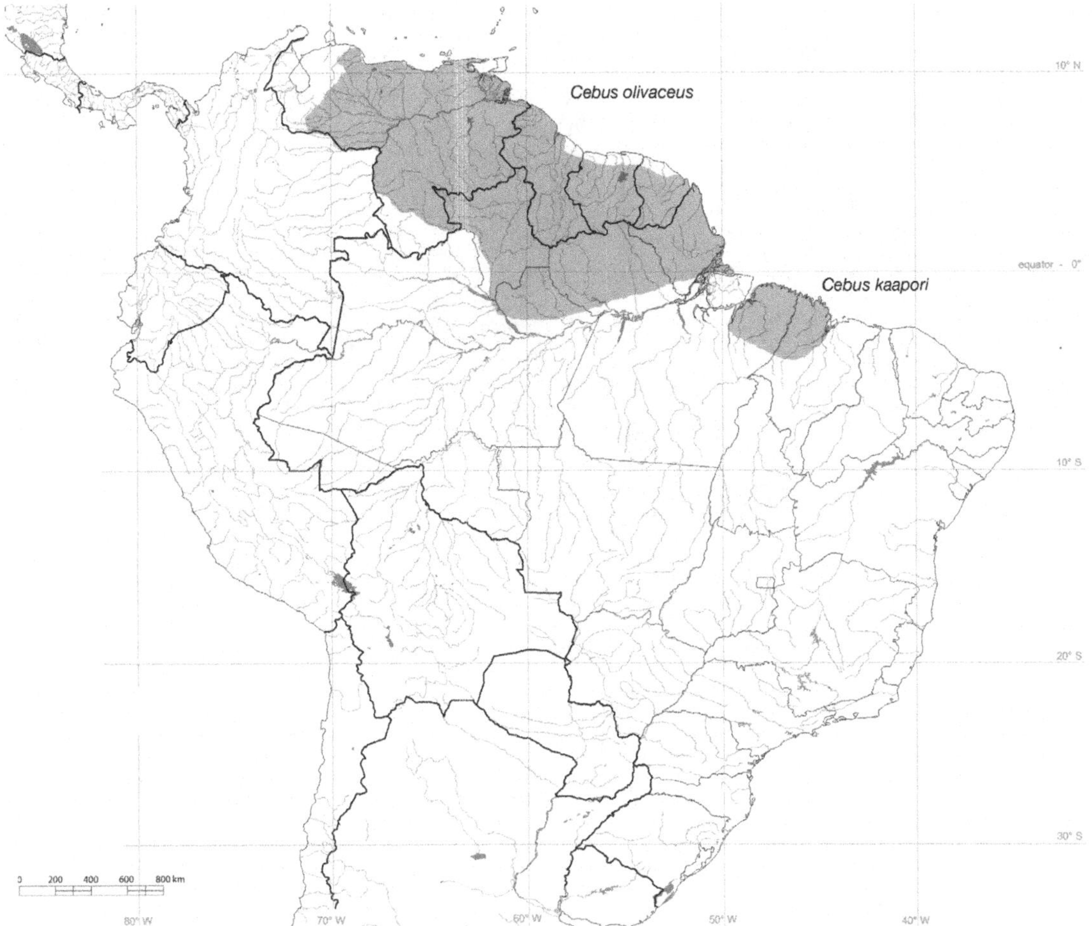

Figure 1.3. The geographic distribution of the weeper or wedge-capped capuchin monkey, *Cebus olivaceus*, in the north-eastern Brazilian Amazon, French Guiana, Suriname, Guyana, and Venzuela, and the Ka'apor capuchin monkey, *Cebus kaapori*, in the eastern Amazon, south of the Rio Amazonas in Brazil. Sources: *C. olivaceus* – Hershkovitz (1949); Hill (1960), Kinzey (1997), Mittermeier (1977a), Bodini and Pérez-Hernández (1987), Eisenberg (1989), Sussman and Phillips-Conroy (1995), Linares (1998); *C. kaapori* – Queiroz (1982), Ferrari and Queiroz (1994), Ferrari and Lopes (1996), Silva Jr. and Cerqueira (1998).

maintaining a toehold of occupation on the South American continent along the western edge of the Colombian Andes, through the Chocó-Darien region to northwestern Ecuador (Hernández-Camacho and Cooper 1976, Rodríguez-Luna *et al.* 1996, Reid 1997, Marineros and Gallegos 1998). Hershkovitz (1949) discussed the taxonomic history of this species, and recognized five subspecies as valid, three in Colombia and two in Central America. They were: *C. capucinus capucinus* (Linnaeus, 1758) from Panama and north-west Colombia; *Cebus capucinus curtus* Bangs, 1905, from the Island of Gorgona in the Pacific Ocean off the coast of Colombia; *C. capucinus nigripectus* Elliot, 1909, from the Cauca Valley in Colombia; *C. capucinus imitator* Thomas, 1903, from Chiriquí and the islands of Coiba and nearby Jicarón, Panama; and *C. capucinus limitaneus* Hollister, 1914, from eastern Honduras. Hernández-Camacho and Cooper (1976) mapped the range of the species in Colombia, but considered that the characters used to identify the subspecies were so variable as to make their designation impossible, at least based on the material available, and considered that the Central

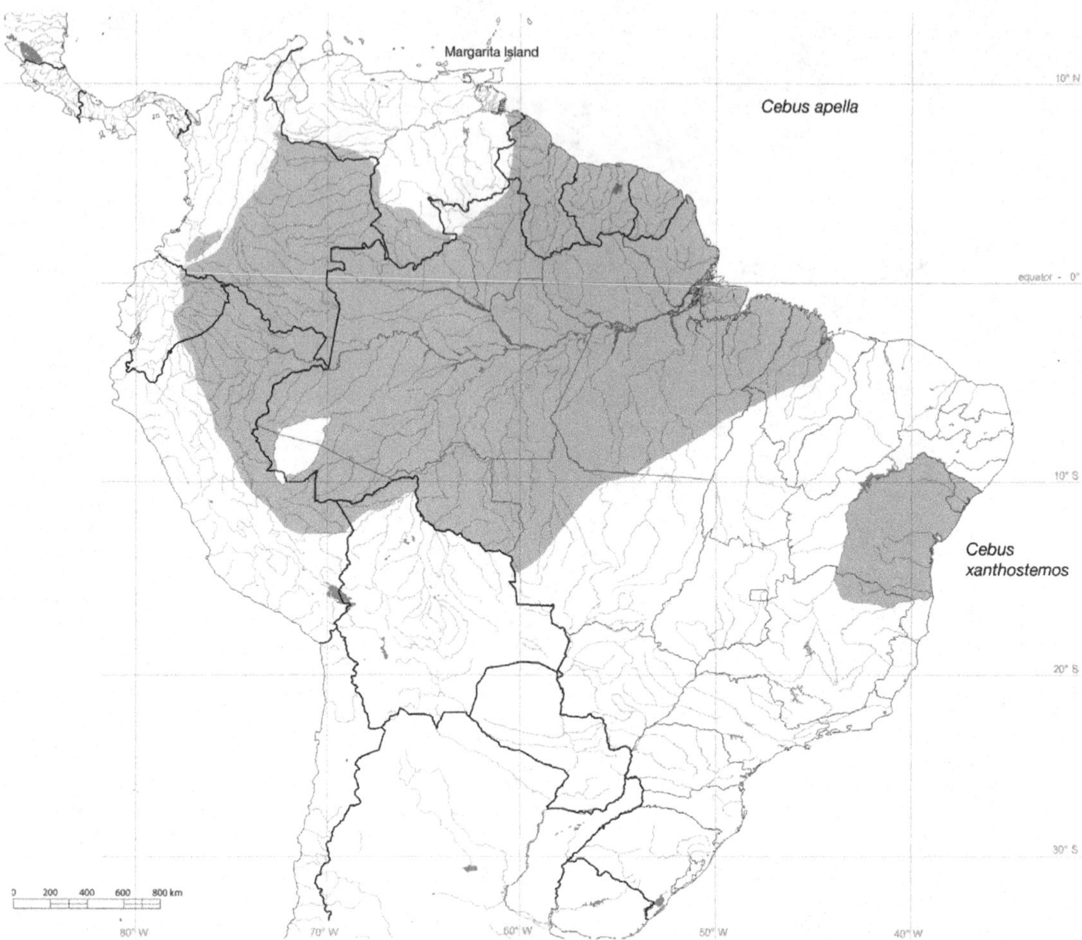

Figure 1.4. The geographic distribution of the tufted or brown capuchin, *Cebus apella*, in Venezuela (including an isolated population on the Island of Margarita – *C. a. margaritae*), Colombia, Ecuador, Peru, Bolivia, Brazil, French Guiana, Suriname, and Guyana, and the yellow-breasted, golden-bellied or buff-headed capuchin, *Cebus xanthosternos*, in Bahia, eastern Brazil. (Drawing by Kimberly Meek.)

Note 1. In the taxonomy of Groves (2001), *C. apella* is comprised of seven subspecies, none of which are recognized by Silva Jr. (2001) except for *macrocephalus* which he considered to be a full species. According to Silva Jr. (2001), *C. macrocephalus* is the tufted capuchin occupying the entire upper Amazon, west from the mouth of the Rio Japurá, and from the Rios Xapuri and Branco in the state of Acre, in Colombia, Peru, Ecuador, and north of the Río Madre de Dios in Bolivia.

Note 2. Groves (2001) recognized the tufted capuchins from the upper Rio Juruá and Rio Envira as *C. libidinosus juruanus*. His belief is that the range of this form (not *C. a. apella* as shown here) extends from the upper Juruá east along the entire southern part of the western Amazon to northern Mato Grosso, and as such is continuous with other *C. libidinosus* subspecies (C. P. Groves, pers. comm. June 2002). We are unable to map this range because the limits are unknown. Silva Jr. (2001) considered *juruanus* to be a junior synonym of *C. macrocephalus*, and the capuchin monkeys from the Mato Grosso as belonging to *Cebus cay* Illiger, 1815, not recognized as valid by Groves (2001). See text for further explanation. Sources: *C. apella* – Hill (1960), Hernández-Camacho and Cooper (1976), Bodini and Pérez-Hernández (1987), Eisenberg (1989), Aquino and Encarnación (1994), Ferrari and Lopes (1996), Linares (1998), Eisenberg and Redford (1999), Boher-Benetti and Cordero-Rodríguez (2000); Groves (2001); Silva Jr. (2001); *C. xanthosternos* – Torres (1988), Rylands *et al.* (1988), Oliver and Santos (1991); Coimbra-Filho *et al.* (1991, 1991/1992), Silva Jr. (2001).

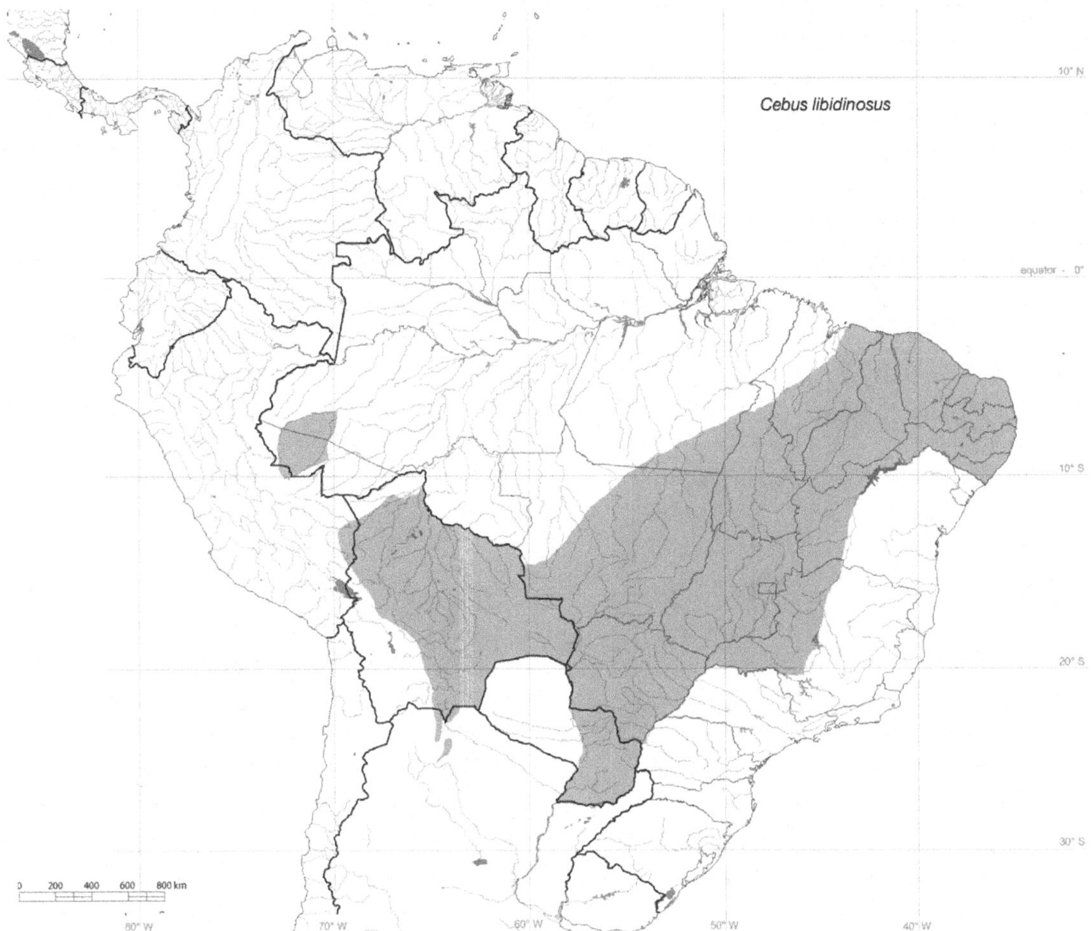

Figure 1.5. The geographic distribution of the black-striped or bearded capuchin monkey, *Cebus libidinosus*, in Brazil, extreme southeastern Peru, Bolivia south of the Río Madre de Dios, Paraguay and Argentina. (Drawing by Kimberly Meek.) Note that Groves (2001) indicates that *C. l. juruanus* extends east from the upper Juruá, along the southern Amazon, presumably including the Pando region of northern Bolivia into Rondônia and northern Mato Grosso in Brazil. See text and Figure 1.4. Sources: Anderson (1997), Aquino and Encarnación (1994), Brown (1989), Groves (2001), Hill (1960), Kinzey (1982), M. Di Bitetti (pers. comm.).

American populations were subject to the same limitation. Rodríguez-Luna *et al.* (1996) mapped the ranges of the three subspecies in Central America. Groves (2001) regarded them all as junior synonyms of *Cebus capucinus*.

The northern limits of this species in historic times are difficult to understand. An early and often-cited paper (Hollister 1914) claimed that a white-faced capuchin had been collected in Belize for the U.S. National Museum. But in spite of second-hand accounts of sightings presumed to be white-faced capuchins (e.g., McCarthy 1982), no recent surveys (Dahl 1986, Hubrecht 1986) have confirmed the report. There have been unauthenticated sightings of capuchins in the Mayan Mountains of western Belize and in Sarstoon National Park on the southern border of Belize (the latter even advertises the presence of capuchins on the park's web site), but these await confirmation. There are no capuchins (and no reports of them) in El Salvador, Guatemala, or Mexico. Why is *C. capucinus* found in Honduras but not in the more northerly tropical forests of Guatemala, Belize and the Yucatán Peninsula

Figure 1.6. The geographic distribution of the black or horned capuchin monkey, *Cebus nigritus*, in Brazil, northern Argentina and Paraguay. (Drawing by Kimberly Meek.)
Note that Groves (2001) recognizes three subspecies, whereas Silva Jr. (2001) recognizes only two of them, and as full species. *C. robustus* occurs between the Rio Jequitinhonha and Rio Doce, and *C. nigritus* throughout the remainder of the range shown here. Sources: Groves (2001), Hill (1960), Kinzey (1982), Oliver and Santos (1991), Rylands *et al.* (1988), Stallings (1985), M. Di Bitetti (pers. comm.).

(Mexico) where spider monkeys and howlers do occur? It is possible that *Cebus capucinus* moved northward from Colombia and Panama more recently than spider and howling monkeys and never made it over the gaps in tropical forests present in historic times between eastern and western Honduras, within El Salvador, and between southern and northern Guatemala (Emmons and Feer 1997). It is also possible that features such as the Río Motagua that cuts laterally across Guatemala from the highlands to the ocean or the Río Sarstoon marking the southern border of Belize, have acted as barriers to their dispersal. However, this is conjecture, as we have almost no evidence to bear on this question, and thus, at present, no clear answer to this puzzle.

Cebus albifrons (Humboldt, 1812) White-fronted capuchin

C. albifrons has a more limited distribution than the Amazonian tufted capuchins, occurring in westerly parts of the South American continent, from Colombia, Ecuador (both east and west of the Andes) to western and southern Venezuela, south through Peru into Bolivia and western and central Amazon in Brazil. The

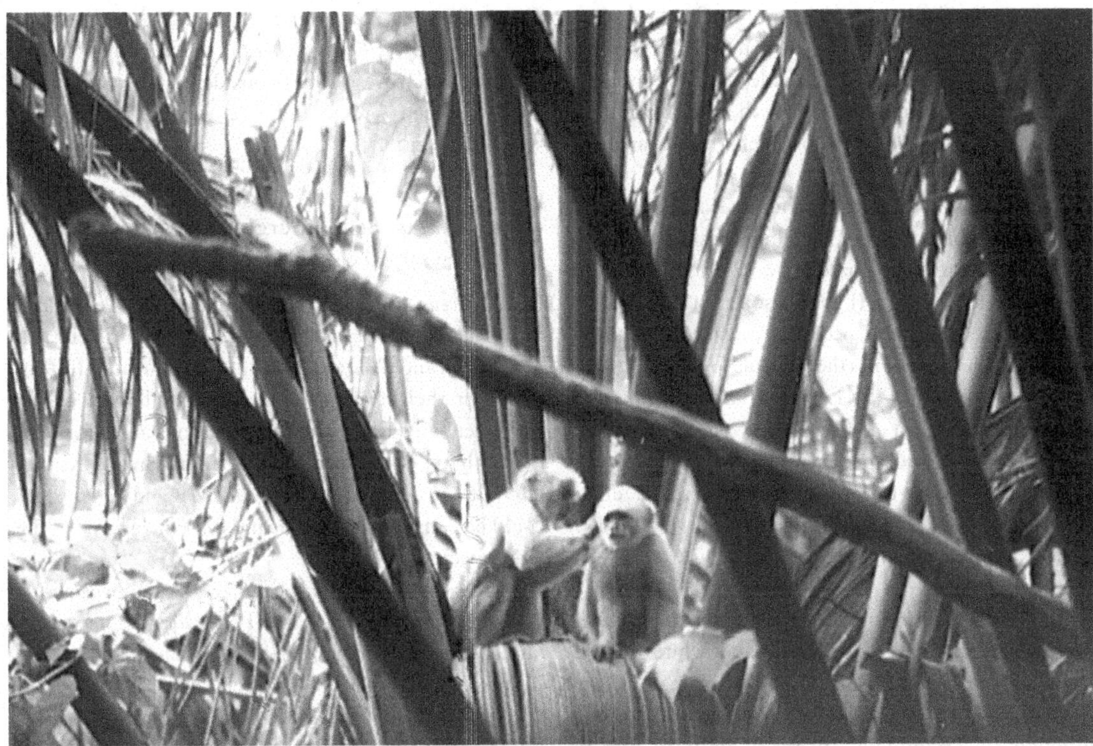

Figure 1.7. One white-fronted capuchin (*C. albifrons trinitatis*) grooming another in Trinidad. (Photograph courtesy of Brian Grafton.)

limit in the east of the range is the Rio Tapajós, south of the Rio Amazonas.

Hershkovitz (1949) listed eight subspecies in Colombia and Venezuela, one in western Ecuador, one on the Island of Trinidad, and three in the Amazon basin (see Figure 1.7). They are as follows: *C. albifrons albifrons* Humboldt, 1812, from the banks of the Orinoco, near the mouth of the Río Ventuari; *C. a. hypoleucus* Humboldt, 1812, from the Río Sinu, Bolívar, Colombia; *C. a. malitiosus* Elliot, 1904, from the Sierra Nevada de Santa Marta, Colombia; *C. a. cesarae* Hershkovitz, 1949, from the Río Cesar, Magdalena valley, Colombia; *C. a. pleei* Hershkovitz, 1949, from Mompós, west bank of the Río Magdalena, at the base of the northern extremity of the Cordillera Central, Colombia; *C. a. versicolor* Pucheran, 1917, from the middle Río Magdalena, Colombia; *C. a. leucocephalus* Gray, 1865, considered by Hershkovitz to be from the Río Lebrija, Santander, Colombia; *C. a. adustus* Hershkovitz, 1949, from the eastern base of the Sierra de Perijá in Venezuela and Colombia; *C. a. unicolor* Spix, 1823, from the Rio Tefé, Amazonas, Brazil; *C. albifrons yuracus* Hershkovitz, 1949, from the Ríos Marañón and Napo, eastern Ecuador and northern Peru; *C. a. cuscinus* Gray 1901, from the upper Río Madre de Dios, Peru; *C. a. aequatorialis* Allen, 1921, from northwestern Ecuador (and probably also the Tumbes region in northern Peru (Encarnación and Cook 1998)); and *C. a. trinitatis* Von Pusch 1941, from Trinidad.

Hershkovitz (1949) and Hernández-Camacho and Cooper (1976) mapped these subspecies in Colombia. Hernández-Camacho and Cooper (1976) examined carefully those proposed by Hershkovitz (1949) and came to the following conclusions:

1. *C. a. malitiosus*, with a dark brownish overall coloration and rather light yellowish shoulders, is a well-defined subspecies inhabiting the deciduous and humid forests of the northern slopes of the Santa Marta Mountains at least as high as 1300 m, although the eastern and southern limits of its range are not well defined.

2. *C. a. cesarae*, light colored, is a well-defined subspecies occurring in the Department of Magdalena, southward from Ciénaga Grande, and the lowlands of the Department of Cesar, north to the deciduous and gallery forests of the Río Ranchería, Department of Guajira.
3. *C. a. versicolor* is a complex of forms from the Cauca-Magdalena interfluvium, including, besides *C. a. versicolor* (intermediate phase), *C. a. leucocephalus* (dark phase) and *C. a. pleei* (light phase).
4. *C. a. adustus* probably occurs in piedmont forests of western Arauca, the northern tip of Boyacá and north Santander, besides the Lake Maracaibo region and upper Apure basin of Venezuela.
5. *C. a. unicolor*, widespread in the upper Amazon, is very similar to the type species, and probably a junior synonym of *C. a. albifrons* (confirmed with further study (Defler and Hernández-Camacho 2002)).

The map of Hernández-Camacho and Cooper (1976) is full of question marks, and white-fronted capuchins are undoubtedly more widespread in the Colombian Amazon than has been documented to date, although this is difficult to ascertain today due to widespread forest destruction in the region.

Bodini and Pérez-Hernández (1987) plotted localities for *C. a. adustus* to the north-west of Lake Maracaibo in northwestern Venezuela, *C. a. leucocephalus* to the west, south and east, surrounding Lake Maracaibo, and *C. a. unicolor* along the upper Río Orinoco, in the Federal Territory of Amazonas.

Hernández-Camacho and Cooper (1976) reported further specimens of dark brown animals from the middle San Jorge valley that had features intermediate between *C. capucinus* and *C. albifrons*, and suggested that further studies may argue that the two are conspecific. Another important area that they pointed out was in western Ecuador, where *C. a. aequatorialis* and *C. capucinus* are both known to occur, although sympatry and intergradation have not been determined. Since capuchin species are known to hybridize in captivity (see Table 1.3), it is probable this may also occur in the wild.

Groves (2001), influenced by Hernández-Camacho and Cooper (1976), reduced the number of subspecies to the following: *C. albifrons albifrons*, *C. a. cuscinus* (*yuracus* as a junior synonym), *C. a. unicolor* (unaware of the study by Defler and Hernández-Camacho (2002) which argued that it is a junior synonym of the type species), *C. a. trinitatis*, *C. a. aequatorialis*, and *C. a. versicolor* (*leucocephalus*, *malitiosus*, *adustus*, *cesarae* and *pleei* as synonyms). Groves (2001) considered that the differences between *cesarae* and *pleei* were insufficient to separate them.

Cebus olivaceus Schomburgk, 1848 Weeper or wedge-capped capuchin

C. olivaceus is widely distributed in Venezuela, the Guianas, and south to the Amazon in Brazil. *C. albifrons* and *C. olivaceus* are mostly allopatric (do not overlap in their ranges), but according to Kinzey (1997) and Eisenberg (1989), their distributions are coincident in southern Venezuela and northern Brazil. Kinzey (1997) even suggested that *C. albifrons* and *C. olivaceus* could be considered western and eastern variants of the same species.

Hershkovitz (1949) recognized five subspecies of *C. nigrivittatus* (= *olivaceus*). They were: *C. nigrivittatus nigrivittatus* Wagner, 1848, from São Joaquim, Rio Branco, Brazil; *C. n. olivaceus* Schomburgk, 1848, from the southern foot of Monte Roraima, Brazil; *C. n. castaneus* I. Geoffroy, 1851, from Cayenne, French Guiana; *C. n. apiculatus* Elliot, 1907, from the Río Caura, near its confluence with the Río Orinoco; and *C. n. brunneus* Allen, 1914, from Yaracuy, northern Venezuela. Hershkovitz (1949) mapped their type localities but not their ranges. Bodini and Pérez-Hernández (1987) mapped localities for *C. n. brunneus* along the Cordillera de la Costa of extreme northern Venezuela, and two localities in western Venezuela. *C. n. nigrivittatus* they restricted to the Venezuelan Amazon, along the north of the upper Río Orinoco, *C. n. apiculatus* to the south and west of the Río Orinoco, north of the Río Ventuari and west of the Río Caroni, and *C. n. olivaceus* to eastern Venezuela, east of the Río Caroni and south of the Río Orinoco, being the form extending east and south into the Guianas and Brazil. Bodini and Pérez-Hernández (1987) believed that the form in north central Venezuela, north of the lower and middle Río Orinoco, is an undescribed subspecies.

In Brazil, *C. olivaceus* is restricted to the north of the mainstream of the Amazon and east of the Rios Negro, Branco and Catrimani, but its range boundaries are unclear. It may be limited in the north by the Serra Parima and Serra Pacaraima, although extending into parts of northern Pará or Amapá, but this is conjecture. Hill (1960) extended the range of *C. n. olivaceus* as far as the Rio Parú, a north bank tributary of the Amazon just above the Rio Jari. In Suriname, *C. olivaceus* is almost entirely restricted to the interior. Like *Chiropotes* and *Ateles*, it just enters the old coastal plain in a small part of western Suriname (Mittermeier 1977a). Sussman and Phillips-Conroy (1995) indicated that it occurs throughout Guyana, and there is no evidence to suggest otherwise for French Guiana.

Cebus kaapori Queiroz, 1992 Ka'apor capuchin

Cebus kaapori occurs in the states of Maranhão and Pará, south of the lower Rio Amazonas, in eastern Amazonia, Brazil. Although the Ka'apor and the Guajá Amerindians hunted it both for food and for pets (see Figure 0.2), it had escaped the notice of scientists until 1992, when Queiroz obtained a specimen from the right bank of the Rio Gurupí, in Maranhão. It occurs between the lower Rio Tocantins, Pará, and the Rio Grajaú in Maranhão. Masterson (1995) reported on a cranial study, and information on its range is provided by Ferrari and Queiroz (1994), Ferrari and Lopes (1996) and Silva Jr. and Cerqueira (1998). It is an untufted capuchin clearly affiliated to *C. olivaceus* and, although separated by the Rio Amazonas, it may more correctly be considered a subspecies (Harada *et al.* 1995, Rylands *et al.* 2000). Groves (2001) and Silva Jr. (2001) maintained its status as a full species.

Cebus apella (Linnaeus, 1758) Tufted or brown capuchin

Only one Amazonian tufted capuchin monkey has been recognized over the last 30 years, *C. apella apella* (see for example, Eisenberg 1989, Eisenberg and Redford 1999, Emmons and Feer 1997, Nowak 1999, Wolfheim 1983). The uncertain distributions of a series of Amazonian tufted capuchins proposed by Hill (1960), and the difficulties in understanding regional patterns with such a high degree of individual variation was largely responsible for this (see Figure 1.4). Of the nine Amazonian tufted capuchins listed by Hill (1960), however, Groves (2001) maintained all but two. *C. apella magnus* von Pusch, 1941, he considered a junior synonym of the form *macrocephalus*, and *C. a. maranonis* von Pusch, 1941 (recognized by Aquino and Encarnación (1994)) a junior synonym of the form *peruanus* Thomas, 1901. Groves listed *C. apella apella* (Linnaeus, 1758), *C. a. fatuellus* (Linnaeus, 1766), *C. a. macrocephalus* Spix, 1823, *C. a. peruanus* Thomas, 1901, *C. a. tocantinus* Lönnberg, 1914, and *C. a. margaritae* Hollister, 1914.

It would seem that the stronghold of the type species is the Guianas, and Brazil, at least east of the Rio Negro. Boher-Benetti and Cordero-Rodríguez (2000) extended it to the southern extreme of the Orinoco Delta, although it is otherwise not recognized as occurring in eastern Venezuela (Bodini and Pérez-Hernández 1987, Linares 1998). In Venezuela it occurs in the Federal Territory of Amazonas, along both sides of the upper Río Orinoco, its precise range being limited by savannas. In the scheme of Groves (2001), the range to the west and the south of the Rio Amazonas is constrained by *C. a. macrocephalus* and *C. a. tocantinus* respectively. Hill (1960) indicated that *C. a. apella* occurred south of the Rio Amazonas from the Rio Xingú east to the Rio Maruím in Maranhão, excluding in his map (between pp. 462 and 463) the lower Rio Tocantins, domain of *C. a. tocantinus*. Hill (1960) also attributed the Rio Irirí, a west bank tributary of the Xingú, to the form *tocantinus*. In the southern Amazon, *C. apella* would be restricted by the transition to *Cerrado*, the bush savanna of central Brazil where *C. libidinosus libidinosus* occurs. Groves (2001) gave the range of *C. a. macrocephalus* as the upper Amazon, east to the Rio Tapajós, south of the Rio Amazonas, but mentioned material from Itacoatiara on the north bank (also listed as a locality for *macrocephalus* by Cruz Lima (1945) and Vieira (1955)). Itacoatiara is east of Manaus, and would therefore be an incursion into the Guianan range of *C. a. apella*. Vieira (1955) gives the locality of Manaus for *C. a. apella*, and Itacoatiara, about 100 km east, for the form *macrocephalus*! Vieira (1955) placed *macrocephalus* throughout the middle Amazon of Brazil, east as far as the Rio Xingú (Redenção), and listed localities such as Codajáz (north of the Rio Solimões, west of the Rio Negro), Jaburú on the Rio Purús, and south through Rondônia to the Rio Guaporé and northern Mato Grosso, north of Cuiabá. Consensus on the range of *C. a. apella* is only on the Guiana Shield east of the Rio Negro and south of the Rio Amazonas, east of the Rio Xingú (Figure 1.4).

Cebus apella fatuellus is from Colombia. According to Hill (1960) the range of this form is eastern Colombia, possibly extending southwards along the lower slopes of the Andes into Peru, where it would meet the range of *C. a. peruanus*. Hernández-Camacho and Cooper (1976)

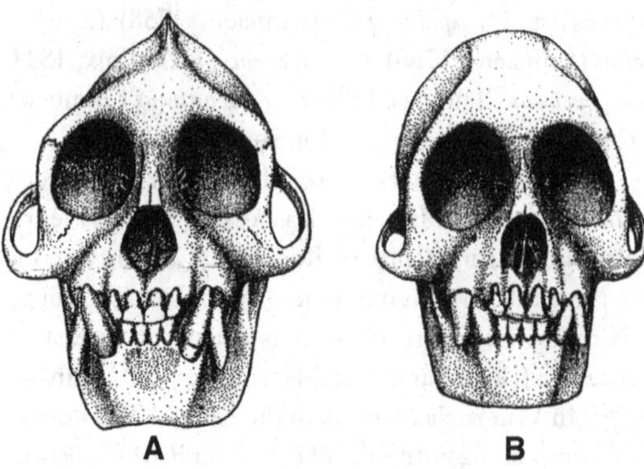

Figure 1.8. Sexual dimorphism in the skulls of male (A) and female (B) capuchins with tufts; that is, subgenus *Sapajus* according to Silva, Jr. (2001) (see Table 1.2). Major differences between sexes include the morphology of the canine teeth and the vomer and the presence in the male of a sagittal crest at the interparietal suture. The sagittal crest evident in this drawing is not seen in other (untufted) *Cebus* according to Silva Jr. (2001).

mapped the distribution of *Cebus apella* in Colombia in some detail, although they did not apply subspecific names. It occurs throughout the Colombian Amazon and foothills of the Andes (to at least 1300 meters). There is also an isolated population in upper Río Magdalena in the Department of Huila up to 2700 m in the region of San Agustin and a small part of the Department of Cauca (Tierradentro), at altitudes of up to 2500 meters near Inzá. It extends north as far as the Río Arauca on the Venezuelan border. Hill (1960) indicated a range covering eastern and north-central Colombia, along with a large part of western Venezuela up to Lake Maracaibo, while Bodini and Pérez-Hernández (1987) and Linares (1998) reported that *C. apella* occurs only in southern Amazonian Venezuela. Presuming that the Colombian populations are those which Groves (2001) is recognizing as *C. a. fatuellus*, the question remains as to the southern limits (Río Putumayo?) where it would meet *C. a. peruanus*, and the western limits, where it would meet *C. a. macrocephalus*, or *C. a. apella* if *macrocephalus* were to be restricted to the south of the Rio Amazonas, as indicated by Aquino and Encarnación (1994). *C. a. macrocephalus* is the form occurring in the upper Amazon, east of the Río Ucayali (Aquino and Encarnacíon 1994). According to Hill (1960), from there it ranges east traversing the lower and middle stretches of the southern tributaries of the Rio Solimões-Amazonas as far as the Rio Tapajós. Vieira (1995) also listed the Rio Xingú.

Aquino and Encarnación (1994) gave a distribution for *C. a. peruanus* as north of the Ríos Madre de Dios and Inambari, east along the right (south) bank of the Rio Purús. Moving north, Aquino and Encarnación (1994) indicated an unknown, but distinct *Cebus apella* subspecies for the Rio Pachitea basin: "The specimens from the montane forests between 800 meters and 1000 meters a.s.l. from the Departments of Huanáco, Pasco and Junín show distinct phenotypical characters and might belong to a fifth subspecies" (p. 36). North of there and west of the Río Ucayali, is the form *maranonis*, extending across the Río Marañon basin to the Río Putumayo. Groves (2001) considered *maranonis* to be a junior synonym of *peruanus*. If he is correct in this, it implies that the unidentified form from the Pachitea basin, separating as it does the distributions of the two, is also attributable to *C. a. peruanus*.

Cebus a. tocantinus from the lower Rio Tocantins in Pará is a dark race, very similar to *peruanus*, but both are recognized as distinct taxonomic entities by Lönnberg (1939) and Groves (2001) due to their geographic separation, south of the Rio Solimões-Amazonas, by the forms *macrocephalus* and *apella*. The range given by Hill (1960) was a small blob around the type locality at the mouth of the Rio Tocantins, otherwise entirely surrounded by *C. a. apella*. Groves (2001) stated that he had seen specimens from the south bank of the lower Amazon as far west as the Rio Madeira. This would take in part of Hill's (1960) ranges for *C. a. apella* (between the Rios Tocantins and Xingú) and *C. a. macrocephalus* (between the Rios Tapajós and Madeira). Lönnberg (1939) indicated that it occurs some way to the southwest of the type locality, presumably between the Tocantins and Xingú at least.

Cebus apella margaritae is restricted to the highlands of the island of Margarita, Venezuela (Sanz and Marquez 1994). Groves (2001) had reservations about the taxonomic status of this subspecies. Silva Jr. (2001) provisionally considered it a synonym of *C. apella*. It is darker than the tufted capuchins in the Federal Territory of Amazonas on the upper Orinoco in Venezuela (Linares 1998), and is well separated from all other tufted capuchin monkeys, the nearest population being the southern extreme of the Río Orinoco delta (Boher-Benetti and Cordero-Rodríguez 2000). Groves (2001; see also Linares 1998) supposed that it was introduced in Pre-Columbian times, and found it to be more closely allied to *C. a. fatuellus* than to *C. a. apella*.

Cebus libidinosus Spix, 1823 Bearded capuchin or black-striped capuchin

Cebus libidinosus, as defined by Groves (2001), is the capuchin monkey of Bolivia, northwestern Argentina, central and north-east Brazil, the Pantanal of Mato Grosso, and eastern Paraguay (Figure 1.5). Its tufts are not very well developed, at least in the nominate subspecies, forming just a fringe along the forehead. It ranges west of the Rio São Francisco, through the *Cerrado* or bush savanna, of central Brazil, being replaced by *C. apella* to the north along the transitions to the Amazon rain forest and the dry forests of Mato Grosso. The range was described by Kinzey (1982), who considered it a subspecies of *C. apella*. Groves (2001) recognized the following subspecies: *C. libidinosus libidinosus* Spix, 1823; *C. libidinosus pallidus* Gray, 1866; *C. libidinosus paraguayanus* Fischer, 1829; and *C. libidinosus juruanus* Lönnberg, 1939.

Cebus l. libidinosus occurs in western Minas Gerais and part of western Bahia, east of the Rio Saõ Francisco, north through the north-eastern states of Sergipe, Piauí, Pernambuco, Natal, and Ceará to Maranhão. *C. l. pallidus* is known from Bolivia and Peru, but there is some confusion as to its distinctiveness from *C. l. paraguayanus*. Mudry de Pargament and Slavutsky (1987, p. 116), studying the chromosomes of the Paraguayan forms (*C. l. paraguayanus*) and those in north-west Argentina, concluded that they "constituted a single karyomorphic population, in spite of the phenotypic differences observed." Aquino and Encarnación (1994) placed *C. apella pallidus* in south-west Peru, south of the Ríos Madre de Dios and Inambari, Department of Puno. Anderson (1997), on the other hand, attributed all northern and central Bolivian tufted capuchin monkeys to *C. apella pallidus*, even those in the Pando region north of the Madre de Dios. Anderson (1997) identified *C. a. paraguayanus* as restricted to the south-east of Bolivia. Mantecon *et al.* (1984) and Brown and Colillas (1984) indicated *paraguayanus* for northern Bolivia and south-east Peru. *C. l. paraguayanus* is the capuchin from eastern Paraguay, east of the Río Paraguai, but absent from the Chaco to the west of river. It might be restricted to eastern Paraguay and the Pantanal of Mato Grosso, extending into south-eastern Bolivia, but this has yet to be confirmed.

Groves' (2001) recognition of the dark *C. l. juruanus* (all other *libidinosus* subspecies are pale) presents an interesting problem. It occurs on the upper Juruá, at least it seems between the Rio Juruá and the Rio Envira (Hill 1960). If, as Aquino and Encarnación (1994) suggested, *pallidus* is restricted to the south of the Río Madre de Dios, and *C. apella peruanus* is the form to the north, with *C. apella macrocephalus* occurring north of the Rio Purús, then *C. l. juruanus* would be completely isolated from other *libidinosus* subspecies by *C. apella*. Groves (2001) indicated a broad distribution for *C. l. juruanus* extending west into Mato Grosso.

Cebus xanthosternos Wied, 1826 Yellow-breasted or buff-headed capuchin

Cebus xanthosternos has been recognized as a good species for some time (Mittermeier, *et al.* 1988, Rylands *et al.* 1995, 1996, 1997). It is restricted to the Atlantic forest of southern Bahia, Brazil, north of the Rio Jequitinhonha, at least as far north as the Rio Paraguaçú near Salvador (Figure 1.4), but probably historically throughout the entire area west of, and north to, the Rio São Francisco (Coimbra-Filho, *et al.* 1991, Coimbra-Filho *et al.* 1991/1992). Torres (1988) identified a distinct phenotype in this area. On the left bank of the Rio São Francisco is *C. libidinosus libidinosus*. It was first recognized because of its distinctive pelage patterns (bright yellow color and large round head with no evident tufts and a very small black cap), but research on chromosome variation by Seuánez *et al.* (1986) showed it also to be distinct karyotypically – capuchins classified as *xanthosternos* have intercalar heterochromatin at chromosome pair 11 that is different from that of *C. apella*.

Cebus nigritus (Goldfuss, 1809) Black or Black horned capuchin

This species occurs in the Atlantic forest of Brazil, south to northeastern Argentina in the Province of Misiones (Figure 1.6). Groves (2001) recognized the following three subspecies: *C. nigritus nigritus* (Goldfuss, 1809); *C. n. robustus* Kuhl, 1820; and *C. n. cucullatus* Spix, 1823, all of which were considered junior synonyms of *C. apella nigritus* by Kinzey (1982). *C. n. nigritus* is a large, dark-colored race, very dark brown to black, with long hair, and in adults two elongated lateral frontal tufts or ridges on the crown which contrast with the whitish superciliary areas and cheeks (Hill 1960). It is the form south of the Rio Doce, in the states of Minas Gerais and

Espírito Santo, Brazil, and extending south along the coast through Rio de Janeiro, São Paulo, Paraná, Santa Catarina and Rio Grande do Sul. *C. n. robustus* is a distinct form, considered a full species by Silva Jr. (2001), and distinguishable by the median conical crest on the crown. The general color is bright red-brown. It occurs from the Rio Jequitinhonha in southern Bahia and northern Minas Gerais, south through Espírito Santo to the Rio Doce (Rylands *et al.* 1988, Oliver and Santos 1991). The westernmost locality in the state of Minas Gerais is given by Pinto (1941), who collected specimens from the headwaters of the Rio Pissarão, in a mountainous region north of the Rio Piracicaba, not far from the town of Presidente Vargas. It is possible that the Serra do Espinhaço of Minas Gerais, running north–south and defining the transition from the Atlantic forest to bush savanna (*cerrado*) in the west, marks the western limits of the distribution of this form. Further west (and west of the Serra do Espinhaço), Kinzey (1982) recorded a specimen from Tomas Gonzaga, near Corinto, Minas Gerais, which he listed as *C. a. robustus* (= *C. a. robustus* × *C. a. libidinosus*). Silva Jr. (2001) identified these animals as *C. robustus*, however. Groves (2001) resurrected the name *cucullatus* Spix, 1823, for the capuchin monkeys which had previously been called *vellerosus* I. Geoffroy, 1851 by Vieira (1955) and Brown and Colillas (1984), for example. It would seem that its range runs parallel to, and west of, that of *C. n. nigritus*, from the Rio Grande on the border between the states of Minas Gerais and São Paulo, south (east of the Rio Paraná) into the Province of Misiones, in northeastern Argentina. North of the Rio Grande is *C. l. libidinosus*. Silva Jr. (2001) regarded the form *cucullatus* to be a synonym of *nigritus*.

PHYLOGENY AND EVOLUTION

For many years the Neotropical primates were grouped into just two families: the Callitrichidae (pygmy marmosets, marmosets, tamarins and lion tamarins) and the Cebidae (*Cebus* and the remaining New World monkeys: squirrel monkeys, night monkeys, titis, sakis and uakaris, howlers, spider monkeys, woolly monkeys and muriquis). However, the morphological studies of Rosenberger (1981) and subsequent genetic research (reviewed in Schneider and Rosenberger 1996) supported the view that callitrichids are closely related to *Cebus* and *Saimiri* and should be placed in the same family (Cebidae), with the remaining New World monkeys belonging to two (Atelidae and Pitheciidae) or three families (Atelidae, Pitheciidae and Aotidae).

Although there is disagreement about whether *Cebus* is the earliest offshoot of the primate radiation (Ford 1986), the second offshoot (Kay 1990), or a sister group of the callitrichines (Rosenberger 1981), there is considerable agreement that capuchins find their closest phylogenetic affinities with squirrel monkeys (e.g., Fleagle 1999, Rosenberger 1981, Schneider and Rosenberger 1996, but see Ford 1990). DNA studies such as those by Schneider *et al.* (1996) and Goodman *et al.* (1998) are beginning to clarify the phylogeny of Neotropical primates. These molecular analyses support the grouping of *Cebus* with *Saimiri* and with the callitrichines (and possibly with *Aotus*, the owl monkey) into one monophyletic clade. The molecular data also suggest they emerged at almost the same time during the New World primate radiation but they do not tell us clearly whether it was capuchins, squirrel monkeys or owl monkeys that might have separated first in the family Cebidae. Goodman *et al.* (1998) proposed that the three extant families of Neotropical primates all emerged around 25 million years ago at the Oligocene–Miocene boundary. Schneider *et al.* (2001) used conjoint molecular analysis and molecular clock estimates to suggest when the genus *Cebus* (tribe Cebini) might have appeared. Using Goodman *et al.*'s estimated emergence date of 25 mya to calibrate their molecular clock, Schneider *et al.* suggested that both capuchins and owl monkeys first branched off from a common cebid ancestor around 22 to 23 million years ago, whereas the marmosets and tamarins split into their current tribes around 13 to 16 million years ago.

Schneider *et al.* (2001) noted that reliable fossil data are still necessary for substantiating and clarifying the branching patterns in the family Cebidae. Unfortunately, the fossil record is, as yet, almost entirely silent on the subject of how and when capuchins evolved. Although discoveries of the past couple of decades have substantially increased the number of platyrrhine fossils known to science (Fleagle and Kay 1997, Hartwig 1994, Tejedor 1998), the fragmentary and sparse nature of this record makes it still very difficult to correlate/interdigitate fossils with the molecular arguments described above, and equally difficult to reliably relate individual fossil finds to modern genera of New World monkeys.

The early platyrrhine fossils from an Oligocene site in Bolivia and from the early to middle Miocene in Argentina and Chile are all very difficult to interpret (Fleagle and Kay 1997). In contrast, a wonderful site in central Colombia called La Venta yielded fossil monkeys from 11–13 million years old that are remarkably similar to modern platyrrhines and are clearly related to living sakis, titis, and tamarins. However, there have been no fossil discoveries there thought to be early relatives of *Cebus*. Disappointingly, it is only from relatively recent Pleistocene sites in Haiti and the Dominican Republic (Hispaniola) in the Caribbean (dated as recent as 3860 years ago) that we find any fossil platyrrhines that can be even tentatively attributed to the genus *Cebus* (Fleagle and Kay 1997, Ford 1986, 1990, MacPhee and Woods 1982, Tejedor 1998). Several fragmentary subfossils from cave deposits at these sites have been assigned to the species *Antillothrix bernensis* (MacPhee *et al.* 1995) and are said to have a body size and dentition reminiscent of living *Callicebus* or possibly *Cebus* monkeys (Ford 1990, MacPhee and Woods 1982). Other Pleistocene fossils from the Caribbean have been argued to be similar to *Alouatta* and *Ateles*, whereas still others represent strange forms completely unknown in modern platyrrhines. There are no native nonhuman primates on Hispaniola or Cuba today, but Fleagle and Kay (1997) noted that the large islands of the Caribbean are close to the Yucatán Peninsula and Venezuela. One explanation for the presence of Pleistocene monkeys on Cuba and Hispaniola is that they were able to disperse over water from nearby parts of Central and South America (where *Cebus* and *Alouatta* and *Ateles* do occur today). Another possible explanation is the common proclivity of humans to take animals with them when they migrate to new locations (MacPhee and Rivero de la Calle 1996). Thus far we have been allowed only this tiny window into the evolutionary past of capuchins and through a glass still so dark that we cannot begin to use the fossil data to substantiate the molecular-based argument that capuchins are most closely related to squirrel monkeys, callitrichines and owl monkeys. As reiterated by many primate paleontologists and systematists, we are still eagerly awaiting a much larger fossil database to clarify the picture of cebid evolution.

HABITAT TYPES

Flexibility, opportunism, and adaptability are the hallmarks of capuchin success. These monkeys occupy virtually every type of Neotropical forest, including humid and dry tropical forests, swamp forests, seasonally flooded forests, mangrove forests, and gallery forests, as well as dry, deciduous forests where rainfall is absent for 5–6 months a year (Freese and Oppenheimer 1981). They range from sea level to premontane forest and montane cloud forest up to 2100–2700 meters a.s.l. in the western Andes in Peru, Bolivia, Colombia and Ecuador (Hernández-Camacho and Cooper 1976). Butchart *et al.* (1995) recorded *C. apella* and *C. albifrons* at altitudes up to 2350 meters in the Cordillera de Colán, Peru. Further, they can survive in isolated forest patches, and early secondary growth areas that require them to cross large open gaps, for example the dry xeric *Caatinga* of northeast Brazil, the scleromorphic *Cerrado* (bush savanna) of central Brazil, dry seasonal forests such as the Chiquitania in Bolivia, the Mato Grosso in central southeast Brazil and Paraguay, and the *Llanos* of Venezuela. Capuchins are mostly seen in the middle-layers of the forest, but typically use all levels from the canopy to the understorey, and will also go to the ground to drink, forage or travel.

There have been some attempts to determine habitat preferences of capuchin species living sympatrically. For example, in Manu, Peru, *C. albifrons* feed more in taller trees and larger canopies in the primary forest, whereas *C. apella* are more often found in smaller, middle- and understorey trees (Janson 1986b). Defler (1985) compared *C. albifrons* and *C. apella* for their preferred habitats in El Tuparro National Park, Colombia. He found *C. albifrons* in seasonally inundated forest along rivers that were seldom entered by *C. apella*. However, in most parts of the park the distributions of the two species were mutually exclusive, even when the forests appeared identical, and Defler postulated that historical precedence and competitive exclusion might be a partial explanation for their contiguous, non-overlapping distributions. In Suriname, Mittermeier and Van Roosmalen (1981) found that *C. olivaceus* preferred high terra firme rainforest and stayed away from forest edge, whereas *C. apella* occurred in all the forest types of the study site, including high and low rain forest, mountain savanna forest, liana forest and swamp forest. Mendes Pontes (1997) also examined habitat partitioning between *C. olivaceus* and *C. apella*. He censused primates at Maracá in northern Brazilian Amazonia and found *C. olivaceus* more often in *Mauritia* palm swamp forest (*buritizal*), whereas both species occurred frequently in terra firme (nonflooded, partially deciduous) forest. Nonetheless, Mendes Pontes indicated considerable overlap in habitat preferences of the two and, in general, the flexible patterns of all capuchins seem to work against strong species-specific habitat preferences. The numerous habitat types that have been described for each of the capuchin species indicate that they all exhibit the ability to exploit almost any type of Neotropical forest to the fullest extent. Differences in the species can be found in the degree of use of different forest types, and the key to understanding the sympatry of the tufted and untufted species is in their resource use and foraging/ranging behavior.

CONSERVATION STATUS

Capuchin populations across Central and South America are affected by the many ways in which humans alter the natural environment for economic development, in particular by the outright destruction of their forests. Mittermeier and others (e.g., Mittermeier and Cheney 1987, Mittermeier *et al.* 1993, 1999, Wallis 1997) argue that habitat destruction is the most significant factor causing the decline of the world's primate populations, and it is obvious that as the tropical forests disappear, so do the animals such as the capuchin monkeys that depend on them. In spite of widely discussed international concerns about deforestation over the past 3–4 decades, recent statistics on forest loss continue to be grim. For example, Chapman and Peres (2001) recently reported that forest loss for 1980–1995 was 9.7% in Latin America. In middle America, home to *Cebus capucinus*, less than 20% of the forests still remain, while the Atlantic forest has been reduced to no more than 7% of its original extent (Myers *et al.* 2000). Destruction of the Atlantic forest began in Pre-Columbian times but was rapid and widespread from the early 1500s (Coimbra-Filho and Câmara 1996, Dean 1995), and it is now considered one of the most threatened tropical forests in the world, along with those in Indonesia and especially Madagascar (Myers *et al.* 2000). The destruction of the Amazonian forests began in earnest only in the late 1960s and 1970s; in Brazil with the advent of major development and colonization projects, and in the Andean countries with migrations from the Andean valleys to the lowlands. In Brazil, this has affected particularly the southern and eastern parts, in northern Mato Grosso, southern Pará, and Maranhão, through logging, mining, hydroelectric schemes, and the advance of the agricultural frontier, at first mainly cattle-farming and then also, in the 1990s, soya bean plantations. Thirty years ago about 1% of the Amazonian forests had been destroyed; current estimates indicate more than 20%.

Neotropical forests continue to be cut down for a variety of reasons – logging to harvest trees for export, for homes, furniture, firewood and charcoal, clear-cutting for crops and cattle ranching, road building, and the flooding of extensive areas for hydroelectric projects. The remaining forests, within their matrix of agricultural land, are highly susceptible to runaway fires that start with the deliberate burning of grasslands for cattle. Serious fires of this sort have occurred in Costa Rica, Nicaragua and Brazil. This deforestation can be documented by satellite imagery, but it is now becoming increasingly recognized that occult damage through logging and understorey, or surface, fires is also highly significant and widespread (Chapman and Peres 2001, Cochrane *et al.* 1999, Nepstad *et al.* 1999). Nepstad *et al.* (1999) estimated that logging crews severely damage 10 000 to 15 000 km^2 of forest per year in the Brazilian Amazon. Logging and relatively minor understorey fires resulting from unusually dry years (for example, El Niño years) open the canopy, and reduce the humidity of the forest floor and soil, increasing the forest's susceptibility to fire damage. Nepstad *et al.* (1999) estimated that 270 000 km^2 of forest became vulnerable to fire in the 1998 dry season in the Brazilian Amazon, and Chapman and Peres (2001) noted as one example, that at least 1 million hectares of intact forest was destroyed by fire in one region of Nicaragua during the 5-month dry season in 1997–98. Many capuchins will have perished as a result.

When only partial, forest destruction results in fragmentation, leaving small forest islands in a matrix of agricultural land, many of which are of insufficient size to sustain viable primate populations (Bierregaard *et al.* 2001, Gascon *et al.* 2001). The resulting proximity to humans entices opportunistic, adaptable monkeys such as capuchins to enter neighboring fields to raid crops such as corn, bringing them into direct conflict with farmers (indeed the local names for capuchins in Suriname and Colombia translate as "corn eater" according to Mittermeier *et al.* (1993)). In providing easy access to hunters, fragmentation increases dramatically the susceptibility of the remaining primates to hunting pressure.

Hunting is a highly significant stress on capuchin populations. Usually monkeys are hunted for food, both at the subsistence and commercial levels, but many species are also hunted for body parts as ornaments or medicine. In South America, as well as in Africa and some parts of Asia, primates are one of the most popular forms of "bushmeat." Hunters prefer to save their arrows or bullets for the larger monkeys, which in the Neotropics are the spider monkeys, woolly monkeys and howlers, but capuchins are always game whenever they are found. Infants are often raised as pets, at least until they are large enough to be eaten. Nascimento and Peres (cited in Chapman and Peres 2001) recorded 203 *C. apella* monkeys killed for food in less than a year in a village of 133 Kayapó Indians in southern Pará, Brazil, and Peres (2000b) found that across Amazonia, vertebrate biomass is highly correlated with the degree of human hunting pressure. See the Prologue for further details on the hunting of, and the pet trade in, capuchins.

The third major threat to primate populations is live capture for export and trade on the pet market, or for biomedical research, pharmaceutical testing or exhibit in zoos or circuses. Unlike the first two major threats, this one is fortunately much reduced and under better control than it was in the 1950s and early 1960s during its peak. Due to international agreements for the regulation of live capture and trade of animals (see below) and due to the changing mission of zoos to become vehicles for conservation and education rather than purely for entertainment, in recent decades many fewer primates have been captured and exported from their native lands and imported into urban centers in North America, Europe and Asia. For example, 7261 *C. apella* were imported to the US from Peru and Paraguay between 1968 and 1973 (Wolfheim 1983), but in the mid 1970s, Peru imposed a total ban on the export of primates (Moya *et al.* 1993). On a larger scale, the US imported approximately 200 000 primates in 1955, and 120 000 in 1965, but primate imports dropped to 15 000 in 1990 (Held and Wolfe 1994) and approximately 10 000 in 1995 (Cowlishaw and Dunbar 2000).

Unfortunately, still largely unregulated is the trade of monkeys within national boundaries. Domestic markets in primates can involve large numbers in some tropical countries. For example, Chapman and Peres (2001) estimated that as many as 45 000 spider, woolly, squirrel, and capuchin monkeys may be held captive as pets in Brazilian Amazonia at any one time.

One of the few bright spots in the gloomy picture of declining primate populations around the world was the immediate decrease in primate trade that resulted from the drafting of the Convention on International Trade in Endangered Species of Wild Fauna and Flora (CITES) in 1973, and its ratification by many countries in 1975. International trade in primates dropped sharply in the late 1970s and continues to decline, primarily due to enforcement of export bans by source countries, import restrictions by user countries, and the resultant encouragement of captive breeding by zoos and biomedical research groups. Countries that are party to CITES (approximately 154 nations presently, including all Latin American countries) agree to regulate and/or ban commercial trade in endangered species of plants and animals. Species that are placed in Appendix I of CITES are considered to be threatened with extinction, and international trade in them is entirely banned, except under special circumstances. Species placed in Appendix II are considered vulnerable, and trade in these species is controlled and monitored to avoid any use incompatible with their survival. All primate species are listed in Appendix II, and 10 species of Cebidae (none of them capuchins) are considered to be sufficiently threatened with extinction that they are placed in Appendix I (see CITES webpage, www.cites.org).

As noted above, zoos in many parts of the world have changed their *raison d'être* over the past few decades, and now perceive the conservation of the animals they exhibit and education of the public as their primary mandates. Indeed, zoos in North America have been reformulated in the past three decades from primarily recreational facilities to substantial contributors to public awareness of nature, participants in conservation endeavors, and locations for scientific research on captive animals. Zoos in countries that are party to CITES are subject to its regulations. To be accredited by the American Zoo and Aquarium Association (AZA), North American zoos must meet standards as to the housing, maintenance, trading, and breeding of the species they exhibit. For example, the AZA established the Species Survival Plans (SSPs) in 1981 and the Population Management Plans (PMPs) in 1994. SSPs are population management and conservation programs for selected species in North American zoos. Each SSP manages the breeding of a species in order to maintain a genetically diverse, demographically stable, healthy, and self-sustaining population. SSPs are directed at species

Table 1.3. *Capuchin monkeys held in zoos monitored by the International Species Information System (ISIS) (Data from ISIS, February 12, 2002).*

	Males	Females	Unknown	Totals for each species	Births < 6 months
C. capucinus	62	62	7	131	3
C. capucinus hybrids	2	1	0	3	0
C. albifrons	27	22	17	66	1
C. albifrons hybrids	1	3	0	4	0
C. apella	193	193	44	430	20
C. apella hybrids	4	16	1	21	6
C. xanthosternos	23	12	4	39	7
C. olivaceus	10	10	0	20	1
Totals	322	319	73	714	38

considered to be endangered and ensure that zoos do not need to take animals from the wild to maintain captive populations (see contributions in Wallis 1997). A list of SSPs in place for Neotropical primates is provided by Rylands, Rodríguez-Luna and Cortés-Ortiz (1996/1997). Capuchins in captivity are currently designated as being managed at the PMP level, which is one level lower in intensity of management than an SSP. Population Management Plans are established for "studbook" populations that do not require the intensive management and conservation action of Species Survival Plans.

"Studbooks" are computerized databases for given species under a management plan, which record the location and complete history of each captive individual. They include pedigrees, and allow detailed genetic and demographic analyses of the entire captive population. According to the International Species Information System (ISIS), there are currently around 714 capuchins in captivity around the world being kept by member zoos (Table 1.3). However, since membership of ISIS is costly, it excludes many private and smaller collections, and the real number is undoubtedly much higher. Studbooks are typically maintained by a single, dedicated volunteer. The *Cebus* studbook was taken over by Mark Warneke of the Brookfield Zoo, Chicago, in 2001, and a complete database on captive capuchins will soon be forthcoming (Cathy Gavillier, Calgary Zoo, pers. comm.; for updates, see the AZA website at www.aza.org).

In terms of monitoring capuchin populations (and indeed populations of all living taxa) in the wild, the primary conservation action and "watchdog" organization is the Species Survival Commission (SSC) of the World Conservation Union (IUCN). Founded in 1948 as the International Union for Conservation of Nature and Natural Resources, the IUCN has a wide range of programs linking conservation, science, society, local action and global policy. The SSC is one of six major organizational units within the IUCN, and is comprised of more than 100 taxonomic, or theme-based, Specialist Groups, of which the Primate Specialist Group (PSG) is one of the largest, with more than 350 members (Mittermeier *et al.* 1999). The PSG works to provide protection to primates, and to minimize the decline of primate populations, through education, research, and the establishment and maintenance of parks and reserves. One of its most important tasks is to evaluate the threatened status of all of the more than 620 primate taxa currently recognized. The Primate Specialist Group produced a *Global Strategy for Primate Conservation* in 1977 (Mittermeier 1977b), describing and prioritizing primate conservation projects, many of which were subsequently funded by the Primate Action Fund of the World Wildlife Fund (WWF), which ran from 1980 to 1986 (Konstant 1996/1997). A similar fund, begun in 1996, is the Margot Marsh Biodiversity Foundation, Virginia, the mission of which is to contribute to global biodiversity conservation by providing support for the conservation of endangered nonhuman primates and their natural habitats (Mittermeier 1996). In the late 1980s and the 1990s, Action Plans for the conservation of African (Oates 1996), Asian (Eudey 1987) and Mesoamerican primates were also published

Table 1.4. *Conservation status of capuchin monkey species and subspecies (Data from the* 2002 IUCN Red List of Threatened Species *(Hilton-Taylor, 2002) and Rylands* et al. *(1997)).*

Category of threat	Species name
Critically Endangered	*C. xanthosternos, C. apella margaritae, C. albifrons trinitatis* (considered Critically Endangered by Rylands *et al.* 1997)
Endangered	No species or subspecies of *Cebus* presently listed in this category
Vulnerable	*C. kaapori, C. apella robustus, C. capucinus curtus*
Lower Risk	*C. apella, C. albifrons, C. olivaceus, C. capucinus*
Data Deficient	*C. albifrons adustus, C. albifrons aequatorialis, C. albifrons cesarae, C. albifrons cuscinus, C. albifrons leucocephalus, C. albifrons malitiosus, C. albifrons versicolor, C. albifrons yuracus*

(Rodríguez-Luna *et al.* 1996). These Action Plans describe conservation activities and primate population status in the relevant regions and provide an indirect measure of the success of various conservation strategies. The PSG works closely with the Conservation Breeding Specialist Group (CBSG), which specializes in the organization of Population and Viability Analysis (PVA) Workshops. Population Viability Analysis is widely applied in conservation biology to predict extinctions risks for threatened species and to compare options for their management (Brook *et al.* 2000, Gärdenfors 2000). Rylands *et al.* (1996, 1997) presented a summary of PSG activities on behalf of the Neotropical primates.

Over 30 years the SSC has maintained a register of threatened species which is now published in an electronic format on the IUCN webpage (www.redlist.org). The system for categorizing the degree of threat to a particular species was proposed by Mace and Lande in 1991, and is now internationally recognized for both regional and global assessments (Gärdenfors 2001). The latest version (3.1), approved in 2000, will be used for all future assessments (IUCN 2001), and is being linked to a global database, the Species Information Service (SIS), to store and organize the information involved in assessing the degree to which a species is threatened. Three categories of threatened status are recognized: Critically Endangered (CR), Endangered (EN), and Vulnerable (VU). Near Threatened (NT) is a category in which a species is close to qualifying for a threatened category, Least Concern (LC) is for widespread and abundant species, and Data Deficient (DD) is assigned to species which have been assessed but for which there is insufficient information available. There are five criteria, or decision rules, for determining the categories: A. Population reduction, B. Restricted distribution, C. Small population, D. Extremely small population, and E. Quantitative analysis showing a probability of extinction. Critically Endangered (CR) taxa face an extremely high risk of extinction in the wild in the immediate future, or a 50% probability of extinction within 10 years or 3 generations, as based on recent population decline, current population estimates, projected continuation of population decline and current geographic spread.

Two capuchins, *C. xanthosternos* and *C. apella margaritae* are listed as "Critically Endangered" on the *2002 IUCN Red List of Threatened Species* (Hilton-Taylor 2002, see Table 1.4). *C. xanthosternos* is today the largest of the primates throughout most of the region where it occurs (howling monkeys, *Alouatta guariba*, and the muriqui, *Brachyteles hypoxanthus*, are already practically extinct there). It is heavily hunted and, with its need for relatively large home ranges, it is also the most seriously affected by the continuing destruction and fragmentation of its forests (Coimbra-Filho *et al.* 1991/1992, Oliver and Santos 1991, Santos *et al.* 1987). As a consequence it is regarded as one of the most endangered of the New World primates (Pinto, *et al.* 1998, Rylands and Pinto 1994) and is facing imminent extinction in the wild if its habitat in the Atlantic forest of southern coastal region of Bahia (Figure 1.4) continues to be destroyed at present rates. An International Recovery and Management Committee was set up for *C. xanthosternos* (and *C. nigritus robustus*) by the Brazilian Institute for the Environment (IBAMA) in 1992 (Santos and Lernould 1993), which oversees a still-incipient captive breeding program, initiated at the Rio de Janeiro Primate Center, but which now has the participation of a number of zoos in Europe, including Zürich, Switzerland; Frankfurt, Germany; Mulhouse, France; and Chester, England.

Only a few, small, populations of *C. apella margaritae* remain on Margarita Island, Venezuela (Figure 1.4). Rylands *et al.* (1997) also list a subspecies of the white-fronted capuchin, *C. albifrons trinitatis* restricted to Trinidad, as critically endangered (Figure 1.2). These two capuchin subspecies occur only in mountainous regions on their respective islands and are hunted as crop pests and for pets (Agoramoorthy and Hsu 1995, Martinez *et al.* 2000). Martinez *et al.* (2000) reported that *C. olivaceus* is a common pet on Margarita Island, and indicated that there may be feral populations competing with the few remaining *C. a. margaritae*. While no capuchins are presently in the Endangered category, one species (*C. kaapori*) and two subspecies (*C. nigritus robustus* and *C. capucinus curtus*) are considered to be Vulnerable, because of their small geographic distributions, deforestation, and hunting pressure. *C. capucinus curtus* is another island population (Gorgona Island, Colombia, Figure 1.1), and *C. kaapori* can only be found at a few locations in an extensively deforested region of eastern Amazonia, Brazil (Figure 1.3). *C. kaapori* "inhabits the region with the highest human population density and the highest level of deforestation and habitat degradation in Amazonia" (Carvalho, *et al.* 1999, p. 42), making it one of the most endangered of the Amazonian primates. *C. nigritus robustus*, although in some places locally abundant, has a relatively small range and has been decimated over a large part of the Atlantic forest where it lives.

Mainly because of their reasonably large geographic ranges in comparison with other taxa, the four "traditional" capuchin species (*C. capucinus*, *C. albifrons*, *C. olivaceus* and *C. apella*) are categorized as Lower Risk. However, while two subspecies of *C. albifrons* are considered Lower Risk by Rylands *et al.* (1995) (one of them, *unicolor*, now considered a junior synonym of the other, *C. albifrons albifrons*, see Defler and Hernández-Camacho (2002)), nine others are listed as Data Deficient (DD). Nothing is known of their status, but their (poorly known) distributions are evidently small and in relatively disturbed areas of Colombia, Ecuador and Venezuela. Although it is possible that a number of subspecies are not in fact valid (Groves 2001, Rylands 2001), the preservation of these populations is vital to protect the genetic diversity of the species.

Much primate conservation monitoring focuses necessarily on population declines. However, the increasing establishment and improved protection of parks and reserves in some Latin American countries, such as Costa Rica, is a bright note in an otherwise gloomy picture of capuchin conservation status. For example, some studies (Fedigan and Jack 2001, Sorenson and Fedigan 2000) demonstrate that capuchin populations have increased substantially in the 30 years subsequent to the establishment of a national park on reclaimed ranchlands. This indicates that attempts to protect, and indeed to regenerate, tropical forests can promote the recovery of capuchin populations in parts of their natural habitats where they have experienced serious decline due to human disturbance.

SUMMARY

The taxonomic status of capuchin species and subspecies is still unsettled. Groves (2001) provided a revised taxonomic scheme for *Cebus* that proposes three new species elevated from subspecies status. He did not, however, publish distribution maps with his revised taxonomic scheme, only brief notes of the geographic ranges of the taxa he listed. Because capuchins show considerable phenotypic diversity at the individual, population and subspecies level that has not been explored genetically, and because capuchin monkey populations over large parts of South America remain undocumented, it is not possible to define the exact ranges of the recognized forms, although the study of Silva Jr. (2001) has now provided a remarkable synthesis of what is known to date. As noted earlier, the vast majority of the existing literature on capuchins refers to these monkeys in terms of the four traditional species rather than the eight new ones, and the following chapters of our book will therefore do the same, except in cases where one of the recently designated species (such as *C. xanthosternos*) has been the specified object of study.

Some recent molecular studies have restricted the composition of the family Cebidae to the capuchin monkeys, squirrel monkeys, marmosets and tamarins. Although molecular clock inferences suggest that *Cebus* might have first appeared 25 million years ago, there is almost no fossil evidence to confirm this, or to verify the hypothesized phylogenetic relationships between capuchins and their congeneric relatives.

The flexible habitat preferences and wide geographic spread of capuchins across Central and South America have helped these monkeys maintain viable populations in many parts of their ranges even as humans destroy and radically alter their forests. At the same time, the capuchin proclivity to respond to agricultural encroachment on their forests with a switch to crops leads to these monkeys being hunted as pests. They are also large enough to be a prominent item in the diet of many native South American groups, and entertaining and endearing enough to be kept as pets when they are young. The continuing destruction of Neotropical forests has led to the isolation and population decline of some subspecies and species of capuchins, which are now placed in the Critically Endangered and Vulnerable categories of the IUCN Red List. We cannot rest complacent that capuchins will continue to thrive in the tropical forests of Central and South America without concerted efforts to conserve and restore their habitats, and to control the capture and trade of these fascinating, adaptable monkeys.

2 • Behavioral ecology: how do capuchins make a living?

There are at least two ways to find capuchins in the forest. You can walk rapidly around their home range searching for them or you can wait for them at one of their favorite spots – a fruiting tree or a waterhole. In dry season, they will come to a water source to drink at least once a day. Sitting quietly by such a waterhole, I hear the familiar twittering vocalizations and whipping of branches under monkey feet that announce the imminent arrival of a capuchin group. A low-ranking subadult male arrives first, rapidly scans the area around the waterhole, takes note of my presence, and descends for a drink among the buzzing insects at the spring-fed waterhole. He leaps up from the water just as I realize that I am now surrounded by white-faced capuchins, like so many elves in the forest, some walking on the ground, others travelling in the lower canopy. The alpha male and female head down to the water with several infants and young juveniles. The subadult male who came in to drink first is now in a *Luehea* tree checking for the fruit that have begun to crack open so that their seeds will be dispersed when the wind shakes the branches. The male selects one of the hard fruits and pounds it against a large branch, cupping his other hand around the fruit to catch the dehiscent seeds in mid-flight. A wrestling, chasing game has broken out among the juveniles on the leaf litter, some of whom are unfurling leaves and rolling over stones to look for insects. A shower of debris falls on me and I look up to see an adult female rummaging in a tangle of vines above my head. Her sister sits in a neighboring *Bursera* tree, tapping twigs and listening for hollow sounds, before she bites them open to lick out the beetle larvae. I keep a wary eye on her juvenile son who is prying pieces off a wasp nest attached to a branch of the tree – sometimes the angry wasps swarm us all. Suddenly, the alpha male, who has been sitting watchfully, gives a loud call signaling that another group of capuchins has been spotted. Followed by the rest of the adult males, he rushes off in the direction of the other group, who are probably coming in for a drink. Infants leap onto the nearest sizable monkey and then the females and young fade into the forest. Like a rambunctious troop of travelling tinkers they are gone as swiftly as they arrived. Apart from the incessant drone of the insects, silence descends on the waterhole again.

If we are what we eat, then capuchins should be tough, ingenious and complex creatures with effective defenses against predators. And so they are. Even though capuchins of all species appear to relish easily accessible and fleshy fruits like figs if available, members of the genus *Cebus* also find ways to obtain food from sources that other monkeys do not use. These foods "less often taken" by other types of primates are embedded in hard shells and tree bark, or protected by spines, thorns or biting/stinging organisms. Indeed, capuchins seem to make a specialty of targeting food that "fights back" – that is, plants, insects and small vertebrates with strong and elaborate defenses against predators. Focusing on capuchins' ability to locate, process and remove embedded foods, Parker and Gibson (1977) referred to them as "extractive foragers," an appropriate appellation. Terborgh (1983) described *C. apella* as "destructive foragers" because of their propensity to pound apart hard nuts, strip bark, break branches and bite open tough fruits – all of which requires strength. Similarly, Fragaszy (1990a) called *C. olivaceus* "strenuous foragers." They can also be characterized as innovative and "extreme" foragers because of their ability to acquire sustenance from a vast array of unlikely and potentially dangerous sources that require special foraging skills and pugnacity – for example, bee hives and wasp nests, army ants, tree holes, vine tangles, palm fronds, bromeliads, uprooted saplings, spider webs, and swollen acacia thorns guarded by ferocious ants. Capuchins are even known to snatch food on occasion directly from the mouths of predators and the live traps of astounded field biologists. No wonder that Costa Ricans refer to them as "monos muy bravos" (bold or belligerent monkeys).

This unique capuchin perspective on where and how to acquire food has at least two types of adaptive value. First, it gives them the flexibility to switch from readily accessible foods such as fruits to more inaccessible ones at times of fruit scarcity. Second, it reduces the degree of dietary overlap between capuchins and other arboreal vertebrates that specialize on fruits or insects, such as spider monkeys, squirrel monkeys, coatis, ant eaters and parrots. Thus, capuchins may sidestep direct food competition with these species by exploiting somewhat different resources.

In this chapter we will look briefly at studies of how capuchins manage to make a living and stay safe. Thus we will first talk about how capuchins distribute themselves and navigate in space. Then we will examine variation in capuchin behavior over time – daily activity patterns, nightly sleeping preferences and seasonal patterns in ranging and feeding. Next we will turn to what they prefer to eat – their diet as well as the flexibility they exhibit in food and feeding patterns between and within groups, and the occasional use of plants and invertebrates for non-nutritional purposes. Finally, we will consider group size and its relationship to feeding competition, predation risk and reproduction.

USE OF SPACE: RANGING PATTERNS

How groups of capuchins use their home ranges

One of the primary areas of interest to primate ecologists is how monkeys use space – what areas does a group cover as it makes its rounds? Does the group exploit this range in a regular and predictable or erratic manner? And does the group defend this area against conspecifics? As human primates who must struggle to find and follow groups of monkeys through the forest, we often consider the routes that capuchins travel through their home ranges to be rather unpredictable. Occasionally over a short period of a few days we can predict that our study group will rise before dawn to rush to a large-canopied tree in fruit where they will spend much of the day. But even large fruit trees are soon out of season and then we are back to trying to guess among the many foraging routes and options that capuchins seem to employ.

Terborgh (1983) envisioned at least four distinctive ways that a group of monkeys could exploit its range. A species that relies largely on insects – a rather evenly dispersed source of food – would be expected to cover its range homogeneously. At the other extreme, a species that relies on highly patchy resources such as fruiting trees would use space in a shifting irregular pattern as patches of food come into production. A species that needs to defend its resources would devote a great deal of time to patrolling the boundaries of its range, and a species dependent on one sleeping site or waterhole or other essential but limited resource would be expected to concentrate its movement in a core area. In practice, capuchins at various study sites use combinations of these models in response to the distribution and availability of the resources they are using at any point in time. Given that capuchins can and do vary the proportions of invertebrates and plant matter in their diet and that they live in forests where fruit and insect abundance varies seasonally, it is perhaps not surprising that the nature of their range use is diverse and complex.

For example, at Manu National Park in Peru, *C. apella* mainly use a core area in a relatively small home range of about 80 hectares (0.8 km^2), whereas *C. albifrons* groups shift the focus of their activity from one spot to another in a larger home range of about 150 hectares (1.5 km^2; Terborgh 1983). *Cebus apella* groups show extensive overlap of their home ranges, but *C. albifrons* are aggressive toward neighboring groups. In El Tuparro National Park in Colombia, *C. apella* also show extensive home range overlap and peaceful interactions between groups, whereas *C. albifrons* defend nearly exclusive territories in large home ranges of 120 hectares (1.2 km^2; Defler 1982). Terborgh concluded that during the rainy season when food is abundant, the patterns of these two capuchin species do not look very distinctive. But when fruit becomes scarce, *C. apella* switch to *Astrocaryum* palm fruits and pith within the central area of their range, whereas *C. albifrons* travel further around their range looking for fruit trees in production. *Cebus albifrons*' use of their home range within any short period of weeks is very uneven, while *C. apella* make more consistent use of their home range. This seems to be a choice between staying close to home and eating whatever one can (*C. apella*) or travelling further and eating something a bit more desirable (*C. albifrons*).

However, in Iguazú National Park in Argentina, near the southern edge of their distribution, *C. apella* conform to a shifting patch model even though they use a core area heavily throughout the year (Di Bitetti 2001). Also, the distribution and productivity of fruit trees are the main factor affecting the shifting patterns of range use by *C. apella*. Range size (mean: 161 hectares (1.61 km^2), range: 81 to 293 ha) is correlated with the number of females per group. Unlike the situation at Manu and El Tuparro where *C. apella* groups interact peacefully, tufted capuchin groups at Iguazú are universally aggressive toward each other wherever they meet, with the losing group retreating toward its core area. Similarly, *C. apella* in French Guiana shift their ranging patterns in response to fruit availability and distribution, while at the same time maintaining a core area of high use (Zhang 1995a).

Janson (1996, 1998b) and Janson and Di Bitetti (1997) have carried out extensive experimental work at their Iguazú site in which they place fruit on feeding platforms in a group's range during the season of fruit scarcity. From these experiments, they have determined that a group of *C. apella* travels from one food site to another on the basis of three variables/principles: they prefer food sites closer to their present location, they prefer sites with a greater abundance of food, and they prefer sites that they have not visited within the previous 24 hours. When these variables are all at play, it seems that distance to be traveled is more important than amount of food reward to *C. apella*. Janson (1998b) inferred from the results of these experiments that the movements of his study group are guided by spatial memory (see Chapter 8).

Cebus olivaceus in Venezuela occupy fairly large home ranges, averaging 275 hectares (2.75 km^2), which they use erratically. When fruit becomes scarce in one part of their range, the group moves to another region in the range with higher food availability (Robinson 1986). Although they move in a forward motion, with little backtracking between areas of their range preferred for feeding and those sites best suited for sleeping, they use their home range very irregularly, spending 75% of their time in 31% of their range. In Robinson's view, *C. olivaceus* essentially forage from one fruit tree to the next, eating insects along the way.

Ranging patterns of *C. capucinus* were briefly described by Oppenheimer (1968, Barro Colorado Island in Panama), Mitchell (1989, Barro Colorado Island in Panama) and Buckley (1983, Honduras) and studied in detail by Chapman (1987a) and Rose (1998), both working in Santa Rosa National Park in Costa Rica. All these researchers reported home range sizes for *C. capucinus* to average 1 km^2, although Rose (1998) found that larger groups with greater numbers of adult males have larger home ranges than do smaller groups. Oppenheimer concluded that the *C. capucinus* of Barro Colorado Island are "territorial," as did Buckley for the capuchins he studied in Honduras. However, Mitchell argued that this species is xenophobic rather than territorial – interacting aggressively wherever groups meet, rather than defending a specific core area. That is also the case at the Santa Rosa field site in Costa Rica, and seems to be true at most field sites where capuchins have recently been studied.

In the dry forests of Costa Rica, capuchins (unlike howlers at the same site) do not make any one food tree their focal point for any sizable length of time (Chapman 1988). Rather capuchins move from one food tree to another in a matter of a few hours, or they make foraging excursions away from a large fruit tree only to return several hours later. Ranging patterns vary over time as different types of foods become available. During the dry season, these capuchins become almost "central place foragers" around the few spring-fed waterholes that can provide them with moisture. Water is a resource rarely considered in ecological studies of primates, but at Santa Rosa it was found to be an important determinant of how capuchins range in the dry season (Fedigan *et al.* 1996). *Cebus capucinus* is the only monkey species in Santa Rosa that descends to drink water almost daily during the dry season. As they do with large fruit trees, capuchins seem to make short excursions away from the water source and return to it repeatedly. Almost immediately after the rains begin and water becomes available from tree holes, the groups vastly increase their travel over their home ranges.

It seems generally true in primate ecology that the more data we collect on a given species, system, or phenomenon, the more we learn to appreciate the flexibility and complexity of that system. During the early days of capuchin field work, researchers tended to make a distinction between tufted and untufted capuchins in their pattern of range use. It looked as if the "tufted" *C. apella* had smaller, overlapping home ranges with core areas and little aggression between groups, whereas the untufted capuchins (*C. albifrons*, C. *olivaceus* and

C. capucinus) had larger home ranges and exhibited "territorial" behaviors (defense of exclusive ranges). Increasing evidence from a number of sites has shown that ranging patterns are more variable and complex, such that it is difficult and probably not appropriate to describe one "typical" home range use pattern for each species. However, we can conclude that at least two major factors are responsible for how capuchins use their home ranges: (1) the way in which their food and other resources are distributed spatially; and (2) their ability and propensity to exploit seasonally available resources. Furthermore, it is probably accurate to conclude that capuchins are not territorial in the sense of actively working to exclude all conspecific groups from their home range (as do the closely related callitrichid monkeys – marmosets and tamarins). Rather, capuchin groups exhibit a tendency to react aggressively to conspecific groups whenever and wherever they meet. This tendency is variably expressed in different habitats, depending on food availability, distribution and abundance.

How monkeys distribute themselves within the group

As well as looking at how a social group of primates distributes itself in its habitat, primatologists are also interested in how individual monkeys deploy themselves in the space within the group. An individual's position in a foraging group relative to its neighbors can affect both its foraging success and its vulnerability to predators. Predation risk exerts pressure on group members to coalesce, whereas feeding competition leads them to disperse. In *C. olivaceus*, the best spatial position for finding and capturing insects is at the front of a group (Robinson 1981). This is because the supply of edible insects is depleted as the foraging group passes over an area. The best position for accessing fruit trees is just behind the leading edge of the group. Dominant individuals in this position can exploit information about location of fruit trees from those in front of them by letting others do the finding and then supplanting them from the choice fruit tree locations. But the best position for safety from predators is at the center of the group. Taking these factors into consideration, one would predict that dominant adults would be found at the front-center of the group, which is the optimal position for both food access and safety from predators. Infants and juveniles would be more likely found in the safe central areas of the group, and subordinate adults would be more likely found on the peripheries – the vanguard or the sides and rear.

By and large these predictions do fit the general pattern of how individuals locate themselves in traveling, foraging groups of *C. apella* (Janson 1990a, b) and *C. capucinus* (Hall and Fedigan 1997). However, there are many complexities brought about by the fact that individuals in capuchin groups often forage in a very dispersed manner and that individuals may locate foods that require some form of processing and extraction. Then, because the group is continually in motion, an individual may start out at the front of the group, locate a tasty morsel that requires extraction and stop to deal with it while the rest of the group moves on, finally ending up at the rear of the group.

Furthermore, in *C. olivaceus*, ranked subgroups may take turns in a fruit tree (Robinson 1981). First the dominant male and female and their young offspring may feed simultaneously while the others wait in adjoining trees or take fallen fruit from the ground. If the tree is large enough, middle-ranked individuals may also feed with the dominants. Then when the first subgroup is satiated they move out of the tree and those "waiting in line" will move in next. Similarly, Phillips (1995) pointed out that *C. capucinus* tend to forage alone in small trees, but in contrast they forage in successive subgroups in medium-sized trees, and they forage simultaneously in large fruit trees. This suggests that the degree of intragroup feeding competition varies with the size of the food patch. In addition, the degree to which the entire group is spread out (or cohesive) during foraging is probably related to the type of food available as well as to the size of the group (Chapman and Chapman 2000, Garber and Rehg 2000).

USE OF TIME: ACTIVITY PATTERNS

Daily activity

As a researcher, the greatest drawback and the biggest motivator to following capuchins is that these monkeys are almost unremittingly active during the day, from shortly before dawn to sometime around dusk. We look enviously at our howler-watching colleagues carrying collapsible chairs and paperback novels into the forest to tide them over the long period of howler inactivity during the middle of the day. Although there is variation among species, groups, individuals and seasons, here is the typical pattern of a capuchin's day:

> When the first fingers of dawn stretch across the sky, well before light has come to the forest floor, our focal monkey starts to stretch and move away from the warm cluster of bodies with whom he has shared a night's rest. He travels quickly into that neighboring fruit tree the group was foraging in last evening just before they entered the sleeping tree. He picks at the leftover fruit from last night and gives a contact call. Once everyone is awake and moving around, he begins to travel in earnest search of his first big breakfast of fruit. Depending on the size of fruit trees in this season, he will visit and feed in one or several trees between 5 and 9 a.m. Around mid-morning, the pace of feeding and traveling slows down and our focal capuchin begins to forage for insects. Throughout the hotter middle part of the day, he will intersperse insect foraging with eating whatever fruit is close at hand and with short periods of rest. And, if it is the dry season, he will visit the waterhole for a drink. If he is a young male, he may join the juveniles in rough and tumble play; if he is an older male he may join adults for a grooming session. When the heat of the day begins to break in mid-afternoon, our focal capuchin once again begins to search intensively for fruit to fill his stomach. From now until dusk the group will travel in search of fruit, sometimes eating a few insects along the way. Dark comes early and quickly in the tropical forest and sometime between 5 and 6 p.m. the group travels in the direction of sleeping trees. Their favorites are tall emergent trees near a good fruiting tree. When it is already dark on the forest floor but there is still light on the canopy, our focal capuchin enters a tree containing a likely sleeping companion or two and tries out a few places before making himself comfortable in the fork of a branch.

These are the main components of an activity or time budget: travel, feed/forage, rest, socialize. (For examples of detailed, quantitative time/activity budgets, see Bernstein 1965, Fragaszy 1986, Fragaszy 1990a, Robinson 1984a, Ross & Giller 1988, Terborgh 1983, Thorington 1967, Zhang 1995a.) What is important about the pattern of such very general categories of behavior as travel, forage, rest, socialize, and why would a primatologist take time to record and compare them with other species? Activity patterns are of interest to ecologists because they inter-relate in multiple ways with diet, body size, range use and social behavior, as well as with external factors such as ambient temperatures and food availability.

What a monkey eats constrains how he uses his time and space. Plant eaters, especially large ones, have to spend much of their time ingesting, in order to get enough food into their systems to meet their metabolic needs. Insect eaters have to spend much of their time locating and capturing prey. The primary difficulty for a frugivore is that fruit trees are scattered in patches between which they must travel. The primary difficulty for an insectivore is that insects are small, scattered and often concealed food items, so they take longer than fruit to locate and eat. Thus, as Terborgh (1983) described, a need to obtain protein in the form of insects puts constraints on the time budget, and the need to obtain ready carbohydrates in the form of fruits puts demands on the use of space.

Most capuchins are classified as omnivores, specifically as frugivore-insectivores. Although these classifications are over-simplifications, and capuchin diets are highly variable across time and space, this classification fits well with capuchin body size and behavior. Omnivorous primates often obtain their carbohydrates and the bulk of their calories from fruit – primate fast food if you will. For their much-needed proteins, they can turn to young leaves (as do howlers), or seeds (as do saki monkeys), or insects, as do capuchins. But if insects are the major source of protein for an omnivore, then its body size cannot be very large. Terborgh (1983) estimated that the weight of an adult male capuchin (about 4 kg) is near the upper limit for an animal that relies on insects for its protein. Terborgh also suggested that capuchins spend a great deal of their time foraging and that, in this respect, they conform to some of the expectations of "energy maximizers" rather than "time minimizers" (Schoener 1971). Because capuchins must devote a lot of time to locating and capturing their protein and a lot of travelling to reach their carbohydrates, there is less leisure for resting and for social interactions than is found in monkey species with other diets that are gum-, nectar-, seed- or leaf-based.

Robinson (1984a) did an extensive analysis of why *C. olivaceus*, as a general pattern, are prone to eat more fruit in the early mornings and late afternoon, and to eat more invertebrates during the middle of the day, especially during the dry season. He found that neither fruit nor insects were differentially available at the times of day that capuchins prefer to eat them. However, he suggested that *C. olivaceus* in Venezuela might prefer to fill their stomachs with sweet, fleshy fruits (like figs) first thing in the morning to quell their hunger quickly and raise their blood sugar. (Terborgh found the same pattern and made the same suggestion for *C. apella* and *C. albifrons* in Peru). Since many of the invertebrates consumed by capuchins are found on the ground, Robinson also suggested that monkeys may wait to hunt these animals until later in the morning when there is less danger from nocturnal terrestrial predators such as the cats and tayras. Freese and Oppenheimer (1981) suggested that capuchins may not hunt invertebrates in the early morning hours because the light is too dim to locate and capture them effectively. The fact that capuchins slow down their foraging efforts in the middle of the day is likely related to increasing ambient temperatures and the need to conserve energy and water at the hottest times of the day (Robinson 1984a).

Nightly sleeping patterns

Most primates occur at tropical latitudes where darkness covers the forest for approximately 12 hours out of every 24 around the year. But even though primates in the wild spend nearly half of their lives sleeping, there has been very little published about their nocturnal patterns and preferences (for a review, see Anderson 1998). We know that arboreal primates sleep in trees, but beyond that how do they select roosting sites? There are probably several factors that influence their choices: safety from predators, comfort, stable substrates that lower the possibility of tumbling during the night (yes, even monkeys fall out of trees), and locations that are large enough for the entire group to sleep fairly cohesively.

From the few published reports available and the unpublished accounts of researchers, capuchins seem to prefer large tall emergent trees with many horizontal branches and most groups have more than one sleeping site (Figure 2.1). For example, based on a 14-month study, Zhang (1995b) reported that *C. apella* in French Guiana preferred tall palms in the center of their home range, which were inaccessible to most predators and afforded the monkeys security, comfort and the possibility of sleeping in social subgroups. Similarly, Di Bitetti *et al.* (2000) found that *C. apella* in Argentina preferred tall emergent trees (although not palms) with a big crown diameter and many horizontal branches that allowed small subgroups of monkeys to sleep together on the ends of branches. Di Bitetti *et al.* (2000) documented 34 different sleeping sites used by the study groups over 203 nights of observation. Although some sleeping sites were used repeatedly, they were used in an unpredictable fashion, which is hypothesized to lower the risk that a predator can key into a consistent sleeping site. Both the Zhang and the Di Bitetti *et al.* studies found that *C. apella* seem to sleep at the closest appropriate site to where they have been foraging in the late afternoon, rather than making a long trek to a favorite site, as do many of the terrestrial primates of the Old World.

Defler (1979a) reported that *C. albifrons* prefer tall trees and sleep near the ends of branches, and Freese and Oppenheimer (1981) described the same pattern for *C. capucinus*, so it is likely that this preference is a generic trait. Most aspects of sleeping site selection are attributed to predator defense – for example, the fact that capuchins like large, tall trees but they do not sleep near the trunk where an arboreal carnivore might be able to reach them. Instead they sleep out on the branches, often in a stable fork of a large branch, which will vibrate if a predator should step onto it in the night. They also like to sleep in areas where a few large trees occur in a clump so that it is easy for a monkey to leap from the branches of one tree into the next, should trouble appear in the original sleeping tree. Buckley (1983) pointed out that capuchins may also like to sleep high in the trees because there is a lower abundance of mosquitoes and biting flies at that level in the canopy. Freese and Oppenheimer had originally reported that *C. capucinus* frequently used one sleeping area, however, many later studies have reported that *C. capucinus*, like the other *Cebus* species, switch their sleeping sites unpredictably from night to night. *Cebus capucinus* in Santa Rosa appear to have favorite sleeping trees in each area of their range that are different from their favorite feeding trees. Even though a given group may have only a small repertoire of favorite sleeping sites, it is rare for it to use the same site on consecutive nights; they seem to prefer to rotate sites.

In sum, all capuchins seem to prefer to sleep in tall, emergent trees with many horizontal branches, but species and even group differences have been found in the number of sleeping sites one group will use and how consistently they use them, as well as whether or not they use palm trees. Di Bitetti *et al.* (2000) suggest that

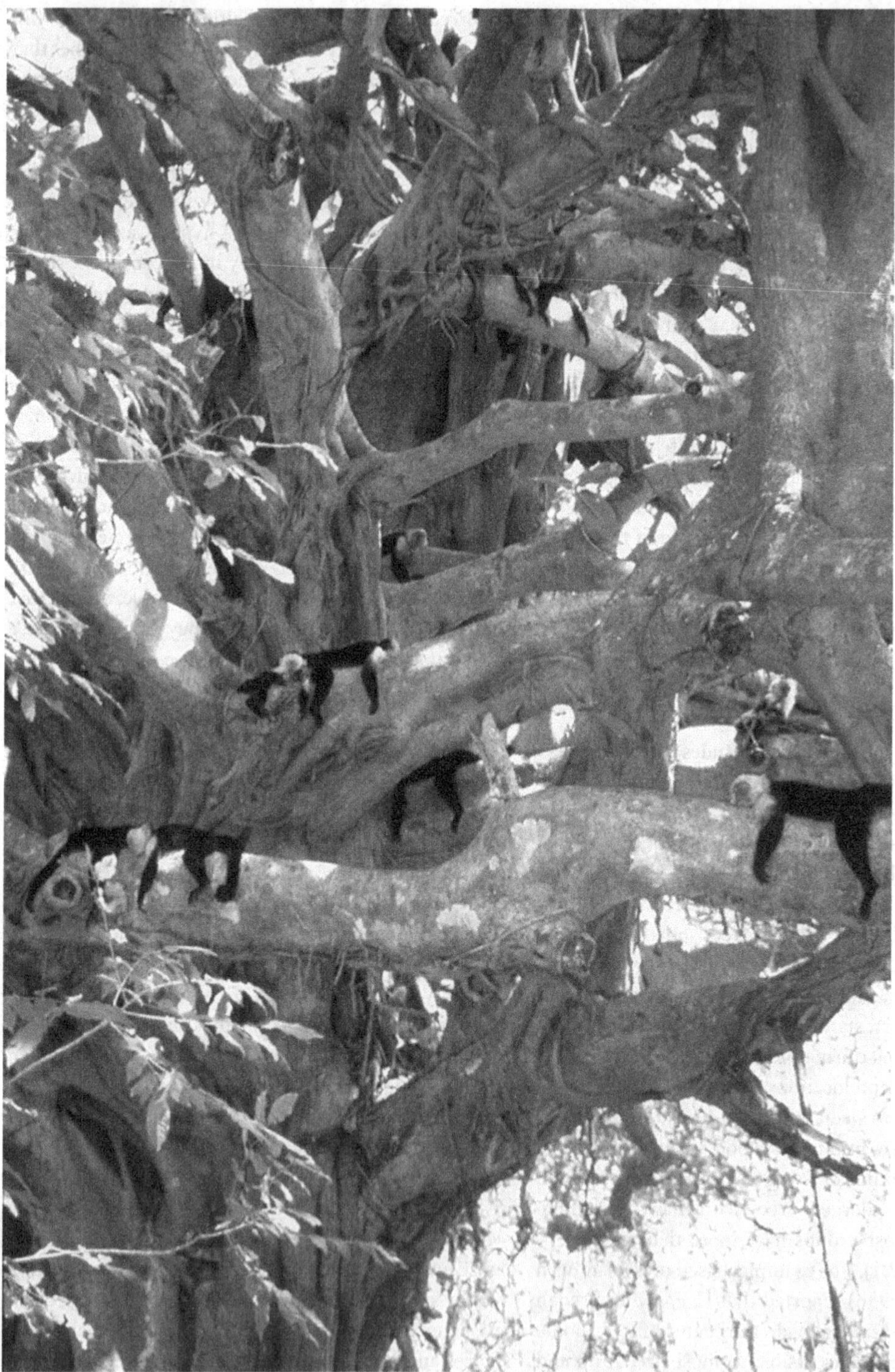

Figure 2.1. A group of *Cebus capucinus* during their siesta in a strangler fig (Photograph courtesy of Katherine MacKinnon).

this variation is due to differences in group size and the availability of trees with desirable features for sleeping.

Seasonal feeding and ranging patterns

Those of us who live in temperate zones far from the equator tend to think of the tropical forests, where most nonhuman primates live, as warm and wet throughout the year, that is, "nonseasonal." What has probably led us to this inaccurate assumption is that in tropical zones the average daylength and daily average temperatures do not fluctuate as strongly throughout the year as they do further from the equator – in other words tropical areas do not experience extreme cold and dark seasons. But rainfall patterns in the Neotropical forests do vary greatly over the course of the year leading to wet and dry seasons that can be as extreme in their own way as our more familiar summer and winter patterns. Tropical flora and fauna respond to wet and dry seasons with fruiting cycles/phenology (Frankie *et al.* 1974) and invertebrate population cycles (Croat 1975, Janzen 1973). This means that the fruits, flowers, leaves and insects on which most monkeys rely for food are differentially available in different months (and different years). Many studies of primates in Neotropical forests have reported that the fruits on which monkeys feed are at their lowest abundance in the late wet season and early dry season, resulting in a resource bottleneck for frugivorous vertebrates around the cusp of the dry season (Janson 1984, Peres 1994, Terborgh 1983). Although other researchers have reported peaks and troughs in fruit production at slightly different points in the annual cycle, it is fair to say that the rainy season is usually associated with more food for capuchins and the dry season with less (Chapman 1988, Miller 1992a, Robinson 1986).

For example, in the tropical dry forests of Guanacaste Province in Costa Rica it is usual for no appreciable rain to fall between December and May, whereas 800–2600 mm of rain falls between the months of June and November. In these forests, more of the fruits that *C. capucinus* prefer are produced during the early rainy season than at other times. Caterpillars (another favorite capuchin food) are also widely available during the early rainy season. Only the asynchronously fruiting fig trees and a few other tree species produce fruit eaten by capuchins during the dry season, and such plant species are vital to capuchin survival through the dry months. During the season of fruit scarcity, *C. capucinus* turn to eating more chitinous insects and embedded hymenoptera larvae, and they prey more readily on squirrel, coati and bird nestlings and eggs. It is also likely that vertebrate predation increases during the dry season because squirrels, coatis and many bird species nest at this time, and young pups or nestlings are easier for capuchins to subdue than are adult prey. During the late dry season, capuchins in Santa Rosa become central place foragers around the few remaining bodies of standing water. In a tropical dry forest with only seasonal streams, water can become a scarce resource, especially for monkeys such as capuchins that like to drink at least once a day in addition to the moisture they obtain from the fruits and insects they consume. As the waterholes dry up, day ranges become smaller and there is more overlap between groups, several of which may be trying to use the same waterhole. Intergroup encounters are more frequent during the dry than the wet seasons and a range of related behavioral patterns (vigilance, alarm calls, male–male aggression, male group transfers, etc.) are more likely to occur in the dry season.

The *decrease* in *C. capucinus* day ranges that we see during the dry season in Santa Rosa is in contrast to seasonal ranging patterns found in other capuchin species and sites. For example, Terborgh (1983) reported that during the low fruit season at Manu, *C. apella* maintain their same home ranges by switching to foraging on the palm nuts and pith located in their core area. However, *C. albifrons* start travelling widely in search of food, thus increasing their home range size. Janson (1984) also described how, during the fruit scarcity season, *C. apella* at Manu turn to small crown fruit trees, especially *Scheelia* palm nut clusters. Only the alpha male is strong enough to break open these nut clusters, but he allows the others to eat from them. Janson argued that a female *C. apella* at this site will invariably choose the alpha male for mating during her conceptive periods in part because the alpha male's tolerance at food trees can strongly influence her feeding success and that of her offspring. Similarly, Peres (1994) noted that *C. apella* in Amazonia switch from being generalist foragers during the wet season to palm nut specialists during the dry season. At other sites where *C. apella* have been studied, such as Iguazú National Park in Argentina (Di Bitetti 2001a) and Nouragues field station in French Guiana (Zhang and Wang 1995), brown capuchins have instead been reported to *increase* their day ranges at the start of the fruit scarcity season, because they travel to distant areas of their range in search of food. Similarly, Robinson (1986) reported that *C. olivaceus* increase their day ranges during the fruit bottlenecks, and Peres (1994) picturesquely described *C. albifrons* groups as travelling "relentlessly and uncohesively" across huge ranges during the season of fruit scarcity.

All this variability begins to make some sense when we consider that there are two basic foraging responses that animals can make to fruit scarcity: they can turn to a less preferred food that is still readily available right in their core area (as *C. apella* does with palms at many sites) or they can travel to distant parts of their range that offer some preferred fruit, even if in low abundance and widely dispersed patches (as *C. albifrons* and *C. olivaceus* are reported to do). And sometimes they can do a little of both. Furthermore, *C. capucinus* in Santa Rosa have to contend with an additional constraint in the form of limited drinking water, so they decrease their ranges when waterholes are drying up and turn to eating non-fruit items within a short traveling distance of water. The list of foods that capuchins resort to at low-fruit times is extensive and instructive, and their ability to use these alternative foods is essential to their success in a variety of Neotropical forest types. We have already discussed palm nuts and pith, which are widely reported to be used by capuchins during the lean season (Gilbert *et al.* 1988, Izawa 1979, Peres 1994, Spironelo 1991, Terborgh 1983). In periods of fruit scarcity, they are also reported to turn to embedded insects, leaf bases of bromeliads, bromeliad fruit, flowers and nectar, seeds of wind-dispersed trees, hard-shelled dry fruits which must be banged open, and vertebrate prey such as nestlings and mammal pups (Brown *et al.* 1986, Brown and Zunino 1990, Chapman 1988, Fedigan 1990, Galetti and Pedroni 1994, Miller 1992a, Mitchell 1989, Peres 1994). Many of the extraction and processing skills of capuchins that have drawn so much attention in research with captive capuchins are best seen in the field when these monkeys are in the process of obtaining and eating such "fall-back" foods. Teaford and Robinson (1989) demonstrated that the reliance of capuchins on hard, dry fruits and chitinous insects during the dry season is even detectable on their teeth due to seasonal differences in microwear on molars.

Tropical trees do not fruit all at once. Studies of fruiting phenology have shown that most Neotropical trees are highly seasonal but have staggered periods of fruit production throughout the year when compared across plant species. It is, however, the case that at many of the capuchin study sites *more* of the trees on which monkeys rely produce ripe fruit during the early and middle parts of the rainy season. In contrast, some trees, such as the ubiquitous and important fig species, have adopted a strategy of fruiting asynchronously or aseasonally so that animals will come to them for food at all times of the year and disperse the tiny fig seeds throughout the forest in their dung. And although rainfall is the key to tropical seasons, the wet and dry seasons do not correspond exactly to the seasons of fruit abundance and scarcity. There is variation across sites in the timing of fruit peaks and troughs. Finally, due to variably prevailing wind patterns and ocean currents, wet and dry seasons occur in different months of the year at the various Central and South American sites.

In Chapter 4, we will address the question of whether these seasonal patterns of foraging and food availability lead to any seasonal patterns in mating and births.

DIET

Use of plant and invertebrate resources

Once we know where our animals occur in space, how they typically move around in it, and the general temporal patterns of their activities, the next thing that a primate ecologist usually wants to know is what they eat. Capuchins are famous for their eclectic diet or (as some researchers have suggested) for trying to eat almost anything remotely edible. But most of their diet is made up of fruits and insects. Hladik *et al.* (1971) did a nutritional analysis of the foods eaten by *C. capucinus* on Barro Colorado Island in Panama in order to determine the total nutritional content of their diet. They found that the capuchin diet consisted of 14.4% protein, 26.3% carbohydrate/sugars, 15.8% fat, 7.6% cellulose and 36% minerals and insoluble sugars. The capuchin diet at Barro Colorado Island was made up of considerably more protein and fat, and fewer carbohydrates and soluble sugars than was consumed by the frugivorous spider monkeys and folivorous howlers studied at the same site.

A number of studies have produced food lists for capuchins, demonstrating the vast array of plants and animals these monkeys eat (see Appendix I; see

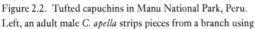

Figure 2.2. Tufted capuchins in Manu National Park, Peru. Left, an adult male *C. apella* strips pieces from a branch using strong arms and teeth. Right, a younger adult male feeds on *Scheelia* palm fruits. (Photographs courtesy of Charles Janson).

also Buckley 1983, Chapman 1987b, Defler 1979a, Galetti and Pedroni 1994, Izawa 1979, Miller 1992a, 1998a, Mitchell 1989, Moscow and Vaughn 1987, Oppenheimer 1968, 1982, Terborgh 1983, Robinson 1986). Freese and Oppenheimer's (1981) list of capuchin foods includes more than 50 families and about 200 species of Neotropical plants; Appendix I expands this list considerably. Janson and Boinski (1992) published a list of animal prey consumed by capuchins that included 17 taxa of arthropods, mainly social hymenoptera and lepidoptera larvae. Terborgh (1983) states that, at his field site in Manu, Peru, the most important families of food trees for *C. apella* and *C. albifrons* are Moraceae (including figs), Annonaceae (including soursop) and Palmae (palm nuts and pith; see Figure 2.2). To these three families of food trees, we would add the following ones that include many species of food trees important to *C. capucinus*: Anacardiaceae (e.g., mangos), Leguminosae (*Inga*, *Acacia*, "bean" type fruits), and Rubiaceae (*Randia*, *Genipa*).

Why do capuchins eat the things they do? For example, why do they prefer the fruits of some plants to those of others also produced in their habitats? Terborgh (1983) has argued that it is not possible to come up with simple criteria that distinguish the fruits of the food trees selected by capuchins – no common color, size or design, except that they are usually fruits made up of soft pulpy material surrounding the seeds, and they are abundant and readily digestible. However, Terborgh then went on to show that there are some features that make it more likely capuchins will eat particular fruits. For example, in Manu, the larger primate species eat larger fruits, and only capuchins can access the hard fruits because they can bite them open with powerful mandibles and molars, after pounding them against branches and rocks. All the monkeys at his site preferred yellow or orange fruit,

Figure 2.3. Young capuchins often eat soft-bodied invertebrates. Left, a young *C. capucinus* holding a wasp nest and feeding on the larvae. Santa Rosa National Park, Costa Rica (photograph courtesy of Katherine MacKinnon). Right, a juvenile *C. olivaceus* eating a caterpillar (drawing by Stephen Nash from a photograph by Dorothy Fragaszy).

which he suggested is related to their visual system. New World monkeys discriminate best among the shorter wavelengths and discriminate colors such as green, blue and yellow (see Chapter 5), leading Terborgh (1983) to conjecture that orange and yellow fruit stand out best against green foliage for capuchins (see also Dominy and Lucas 2001). Furthermore, Janson (1983) demonstrated that capuchins prefer large orange, yellow, brown or green fruits with a husk, whereas birds exclusively eat small red, black, white or purple colored fruits without a husk. Recent research indicates that male capuchins, unlike howling monkeys, do not have the trichomatic color vision that would allow them to discriminate red fruit against a green foliage background (although a majority of females do; Jacobs 1998, see Chapter 5). Finally, Terborgh (1983) concluded that the size and height of trees producing the fruit is very important to whether it will be included in a monkey's diet, however capuchins are the most flexible of all the species at his site in their use of canopy height and tree size.

Why do capuchins prefer to obtain their protein from arthropods and vertebrate prey rather than from readily available sources used by other primates such as young leaves and seeds? And why do capuchins prefer certain kinds of insects, especially embedded insects? The answer is no doubt related to their body size. Dietary patterns are highly related to body size in primates and other vertebrates. Larger species tend to be folivores and frugivores, thereby relying more on carbohydrates to meet nutritional needs, whereas smaller species tend to be more insectivorous, relying more on calories obtained from protein and fat, although still consuming fruit for ready energy (see Richard 1985). Janson and Boinski (1992) looked at the role of morphology in squirrel monkey and capuchin diets and concluded that the bigger the cebine, the less reliance on mobile insect prey and greater reliance on hidden or well-protected insects, such as social hymenoptera and embedded larvae (see Figure 2.3). For example, the smaller-bodied squirrel monkeys obtain most of their prey (typically grasshoppers, cockroaches, beetles, spiders) from the surface of substrates or from unfurling leaves, whereas capuchins often search for prey hidden in tough substrates that require force to remove.

Janson and Boinski (1992) also noted that there are extensive behavioral adaptations for capuchin foraging and dietary patterns. For example, capuchins have the largest brain-to-body ratio of any nonhuman primate and complex convolutions on the surface of the brain (see Chapter 5). This neural complexity is likely to be related to their foraging skills – both their ability to access food that other monkeys cannot or do not eat, and their ability to map their food sources in time and space (King 1986).

Many authors have noted that capuchins use behaviors in foraging that are most unusual in monkeys. To put

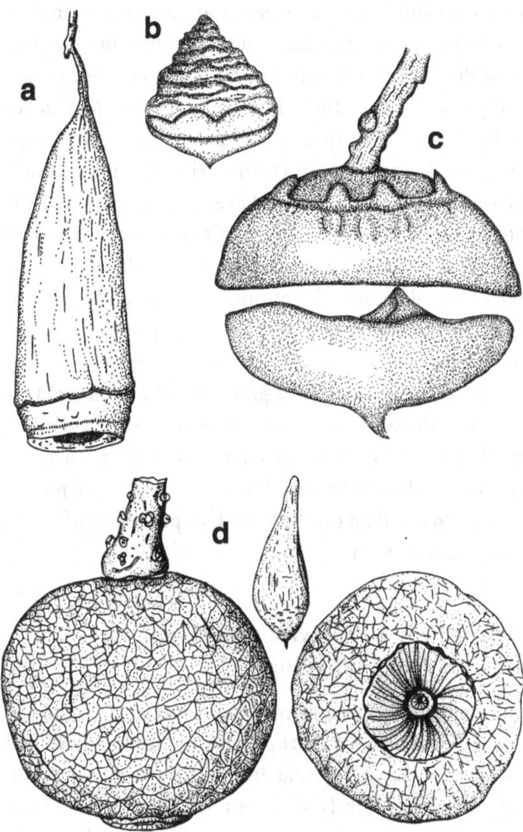

Figure 2.4. Composite of hard-shelled fruits exploited by capuchins. (a) *Couratari oblongifolia* (9 × 2.8 cm); (b) *Lecythis corrugata* var. *corrugata* (previously named *Eschweilera corrugata*) (2.5–5 × 2.5–6 cm); (c) *Eschweilera poiteaui* (3.5 × 9.5); (d) *Bertholletia excelsa* (9–12 cm diameter). Drawings by Betty van Roosmalen-Blijenberg, from van Roosmalen, M. G. M. 1985. *Fruits of the Guyanan Flora*. Utrecht-Institute of Systematic Botany. (Courtesy of Betty van Roosmalen-Blijenberg and Marc van Roosmalen).

it in a nutshell, capuchins forage actively and destructively, using their hands and teeth, often simultaneously, to pull, bite and smash open potential food items. Field observers from sites spanning Central America to Argentina (Boinski *et al.* 2000, Fragaszy 1986, Fragaszy and Boinski 1995, Izawa 1979, Panger 1998, Terborgh 1983, see Janson and Boinski 1992 for review) emphasize that capuchins commonly bang food-containing objects on a substrate to break them open (snails, hard nuts, well-protected fruits and seeds, see Figure 2.4). This kind of behavior is called "combinatorial," because it combines an object with a substrate. Vigorous actions combining an object with a surface are surely a signature feature of capuchins' foraging activities, although the kind and quantity of foods obtained from these actions apparently vary considerably across sites, and across seasons at the same site. Combinatorial activities occupied a small amount of adult and juvenile capuchins' (*C. capucinus*) time (< 1%) at Panger's site in Costa Rica, and about the same amount (0.7% of all samples; 2% of foraging samples) in Fragaszy's *C. olivaceus* site in Venezuela (Fragaszy unpubl. data).

After an object is opened, capuchins use finely controlled actions to extract bits of food. Capuchins use dexterous actions to process softer, exposed food items as well, for example, when they pull parts of insects away and discard them. In Chapters 6, 7, 9 and 10, we take up the fascinating story of capuchins' unusual manipulative proclivities in more depth, in relation to the anatomical equipment and motor control that enable it, how it develops ontogenetically, how it is used in solving problems, and how it can even lead to the spontaneous discovery of how to use an object as a tool. Despite the ecological significance and phylogenetic uniqueness of capuchins' dexterous, destructive and combinatorial foraging, this aspect of their behavior is not well-studied in field settings. We suggest that studies of extractive activities and dexterous processing of foods will be very useful to understand the evolution and adaptive context of capuchins' marvelous manipulative skills.

Dietary flexibility – between groups

We have seen that capuchin dietary preferences differ among species, and vary according to time of day and season of the year. However, capuchin diets may also differ between groups of the same species, even between neighboring groups or groups separated by very little distance. For instance, Miller (1998a) documented that two populations of *C. olivaceus* separated by 60 kilometers in Venezuela consumed only 15 out of 79 plant species in common (19%). Chapman and Fedigan (1990) found three neighboring groups of *C. capucinus* in Santa Rosa National Park in Costa Rica exhibited equally radical differences in diets to the extent that one of the three groups would probably fall into a dietary classification of frugivory (83% of their diet being fruit) and another as faunivory (nearly 50% of their food being insect and vertebrate prey). Brown and Zunino (1990) studied dietary variability in *C. apella* diets at different sites in the northwest and northeast corners of Argentina, the southern edge of *Cebus* distribution. They found that bromeliad leaf bases (a nutritionally poor but easily obtained food) made up 73.6% of food at one site, but only 2% of the diet at another site.

Why do groups of the same species show such dramatic differences in diet? The most obvious answer is that the foods available to them must differ and that is indeed a major explanatory factor. A given plant species may grow at one site and not another, or it may grow at both sites but in variable abundance. For example, bromeliads are much more abundant at El Rey National Park where they are frequently consumed by *C. apella* than at Iguazú National Park where they are seldom consumed (Brown and Zunino 1990). Miller (1998a) concluded that much of the divergence she found between *C. olivaceus* diets at two sites was explained by differential resource availability, and Chapman and Fedigan (1990) found that about 30% of *C. capucinus* dietary divergence was explained by availability. As was noted in our discussion of seasonality in the tropics, we often assume that tropical forests are homogeneous when they are in fact quite variable across space as well as time. Miller censused trees at two sites separated by a mere 60 kilometers in Venezuela and found that only 29% of the tree species occurred at both sites. However, even if local differences in food availability explain some part of dietary divergence between groups, that still leaves some variation to be explained. Chapman and Fedigan (1990) found that 69% of the time, the use that *C. capucinus* made of a food tree did not correspond to its availability. What other factors may be at work? It is difficult to quantify the role of other variables that may lead capuchins in different groups to diverge in diet, but we hypothesize that both local traditions and food profitability may be involved. When different macaque groups and chimpanzee communities show intraspecific variability in diet and/or food processing technique, this is often said to be due to "cultural" differences or to local traditions (Boesch *et al.* 1994, de Waal 1999, McGrew 1998, Watanabe 1994, Whiten *et al.* 1999; see Chapter 13). That is, individuals in different groups of the same species may have learned to recognize different sets of plant items as food and only eat items in this set, and they may have developed different ways of processing these foods. Profitability is as difficult to document as local traditions, but it relates to a situation where a given food may be present in the ranges of two groups, but another, more desirable food is also present in the range of one group and not the other. Thus, acacia beans may occur in the ranges of all three study groups of *C. capucinus*, but the only group to eat them during the dry season is the one group that does not have access to fleshy fruits during this time period. In order to evaluate the role of relative profitability in this case, we would need to be able to calculate nutritional value, toxicity and availability of all the potential foods in a group's home range, something that no one has yet undertaken to do.

In summary, we have a good idea of the factors that play a role in dietary divergence between groups – differential availability, abundance and profitability of foods and possibly locally learned traditions – but we do not yet have extensive data on their relative power to influence what capuchins eat.

Dietary flexibility – within groups (sex, age, rank, individual)

Capuchins living in one group are of course not all the same size or age, nor do they have the same nutritional needs. It makes sense that individuals within a group will forage in slightly different ways, reflecting their own preferences for substrates or activities, their own assessments of the risks of moving into particular areas, their own foraging skills and their own tastes in food. We cannot tease apart the contribution of these different factors (and perhaps others we have not mentioned) to the differences we see across individuals in foraging activities in natural habitats. However, we can document differences within groups in foraging, and we can consider which of the several plausible sources of variation seem the most influential in a given case.

Two researchers have attempted to do this for capuchins. Fragaszy (1986, 1990a, Fragaszy and Boinski 1995), working with weeper capuchins at Hato Masaguaral in Venezuela, and Rose (1994a, b, 1998), working with *C. capucinus* in Santa Rosa National Park, Costa Rica, have considered within-group variation in activity budgets, foraging activities and diet. In both studies, sex differences are the largest source of within-group variation. Both males and females rely principally on ripe fruit for the bulk of their diet, but both also complete their diet with a variety of other food types, particularly in seasons when ripe fruit is less available (typically the dry season at Santa Rosa and Masaguaral). Although they spend the same amount of time ingesting food and the same amount of time is devoted to eating fruit, adult males and females allocate differing proportions of each day to foraging. Females forage for more of the day than males. Overall, monkeys devote the overwhelming majority of their foraging effort working on plant materials, and consume fruit more often than any other kind of food. However, males devote more time to consuming animal prey than do females (roughly a third of consumption, for males, versus a quarter of consumption for females at Masaguaral).

In addition to differences in the amount of time devoted to eating animal material, males and females differ in how they obtain animal foods, and the kinds of animal foods they eat. Males are more likely than females to chase mobile prey on the ground, to capture larger vertebrate and invertebrate prey, and to take risky actions during prey capture (for example, to grab a paper wasp nest and run with it, with wasps flying after the thief, stinging its face and hands for tens of meters as the monkey flees). The culinary reward for stealing a wasp nest: the soft larvae, scooped out of the hive one by one and eaten with apparent relish (see Figure 2.3). This might be called the original "fast-food" strategy! At Masaguaral, only males run off with wasp nests. Females are more likely to remain a few meters above ground to forage, rather than descend to the ground, and they are more likely to work on sites that reliably produce animal foods in small packages. For example, at Masaguaral, females frequently foraged for long periods of time (half an hour or more) in the crowns and along the trunks of palm trees, sifting through debris that collects in these places for a variety of animal prey (such as roaches, beetles, grubs and small frogs). Palm trees can be likened to grocery stores for female capuchins, where shoppers can find what they want if they take the time to look carefully. Males at Masaguaral rarely even entered a palm tree, let alone spent time there foraging. As a result of their tendencies to search for animal prey in different ways and in different places (immediate capture, on the ground or on larger supports, *vs.* careful collection from plant substrates, above ground), females at Masaguaral devoted more of their foraging time, and more of their total day, to strenuous foraging activities such as banging, pulling and tearing than did males. Rose (1994 a, b) suggests that the larger size of adult male capuchins compared with females (about 30% larger; about 3.5–4 kg *vs.* 2.5–3 kg) is the best explanation for the sex differences we have found in foraging and diet. Larger animals are less vulnerable to predation, and therefore more likely to accept the risk of foraging on the ground (see Figure 2.5). Males might also be more efficient at actions benefiting from strength (such as pulling up bark), or better able to capture well-armed, mobile animal prey. They might also be more opportunistic, taking foods that are more common at that moment and place, whereas females concentrate on reliable food sources and forage more strategically.

Although sex differences are the most obvious source of variation in foraging within a social group, age also contributes to variation. Immature animals in a group forage differently than do their elders. Infants up to 12 months can hardly be said to forage – their activity more resembles manual and locomotor play, and exploration of anything that can be chewed. Juveniles, on the other hand, are largely or completely responsible for feeding themselves. They seek the same foods in the same places as adults (especially adult females; juveniles also use the ground less than do adult males; see Figure 2.5), they use the same actions as adults to capture or process food items, and they achieve a largely similar diet as do adults in terms of types and quantities of food. However, compared with adults, juveniles are less efficient (in terms of the success rate per time devoted to foraging, especially when foraging quietly – that is, when inspecting, tapping, gently opening, or using their fingers to extract something, for example), spend more time foraging, and engage in strenuous actions in a greater proportion of their foraging activity. Altogether, this combination adds up to a far greater expenditure of energy to feed themselves for juveniles than for adults. Size and skill probably both contribute to this outcome. Inexperienced juveniles are not effective, for example, at pulling up bark to uncover tasty invertebrates such as small snails or grubs, or at ripping out juicy palm pith from the base of a growing frond. Perceptual learning may also be involved: juveniles can spend considerable effort mining "dry holes," earning nothing edible for their work. Adults are apparently better at selecting sites to search, or alternatively, at finding small things while searching. Juveniles also have less knowledge than older individuals about seasonally

available foods, or seasonal shifts in the location of foods (for example, that small snails that were visible on tree trunks in the wet season can be found under loose bark in the dry season).

Foraging places and modes of foraging in juvenile males and juvenile females already differ in the direction seen in adults. At Masaguaral, juvenile males are more likely to be on the ground, more likely to capture animal prey directly, and less likely to use strenuous foraging actions than are juvenile females. Thus, behavior becomes organized in sex-typical ways before the onset of reproduction, when males and females face different metabolic needs. Sources of variation between the sexes, whatever they are, are not linked tightly to size or individual experience. Social learning is not a likely explanation for these differences in capuchins, although affiliative preferences for same-sex partners may support emerging similarities between adults and juveniles of the same sex by aiding homogeneity in use of space and timing of activity. Social status is not a likely explanation either, as adults of both sexes are dominant (in terms of access to resources) to juveniles of either sex. The organizing effects on the brain of prenatal exposure to gonadal hormones remains as the most likely contributing factor in the emerging sex differences in foraging in capuchins.

Variations in foraging in a group of capuchins reflect more than sex and age differences, however. A considerable portion of variation reflects individual differences. Individuals experience their own unique history of success and failure in foraging, and possess a unique set of social and physical characteristics that affect their activity. Thus one might expect that, within a robust species-typical framework, individuals would work out their own strategies for foraging, for example, developing idiosyncratic food preferences or methods of processing foods. At Masaguaral, individual differences in selection of foraging substrates predicted much else about foraging: the kinds of actions to be used, the efficiency of foraging; the kinds of foods obtained. However, juveniles and adults exhibited equivalent variation within their own age-sex class, suggesting that cumulative individual experience (in older animals) did not lead to larger differences among adults than among juveniles. One could envision environments with lesser consistency where variation among adults might accumulate, but at Masaguaral at least, they apparently do not, at least at the resolution of measurement employed in Fragaszy's (1990a) study.

GROUP SIZE AND ITS RELATION TO FEEDING COMPETITION, PREDATION RISK AND REPRODUCTION

All behavioral ecologists agree that primary benefits of living in social groups relate to mechanisms of predator detection and defense (to be discussed in Chapter 3) and to enhanced discovery and defense of resources (van Schaik 1983, Wrangham 1980). Less agreed upon is the recent suggestion that group living in primates may have evolved as an anti-infanticide strategy, that is, as a defense against conspecific threat (Janson 2000 and see Chapter 11). For many years, there has been lively dispute about whether predator or resource defense is a more important reason for sociality, and resource competition itself is a complex phenomenon made up of both direct and indirect forms of competition and intra-group and inter-group levels of competition. In primates, the issue is almost never *should* they live in a social group or not, since almost all primate species are social. Rather the issue is in what *sort of group* should they live, in order to be maximally successful? There have been a number of capuchin field studies that compared the foraging and reproductive success of small groups versus large groups, so we will briefly focus on these as a way of examining feeding competition in this genus of monkeys.

First of all, let us consider the hypothetical benefits of living in a large social group of primates: (1) increased ability to locate good food sources in the range and to defend these resources from other groups by supplanting smaller groups from desired

←

Figure 2.5. Sometime capuchins go to the ground. Above, white-faced capuchins drinking from a cattle trough, one of the few permanent water sources during the dry season in Santa Rosa National Park, Costa Rica (photograph courtesy of Katherine MacKinnon). Below, an adult male wedge-capped capuchin on the forest floor sifting the leaf litter in search of invertebrate prey in Hato Masaguaral, Venezuela (photograph by Dorothy Fragaszy).

resources; (2) reduced predation risk through the three mechanisms of mobbing (strength in numbers), the selfish herd (lowered probability of being chosen as prey) and many eyes (to share vigilance behaviors); and (3) better reproductive success resulting from enhanced fecundity (due to increased access to food) and enhanced survivorship of young (due to increased predator defense). On the other hand, there are also theorized costs of living in a large social group: (1) greater intragroup feeding competition that is expressed as higher rates of agonism and the need to spend more of the daily budget in traveling and searching for food to feed everyone; (2) less time for resting and socializing that may lead to lowered fecundity; and (3) exposure to parasites and diseases. So, we already have a pretty complex set of checks and balances here, and we still have to factor in the food – the abundance, availability and distribution of food sources is a determining factor in which strategies work best for foraging success (Chapman 1990). For example, if resources are distributed in small scattered patches such that a large group has to travel long distances between feeding sites during all the daylight hours, then the other advantages of large groups may not be enough to outweigh this very large energy cost and in this case, small groups would be more efficient. Or if the size of each food patch is so small (one fruit tree is usually considered to be a "patch"), that only one or two of the group members can eat at a given time, then there will either be high frequencies of intra-group aggression or many group members will have poor access to food. Here again, it would be more advantageous to forage in small groups in this case. And as we have already discussed in the section on seasonal patterns, the pattern of food resources changes from month to month, so that larger groups may have greater foraging success in some months and smaller groups in others.

Now that we know just what a complex issue we are examining when we study foraging competition, what can we say about this phenomenon in capuchins? Are large groups indeed better at locating and defending food resources as hypothesized? If large groups can reliably supplant small groups from food (enhanced intergroup competition) does this outweigh the costs of more agonism within the group (increased intra-group competition)? Are large-groups better at predator detection, and does this outweigh the energetic costs of having to travel further in search of food? Do larger groups have greater reproductive success than small ones? These are the sorts of questions that have been studied by capuchin researchers.

Janson (1988) tested the hypothesis that larger groups (averaging 14 members) may benefit from sharing newly found food patches and found no supporting evidence for this in tufted (brown) capuchins at Manu. On the contrary, he found that the foraging efficiency (food ingested per unit distance traveled) of small groups made up of only five members was up to four times greater than that of large groups. Larger groups spent more time foraging, they traveled further and visited more trees per day than small groups, but they had lower food intake per tree visit. While large-crowned fruit trees were virtually "competition free," feeding in small-crowned fruit trees led to aggressive interactions in large groups. Large groups did not rest during the dry season, instead they used all their daylight hours in search of food. There was also differential access to food within the group, with the alpha male consuming up to five times more of a favorite food than did subordinates. Janson calculated that the lowest-ranking individual in a group of 12 could improve its energy budget by 15% if it changed groups and became the lowest ranking individual in a group of six monkeys. Taken together, these findings indicate that the cost–benefit ratio does not favor large groups in tufted capuchins. Indeed, the only evidence of foraging benefits for large groups in this study comes from the finding that larger groups can supplant smaller ones from food. Is this enough of a benefit to outweigh all the energetic costs of large groups just outlined? Janson's answer is that if the group size is sufficiently large (10 or more individuals in this case) then inter-group competition could play a role in favoring large group size. But the energetic costs of competing with conspecifics for food (intra-group competition) greatly outweighs any advantages one might obtain from living in a social group that can supplant other groups (inter-group competition). Indeed, it is so much more efficient to eat as a solitary or in a very small group, that predator avoidance and not enhanced food competition is likely to be the primary benefit of sociality in brown capuchins.

Thus, the bulk of Janson's findings indicate that there are many foraging advantages of small groups for tufted capuchins in Manu. In contrast, Miller (1992a, b) found for *C. olivaceus* at Hato Piñero that females in small groups experienced a foraging disadvantage relative to those in the large groups, as shown by seasonal fluctuations in foraging and food intake. Whereas females in large groups showed consistent foraging and food intake patterns, females in small groups ingested less food and rested more to conserve energy during the dry season when fruit was scarce. Although female *C. olivaceus* in small groups apparently try to mitigate these disadvantages by eating more and traveling/foraging further during the wet season when food is abundant, such food restrictions during one half of the year may be sufficient to depress the fecundity of females living in small groups. Miller argued that the low food intake of small-group females during the dry season was not the result of reduced foraging effort, but an apparent inability to maintain access to food resources due to supplantation by larger groups.

What might explain this discrepancy between Janson's argument that individuals have greater foraging success in small groups and Miller's conclusion that they do better in large groups? For one thing, Janson and Miller studied different species with different diets and patterns of food distribution. For another, it is clear that "large" and "small" are relative terms – Miller's small group of *C. olivaceus* consisted of 15 individuals which would be a large group in Janson's study of tufted capuchins.

Do larger groups enhance predation avoidance? De Ruiter's (1986) study of *C. olivaceus* at Hato Masaguaral showed that members of small groups spent more time in vigilance behavior and stayed at higher levels in the canopy than did members of large groups, who often came to the ground to forage. It is a common finding in many social species of animals that rates of individual vigilance are inversely related to group size – this is thought to be the case because the more eyes there are to scan for predators, the less often any given individual has to do so (the "many eyes hypothesis"). Thus one clear advantage of living in a large group is that the individual can spend less time looking for predators and more time feeding and resting. In a study of vigilance in *C. capucinus*, Rose and Fedigan (1995) found that individuals had lower rates of vigilance in groups with more adult males, and that the alpha male of each group was the most vigilant. Thus it seems that one of the benefits of living in a white-faced capuchin group with more males is that each individual can spend less time being vigilant. It also seems that whereas one of the benefits of being alpha male is the ability to eat more of the preferred foods, there are also costs to being the top-ranking male, such as spending more time in vigilance than do subordinates.

If members of larger groups do have feeding advantages and better predator avoidance than members of small groups, then it is also theorized that large groups will exhibit enhanced reproductive success – both greater fecundity and survivorship. Is this the case? Robinson's (1988a, b) 10-year study of 12 groups in the Hato Masaguaral population of *C. olivaceus* indicated that reproductive success was much higher in larger groups. This was due to greater fecundity rather than survivorship, since survival rates were similar in groups of all sizes. Both Robinson (1988b) and Miller (1996) argued that females in larger groups have higher fecundity because they experience less variation, especially seasonal variation, in access to food. Large groups predictably supplant small groups from food, and members of small groups experience a reduced food intake during the dry season.

What can we conclude about group size and feeding competition in capuchins? Certainly that we cannot expect to find simple answers to such complex issues of how resources, group sizes, foraging strategies and reproductive effort interact to create the patterns that we observe in our field studies of capuchin behavioral ecology. Also, that many more careful studies of feeding competition would help us to begin to make sense of the patterns, because for now we often have only very piecemeal evidence, a bit from each of the three better-studied *Cebus* species (*C. apella*, *olivaceus* and *capucinus*) and nothing at all on *C. albifrons*.

SUMMARY

It is appropriate to describe capuchins as both opportunistic omnivores and extractive, processing specialists. According to Milton (1984), primates choose their food on the basis of two principal factors: nutritional value and relative availability in time and space. The internal morphology of the digestive system also constrains diet in primary consumers such as plant-eating monkeys, and external features (dentition, body size, robusticity) may constrain food choices of secondary consumers such as insect-eating and meat-eating primates. Why do capuchins exhibit the types of foraging patterns described in this chapter? First of all, because they can. They have the morphology (body size, jaw strength, hand-eye coordination), physiology (gut capacity) and cognitive ability to target successfully hard-to-process foods and foods with elaborate defense systems. Second, they do so because capuchin foraging patterns are highly adaptive. Their foraging strategies allow them to minimize direct competition at several levels (within and between groups, between species), to be flexible in times of fruit shortages by falling back on alternate sources of food, and to exploit secondary and disturbed forest habitats that frugivorous primates do not.

3 • Community ecology. How do capuchins interact with their local communities and influence their environments?

In March the bands of female coatis (a relative of the raccoon) seem to vanish from the forest floor and from the shade trees by the waterhole. No longer do the juvenile capuchins tug the tails of sleeping coatis or challenge them to move aside in the fig trees. With their groups temporarily disbanded, the female coatis build bulky leafy nests in the tops of tall trees and after giving birth to a litter of pups, watch over their helpless offspring, descending the trees only occasionally to eat and drink. Capuchins have learned to take advantage of those rare occasions and may snatch an infant coati from its nest while its mother is away. Sometimes they even try to take a coati pup when the mother is present. On one such occasion we watched a juvenile male travelling around the periphery of his foraging group repeatedly checking nests in the canopy. So far he had come up with no vertebrate prey but had pounced on several large insects hiding in the foliage and dislodged a lot of leafy debris as he tore the nests apart. Suddenly he gave an aggressive call and we saw him facing off with a female coati on her nest. Other juvenile capuchins materialized from the nearby trees and they surrounded the coati, making threatening sounds and faces, and tugging at her tail. The coati for her part was making short lunges at these small monkeys but remaining close to her nest and staying physically on top of her pups. The contest seemed a stand off until two adult male capuchins from the group bounded into the tree. They attacked the female coati much more forcefully and effectively, one in front of her and the other behind and the mother was dislodged from her nest. She watched while each adult male capuchin scooped up one pup per hand and hobbled off with her offspring – their prey. We braced ourselves for the sights and sounds of infant mammals being eaten alive and I recited to myself an ecosystem mantra: coatis eat crabs and capuchins eat coatis and cats eat capuchins. Each of the adult males quickly let go one of his prizes from one hand in order to eat the other. One pup fell to the forest floor where it was instantly scooped up by a juvenile capuchin that carried it away to a nearby tree for consumption. An adult female picked another pup off a branch. All of the monkeys began by eating the soft parts – the intestines, the tail and the feet, holding the body of their prey against a branch and tearing upward with their teeth. Quickly the squealing sounds of dying coati pups stopped and bits of coati flesh were making their way through the capuchin group. Individuals without meat watched carefully for discarded and dropped bits, and infants approached the meat-eaters closely, peering at their hands and mouths. One of the adult males briefly allowed an infant to chew on the carcass in his hand and another abandoned the remains of his prey on a branch where it was quickly snatched up by a subordinate male. After two hours almost every member of the monkey group had eaten at least a scrap of meat and nothing remained of the coati feast.

One common complaint of ecologists about primatologists is that we are too organism-focused (as opposed to ecosystem-focused) and this book is no exception, being a general introduction to the behavior, biology and natural history of one genus of monkeys, the capuchins. There are, however, a few studies that place capuchins into a larger picture by examining how these monkeys affect and are affected by their ecological communities. In this chapter, we address some of these community-level ecological issues, for example, predator–prey relations, polyspecific associations and seed dispersal by capuchins.

CAPUCHINS AS PREDATORS

Most of the protein in capuchin diets comes from invertebrates, primarily insects and other arthropods like snails. Other primates, such as the squirrel monkeys (*Saimiri* sp.) that live alongside capuchins in many parts of their range, also consume insects as an important part of their nutritional intake. But only capuchins specialize in finding and extracting hidden and embedded insects, e.g., larvae of beetles and hymenoptera, scorpions, centipedes, millipedes, cryptic walking sticks, grasshoppers, cicadas and katydids. Terborgh (1983) pointed out that as predators, monkeys are searchers rather than pursuers. That is, much of their time and effort is devoted to finding the prey, which when located and/or extracted, is quickly captured and consumed. Most of the invertebrates targeted by capuchins do not try to escape by flight but rely on being hidden and/or having noxious defenses such as biting and stinging.

Here are some examples of the many types of behaviors capuchins employ to obtain hidden and embedded insects and other prey (e.g., snails and frogs). They may place their ear to a branch, and tap with a finger while listening – if an embedded insect or frog moves in response, or perhaps merely if the branch sounds different from the others (hollow or extra full), the monkeys then break off the branch and bite it open (illustrated in a sequence of photographs in Izawa 1978). They also strip away bark to reveal the insects hidden underneath, they unfurl leaves, tear apart nests, vine tangles and debris at the base of palm fronds, they stick their hands and arms in tree holes and crack open hard fruits (such as *Hymenea* pods that are infested with beetle larvae). They commonly open such fruits by banging them on hard surfaces, but may also crack them apart with two hands by using a stone or log as a fulcrum (Panger 1998, Panger *et al.* 2003). On the ground, they visually search the leaf litter and sweep the leaves around with their hands as well as rolling over stones to reveal the insects hidden underneath. They will also stand bipedally on the edge of an army ant colony on the march and grab the insects that are fleeing to escape the ants.

While hunting for invertebrates in these ways, capuchins sometimes come across vertebrates that they also capture and consume. For example, nests and tree holes may contain birds, eggs or nestlings. Capuchins are at least partially prepared to deal with the adult birds and mammals that defend their young because these monkeys are accustomed to overcoming the noxious defenses of biting and stinging insects. Whether or not invertebrate predation is the sole reason that capuchins have also become successful vertebrate predators, they are unusual among the nonhuman primates in that they are widely reported to capture and consume a variety of relatively large vertebrate prey that may weigh up to one-third of capuchin body weight, and may constitute up to 3% of their feeding time (Rose 1997, 2001). The types of vertebrate prey that capuchins have been reported to consume include: birds and their nestlings and eggs, lizards, frogs, rodents, bats, squirrels, squirrel nestlings and coati pups (see Figure 3.1).

Because it is rare for nonhuman primates to prey extensively on vertebrates and because anthropologists have focused much attention on the evolution of hunting in humans, there has been great interest in how capuchins locate and capture prey. There are anecdotal and short reports of vertebrate predation for all the species of capuchins (e.g., *C. olivaceus*: Fragaszy 1986, Robinson 1986; *C. apella*: Galetti, 1990, Izawa 1978, 1990b, Olmos 1990; *C. albifrons*: Defler 1979a; *C. capucinus*: Freese and Oppenheimer 1981, Oppenheimer 1982), but most of the data and analyses come from *C. capucinus*, especially from Santa Rosa in Costa Rica, where Rose (1997, 2001) has focused on this behavior pattern (see also Fedigan 1990, Newcomer and DeFarcy 1985).

At Santa Rosa, most of the vertebrate predation is focused on squirrels (*Sciurus* adults and nestlings) and coati pups (*Nasua narica*), which together make up half of the prey taken (Rose 1997) (see Figure 3.2). Approximately 5.4 prey items are taken every 100 hours and there is a successful predation event every 2.3 days, although the rates of vertebrate predation are higher in the fruit-scarce dry season than during the rainy season. Such high rates of hunting can result in 100% destruction of local populations of coatis and magpie jays. Adult male capuchins do most of the hunting (accounting for 52% of all prey taken) and they specialize in hunting adult squirrels and coati pups. Adult females and juveniles focus more on taking eggs and bird nestlings.

Much of the vertebrate predation that occurs in Santa Rosa consists of a single individual coming across prey in a seemingly opportunistic way and removing it from its nest. The finder may start to consume the prey on the spot or move to a large horizontal branch, where she or he rubs the carcass against a branch and starts

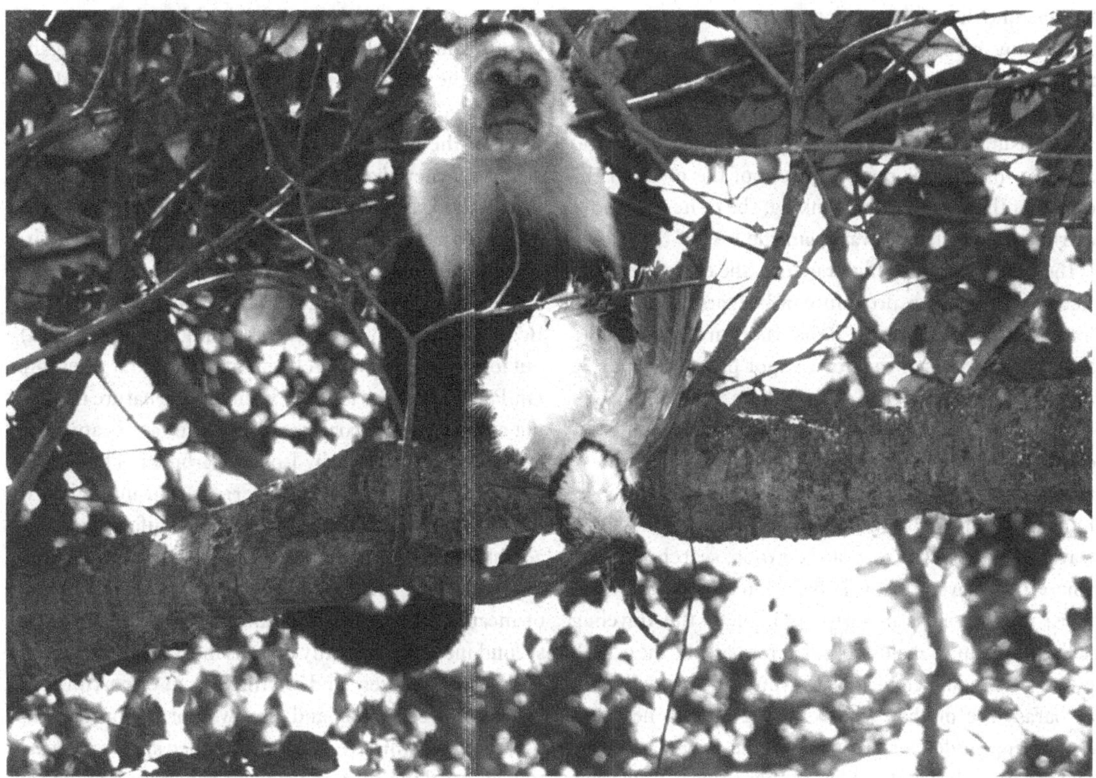

Figure 3.1. Capuchin monkeys capture vertebrate prey. This white-faced capuchin has caught a magpie jay hen (*Calocitta formosa*). Santa Rosa National Park, Costa Rica (photograph courtesy of Katherine MacKinnon).

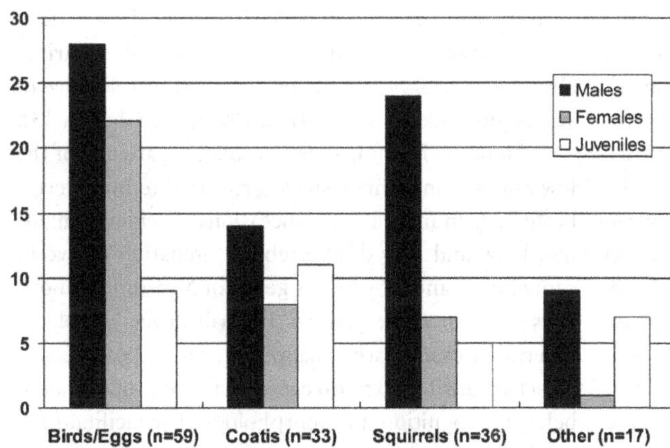

Figure 3.2. Number of prey items taken by adult male, adult female and juvenile capuchins at Santa Rosa National Park during 2929 observation hours. Values include only prey for which the initial captor could be clearly identified. From Rose (2001).

to consume the soft parts. Some vertebrate prey require processing before eating, for example, capuchins partially pluck parrots and magpie jays before consuming them. If the prey is large and takes a while to process and consume, other individuals may crowd around and attempt to touch or remove pieces of the meat. If the owner of the carcass is low ranking, she or he may move or run some distance away from the rest of the group. If the carcass owner is high ranking, she or he is usually tolerant of the close proximity of beggars. The carcass may be dropped entirely, so that there are two or more sequential owners of each carcass, or parts of it may be dropped and then picked up by others.

Hunts for adult squirrels and sometimes hunts for coati pups are somewhat different from the solitary and almost incidental predation acts described above. During mammal hunts, adult male capuchins often travel away from the rest of the group, searching spots where prey have previously been found and where the density of squirrels or coatis is high. If the traveling male or group of males comes across a coati nest with pups, or a squirrel nest with young, they will start to harass the mother who is sitting on the nest and attempting to defend her young. One may slap at the mother from the front, while another reaches under her from behind to grab the nestlings. Sometimes, when white-faced capuchins come across an adult squirrel they may also start to chase it around in the trees. We are not certain how they decide when to chase an adult squirrel and when to leave it alone – probably the size, behavior and location of the squirrel, the hunger of the monkeys and the proximity of other monkeys all play a role. There has been much interest and discussion about whether or not nonhuman primates ever "cooperate" in a hunt, that is to say, whether they act together toward a common goal. Since we cannot get into the minds of the capuchins, it is difficult to document true cooperation (see Chapter 9). However, sometimes during a squirrel hunt, group members move in such a way as to block all possible arboreal escape routes for the squirrel. Rose (1997) has referred to this pattern as collaboration – the adoption of different but complementary goals. They may also "relay" chase the squirrel in turn until they have exhausted it. The squirrel then drops (or is knocked) to the ground where one or more monkeys race up to it and begin trying to subdue it by holding it down and biting it, often being bitten themselves by the squirrel in the process. Only adult male capuchins seem to be able to subdue an adult squirrel, although female and young capuchins do participate in squirrel hunts. Again, as with the nestling coatis, the squirrel carcass takes a while to consume, and beggars may be able to obtain a piece of the meat, or even the entire carcass when the original owner abandons it.

As with "cooperation" there has been much interest in "sharing" of meat. Tufted capuchins in captivity allow other group members, especially infants, to take food readily from them (de Waal 1997, de Waal and Berger 2000, de Waal *et al.* 1993, Fragaszy *et al.* 1997a), but food transfers are rarely reported in field studies thus far. Perry and Rose (1994) examined the nature of sharing and transfer of coati meat in white-faced capuchins at Santa Rosa and nearby Lomas Barbudal. They found that coati meat is often transferred from individual to individual, especially from mother to infant, but that actual direct giving of meat is rare. That is, it is exceedingly rare for a white-faced capuchin to break off a piece of meat and hold it out for another to take. Instead, a second individual usually obtains meat from the first by approaching closely and waiting for an opportunity to "steal" a piece rapidly and run away, or by approaching with overt submissive gestures and then slowly and cautiously breaking off a piece from the carcass while the owner is pre-occupied with biting into and chewing up pieces of meat. Owners often tolerate the close proximity of beggars and frequently are satiated before consuming all of a coati pup and will drop the carcass close to one or more of the beggars – Perry and Rose (1994) found an average of 2.89 owners per coati carcass. This pattern falls somewhat short of our notion of active sharing, although it could certainly be considered "tolerance of scrounging" (see Fragaszy *et al.* 1997a; and Chapter 13).

Along with chimpanzees, capuchins are one of the few nonhuman primate species reported to hunt vertebrate prey in more than an occasional, incidental manner. How and why did vertebrate predation evolve in capuchins – and why in this genus of Neotropical monkeys and not other genera? We will never be able to determine exactly what happened, but we can reconstruct a plausible scenario based on features of capuchin behavior, cognition and morphology that facilitate the finding, capture and consumption of vertebrate prey. As noted above, capuchins regularly search crevices and other hiding places in such a way that vertebrates as well as invertebrates are occasionally flushed. And by specializing in insects that are cryptic, embedded and/or equipped with noxious defenses, capuchins are already well prepared to solve the problems of exploiting foods that hide, run and fight back. They use cognitive and manipulative skills to deal with invertebrate prey that are readily applicable to vertebrate prey. Furthermore, as noted by Rose (1997, 2001), capuchins show a strong tendency to be "monos muy bravos" (i.e., belligerent to other animals), especially toward predators and competitors, and they can apply those assertive tendencies to the capture of prey, such as adult squirrels, that can bite back. Finally, one of the ways capuchins cope with the highly seasonal environments in which they live is by switching to alternative foods when fruits and insects are scarce. And one of the alternative foods they may switch to is vertebrate prey. In the tropical dry forests of Santa Rosa in Costa Rica, it happens that the main nesting periods for parrots, squirrels and coatis are during the dry season, when fruit and caterpillars are less available to white-faced capuchins. It is possible that such a combination of seasonal decline in fruit with rise in nestling availability played a role in the development of what has become systematic rather than merely incidental vertebrate predation by capuchins at this and several other sites where they have been studied.

CAPUCHINS AS PREY

Capuchins prey on vertebrates smaller than themselves and in turn they are preyed upon by those Neotropical carnivores that are large enough to take a monkey. Although most of the evidence is circumstantial, it is widely believed that capuchins, especially young ones, are subject to predation by animals such as jaguars, pumas, jaguaroundis, coyotes, tayras, venomous and constricting snakes, caimans and crocodiles, and raptors such as eagles and the larger hawks and owls. Scientists believe that the risk of being killed by one of these carnivores is such a significant factor in the life of a primate that enhanced predator detection and defense has frequently been proposed as a primary benefit of group living. Improved resource defense is also proposed as a major benefit of group living, and this hypothesis is fairly straightforward to test through the study of inter-group feeding competition.

Unfortunately, predation patterns are quite difficult to document. This is because it is so rare to observe actual events in which predators capture and consume monkeys. There is only one published observation of a predator taking a capuchin and that is Chapman's description of a boa constrictor capturing a juvenile white-faced capuchin at the Santa Rosa field site in Costa Rica (1986). On the other hand, researchers have frequently observed carnivores attempting to prey on monkeys. And Rettig's (1978) study of the nest of a pair of harpy eagles in Guiana found that they brought 13 capuchins to their young to eat over a period of 328 days, that is one capuchin every 25 days. It is likely that our presence in the field close to our study monkeys inhibits the actions and success of predators, which are almost all wary of humans. Even if successful captures of capuchins by predators are rare events in nature, each monkey has only one life to lose and if a group loses only one individual per year to predation that loss can still play an important role in the demographic profile of the population.

Monkeys can thwart predation attempts in several ways: they can be cryptic and silent; they can run away from predators or onto substrates (e.g., high thin branches) where predators cannot follow; they can be vigilant and give loud alarm calls to warn their conspecifics; they can actively confront predators, or they can use a combination of all of these. Capuchins seem to have specialized in the third and fourth strategy: they try to pre-empt attacks through vigilance, early detection and alarm calling, and they frequently mob predators. These strategies appear to work well with "sit and wait" predators, such as snakes and Neotropical felids, for whom the game is up once the predator is detected and noisily harassed. The mobbing of a predator by capuchins is quite a spectacular event, led mainly by adult and subadult males who vocalize loudly and repeatedly with threats and alarm calls, while breaking large branches and dropping them onto the predator (see Figure 3.3). Although it would be an exaggeration to say that capuchins actively throw sticks with aim and accuracy, they do seem to have acquired the martial art of "bombing" the object of their mobbing attack with dead boughs and other debris. They do this by locating themselves immediately over the predator (repeatedly if necessary) and looking for dead branches to break that will fall directly on the target. They usually break branches by bouncing on them with their feet or shoving on them with their hands; occasionally they toss them. Boinski (1988) described how a group of white-faced capuchins in Corcovado National Park, Costa Rica, managed to kill a large venomous snake (*Bothrops asper*, the "fer de

Figure 3.3. Capuchin monkeys respond jointly to potential predators. These two white-faced capuchins react to a *Boa constrictor* snake by making alarm calls. Santa Rosa National Park, Costa Rica (photograph courtesy of Katherine MacKinnon).

lance") after they first pinned it to the spot by dropping a heavy branch on it, and then approached the snake on the ground to flail at it with another stick. One adult male in particular came to the ground and rained 55 blows onto the head of the snake (and the ground nearby) with a large dead stick that he wielded (somewhat clumsily) like a club. The dead snake was later found to have broken bones as well as severe lacerations to its body from the capuchin attack (see Vitale, *et al.* 1991 for capuchins' behavior toward a snake model in captivity). Unhabituated capuchins react to humans as predators, and researchers have been hit more than once by a dropped branch. In fact, one field researcher was knocked unconscious, felled by a branch broken over her head by a capuchin (M. Panger, pers. comm.). Census takers learn to stop taking data and move out of the way when an unhabituated, alarm-calling male scans the canopy directly above them for a likely branch to break and drop. These examples demonstrate that capuchin mobbing is a very effective way to discourage a large predator on the ground from attempting to pursue them.

The other effective strategy used by capuchins is early detection. These monkeys frequently cease whatever activity they are engaged in, sit or stand upright in an alert posture, and visually scan beyond the immediate vegetation – a behavior that is commonly referred to as "vigilance." Rose and Fedigan (1995) found that white-faced capuchins spend anywhere from 1% to 6% of their waking hours being vigilant. They also found that adult males are more vigilant than adult females and that alpha males are the most vigilant of all. Fragaszy (1990) found that adult male *C. olivaceus* were vigilant in 18% of samples, as compared with 9–13% in adult female and juvenile male samples. Van Schaik and van Noordwijk (1989) proposed that females choose protective males, and that males can effectively protect the group through vigilance and alarm calling. Using models rather than live predators, they found that both *C. albifrons* and *C. apella* males are more vigilant than females and are better at detecting predators and more active in facing them down. Although the findings on *C. capucinus* confirm van Schaik and van Noordwijk's proposal that male capuchins are more vigilant and more active in confrontations with predators, Rose (1997) did not find that males were better at early detection of predators in terms of giving the initial alarm calls. More research on different populations and species of capuchins to see who gives the first alarm calls when a predator is spotted would help us to better understand the pattern.

It is important to note that almost all members of a capuchin group are active in the predator detection and defense system. All monkeys except infants show some vigilance behavior and give alarm calls and participate in a mobbing. It is just that adult males spend more time in vigilance and are more active in confrontations with predators than are females and juveniles. There are probably several reasons why males are more vigilant than females. For one thing, there is a cost to vigilance – vigilant individuals spend less time eating, resting and being social. Adult females, most of whom are lactating or pregnant, spend more time eating and less time being vigilant than do adult males. For another, males are found more often on the ground and on the periphery of the group, where most vigilance behavior is needed. And perhaps most important, adult male capuchins seem to have more than one reason to be vigilant (see Fragaszy 1990) – they not only stay on the look-out for possible predators, they also watch for the approach of other groups, especially non-group males. Rose and Fedigan (1995) found that male white-faced capuchins are more vigilant in areas of home range overlap, and that they are the first to detect and run off novel or neighboring males. There is even a particular chucking vocalization that is given almost exclusively by males when they spot non-group males in the vicinity. Fragaszy (1990) found that *C. olivaceus* males are more likely to locate possible prey items during vigilance. Whatever multiple reasons male capuchins may have for staying so alert to their surroundings, it clearly benefits the females and the young of the group to have these males act as a finely tuned alarm system that quickly detects and aggressively deters intruders. Perhaps the attentive and belligerent approach that capuchins take to interlopers is another reason we see so few successful predatory attacks on these monkeys.

CAPUCHINS AS AGENTS OF DISPERSAL

Pollination

Although capuchins do not commonly consume flowers (Chapman 1988, Galetti and Pedroni 1994, Janson, *et al.* 1981), many studies report that *Cebus* monkeys insert their faces deeply into the flowers of certain vine and tree species to obtain the nectar therein (e.g., Oppenheimer 1968, Prance 1980, Torres de Assumpção 1981). At Santa Rosa, researchers may first notice that the white-faced capuchins have been nectaring by the pollen dust that clings to their facial fur. A few primatologists have suggested that capuchins serve as pollinators, but no one has yet conducted a study directly focused on this topic.

Capuchins are described as visibly lapping or licking the nectar out of flowers, during which time their muzzles rub against the pollen-bearing anthers so that their faces become covered in pollen. They then travel to another inflorescence or plant to obtain nectar and likely contact the stigmas with the pollen they have transported from the first plant. The monkeys may destroy some of the flowers by breaking them off the stem – Prance (1980) even describes *C. apella* as picking the cup-shaped flowers and then holding them up to their mouths to drink. However, most reports emphasize the lack of severe damage to the flowers, because the monkeys bring their faces into the plant rather than vice versa. Janson *et al.* (1981) have argued that the use of flower nectar without destroying the inflorescences, along with the heavy uptake of pollen on facial fur and the traplining behavior of moving from flower to flower, all suggest the possibility of a pollination role for monkeys. Furthermore, Janson *et al.* (1981) pointed out that the plant species which monkeys visit for nectar share some characteristics making them particularly attractive to non-flying mammals: these vine and tree species (e.g., *Ceiba* and *Ochroma* in the family Bombacaceae, *Mabea* in the family Euphorbiaceae, *Combretum* in the family Combretraceae) tend to flower in the dry season when trees are leafless and fruit is scarce; their flowers are large, tough, and often cup-shaped in long inflorescences with exerted stamens; and the conspicuous flowers open simultaneously.

Sussman and Raven (1978) argued that non-flying mammals such as primates and marsupials have been competitively excluded from significant pollinator roles by bats, and indeed reports of capuchins as pollinators

Figure 3.4. Capuchins may disperse seeds. Here a white-faced capuchin extracts winged seeds from *Luehea* pods. In the right upper corner: a close view of the pod and its seeds. Santa Rosa National Park, Costa Rica (photographs courtesy of Katherine MacKinnon).

often suggest they are secondary to bees, bats and birds in significance to the plants. However, nectar is one of the important alternative foods for Neotropical monkeys when fruits are scarce, and we now have more reports of capuchins using nectar and carrying pollen on their fur than we did when Sussman and Raven published their paper. It seems that the time is right for field researchers to examine the extent and significance of pollination behavior by capuchin monkeys.

Seed dispersal

As with flowers, capuchins seldom deliberately consume seeds for nutritional purposes. But they often eat fruits in which the seeds are tenaciously embedded in the nutritious pulp they seek. Small seeds are usually swallowed whole with the fruit pulp whereas large seeds may be scraped cleaned of pulp and then discarded. Just as a plant may use flower nectar to entice animals to spread its pollen, so it may employ fruit to inveigle animals to disperse its seeds. Some primates make seeds an important part of their diet, and destroy them by chewing them up before ingestion. Even capuchins may consume seeds as a source of food during the fruit-scarce dry season, thereby acting as seed predators. For example, the protein-rich winged seeds of *Cariniana micrantha* are consumed by *C. apella* in Brazil (Peres 1994) and winged seeds of *Luehea candida* are eaten by *C. capucinus* in Costa Rica (Chapman and Fedigan 1990, O'Malley 2002; see Figure 3.4). But for the most part, capuchins are more likely to disperse than destroy the seeds they contact while foraging.

Some researchers have mentioned in passing that capuchins disperse seeds (Hladik and Hladik 1969,

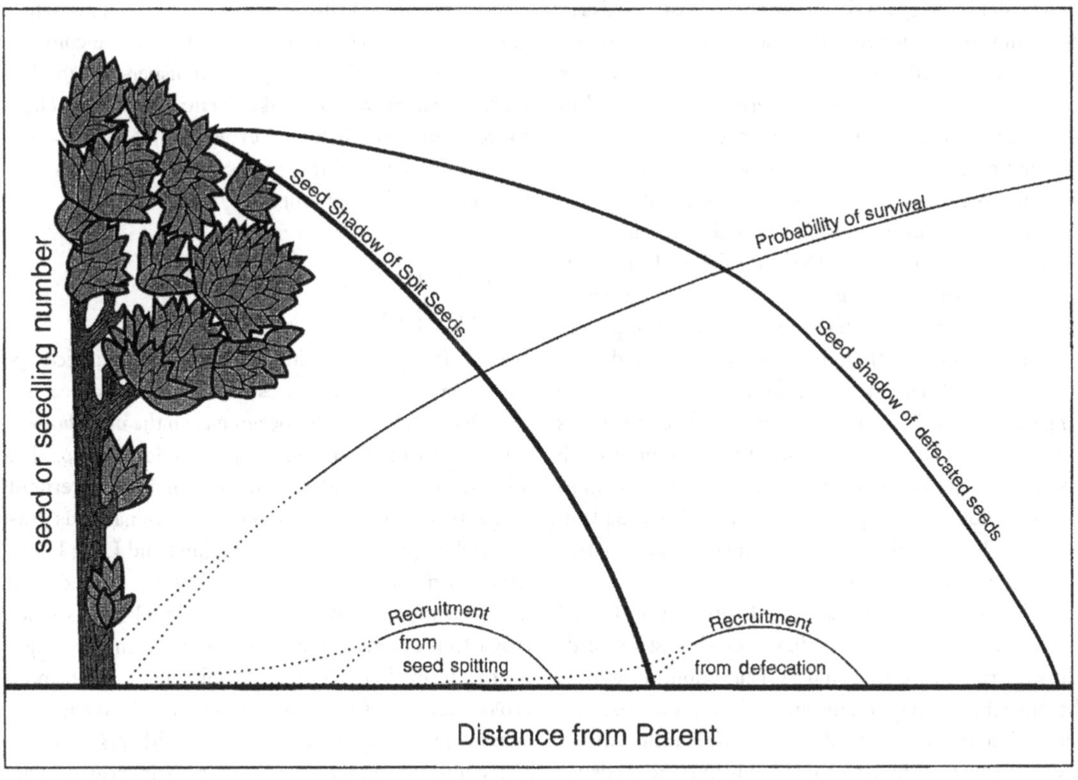

Figure 3.5. A hypothetical illustration of the seed shadow, the probability of seed survival, and the recruitment curves that would result from seeds that are primarily spit out, as by Old World monkeys, and seeds that are primarily defecated, as in New World monkeys. The dotted lines with respect to the probability of survival and recruitment curves illustrate the reason for debate about the likelihood that seeds can survive under the parent's canopy. From Chapman 1995. Copyright © 1995 Wiley-Liss, Inc. Reprinted by permission of Wiley-Liss, Inc., a division of John Wiley & Sons, Inc., and by permission of the Author.

Janson *et al.* 1981, Oppenheimer 1968) and three researchers have carried out analyses of this phenomenon. Rowell and Mitchell (1991) compared seed dispersal in white-faced capuchins and guenons (Old World monkeys in the genus *Cercopithecus*). They found that capuchins tend not to crush or damage seeds from the fruits they ingest and that capuchin feces is composed largely of intact seeds held together by small amounts of unstructured organic matter. This is quite different from the guenons who crush seeds before swallowing them and excrete a firm paste with no intact sizable seeds in their feces. In 150 capuchin fecal samples, researchers found seeds from 39 different species, and they planted seeds from 35 species, from which seedlings of 31 species germinated. The successful germination rate for seeds that had passed through a capuchin's gut was 60.5%. They also found that capuchins carry fruit with seeds in their hands 10–20 meters and they deposit seeds in their feces 200–1000 meters away from the original plant. Similarly, Zhang and Wang (1995) documented that *C. apella* deposited seeds on average 390 meters from the original plant where they had eaten the fruit and they found that seeds passed through capuchin guts germinated well, although not any better than seeds not so treated. Capuchins defecated fewer seeds per deposit than did spider monkeys but the seeds they deposited had greater probability of surviving to the seedling stage than those deposited by spider monkeys.

Chapman (1989, 1995) carried the study of capuchin seed dispersal one step further by examining the fate of dispersed seeds (see Figure 3.5). He collected seeds from the feces of white-faced capuchins, howlers and spider monkeys in Santa Rosa and constructed artificial dung piles containing these seeds on the forest floor. His study showed that within 5 days of being placed in the forest, 51.8% of the seeds were either removed by secondary dispersers such as dung beetles, or destroyed by seed predators such as mice (there was great variation across plant species). Out of 793 seeds placed out in the forest, only 11 seeds remained at the locations where they were placed and had germinated by the time the study terminated 6 months later, and only one of the seedlings became established to a height of greater than 5 cm. Thus, although the majority of seeds that passed through a monkey's gut were found to be capable of germination, this study indicates that from the tree's perspective, the rate of successful propagation through monkey-dispersed seeds may be low. In order to evaluate the value of this type of seed dispersal to the plant, it would be helpful to know the success rate for tree propagation by other means.

We can say capuchins are well situated to be seed dispersers by the fact that they seldom destroy seeds during fruit processing and consumption, as well as because they travel long distances during the relatively short time it takes for seeds to go through their digestive systems (1–4 hours), thus depositing the seeds far from the parent tree. Seeds that have passed through a mammal's gut have enhanced germination rates (Howe 1980, 1984, Janzen 1983). But just how important and effective are capuchins as seed dispersers? Howe (1980) thought primates were not important compared with birds because monkeys deposit seeds in larger clumps that may lead to increased competition at the seedling stage, and/or attract seed predators, such as terrestrial rodents. Primatologists have counter-argued that at least some species of monkeys are good seed dispersers (Garber and Lambert 1998, Gautier-Hion *et al.* 1985, Zhang and Wang 1995). After all, nonhuman primates constitute between 25 and 40% of the frugivore biomass in tropical forests (Chapman 1995) and they defecate large numbers of viable seeds. The effectiveness of primates as seed dispersers really depends on the type of plant being propagated and the behavior, morphology and physiology of the animal doing the dispersing. And, as Chapman (1995) and Garber and Lambert (1998) have pointed out, the seed-dispersing efficacy of primates also depends strongly on the postdispersal survival of the dispersed seeds, and how that survival is influenced by the frugivore's actions. Thus, we now recognize that the traditionally simple distinction between seed predators and seed dispersers is, in fact, a much more complex set of ecological relationships. Seed dispersal is another issue on which we would like to encourage researchers to carry out more field and experimental work to help us better understand the interactions of capuchins with their ecological community.

CAPUCHINS AS ECOSYSTEM "ENGINEERS"

Back in 1968, Oppenheimer first noted that an ecological description of a species should take into account the effect that the study species has on the morphological structure of the habitat. He further documented that white-faced capuchins altered the branching patterns of one of their food trees by removing terminal buds during feeding (1968, 1977, Oppenheimer and Lang 1969). More recently, Jones *et al.* (1994, 1997) described such physical modification of the environment by one species that affects others as "physical ecosystem engineering." By this term, Jones *et al.* were referring to nontrophic ecological effects such as beavers building dams and such as trees providing shelter, shade, humidity, safety and soil stability to other organisms in their local habitat. Although no capuchin researcher has followed up on Oppenheimer's original study, it seems likely that these monkeys do affect the structure of the woody vegetation in their habitat through their frequent "tree pruning" behaviors. Capuchins actively remove tree branches and sections of vines for at least three reasons: extractive foraging, agonistic display and foraging on apical meristems. Some branches are broken off inadvertently when the monkeys move through the forest and the vegetation gives way underneath their weight. However, it is mainly during foraging for embedded insects that capuchins shower the forest floor with dead and infested branches that they have detached in search of food. Often researchers are first alerted to an infestation of a tree with insects such as beetle larvae when a capuchin group swarms the tree, tapping the branches while listening closely for the sounds of embedded insects, then breaking off branches to open with their teeth and lick out the grubs. Oppenheimer and Lang (1969) argued that such intense feeding bouts reduce the insect populations, and thereby the amount of damage done to the vegetation during that infestation. No one has yet verified the significance of these effects, but "tree pruning" is sufficiently common during capuchin foraging (and agonism and predator defense) that researchers on the ground must be constantly alert to the dangers of falling branches.

CAPUCHINS AS CROP RAIDERS

Crop-raiding is much less common in New World than Old World monkeys (Cowlishaw and Dunbar 2000), probably because of the smaller body size and lack of terrestriality in Neotropical monkeys, and perhaps because there is a lower incidence of agricultural fields abutting highly disturbed forests in the Neotropics than the African and Asian forests. Nonetheless, several characteristics of capuchins fit the profile of the primates that are more likely to raid crops: omnivorous, opportunistic monkeys that will come to the ground for at least part of their food. Several authors have mentioned in passing that capuchins may feed on cultivated foods from fields and gardens that border their forest habitats (Baldwin and Baldwin 1976, Defler pres. comm, Freese and Oppenheimer 1981, Hernandez-Camacho and Cooper 1976, Hill 1960, Klein and Klein 1976, Rylands *et al.* in press). But only Galetti and Pedroni (1994) have calculated the extent of human-produced foods in the diet of one population of *C. apella* on a reserve in southeast Brazil. These two authors described tufted capuchins as eating corn, potatoes and manioc from plantations surrounding the Santa Genebra Reserve. They accomplished this feat by harvesting the food quickly and running back into the forest with it, before dogs could attack them. Corn constituted up to 25% of their diet, depending on the season, but potatoes and manioc were less commonly consumed. Defler (pers. comm, March 2003) reported that *C. apella* in Columbia feed on corn, sugar cane, cacao and fruit trees. Although capuchins are clearly less frequent crop pests than Old World species such as baboons and vervets (see Naughton-Treves *et al.* 1998), at least two articles make reference to the hunting of capuchins in order to control their numbers and minimize crop-raiding (Ferrari and Diego 1995, Gonzales-Kirchner and Sainz de la Maza 1998). Furthermore, if humans continue to log and otherwise disturb the forests where capuchins occur and to cultivate agricultural fields in close proximity to these forests, this adaptable, inventive genus of monkeys has a strong probability of coming into increasing conflict with humans through crop-raiding.

CAPUCHINS AS HOSTS FOR PARASITES

Parasites are organisms that exist at the expense of host species without benefiting them in return; their effects are either neutral or they harm their host. They are of interest to ecologists because of the coevolutionary relationships that parasites have established with their hosts – acting as selective pressures on host species while simultaneously being faced with their own ecological problem of getting from one host island to the next.

Relatively little is known about the ecology and life histories of parasites that live on or in free-ranging New World monkeys, although there is a fair amount of literature on the parasites of howlers (see references in Stoner 1996, Stuart and Strier 1995, Stuart *et al.* 1990, 1998) and some on squirrel monkeys (Dunn 1968), muriquis (Stuart *et al.* 1993) and callitrichids (Araujo-Santos *et al.* 1995). Like other primates, Neotropical monkeys can be infected with blood parasites, such as those that cause malaria; they can suffer from ectoparasites, such as bot flies, mites, ticks and lice; and they frequently are hosts to endoparasites such as tapeworms and roundworms. Most of the recent research has focused on intestinal parasites of Neotropical monkeys, because these types of parasites and their ova and larvae are excreted in feces, which can be collected without capturing the animals. Recent studies (reviewed in Stuart and Strier 1995) have shown that moist conditions (wet forests) are conducive to higher parasite loads in Neotropical monkeys than are dry forests, and that repeated pathway use and small home ranges also lead to higher rates of parasites. They also showed that howler and muriqui groups that consume plants known to contain anthelminthic agents had lower parasite loads than groups without access to these plants, suggesting that monkeys may eat some plants to control intestinal worms (Strier 1998, Stuart and Strier 1995).

Studies of fecal samples from captive *C. apella* in Argentina (Santa Cruz *et al.* 1998, 2000) found that these monkeys are hosts to endoparasites such as protozoans (*Trichomonas* spp.), roundworms (*Strongyloides* spp., *Filariopsis arator*) and tapeworms of the family Anoplocephalidae. Necropsies of six of these captive monkeys that died of causes unrelated to parasites found high levels of *F. arator* (a nematode or roundworm that dwells in the lungs and peritoneal cavity). Dunn's (1968) review of platyrrhine parasitism lists many other types of parasites found on or in monkeys of the genus *Cebus*, including intestinal protozoans, trypanosomes (Chagas' disease), tapeworms, thorny-headed worms, flukes and roundworms. Most of the studies described in Dunn's review are quite old and were based on examination of deceased monkeys. Only Stuart *et al.* (1998) appear to have recently examined the endoparasites of living, free-ranging capuchins, which they compared with the parasite loads of sympatrically living howler and spider monkeys in Costa Rica. The fecal samples that Stuart *et al.* (1998) collected from *C. capucinus* indicate that 91% of the capuchins sampled were parasitized as compared with only 49% of the howlers and 51% of the spider monkeys from the same Costa Rican sites. The fecal samples from these capuchins also contained several types of unidentified nematode larvae not found in the howlers and spider monkeys. We suggest that at least three characteristics of capuchin behavioral ecology may lead to higher parasite loads in these monkeys than in sympatric howlers and spiders: only capuchins drink from water holes (isolated standing water holes are well-known sources of parasite contamination); capuchins frequently forage on the ground as well as in the trees, which is postulated to lead to higher parasite loads (contaminated soil; see Dunn 1968); and they eat a wider variety of foods, which may bring them into contact with a greater variety of parasites. All of these suggestions, as well as Stuart *et al.*'s finding of high parasite loads in Costa Rican capuchins, need to be tested and further documented for generalizability.

Almost nothing has been published about the ectoparasites of capuchins, except that they appear to be less subject to botfly infestation than are howler monkeys living at the same sites (Milton 1996). It is possible that capuchins control levels of ectoparasite infestation through frequent social and self grooming (which is almost entirely absent in howlers), and through the capuchin practice of rubbing their bodies with plants known to contain insecticidal properties (Baker 1996, Valderrama *et al.* 2000; see Chapter 5).

As with several other community ecology issues discussed in this chapter, the relationship between capuchins and their parasites is an area that is clearly much in need of further research. We need to know not only which species of parasites are commonly found in or on capuchins, but also how the life cycles, behavior and ecology of these organisms are inter-related. Addressing such issues will require not only the incidental collection and preservation of a fecal sample here and there, but an in-depth longitudinal set of studies that collect parasite information on known individuals over time in variable ecological circumstances.

CAPUCHINS AS PART OF LARGER PRIMATE COMMUNITIES

One central question in the study of primate community ecology is how many sympatric species a given habitat can accommodate. The answer for New World monkeys is that researchers have located from one to 14 species living in any particular Neotropical forest. In the case of capuchins, because of their widespread distribution across Central and South America, and their ability to exploit various habitats, they can be found living sympatrically with almost every type of New World monkey, except perhaps *Alouatta pigra* of Mexico, Guatemala and Belize (the northern end of Neotropical primate distribution does not include capuchins in historic times). Furthermore, capuchins are seldom found as the only type of primate in the forest – they usually live sympatrically with other types of monkeys (see Figure 3.6). It is very rare in the Neotropics for primate species within the same genus ("congeners") to co-exist but capuchins form one of the exceptions – the South American species of *C. olivaceus* and *C. albifrons* each live sympatrically with *C. apella* in some of their habitats. However, Lehman (2000) found in a survey of 16 sites in Guyana, that sightings of *C. olivaceus* were reduced in areas where they co-occurred with *C. apella*, and he suggested that this negative pattern of site association (sightings of *C. olivaceus* go down as sightings of *C. apella* go up) is due to scramble competition between the two species (see also Sussman and Phillips-Conroy 1995).

The finding that capuchins of different species can co-occur at the same sites is due largely to the very widespread distribution of *C. apella* (see Chapter 1). Peres and Janson (1999) calculated that *C. apella* occurs in at least 50% of Neotropical primate communities. These authors also investigated the richness of primate assemblages across the Neotropical sites that range from latitudes of 20 degrees N to 30 degrees S and they found that primate communities become more species-impoverished with increasing distance from the equator. This substantiates a nearly universal pattern of higher species diversity at lower latitudinal levels.

Figure 3.6. Seven sympatric platyrrhine species in a Surinamese rain forest, showing typical locomotor and postural behavior as well as use of different heights in the forest. At the highest levels are the bearded saki (*Chiropotes satanas*) and the black spider monkey (*Ateles paniscus*); below them are tufted capuchins (*Cebus apella*) on the left and red howling monkeys (*Alouatta seniculus*) on the right; in the lower levels are squirrel monkeys (*Saimiri sciureus*) on the left, a capuchin in the center, and golden-handed tamarins (*Saguinus midas*) and white-faced sakis (*Pithecia pithecia*) on the right. Capuchins also travel, forage and play on the ground occasionally. (With permission of Stephen D. Nash and John G. Fleagle.)

Some studies of primate assemblages in South America (Cowlishaw and Dunbar 2000, Reed and Fleagle 1995) have found a correlation between the number of primate species in a given area and the amount of annual rainfall for that area, but Peres and Janson found only a weak correlation between rainfall and primate species richness. Instead, they argue that there is a gradual decline in primate richness from the central equatorial zones of South America to the northern and southern frontiers of Neotropical primate distribution for the following reasons: greater resource seasonality, lower floristic diversity and decrease in available area of tropical forests at increasing distances from the equator. In some contrast, Kay *et al.* (1997) argued that variable primate species richness across South America is determined by differences in plant productivity and by historical biogeography (they found more primate species per site in larger than in smaller geographic regions).

Two of the sites with a large number of Neotropical primate species that have been studied thus far

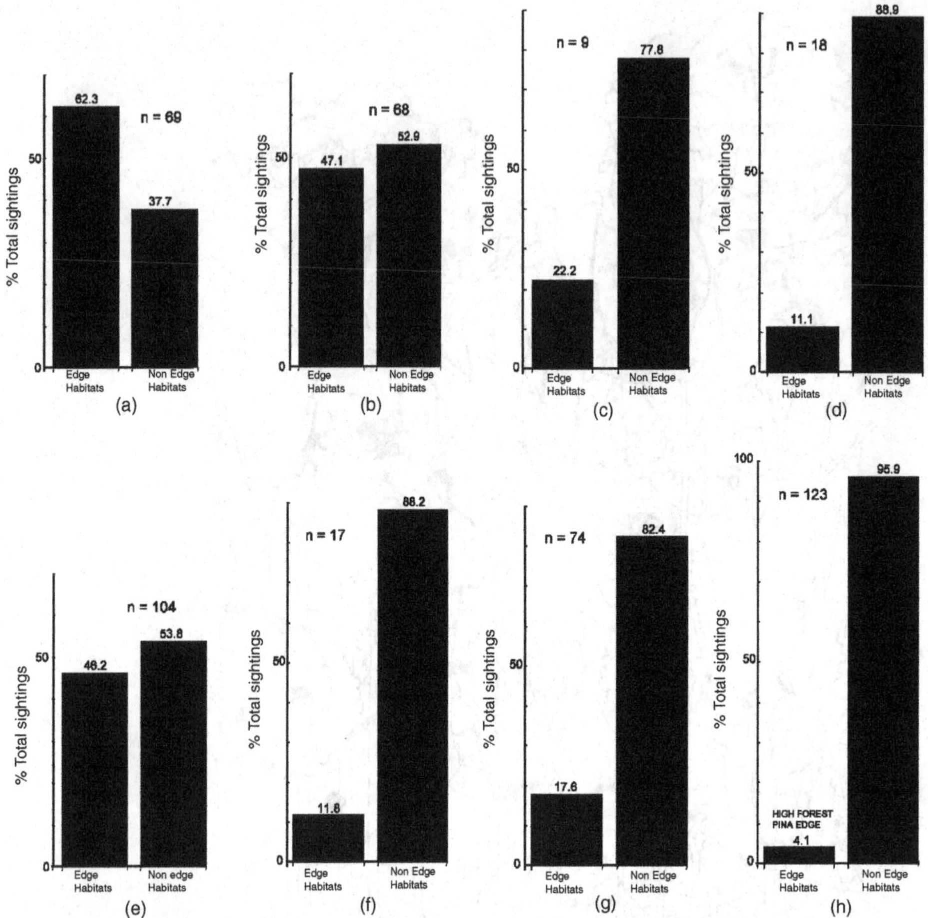

Figure 3.7. Utilization of edges in the Voltzberg study area. (a) *Saguinus midas midas*; (b) *Saimiri sciureus*; (c) *Pithecia pithecia*; (d) *Chiropotes satanas chiropotes*; (e) *Cebus apella apella*; (f) *Cebus olivaceus* (= *nigrivittatus*); (g) *Alouatta seniculus*; (h) *Ateles paniscus paniscus*. (Adapted with permission from Mittermeir and van Roosmalen 1981.)

are Manu National Park in Peru with 13 species of monkeys (Terborgh 1983) and the Raleighvallen-Voltzberg Nature Reserve in Suriname with eight species (Boinski *et al.* 2001, Mittermeier and van Roosmalen 1981, Fleagle *et al.* 1981). In addition there are a number of sites in the Amazonian forests of Brazil that have been surveyed by Peres (1999) in which he located up to 12 species of monkeys at any one site.

Mittermeier and van Roosmalen (1981) argued that the eight species of monkeys in Raleighvallen effectively divided up the available habitat and food resources through a combination of dietary divergence, forest-type preferences and canopy-height predilections. For example, they reported that *C. olivaceus* (called *C. nigrivittatus* at that time) was almost entirely found in the non-edge habitats of high rain forest, whereas *C. apella* was the most adaptable of the eight species in this reserve and occurred in all five forest types in both edge and non-edge habitats (Figure 3.7). At the same site and around the same time, Fleagle *et al.* (1981) found that *C. apella* differed from *Saimiri sciureus* (squirrel monkeys) in that capuchins were more frugivorous, more likely to be found in the middle canopy and moved quadrupedally on medium-sized support branches. The squirrel monkeys were more insectivorous, more saltatory and moved on the smallest of arboreal supports. This differential habitat use occurred even when capuchins and squirrel monkeys foraged together in polyspecific associations (see below). A comparison of capuchins to howlers at the Curu Refuge in Costa Rica (Tomblin and Cranford 1994) also found ecological niche differences in how these sympatric species used the habitat – the capuchins ate a wider variety of foods (many of which require extraction and processing) and they used a wider range of the arboreal habitat than did the howlers.

Looking more closely at how the diets of capuchins diverge from those of sympatrically living primates, Stevenson, *et al.* (2000) found that during times of fruit scarcity in Tinigua National Park, Colombia, the four monkey species studied resorted to different sets of supplementary or "fall back" foods. Although 90% of interspecific aggression still occurred in fruiting trees, during times of low fruit abundance, howlers ate more leaves, spider and woolly monkeys ate more unripe fruit and young leaves, and capuchins ate more stems and insects. The authors postulate that switching to a variety of vegetative parts of plants during periods of fruit scarcity is a resource-partitioning mechanism that reduces competition and allows co-existence among the four sympatric species. Habitat partitioning was also briefly described by Julliot and Simmen (1998) for the Nouragues station in French Guiana, and in more detail by Mendes Pontes (1997) for the five species of monkeys that occur on Maraca Island, Northern Brazilian Amazonia. Mendes Pontes found that the five species at his site all formed some polyspecific associations that involved sharing the food resources, especially co-foraging in the aseasonal fig trees during times of fruit scarcity.

CAPUCHINS IN POLYSPECIFIC ASSOCIATIONS

Primates of one species are sometimes found foraging and traveling together with members of other species. These assemblages are variably referred to as mixed species, polyspecific or heterospecific groups, and they designate nonrandom and lengthy associations between two or more species (Norconk 1990). When such groupings occur in the platyrrhines, they are usually called mixed species groups, and very common participants in such groups are *Saimiri sciureus* and *Cebus apella*. *C. albifrons* is sometimes reported to forage in mixed species groups, but not nearly as frequently as the associations between squirrel monkeys and tufted capuchins. These associations are not just chance aggregations at common resources, but two or more species that truly forage and travel together.

Why would one species ever form a group with another species? Norconk places the many suggested benefits of mixed species groups into two large categories: group-size hypotheses and information-sharing hypotheses. The former refers to suggested benefits such as diluted/reduced predation risk (accomplished through safety in numbers, the selfish herd phenomenon, and/or confusing the predator) and reduced vigilance time per individual (accomplished through many eyes and ears, and/or by sentinels). Basically these are all forms of the argument that an individual has a lower probability of being taken by a predator in a larger, mixed-species group than it does in a smaller, single-species group. The second category of hypotheses, information-sharing, refers to the idea that individuals may experience improved feeding efficiency by pooling their knowledge of where short-lived resources are located, or by one species using the second as a "guide." In a variant of this hypothesis, one species may flush invertebrate prey that are taken by the second species. For example, capuchins at many field sites are reported to be followed by birds such as kites and hawks that feed on insects, snakes and small mammals (Boinski and Scott 1988, Fontaine 1980, Freese and Oppenheimer 1981, Greenlaw 1967, Oppenheimer 1968, Ridgely 1976, Stott and Selsor 1961, Terborgh 1983, Warkentin 1993, Zhang and Wang 2000). Individual birds may travel with a given group of capuchins for up to 20% of the time that the monkey group is observed. Presumably the monkeys are little affected by the presence of the birds, but the birds benefit by feeding on the animals stirred into visibility by the foraging actions of the capuchins. Less commonly, other species may benefit from the messy eating habits of capuchins, who frequently test and drop fruit while foraging. At Santa Rosa in Costa Rica, peccary sometimes gather under large fig trees in which *C. capucinus* are consuming fruit and the peccary eat the dropped fruit as well as becoming covered in them. Sabino and Sazima (1999) report that characid fish, *Brycon nicrolepis*, gather under vegetation along river banks where *C. apella* are feeding, and follow the monkeys along the river course, feeding on the fruits that they drop.

More commonly, however, polyspecific associations involve two species of monkeys traveling and foraging together. Both Terborgh (1983) and Podolsky (1990) studied mixed species groups of squirrel monkeys and tufted capuchins at Manu National Park in Peru. Terborgh found that when traveling at a slow pace to capture insects, *Saimiri* seemed to lead the way, but when travelling rapidly to a fruit tree or new foraging location, *Cebus* led the way. He concluded that squirrel monkeys benefit from the tufted capuchins' knowledge of where ripe fruit trees are located, as well as from capuchin alarm calls and from the capuchins' superior strength which facilitates opening tough *Scheelea* palm nuts. But, in Terborgh's view, the capuchins were not benefiting much from the association and so seldom actively sought out squirrel monkeys, whereas the reverse was often true (squirrel monkeys sought out the capuchins and benefited from the association). Podolsky suggested several reasons why tufted capuchins may tolerate the squirrel monkeys joining them for periods of up to 12 days of foraging – there is more than enough ripe fruit for everyone, individual capuchins can supplant individual squirrel monkeys from food whenever they need to, a larger area can be searched with a mixed species group, and capuchins are less vigilant when surrounded by many small squirrel monkeys. The latter two parts of Podolsky's argument suggest that capuchins do also benefit from their association with squirrel monkeys.

Thus far we have talked about the possible benefits of mixed species groups, but there are also costs. For one thing the levels of feeding competition increase with group size, and that is probably why a tufted capuchin group travels faster and further in search of food on days when it is in association with squirrel monkeys. That is also probably why two species of capuchins, such as the sympatric *C. albifrons* and *C. apella*, are so seldom found in mixed species groups (or seldom even found at the same sites, see Sussman and Phillips Conroy 1995 and Lehman 2000) – they are just too similar in body size, group size and foraging habits.

Furthermore, Boinski (1989) argued that squirrel monkeys (*Saimiri oerstedi*) in Corcovado National Park, Costa Rica seldom travel in association with white-faced capuchins because in this case (unlike the situation in South America) squirrel monkeys obtain neither foraging nor anti-predator advantages from associating with capuchins, and possibly experience a reduction in foraging efficiency when in this mixed species association. The lack of foraging advantages from polyspecific associations in Costa Rica is because squirrel monkeys and capuchins at this site have minimal dietary overlap. The possibility of reduced foraging efficiency is because squirrel monkeys in Corcovado stop feeding and either monitor or avoid capuchins when they hear them coming. Indeed, as noted by Boinski (1989), squirrel monkeys in Panama and Costa Rica are reported to actively and regularly withdraw at the approach of capuchins (Baldwin and Baldwin 1972, Moynihan 1976). Such avoidance could be related to the high rate with which white-faced capuchins prey on small vertebrates, including one unverified report of capuchins killing a squirrel monkey (Freese and Oppenheimer 1981), or to their tendency to supplant other primate species from the trees.

Terborgh (1983) pointed out that in comparison with birds, mixed species groups in primates are relatively rare and are concentrated in parts of the world where the largest monkey-eating raptors occur (Africa and South America). Thus he suggests that the threat of predation is the primary impetus for the formation of mixed species groups in primates, but that the costs of association are relatively high – at least higher than in bird species where mixed species flocks are much more common than in primates.

SUMMARY

Although we have only begun to conduct research on those aspects of ecological relationships and primate communities covered in this chapter, it is clear that capuchins are an integral part of their environment, both influencing and being influenced by the organisms around them. This genus of monkey preys on both invertebrates and small vertebrates, and in turn, is preyed upon by a wide variety of carnivores. They are hosts to many forms of parasites. While their preferred foods are fruits and insects, they also occasionally eat seeds and flowers. At other times, capuchins disperse seeds away from the parent fruit tree, and they carry pollen from one flower to another. However, the effectiveness of their role as seed dispersers and pollinators remains to be tested. Furthermore, it has been argued that capuchins lower the numbers of certain insects that infest trees, and change the morphology of some plant species. They occasionally take agricultural foods from fields and orchards, leading to conflict with farmers, who may hunt them in retribution. They sometimes forage in mixed species groups, benefiting the other types of monkeys that travel with them, whereas at other times they supplant other species from food. Clearly, capuchins participate fully in the many forms of competition and cooperation that characterize the dynamics of ecological communities.

4 • Life history and demography

The male that we call Winston is not having a good day. Sometime between yesterday evening and today, he appears to have been in a serious fight. His right hand and foot are slashed, his upper arm has a gaping wound and his face is bleeding from cuts to his eyebrow and upper lip. He limps and rather than eating, stops constantly to rest and lick his cuts. Sometimes females and juveniles approach and groom him, but mainly they go about their normal day of foraging and traveling. The young beta male has a wound to his tail, suggesting he was facing away, perhaps running away, when bitten by an unknown aggressor. We have followed Winston over a ten-year period during which time he has resided in two other groups before taking up the alpha position in this one, but we have never seen him wounded so severely. We estimate him to be an old male because of his haggard and scarred face, thin frame and coarse hair, and because he already looked past his prime when we first saw him a decade ago. But we don't really know how old he is, and it is difficult to verify the life span of these peripatetic males. Suddenly the beta male gives the call that precedes the appearance of non-resident monkeys. I move in the direction of the beta male's gaze and make out the presence of three adult males headed our way through the trees. I have never seen them before. Two are wounded, but neither so seriously as Winston. I look back at the group and see that the females with infants have all faded out of sight. A few females, several large juveniles and the beta male have all stayed close to the alpha male, who is now standing on his three good legs and looking in the direction of the oncoming males. The newcomers arrive in a rush, breaking branches and giving aggressive calls. For a short while, the remaining members of our study group line up shoulder to shoulder and face off with the invading males like human soldiers in the ritualized battle lines of prior centuries. Then chaos breaks out with monkeys running in all directions. We try to follow on the ground but cannot keep up with the fast pace of the encounter, even though we catch glimpses of monkeys hurtling through the trees and hear crashing and vocalizing all around us. A large juvenile runs past us with wounds on his leg and tail. Finally, we find the group sitting very quietly and cohesively near one of their sleeping trees. Everyone is present except the alpha male and there is no sign of the attackers. We re-trace our steps and find Winston lying on the ground on his back. He has even more severe wounds and does not seem to be able to get up. We sit nearby as unobtrusively as possible until it is dark but the aggressive males do not return to this spot this day. As researchers who do not interfere in the lives of the monkeys but inevitably become attached to them, we are not having a very good day either. The next morning when we arrive before dawn, Winston is nowhere to be found and two of the three invading males are shadowing the group as it begins its daily round.

Capuchins, like parrots and tortoises and a few other mammals, live a very long time. Just how long do capuchins live? And is their span of life unusual among monkeys or among primates in general? How is their long life span related to other features of their biology? These are all questions that have occupied primatologists and evolutionary biologists for some time. At the moment, we have many more theories than data. To conduct a complete analysis of the life history of a taxonomic group, one wants to know how long animals live in natural environments, when they die and the causes of death (which are likely to vary with age), and many details of reproduction (for example, age at first conception, inter-birth intervals in females; age at fertility in males; and for both sexes, if there is a period of reproductive senescence at the end of the life span.) It is safe to say that the science of life history analysis, in primates at least, is in its infancy. Data on longevity, causes of death and reproductive parameters for animals living in natural settings are hard to come by – they require good record-keeping on a large number of animals over many years. We have good records for a very few species of nonhuman primates, principally Japanese macaques and chimpanzees. For capuchins, such data are currently available from only a few sites. Information from captive populations is easier to collect, but even this is surprisingly scarce, and its relation to life history parameters in natural environments is not yet clear.

Accordingly, this chapter is somewhat of a promissory note. We tell you what little we know, and then we tell you what we would like to know more about. We start with life span and how life span correlates with physical features (especially how brain size relates to life span). Then we turn to individual reproductive parameters. Next, we consider a topic where we have begun to accumulate a reasonably good body of data: age/sex compositions of groups. Finally, we present what is known about demographic processes such as natality, mortality, dispersal and group fission/fusion patterns.

LIFE SPAN AND CORRELATES WITH BRAIN SIZE

Primates as a whole exhibit long life among mammals (Austad and Fischer 1992). There is a general positive relation in mammals between body size and longevity, and primates, with their relatively large body size among mammals, are appropriately long-lived for their body size. Within primates, but not across mammals as a whole, longevity is correlated with relative brain size (brain size corrected for body size). This trend has been of great interest to evolutionary biologists and primatologists, as both groups seek explanations for the evolution of anomalously large brains in our own species. The correlation between relative brain size and longevity in primates is only modest, however, about +.5, when apes and humans are removed from the data set (Austad and Fischer 1992). Lest we jump to the conclusion that there is some special link between brain size and longevity, we note that the correlation between life span and the mass of other organs (heart, liver, kidney and spleen) is also positive across mammals, and of the same magnitude as the link between relative brain size and longevity in primates. Within primates, there is no evidence that life span increases across the order the more recently the taxonomic group diverged from the line leading to humans. One idea currently receiving attention by life history theorists is that the positive relation between body size and life span reflects vulnerability to environmental hazards or mortality risk. According to this theory, larger animals are less vulnerable to predation and may be more resistant to food or water shortages (because metabolic rate correlates negatively with body size – large animals have lower metabolic rates than small animals). One might extrapolate from this idea that perhaps the link between relative brain size and longevity in primates reflects a difference in vulnerability to environmental hazards that relates to brain size – larger brains are associated with some sort of defense against hazards. The most interesting potential basis for this relationship, to a behavioral biologist, involves behavioral solutions to life's dangers. However, non-behavioral mechanisms (such as better defenses against oxygen free radicals) are equally likely (Austad and Fischer 1992). All we can say at present with confidence is that capuchins live an anomalously long time, even for primates, and they have anomalously large brains as well. These two anomalies may be linked in a biologically and evolutionarily interesting way. Then again, they may be linked only through a (relatively uninteresting) third process that we have not as yet identified.

So how long do capuchins live? The current record in captivity is nearly 55 years old (Hakeem *et al.* 1996), not too far off the record for chimpanzees (Dyke *et al.* 1995). The record life span documented by Hakeem *et al.* is 41 years or older for each of the four traditional species in the genus (see Table 4.1). A reasonable life span for captive capuchin monkeys, maintained with good nutrition and healthy living conditions, is 40 years. Most elderly capuchin monkeys presently in captivity were captured, not born in captivity, and thus we do not know their ages. An elderly male captured as a "young adult" in 1963 was still healthy and active (and the central male and the females' preferred sexual partner) in Fragaszy's colony in early 1997 (he died that year as a result of socially inflicted trauma). Given his date of capture and an estimated minimum age of 5 (when permanent canines emerge) for designation as an adult at that time, he was at least 39 at the time of death. As adult primatologists with groups of capuchins in our

Table 4.1. *Life history parameters in Cebus*

Species	Gestation (days)	First birth	Interbirth interval (mos)	First siring	Max life span (years)
C. albifrons (nature)	–	–	–	–	–
C. albifrons (captive)	–	–	–	–	44[6]
C. apella (nature)	151–155[1] 149–158[10]	7 yr Mode[1] 5 yr E[9]	19.4[1] 22[9]	–	–
C. apella (captive)	160[4]	5 yr 7 mo[4] 5 yr 8 mo A[11] 4 yr 7 mo E[4] 3 yr 10 mo E[11]	20.6[4] 20.4[8]	4 yr 5 mo[4]	45.1[6]
C. capucinus (nature)	157–167[5]	7 yr A[2]	26.4[3] 23.7[7]	6 yr (mate with ejac.)[2]	–
C. capucinus (captive)	–	–	–	–	54.75[6]
C. olivaceus (nature)	–	6 yr E[9]	–	–	–
C. olivaceus (captive)	–	–	–	–	41[6]

A = Average, E = Earliest, Mn = Minimum, – = no data available.
Sources: (1) Di Bitetti and Janson 2001; (2) Fedigan unpubl.; (3) Fedigan and Rose 1995; (4) Fragaszy and Adams-Curtis 1998; (5) Freese and Oppenheimer 1981; (6) Hakeem *et al.* 1996; (7) Perry, Manson & Gros-Louis, unpubl.; (8) Recabarren *et al.* 2000; (9) Robinson 1988a; (10) Robinson and Janson 1987; (11) Zunino 1990.

laboratories now, we will be looking for homes for them when we retire – and most of them will still be in the prime of life!

Capuchins living in natural conditions are unlikely to live as long as monkeys living in healthy conditions in captivity. Nevertheless, in view of the life spans noted above for captive animals, Robinson's (1988a) estimate that *Cebus olivaceus* at Hato Masaguaral live at least into their mid-thirties seems plausible. Long-term studies of capuchin monkeys promise more data on this unusual aspect of capuchins.

INDIVIDUAL REPRODUCTIVE PARAMETERS

The bare facts about reproduction in relation to life history are presented in Table 4.1. A cursory glance tells us three things: that the values across captive and wild populations are quite consistent; that capuchins start producing offspring quite a long time after they have become fully nutritionally independent of their mothers (i.e., for females, about 4 years after the birth of their next sibling); and that we lack even basic data on reproduction for most species. When looking at this table it is wise to keep in mind that some commonly used terms to identify stages of life history (e.g., ages at sexual maturity, weaning etc.) are actually very difficult to define (Lee 1996). For example, fertility in males increases gradually in adolescence and is not synonymous with sexual maturity (which many authors put at >7 years), nor is it synonymous with full physical maturity, which males may take as long as 10 years to reach (Jack and Fedigan in Press a, b). Although we have a documented instance of a male capuchin siring offspring when less than 5 years old, this occurred in a captive group with no older males present.

Figure 4.1. This tufted capuchin male, about 30 years old, dipped the stick inside the holes of the box and is licking honey off the stick. (Photograph courtesy of Jim Anderson).

Comparing the values in this table to those for other genera reported elsewhere (Hartwig 1996, Robinson and Janson 1987, Ross 1991), capuchins appear anomalous compared with most of their platyrrhine counterparts, particularly when body size is taken into account, in the duration of inter-birth intervals (long), the age at first birth (old) and maximum life span (long; see Figure 4.1). This combination of features translates into a relatively long period of juvenescence in capuchins – those years when the youngster is nutritionally independent of the mother, but not yet reproductively mature – coupled with a long reproductive life. The ecological utility of this pattern may be that young capuchins are buffered against starvation by their slow growth pattern (Janson and van Schaik 1993). Although some decline in fecundity is likely as animals age, we have no data yet indicating a period of reproductive senescence in capuchins.

DEMOGRAPHY

Age/sex composition of groups

As with life history analysis, the study of demography in capuchin populations is in its infancy. We have, however, collated all available field data on age/sex composition of groups (Table 4.2a, b). From this table, we can see that average group sizes range from 12–27 individuals, with an overall average capuchin group size of 18 members. *C. capucinus* have the smallest group sizes, averaging 16 individuals, and *C. olivaceus* have the largest groups, with 21 members on average.

The proportion of adult males in capuchin groups is quite variable, ranging from as low as 16% to as high as 31%. The average proportion of adult males for all capuchins is 25%, or one-quarter of the group. *C. olivaceus* have the fewest males per group (mean = 17%) whereas *C. apella* and *C. albifrons* have the most adult males per group (mean = 33%). The proportion of adult females in groups remains remarkably stable across the four better-known species at 30–33%, or one third of the group. The proportions of juveniles in groups range from 11–47%, with an overall average of 30%, indicating that juveniles constitute on average one-third of the group. *C. apella* and *C. albifrons* have the lowest proportions of juveniles (means = 23% and 22%), and *C. olivaceus* have the highest proportions of juveniles (mean = 41%). The latter value is interesting because *C. olivaceus* also have the lowest proportion of infants in groups (mean = 9%), suggesting that infancy and juvenescence may have been differently defined in census studies of this species compared with others. The proportion of infants in groups ranges from 6–23%, with both ends of the range occurring in *C. albifrons*. Averaged overall, infant capuchins comprise 13% of the group.

In the majority of capuchin groups censused, the adult sex ratio is biased toward females. Exceptions occur in some *C. apella* and *C. albifrons* groups where there were more adult males than adult females. In contrast, *C. olivaceus* groups contain nearly two adult females for every adult male. Overall, the adult sex ratio in capuchins averages 0.81, or 1.2 females for every male. The immature to adult female ratio is an indicator of the age structure and health of a primate population – high ratios indicate good reproductive and survival rates. Overall, capuchin groups include 1.35 immatures to every adult female. *C. apella* have the lowest ratios of immatures to adult females (mean = 1.06) and *C. olivaceus* have the highest ratios (mean = 1.56). The ratio of infants to adult females can be used as a rough indicator of the birth rate, although comparisons are problematic due to different methods of assessing infancy and the possibility of birth seasons (see below). Overall, capuchins exhibit a ratio of 0.44 infants to every adult female, indicating that nearly one in two adult females

Table 4.2a. *Age / sex composition of capuchin groups at various field sites*

Site and year(s) of study	Mean no. groups counted	Mean group size	Proportion adult males	Proportion adult females	Proportion juveniles	Proportion infants	Adult M:F ratio	Immature: adult female ratio	Infant: adult female ratio
Cebus capucinus									
Santa Rosa 1972[1]	1	17	0.29	0.35	0.24	0.12	0.83	1.00	0.33
1983–99[2]	20	15.8	0.21	0.31	0.34	0.13	0.71	1.39	0.41
BCI 1966–1970[3]	2	15	0.17	0.31	0.36	0.16	0.54	1.70	0.52
1986–1988[4]	4	20	0.19	0.32	0.37	0.12	0.60	1.59	0.42
Trujillo 1980[5]	1	14	0.21	0.24	0.39	0.16	0.88	1.59	0.65
Cebus olivaceus									
Hato Masaguaral 1977–85[6]	10	20	0.18	0.33	0.35	0.14	0.54	1.50	0.43
Hato Piñero 1989–91[7]	8	22	0.16	0.33	0.47	0.04	0.51	1.61	0.13
Cebus apella									
Iguazú Nat'l Park 1993[8]	6	14.4	0.19	0.34	0.28	0.19	0.56	1.38	0.56
La Macarena 1986–98[9]	1	18	0.28	0.28	0.33	0.11	1.00	1.60	0.40
Cocha Cashu 1976–77[10]	1	12.8	0.44	0.36	0.11	0.09	1.20	0.57	0.26
Caratinga 1995–97[11]	1	27	0.22	0.33	0.33	0.12	0.68	1.38	0.38
C. albifrons:									
Cocha Cashu 1975–77[10]	1	8.5	0.47	0.35	0.12	0.06	1.3	0.40	0.17
Cocha Cashu 1985[12]	1	16	0.31	0.25	0.25	0.19	1.25	1.4	0.75
El Tuparro[13]	1	35	0.20	0.29	0.29	0.23	0.70	1.80	0.80

Sources: (1) Freese 1976; (2) Fedigan and Jack 2001; (3) Oppenheimer 1982; (4) Mitchell 1989; (5) Buckley 1983; (6) Robinson 1988a, b; (7) Miller 1992a, b; (8) Di Bitetti and Janson (2001 and pers. comm) Based on the composition of six well-studied groups before a group fission occurred; (9) Izawa 1988, 1990a, 1992, 1994, 1997, 1999; (10) Terborgh 1983; (11) Lynch and Rimoli 2000; (12) van Schaik and van Noordwijk 1989; (13) Defler 1979a, b.

(Note that for multiple groups studied over several years, the values are averages across all the groups counted in each year, then averaged for all the years of study).

Table 4.2b. *Averaged group compositions for the four better-known capuchin species*

Species name	Group size	Proportion adult males	Proportion adult females	Proportion juveniles	Proportion infants	Adult M:F ratio	Immature: adult female ratio	Infant: adult female ratio
C. capucinus	16.36	0.21	0.31	0.34	0.14	0.71	1.45	0.47
C. olivaceus	21	0.17	0.33	0.41	0.09	0.53	1.56	0.28
C. apella	17.17	0.33	0.31	0.23	0.12	1.08	1.06	0.43
C. albifrons	19.83	0.33	0.30	0.22	0.16	1.08	1.20	0.57
All four species averaged	18.25	0.25	0.31	0.30	0.13	0.81	1.35	0.44

Table 4.3. *Natality, sex ratio at birth and infant survivorship in capuchins*

	Number of infants born	Died < 6 months	Died < 1 year	Survived > 1 year	Sex ratio at birth M - F - Unknown
Cebus capucinus[1] Santa Rosa 1986–2001	85	20 (24%)	27 (32%)	58 (68%)	52 -18 -15 (Est'd M:F = 2.8)
Cebus capucinus[2] Captive (zoos)	149	24 (16%)	–	–	85–23–41 (Est'd M:F = 3.7)
Cebus olivaceus[3] Hato Masaguaral	133	24 (18%)	–	–	29-55-49 Est'd M:F = 0.53
Cebus apella[4] Iguazú 1992–99	22	–	10 (45%)	12 (55%)	12 -13 -4 (Est'd M:F = 0.92)
Cebus apella[5] Captive	61	8 (13%)	–	–	68-58-15 (Est'd M:F = 1.17)
Cebus apella[6] Captive	36	13 (36%)	0	23 (64%)	20-15-1 (Est'd M:F = 1.33)
Cebus apella[2] Captive (zoos)	863	155 (18%)	–	–	384–137–342 (Est'd M:F = 2.8)

Sources: (1) Fedigan, unpubl. data; (2) Debyser 1995; (3) Robinson 1988a; (4) Di Bitetti and Janson 2001; (5) Fragaszy and Adams-Curtis 1998 and Fragaszy, unpubl. data; (6) Visalberghi, unpubl. data; most infant deaths in this lab occurred when the laboratory was newly established.

has an infant at any given time. We hope that this necessarily limited discussion of age/sex composition of capuchin groups will stimulate others to supply further data that will allow better comparisons and more secure generalizations.

Natality and mortality

We only have capuchin birth data from three field sites, two captive research colonies and one analysis of zoo records on two species (Table 4.3). The field data indicate that M:F sex ratios at birth can vary from being nearly one male for every female (Iguazú: 0.92) to being highly skewed toward female births (Hato Masguaral: 0.53) or highly skewed toward male births (Santa Rosa: 2.8). It is difficult to sex infant capuchins in the field since the females possess a clitoris that is nearly indistinguishable from the males' penis (see Chapter 6). Therefore, at Fedigan's field site in Santa Rosa, we only decide on the sex of an infant when we can detect the presence or absence of testicles, resulting in a high percentage of infants that die before we can definitively sex them (18%). Even with this conservative method

Figure 4.2. Mother and infant dyads. Left, *C. apella*, right, *C. capucinus*. The newborn infants cling to the mother in a cross-neck position, as shown on the right. Older infants cling parallel to the mother's body, as shown on the left. (Drawing by Stephen Nash).

of determining that an infant is a male, and even if we assume that all "unknown" cases are females, the study groups at Santa Rosa have produced many more male than female infants since 1985. This male-biased sex ratio persists through the juvenile years, for which we have recorded sex ratios of 2.8 males to every female in younger juveniles and 1.6 males to every female in older juveniles (Fedigan *et al.* 1996). Why we have sex ratios skewed heavily toward males in Santa Rosa is not yet known, but male-biased sex ratios at birth are known to occur in some species of Old World monkeys (Fedigan and Zohar 1996).

The data summarized in Tables 4.2a and 4.3 allow us only a very incomplete comparison of sex ratios at birth and adult sex ratios from three field sites (Santa Rosa, Hato Masaguaral and Iguazú), but what limited data there are suggest that the ratios of males to females born into study groups can vary considerably from the ratios of adult males to females found in these groups. In two of the three cases, there are much lower ratios of adult males to females than infant males to females (Santa Rosa: 0.71 vs. 2.8; Iguazú: 0.56 vs. 0.92), indicating that capuchins may be similar to many Old World monkey species, in that males are more subject to mortality throughout life (Fedigan and Zohar 1996). It is common for the dispersing sex to experience higher mortality rates than the philopatric sex (see below).

There is a lamentable scarcity of information on mortality in capuchins, either from the field or from captivity. As Robinson (1988a) noted some time ago, deaths are uncommonly observed in a long-lived genus such as *Cebus*. At Santa Rosa, we have observed monkeys dying and found cadavers of our well-known study animals. We have also, albeit rarely, seen a capuchin taken by a predator (e.g., Chapman 1986). However, except in those cases where we have witnessed aggressive encounters leading to severe wounding and subsequent death, we are seldom able to determine the cause of death. Usually individuals simply disappear. When we find the body of an individual that had looked thin, frail or weary in the previous days, we assume that it died of some illness, but without autopsies, we cannot specify the disease.

We do know that capuchin mortality rates are highest in the first year of life (Fedigan *et al.* 1996). Table 4.3 shows that 18–24% of capuchin infants die in the first 6 months of life in nature and, in captive groups, 13–36% of them die in the first 6 months. This table also shows that only 55–68% of capuchins survive past their first year. Survival analysis of the Santa Rosa data showed that only 39% of infants born survive to the age of 5 years (Fedigan *et al.* 1996).

Seasonality of births

As noted by Di Bitetti and Janson (2000), most Neotropical primates show some degree of birth seasonality. Capuchins are no exception, although the extent of

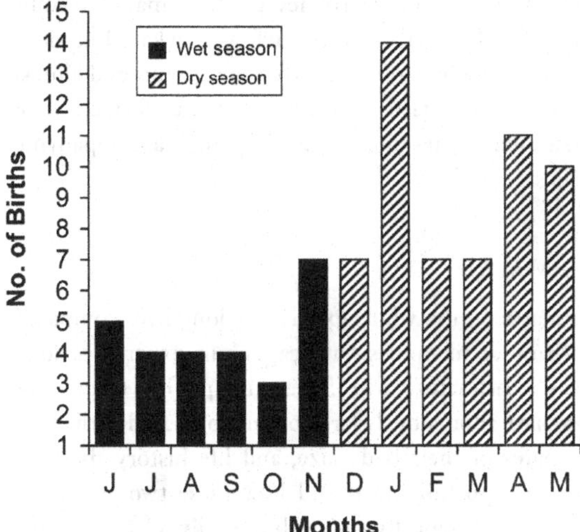

Figure 4.3. Birth seasonality of *C. capucinus* at Santa Rosa National Park, Costa Rica, based on 83 births to individually known mothers on known dates.

strict seasonality varies from site to site. The Central American capuchin, *C. capucinus*, has been documented to give birth in every month of the year, but exhibits a notable birth peak in the dry season and early rainy season (Costa Rica: Fedigan *et al.* 1996, and unpubl.; see Fig 4.3; Panama: Oppenheimer 1968, Mitchell 1989). Somewhat similarly, *C. olivaceus* at Hato Masaguaral (Venezuela: Robinson 1988a) and *C. albifrons* in El Tutarro (Colombia: Defler 1979b) produce most of their infants in the late dry and early wet season. Populations of *C. apella* at La Macarena exhibit a birth peak in the early wet season (Peru: Izawa 1990a, 1992), whereas *C. apella* residing below the equator at Iguazú have an actual birth season in December–January (Argentina: Di Bitetti and Janson 2000). Di Bitetti and Janson (2000) argue that capuchins follow the typical pattern for New World monkeys of giving birth in the period coinciding with or immediately preceding the peak in food availability. The availability of food is a complex issue for primates such as capuchins that can switch to eating insects when fruit levels are low (see Chapter 2), and it is also possible that birth seasons are driven at least in part by optimal physiological conditions for conception (e.g., Santa Rosa capuchin females become thin during the dry season when little water is available and put on a noticeable amount of weight by the middle of the rainy season at the time when most of them conceive).

Dispersal patterns

Capuchins are unusual among New World monkeys in being primarily male-dispersed (Strier 1999). Female capuchins may also leave their natal groups and join another, but much more rarely than males. Males may emigrate from their groups as early as 2 years, but the probability that they will leave their birth groups is highest when they are juveniles of about 3–6 years of age (Robinson 1988a, Jack and Fedigan, in press a). Jack and Fedigan examined nearly 20 years of data on dispersal among *C. capucinus* at Santa Rosa and found that emigration from natal groups does not coincide with the conception season (see above), nor does it result from eviction by resident adult males (the putative fathers) or eviction by adult males who have recently taken over the groups (Jack and Fedigan in press a). These proximate circumstances for emigration suggest that juvenile males leave their natal groups not because they are attracted to estrous females elsewhere or because they are forced to do so by adult males. Rather natal male dispersal coincides with periods of intense inter-group interaction (during which time juveniles "get to know the neighbors"), group fissioning (see below) and occasions when older adult males are wandering and dispersing themselves. Indeed, parallel emigration of multiple males is common in this population, and 82% of confirmed natal male transfers between study groups were in the company of older maternal siblings, or members of the same age cohort (possibly paternal sibs), or into groups containing older familiar males. This indicates that natal male emigration occurs primarily due to attraction to extragroup males or dispersing group mates. Parallel dispersal may confer inclusive fitness benefits (if co-dispersing males are usually related) and provide added protection for young, vulnerable males traveling between groups. A considerable body of primate data shows that dispersing males are subject to high mortality rates.

We also examined the proximate circumstances of secondary dispersal (emigration subsequent to natal dispersal) in the male capuchins of Santa Rosa (Jack and Fedigan in press b). Again we found that the vast majority of emigrations appear to be voluntary (rather than resulting from eviction or takeovers) and that familiar and related males are likely to disperse in parallel. However, in these cases of secondary dispersal, males are older and more likely to enter a new group using aggression and force. Given that male white-faced capuchins face many risks in dispersing between groups, and that these males lead a life of continual movement between groups, there must be some benefits to repeated dispersal. We found that adult males usually increase their dominance rank and improve their odds of finding available females (a lower male to female sex ratio) in their new groups. Furthermore, the median length of time that an adult male remains in a group is about 4 years and females first give birth at 6–7 years of age. This suggests that adult males, by continuing to move between groups, are unlikely to mate with their maturing daughters.

GROUP FISSION AND FUSION

Social groups, like individual capuchins, have life histories and experience change over time. Although the formation of new groups is rare, there have been observations of group fissioning, which is distinct from the dispersal of one or a few males from a group. Such fissioning usually occurs when a small subgroup, consisting of one to two adult males, and one to three adult females with their dependent offspring buds off from the original group. These small subgroups may establish a range of their own or fuse with a neighboring small group. Oppenheimer (1968), Robinson (1988a), Izawa (1994) and M. Di Bitetti and C. Janson (pers. comm.) have all described cases of subgroups fissioning and later fusion with neighboring groups, for *C. olivaceus*, *C. apella* and *C. capucinus*, respectively. At Santa Rosa, Fedigan and colleagues have observed several cases of fissioning in which a small subgroup of adult males, adult females and their offspring departed from one of our study groups and were subsequently seen or inferred to join another group. They have also observed the extinction of one study group that experienced a series of deaths from unexplained illness followed by a male take-over during which infants and resident adult males and females were wounded and subsequently died. It is unlikely that capuchin groups ever come into being as entirely "new" through the joining together of solitaries of both sexes. This pattern of new group formation is common in other platyrrhines such as howlers, but since female capuchins seldom leave their groups and almost never travel solitarily, new groups are only formed when fissioning occurs along matrilines with accompanying males.

SUMMARY

Primates on the whole experience long lives compared with other mammals, and capuchins experience long lives for primates (up to 55 years in captivity). Capuchins also have anomalously large brains compared with other primates of their body size, and life history theorists are busy postulating if and how these two anomalies might be biologically and evolutionarily related. Unfortunately, we lack data on many of the basic life history and demographic parameters that might help us to fully examine this issue. Nonetheless, this chapter summarizes the available records on capuchin gestation lengths, ages at first birth and first siring and interbirth intervals. This information points clearly to a pattern of long, slow life histories. We also summarize available data on age/sex composition of groups and natality patterns, which show us that the adult sex ratio is usually biased toward more females than males, whereas the sex ratio at birth is quite variable. In nature, capuchin females give birth about every 2 years, and show some degree of birth seasonality, either a birth peak or an actual season of births when food is abundant. There is almost no information from the field on the causes and frequencies of death, but we do know that mortality rates are highest in the first year of capuchin life, as is true of other primate species. Male capuchins emigrate from their natal groups around the age of sexual maturity and join other groups, a pattern that they then repeat throughout their lives. Female capuchins do not disperse from their natal groups, but occasional group fissions occur, during which several adult females and their young depart their groups with accompanying males to form new groups or fuse with neighboring groups.

Part II
Behavioral biology

5 • The body

A female wedge-capped capuchin moves toward the trunk of a large palm tree along leafless branches and lianas swaying in the brisk breeze of the dry season. Bracing herself with her feet and tail just below the base of the palm's crown, she begins to search for food. She thrusts her hands quickly here and there into the debris caught at the base of the fronds, and pulls out handfuls of decaying leaves, small twigs, abandoned cocoons, and other bits of plant and animal remains in a quick rhythm. Each time, she looks briefly at the loose material in her hands, then releases it a second later and starts over. Suddenly she feels something softer with her fingertips; she pauses, eyeing the current handful. Here is a tasty morsel for her, a larval insect that had remained hidden from view underneath a dead leaf but was obvious to her sensitive fingers, and now, her sharp eyes. Taking it with her other hand, she pops it into her mouth.

PLACING ANATOMY IN COMPARATIVE CONTEXT

This vignette highlights just a few features of capuchins' anatomy that are fundamental to their behavior (hands, tail, eyes); we will treat these and several others in this chapter. The physical characteristics of capuchins must be seen against the shared features of their fellow platyrrhine taxa to appreciate what is distinctive about them. For example, all platyrrhines share three premolars, and the tympanic ring of the skull is fused to the side of the auditory bulla but does not extend into a bony tube (see Figure 5.1). These are both considered primitive (ancient) features in the primate order that have been lost in catarrhine monkeys. In addition to retaining anatomical features of ancient primates, platyrrhine monkeys possess features that evolved after their most recent common ancestry with Old World monkeys, including the shapes of the lower premolars and the form of the lateral wall of the skull (see Fleagle 1999 for the major features of anatomy and morphology in primates, including some of the major differences between platyrrhine and catarrhine primates). Hill (1960) remains the most detailed comprehensive description of the morphology and anatomy of the genus, down to the number of vibrissae on the upper lip (four). In most domains, the capuchin's anatomy and morphology are representative of platyrrhine monkeys.

Within the shared platyrrhine body type, the genus- and species- distinctive details of the capuchin's anatomy and morphology reflect its body size and behavioral ecology. After reviewing general morphology (size and shape, pelage and skin), we will focus on those anatomical and morphological features that are particularly distinctive to capuchins, and that contribute in obvious ways to their behavioral ecology. These include the skull (cranium and mandible) and teeth, tail, hands, external genitalia, gut and brain. We will also discuss sensory acuity, self care and positional behaviors, and locomotion.

SIZE AND SHAPE

New World monkeys range in size from just over 100 g (for the pygmy marmoset, *Cebuella*) to over 10 000 g (spider monkeys and muriqui, *Ateles* and *Brachyteles* respectively), with a modal size of about 1000 g (Ford and Davis 1992, Smith and Jungers 1997). Capuchins are intermediate in this size range. Wild adult female capuchins weigh on average 2.3 kg, and adult males 3.1 kg (Ford and Davis 1992). Species of capuchins vary somewhat in weight, with *C. albifrons* the lightest at 2.1 kg, followed by *C. olivaceus* and *C. apella* at 2.7 kg, and *C. capucinus* the heaviest at 3.3 kg (average of male and female weights; Ford and Davis 1992). Fleagle (1999)

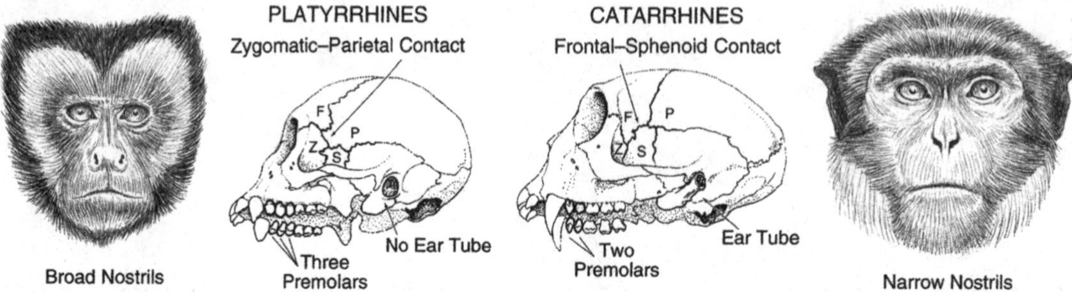

Figure 5.1. Face and skull of a New World monkey (left) and of an Old World monkey (right). The illustration points out several major differences between them (see text). (Drawing by Stephen Nash.)

provides somewhat higher weights for each species, with *C. apella* and *C. capucinus* both at 3.65–3.7 kg, males, and 2.5 kg, females, and *C. albifrons* and *C. olivaceus* at 3.2–3.3 kg, males, and 2.3–2.5 kg, females (see also Smith and Jungers 1997). Animals living in captivity are likely to weigh more than wild specimens. The two largest capuchins known to us each weigh more than 6 kg (both adult male *C. apella*; S. Brosnan pers. comm. and Visalberghi pers. obs.).

Capuchins have moderately proportioned bodies, in accord with their largely quadrupedal style of locomotion (see Figure 5.2). The torso measures 36 cm or longer in adults (Ford and Corruccini 1985). Napier and Napier (1967) reported the maximum head and torso length as 56.5 cm for male capuchins and 48.0 cm female capuchins (all species pooled), and maximum tail lengths as 56 cm (males) and 51 cm (females). Capuchins' tails are proportionally short for platyrrhine monkeys (Rosenberger 1979, 1983).

Capuchins have an inter-membral index (calculated as [length of humerus + radius/length of femur + tibia] × 100) of 81, which is an intermediate value for platyrrhine monkeys (Napier and Napier 1967). The forelimb and hindlimb are proportionately shorter, compared with the length of the torso, in capuchins than in the Atelinae, but are proportionately longer than in *Callicebus*, *Aotus* and *Pithecia*. The length of the arm and forearm (humerus + radius) of adult *C. apella* measure on average 24.9 cm, male (n = 3) and 22.7 cm in females (n = 11). The hands (length of the palm and third digit) measured 9.2 cm and 6.9 cm, males and females respectively (Fragaszy *et al.* 1989).

Cebus apella (the tufted capuchin) stands out as the most robust species in the genus with respect to cranial and dental characteristics (Ford and Corruccini 1985, Jungers and Fleagle 1980). Tufted capuchins are also more sexually dimorphic than the non-tufted species with respect to cranial characters (Masterson 1997, Silva Jr. 2001). Male tufted capuchins have a distinctive bony crest along the top of the skull (an anterior sagittal crest), whereas females of the same species and both sexes of non-tufted species lack this crest (Silva Jr. 2001; see Figure 1.8). We have no data on body scaling, and few anatomical measures of any kind, for *C. olivaceus*.

On perusing the illustrations throughout this volume the reader will see that if all the species were similarly colored and shared similar coiffures it would be impossible to differentiate them from an image alone. However, there are some consistent anatomical differences among the species that suggest species-specific adaptations to a complex of locomotor and foraging activities. For example, habitually moving along the ground, climbing, or using small supports is associated with characteristic skeletal differences. Ford and Hobbs (1996) report that 26 of 78 measures and 11 indices of limb bones varied significantly across the four historical species of capuchins. *C. apella* stands out in their sample as having short forelimbs and relatively shorter hands and feet compared with the other species. Moreover, *C. apella* has a short deltopectoral crest, the attachment area for the M. deltoideus, a muscle that participates in quadrupedal locomotion, and smaller joint surfaces in the limbs than the other species. This latter difference may reflect differences in joint mobility, more than the ability of the joints to bear mechanical loads (Rafferty and Ruff 1994). *C. capucinus* has longer limbs and larger joint surfaces, and *C. albifrons* is more similar to *C. capucinus* than to *C. apella*. The two skeletons of *C. olivaceus* examined by Ford and Hobbs (1996) were surprisingly distinctive, and more similar to *C. apella* than the other two species.

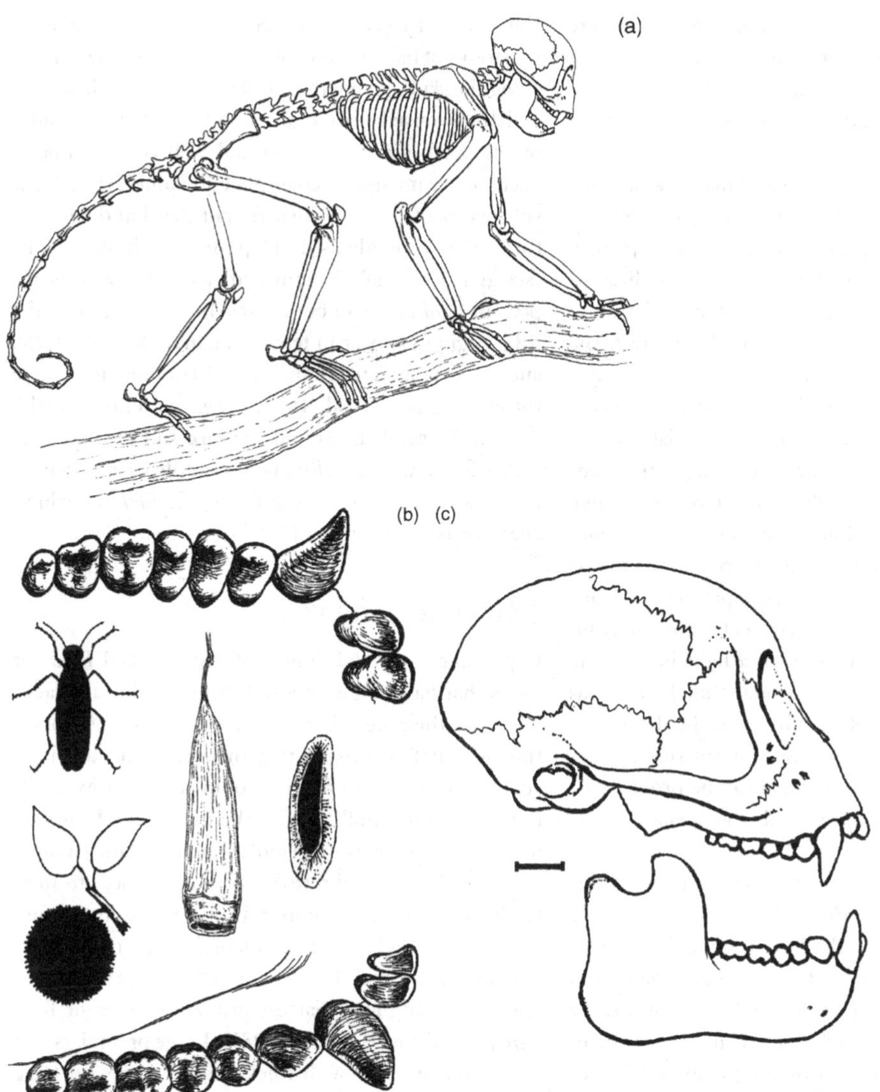

Figure 5.2. (a) Skeleton of a capuchin; (b) teeth of the upper jaw (top) and of the lower jaw (bottom) with typical foods of capuchins and a lateral view of the skull (right). Note the large mandible which reflects adaptation for incisal extraction and for eating hard foods, i.e., durophagy. Scale is indicated by a one centimeter bar below the skull. (Drawing by Stephen Nash.)

PELAGE AND SKIN

Capuchins' furs range in color from nearly white through blonde and red tones to deep brown and black (see color plates). Usually the skin of the face is paler than skin on the other parts of the body, the fur on the chest is lighter, while the forearms, back, and tail are the darkest color in animals that have variegated coloration. White-faced capuchins typically have pink skin on the face, although older individuals can develop brownish pigmented patches on the face, and occasionally an all-brown or nearly gray skin tone on the face. In other species of capuchins, the skin of the face ranges from pinkish-beige to black. Tufted capuchin monkeys housed indoors may have light brown or gray pelage and pinkish skin on the face, but both skin and pelage can become much darker after exposure to the sun. The lighter fur on the chest and face in particular often takes on a distinctly bronzed/reddish hue in animals that were nearly mouse-gray in indoor housing. Other species of capuchins may also "tan" if they move from indoor to outdoor housing, but we do not have evidence of this at present.

Capuchins have distinctive facial markings and fur coloration that, in principle, make them easy to identify at the individual and species level (see color plates). Among tufted capuchins, patterns of white or lighter fur along the sides of the cheeks, on the chin, and around the eyes, and the pattern where the black cap meets the lighter hair on the forehead all combine to give individuals a distinctive appearance. Older *Cebus apella, olivaceus* and *capucinus* can have mottled patches of facial skin and less facial hair while at the same time having longer hair at the crown (see Rowe 1996 for color photographs). However, individuals within species vary greatly, both across geographic regions (as described in Chapter 1) and within the same group. For example, within a single group of *C. olivaceus* in Venezuela, one subadult male was almost fully blonde with a light brown cap, similar to a typical *C. albifrons*; one adult female was uniformly dark brown/black, similar to the darkest *C. apella* (Fragaszy, pers. obs.). Thus, in practice, species identification of a single individual can be problematic unless one is experienced with all the species in the genus.

In general, the monkeys become more individually distinctive with age. Field observers typically use individual details around the face, such as black sideburns, clumps of white or lighter fur, slight variations in the shade of the fur, the presence and shape of tufts or crowns of fur along the skull and jaw, and the shape of the dark cap to identify individual capuchins. Younger animals are distinctively different from adults but often more difficult to tell apart individually. For example, infant white-faced capuchins are more uniformly colored than older animals. Newborn tufted capuchins typically have a uniformly dark face but within a few weeks develop markings with high contrast between light and dark areas. In *C. olivaceus*, younger individuals have darker black or brown caps and lighter faces than adults.

In most, but not all, subspecies of *C. apella*, gradual lengthening of the hair at the crown and along the cheeks throughout adulthood results in the prominent tufts of hair that give the species its common name in English, "tufted capuchin." Some of the more dramatic variations of hair contours include a central ridge of hair along the skull, giving the monkey a punk-like hair style and thick raised tufts that extend far along the sides of the skull, giving a robust and well-coiffed appearance (see Frontispiece; color plates C2 and C3). In those subspecies of *C. apella* that do not develop distinctive tufts, older individuals develop longer hair around the face giving a "halo" or "crowned" look, which also happens in *C. olivaceus* of both sexes. In tufted animals, the tufts begin to appear in females at about 6 years of age, and a few years later in males, and continue to enlarge throughout adulthood. In *C. capucinus*, older females develop longer hair around the face so that they look rather fuzzy, whereas older males simply become balder in the face. We do not know if aging *C. albifrons* exhibit characteristic changes in fur.

SKULL AND TEETH

Capuchins eat a wide range of hard-husked foods or foods that have to be removed from woody substrates. They use their teeth to obtain and process these foods; thus it is not surprising that their jaws and teeth are adapted to resist strains that would occur, for example, during vigorous pulling with the incisors and canines, or crushing with the premolars and molars (Anapol and Lee 1994; see Figure 5.2b). With respect to their teeth, capuchins have more robust canines, longer and thickly enameled incisors and large, square molar teeth with thicker enamel than other nonhuman primate genera, especially those that eat primarily leaves or fruit (Anapol and Lee 1994, Kay 1981). Large premolars and square molar teeth with thick enamel and with large occlusal surfaces and low cusps are useful in cracking and grinding hard tough plant material (such as wood) and nuts (Anapol and Lee 1994, Kay 1981). The mandible (lower jaw) shows adaptations for vigorous pulling and mastication: the symphysis (frontal center of the lower jaw) is both robust (thickened), to counter strong torsional and lateral bending strains, and vertically deep, to counter vertical shear and transverse bending strains (Daegling 1992, Hylander 1979, 1984, 1985, 1988). These strains result from biting and pulling with the incisors. The mandible in capuchins also has long, deep corpora (the horizontal portions of the bone) that are characteristics of platyrrhines in general (Bouvier 1986).

However, unlike most platyrrhines, capuchins and squirrel monkeys have a relatively large ratio of masseter muscle to temporalis muscle, an arrangement that produces a strong incisal bite. Their mandibular corpora are thickened laterally to counter the higher torsional strains that result from this activity (Anapol and Lee 1994, Bouvier 1986, Daegling 1992).

Tufted capuchins feed more than the other capuchin species on nuts and other hard or well-protected foods that they pull and bite open. For example, Terborgh (1983) reports that *C. apella* searched tough substrates such as the bases of palm fronds, bamboo canes, dead branches and termite nests on 44% of its searches. *C. albifrons* searched similar substrates during a third of its searching activities (Terborgh 1983) and *C. olivaceus* took about a fifth of its animal prey from these sorts of substrates (Robinson 1986). Thus it is not surprising that the anatomical adaptations for eating hard foods ("durophagy") are most evident in tufted capuchins. As noted earlier, male tufted capuchins possess an anterior sagittal crest (Bouvier 1986; Kinzey 1974; Silva Jr. 2001; see Figure 1.8); this feature provides a greater area for the attachment of the *M. temporalis*, a muscle that participates in resisting pulling against the incisors. Compared with other capuchins, *C. apella* has thickened corpora that counter torsional strains from both bite force and muscle forces (G. Jones, pers. comm.), the ascending ramus is generally larger than those of *C. capucinus*, but similar in shape (Bouvier 1986, Daegling 1992), and the mandible is relatively shorter than in other capuchins, which may decrease forces on other parts of the jaw when the monkey crushes hard objects with its molars (Bouvier 1986). Tufted capuchins also have a larger zygomatic arch than the other *Cebus* species; a larger arch reflects a larger *M. temporalis* and provides a greater area for the *M. masseter* to attach (G. Jones, pers. comm.).

The differences in morphology and diet across the species of capuchins are illuminating, but we should emphasize that all species of *Cebus* possess the same suite of cranial characteristics that supports reliance on encased sources of food (Janson and Boinski 1992). Such foods require both strong crushing bites and vigorous pulling and biting with the incisors. Janson and Boinski point out that these morphological specializations will be most important to capuchins during periods of food scarcity, and that capuchins also have behavioral adaptations to cope with the challenges of obtaining animal prey or well-protected plant foods.

TAIL

Capuchins possess prehensile tails; i.e, they can use their tails to grasp an object or a surface (such as a branch), and to support weight. The ateline group of platyrrhine monkeys (howling, woolly, muriqui and spider monkeys, following Schneider's classification; Schneider and Rosenberger, 1996) also have prehensile tails. (No Old World monkeys have prehensile tails.) Using the tail to bear weight is associated with many anatomical characteristics, described thoroughly by Bergeson (1996). However, capuchin monkeys' tails are quite different physically and functionally from the prehensile tails of the atelines. For example, atelines have an area of glabrous (hairless) skin near the tip of the tail with papillary ridges and sensory bodies like those of the fingertips. Capuchins' tails are furred for their full length and no area of their tails has specialized skin or sensory bodies. The anatomical differences are highlighted in Table 5.1 (reprinted from Bergeson 1996, pp. 24–26).

Overall, the tail of capuchins is shorter and not as flexible or dexterous as the tail of atelines. Adult capuchins rarely suspend themselves solely by the tail, although youngsters often do so. Sometimes this difference is highlighted by describing the tail of capuchins as "semi-prehensile" (e.g., Napier and Napier 1967). Nevertheless, the prehensile tail is an extremely important apparatus for capuchins. Capuchins use their tails to grasp supports when moving quadrupedally, particularly after leaping, but even more often they use their tails when foraging (Bergeson 1996, Garber and Rehg 1999, Youlatos 1999). Using the tail to support the body enables a variety of postures and frees the hands for other pursuits. For example, the tail is often used to maintain an "inverse bipedal" posture – that is, with the body facing downward, feet thrust against a surface or holding onto supports, and the tail anchoring the body from above (Bergeson 1996). In this position both hands are free, enabling the monkeys to feed on fruits on terminal branches or search for invertebrates hidden along the trunks of trees. When seated with the tail wrapped around a branch, the monkey can exert strong forces, as when pounding a hard object against a tree

Table 5.1. *The morphological complex of the prehensile tail in* Cebus *and the* Atelines *(howling, spider and woolly monkeys). Taken from Bergeson (1996).*

Atelines	*Cebus*
A. Tail length	
1. Although the tail of *Alouatta* is only moderately elongate when compared with other platyrrhines, the tails of the other atelines are relatively very long (Lumer and Schultz 1947, Rosenberger 1979, 1983).	The tail of *Cebus* is relatively short when compared with other platyrrhines. In terms of length the tail of *Cebus* appears to have evolved in an opposite direction than the ateline tail (Lumer and Schultz 1947, Rosenberger 1979, 1983).
B. Skin	
1. Atelines possess a tactile pad on the ventrodistal portion of their tails. This pad is hairless and contains hypertrophied sweat glands and an abundance of specialized tactile receptors known as Meissner's corpuscles (Biegert 1961). Meissner's corpuscles are specialized nerve cells that among non-atelines are found only under finger or toe pads. Meissner's corpuscles give the caudal tactile pad of the atelines all the tactile sensitivity of an extra hand or foot. V-shaped epidermal ridges on this tactile pad help to distribute the shearing forces generated during tail use by the friction between pad and support (Biegert 1963, Niemitz 1990).	*Cebus* does not possess a tactile pad on its tail.
C. Muscle	
1. In atelines, the dorsal muscle bundles are thicker than the ventral ones in the proximal portion of the tail; distally, the reverse is true (Ankel 1972, Chang and Ruch 1947, Hill 1960, Lemelin 1995).	*Cebus* shares this condition with the atelines, though not to the degree seen in the atelines (Hill 1960–1964, Lemelin 1995).
2. In atelines, the long tendons of *m. flexor caudae lateralis* and *m. extensor caudae lateralis* cross a relatively small number of vertebrae before insertion (Lemelin 1995). This provides a greater range of flexion and extension to the tail.	*Cebus* shares this condition with the atelines (Lemelin 1995).
D. Bone	
1. In the atelines the lumbar transverse processes arise from the roots of the neural arch as opposed to from the widest part of the vertebral body. This corresponds to the increased innervation and vascularization of the ateline tail (Ankel 1972).	In *Cebus* and nonprehensile-tailed primates, the transverse processes arise from the widest part of the vertebral body (Ankel 1972).
2. The neural arches of the caudal vertebrae are elevated in the ateline tail (Ankel 1972; also see Jenkins and Krause 1983).	The neural arches of the caudal vertebrae are elevated in the *Cebus* tail, but less so than in the atelines (Ankel 1972).

Table 5.1. (*cont.*)

Atelines	*Cebus*
3. In atelines, the sacroiliac joint is expanded in order to better withstand the tensile forces transmitted through the tail during tail suspension (Ankel 1972, Leutenegger 1970). Straus (1929) noted that in all primates except the atelines, one or occasionally two sacral vertebrae participate in the sacroiliac joint. In the atelines, three vertebrae normally participate in the sacroiliac joint.	The sacroiliac joint of *Cebus* does not appear to be expanded, and may in fact be reduced. Straus (1929) noted that in *Cebus*, 1.5 sacral vertebrae participate in the sacroiliac joint.
4. Atelines have large sacral indices. The sacral index is a measure of the relative sizes of the cranial and caudal apertures of the sacrum; the caudal aperture is relatively large in atelines (Ankel 1965, 1972). The large caudal outlet corresponds to the increased vascularization and innervation of the ateline (prehensile) tail.	*Cebus* has a moderately enlarged sacral index. The magnitude of this index is intermediate between nonprehensile-tailed primates and the atelines, but approaches that of several platyrrhine genera that do not have prehensile tails (Ankel 1965, 1972).
5. The distal caudal vertebrae of atelines are relatively wide in the middle. This functions to withstand the relatively large stresses that are transferred to caudal vertebrae during tail suspension or grasping (German 1982).	The distal caudal vertebrae of *Cebus*, like those of the atelines, are relatively wide in the middle (German 1982).
6. In atelines, the proximal transverse processes of the distal caudal vertebrae are relatively wide. This expansion is due to the attachment of specific muscles that rotate the tail (*m. flexor caudae lateralis, m. intertransversarii caudae*) German, 1982, Meldrum and Lemelin 1991).	In *Cebus*, the proximal transverse processes of the distal caudal vertebrae are relatively wide, but not as wide as in the atelines. *Cebus* is intermediate between the atelines and nonprehensile-tailed primates in this respect (German 1982).
E. Brain – Neurology	
1. In most primates, the cortical representation of the tail is buried deep within the great sagittal sulcus of the brain. In the atelines, the caudal sensorimotor representation is greatly expanded and has migrated out of the great sagittal sinus (Fulton and Dousser de Barenne 1933, Hirsch and Coxe 1958, Petras 1979). This has caused a lateral shift of the remaining somatotopic cortex and usually results in the confluence of the Sylvian and intraparietal sulci.	The brain of *Cebus* is relatively large and elaborate, but does not contain a confluence of the Sylvian and intraparietal sulci. The caudal sensorimotor representation is expanded, but not to the degree seen in the atelines (Falk 1980, Hirsch and Coxe 1958, Radinsky 1972).

trunk, without losing its balance. Figure 7.1 illustrates these positions.

The capuchins' prehensile tail is not a general adaptation to support a large body in the trees, but an adaptation for their style of locomotion and feeding (Bergeson 1996). Rosenberger (1992) and Rosenberger and Strier (1989) suggest that the evolution of the prehensile tail in capuchins corresponded with the continued pursuit of a relatively insectivorous and faunivorous diet despite their relatively large size for this kind of diet. In a very

Table 5.2. *Forelimb indices (data taken from Fragaszy et al. 1989).*

	N	D1/D3[a]	D3/hand[b]	Hand/arm[c]	Palm length/palm span[d]
Cebus					
Adult male	3	74	48	27.0	1.5
Adult female	11	69	51	28.2	1.6
3 yr	4	60	51	28.6	1.6
2 yr	2	57	51	30.2	1.7
16 mo	4	59	50	30.0	1.9
26 wk	4	60	49	30.6	1.8
16 wk	5	60	50	31.7	1.8
7 wk	5	56	50	32.8	1.5
Neonates	2	52	55	47.9	1.3
Saimiri					
Male	7	56	51	25.6	1.2
Female	7	50	52	26.2	1.2

[a] $\text{Index} = \frac{D1}{D3} \times 100.$

[b] $\text{Index} = \frac{D3}{\text{Palm length} + D3} \times 100.$

[c] $\text{Index} = \frac{\text{Palm length} + D3}{\text{Palm length} + D3 + \text{forearm} + \text{upper arm}} \times 100.$

[d] $\text{Index} = \frac{\text{Palm length}}{\text{Palm span}}$

important sense, then, capuchins' prehensile tail enables their distinctive manual activities that differentiate them from other arboreal monkeys. Despite its importance to their lifestyle, however, capuchins' use of their tail is relatively crude in comparison with the other New World monkeys with prehensile tails, the atelines. Freese and Oppenheimer (1981, p. 385), comparing the use of the tail between a spider monkey and a capuchin, liken the contrast to "the difference between driving a sports car and a utility helicopter." Nevertheless, with sufficient motivation and practice, capuchin monkeys can use the tail for rather delicate prehensile maneuvers, such as picking up a small object. We sometimes see actions of this sort in captive capuchins, such as when they reach through wire mesh with the tail to retrieve something.

HANDS

Superficially, the hands of capuchins are not particularly distinctive from the hands of other platyrrhine monkeys. All these genera have relatively slender palms and long fingers, and their thumbs (except spider monkeys that have vestigial thumbs) are proportionally almost as long relative to the other digits as the capuchins; capuchins' thumbs are relatively about as long as humans' thumbs (Napier and Napier 1967, see Table 5.2). Nevertheless, despite the superficial similarities of their hands to the hands of their platyrrhine relatives, capuchins use their hands quite differently.

To understand what is special about the hands of capuchin monkeys, we need to step back a moment and think about how we use our hands to manipulate objects. The thumb is critically important to how humans achieve precise control of objects: very fine movements as well as very strong grips can be achieved by opposing the thumb and the index finger (Napier 1980, Susman 1998). The precise opposition of the tips of the thumb and index finger that humans achieve is enabled in part by a particular kind of joint, called a "saddle joint", at the junction between the carpal and

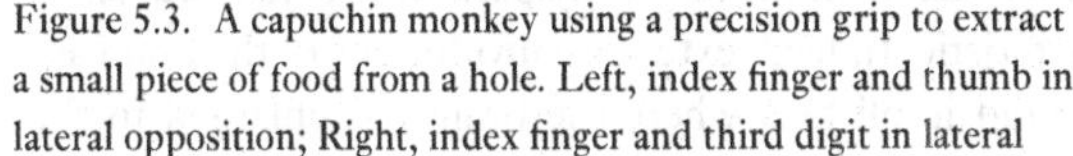
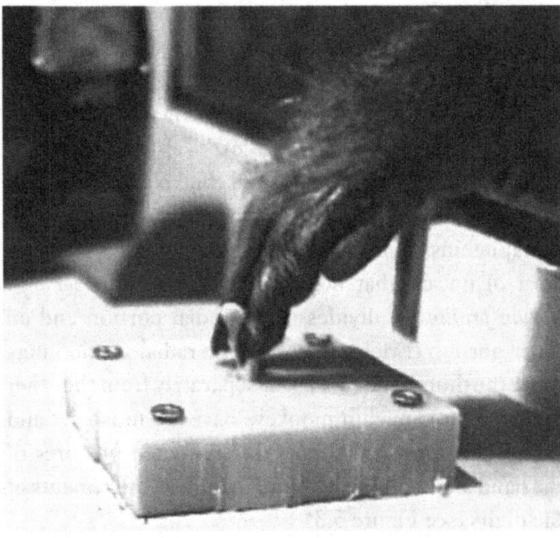

Figure 5.3. A capuchin monkey using a precision grip to extract a small piece of food from a hole. Left, index finger and thumb in lateral opposition; Right, index finger and third digit in lateral opposition (scissor grip). (Photographs courtesy of Giovanna Spinozzi.)

metacarpal bones (the carpometacarpal joint). The saddle joint allows the thumb to rotate so that its tip faces the tips of the other digits. In addition to joint morphology, the muscles of the hand and thumb act in ways that produce rotation at the joints (Rose 1992). Moreover, pad-to-pad opposition is easy for us because of the relative proportions of the digits and their position on the palm. Humans have relatively long thumbs so that with slight flexion, the index finger comes to rest on the thumb in the pincer grip. Humans can also roll digits obliquely across the palm toward the thumb, enhancing the precision grip (Wilson 1998). For example, we can touch the tip of our little finger to the tip of our thumb.

Capuchin monkeys, like other platyrrhine monkeys, do not have a saddle joint at the carpometacarpal joint; for this reason, they were classified by Napier as having pseudo-opposable thumbs that flex in parallel with the other digits (Napier 1961). With a pseudo-opposable thumb, when the hand is used to pick up something small, the thumb closes in parallel to the other digits in a palmar, or power, grip. In this grip, the whole hand must move to move the object. In a precision grip, the fingers support the object and can move it even if the rest of the hand is still (see Figure 5.3).

However, Napier misjudged the dexterity of capuchin monkeys. These monkeys, like Old World monkeys and apes, are now known to achieve precision grips in a variety of ways (Christel 1993, Christel and Fragaszy 2000, Costello and Fragaszy 1988, Tonooka and Matsuzawa 1995), and they are the only platyrrhine taxon known to do so. Given the manner in which capuchins use their hands during feeding and other manipulative activities, both strength and precision grips seem important to their normal manual function. What features of anatomy support capuchins' relatively enhanced dexterity and manual strength?

We can identify two features of skeletal anatomy that may be implicated. First, Rose (1992) reports that capuchins are more similar to Old World monkeys than are *Alouatta* or *Lagothrix* in the range of conjunct rotation that occurs at the carpometacarpal joint because of the geometry of the joint surfaces. This characteristic may support some degree of opposability between thumb and other digits. Conjunct rotation results from combined flexion and abduction during opposition of the thumb with other digits. Second, capuchins have a relatively deep carpal arch, which is associated with an arrangement of bones in the hand that would enhance opposition as well as the strength of digital flexors (Napier 1960, 1961). Both opposition and strong flexors support a strong grip on objects.

In addition to the features of skeletal anatomy supporting opposition of the thumb and strong grips, capuchins have muscular and neural characteristics that relate directly to their elaborated dexterity compared with other platyrrhine genera. First, capuchins can oppose their thumbs to the other digits in part through rotation of the proximal phalanx by the *abductor pollicis brevis* muscle (Rose 1992). Flexors also contribute to opposition. Second, individuated control of the digits in flexion requires separate muscle activation in each digit. In capuchins, as in most simian primates, the deepest layer of muscle that flexes the fingers (*M. flexor digitorum profundus*) divides into a radial portion and an ulnar portion (Erickson 1948). The radial portion may move the thumb and forefinger separately from the other digits. Thus capuchin monkeys have the muscular and skeletal anatomy to support differentiated postures of the hand and individuated (independent) movements of the digits (see Figure 5.3).

The muscular and skeletal equipment of the hands in capuchins has counterpart neuronal equipment in the spinal cord. To move one finger independently of another, the muscles of the two fingers must be innervated by different neurons. Neurons innervating the muscles of the fingers originate in the spinal cord. Hence, independent movements of the digits on the same hand require these neurons in the spinal cord (called motoneurons) to innervate those digits selectively. Direct synapses between neurons that originate in the primary motor cortex of the brain (corticospinal neurons) and motoneurons support independent movements of the digits (Kuypers 1981). Direct synapses of corticospinal neurons with motoneurons are prevalent in the spinal cord of Old World (catarrhine) primates, all of which typically show independent control of the digits. Platyrrhine primates, in contrast, were thought not to have direct synapses between corticospinal neurons and motoneurons innervating the digits (squirrel monkeys, *Saimiri*, for example; Bortoff and Strick 1993). Thus it was quite surprising to find that direct synapses between corticospinal neurons and digital motoneurons in cervical spinal cord are more prevalent in capuchins than in macaques (Bortoff and Strick 1993). (Macaques routinely exhibit fine precision grips by opposing the thumb to the other digits.) Capuchins have relatively large spinal cords for their body size, quite possibly reflecting abundant sensorimotor fibers (Rilling and Insel 1999).

This is the only known instance, to our knowledge, of a substantial organizational difference in the central nervous system among taxa in the same subfamily. It stands as an important example of evolutionary convergence of neural organization between capuchins and Old World primates, and it piques our interest as to whether there are related substantial differences between capuchins and other platyrrhine monkeys in the neural organization of manual activity at other levels of the neuroaxis.

BRAIN

Absolute and relative size

Anatomists have long considered capuchins' brains to be rather large for their body size, and otherwise have found them interesting on account of the extensive convolutions in the surface of the neocortex (Hill 1960; see Figure 5.4). But appearances are deceptive in this instance. Convolutions of the brain's exterior surface are not a strong indicator of any particular behavioral characteristic. Indeed, exterior convolutions of the brain are found in all taxa where the brain has enlarged to the point where more material must be packed into a limited cranial capacity. The extent of folding varies with brain size in an orderly way (Zilles *et al.* 1989).

To appreciate what is special about the brain of the capuchin, we must adopt a different comparative metric than convolutions or absolute brain size. The story of absolute and relative brain size, as well as the most intriguing part of the mystery, the relation between the capuchins' unusual brains and their unusual behavior, requires more than an examination of appearances for a clear understanding.

Brains vary across taxa, but so do body size, metabolic rate and many other characters. How should we take these other variables, especially size, into account in our comparisons of brains? This question has occupied excellent minds for a very long time (Deacon 1990a, b, Gould 1977, Harvey and Krebs 1990, Holloway 1979, Jerison 1973, Martin and Harvey 1985). How to interpret the variations in relative brain size and structure with respect to the evolution and function of the brain, particularly the primate brain, has also been the subject of continuing debate (Aboitiz 1996, Aiello and Wheeler 1995, Armstrong and Falk 1982, Barton 1999, Barton and Harvey 2000, Barton *et al.* 1995, Falk and Gibson 2001, Finlay and Darlington 1995, Reader 2003). In this larger debate, capuchin monkeys hold our interest as a point of extreme variation in the evolutionary radiation of platyrrhine monkeys because of their large relative brain size and corresponding enlargements of specific areas of the brain (Stephan *et al.* 1970). In what

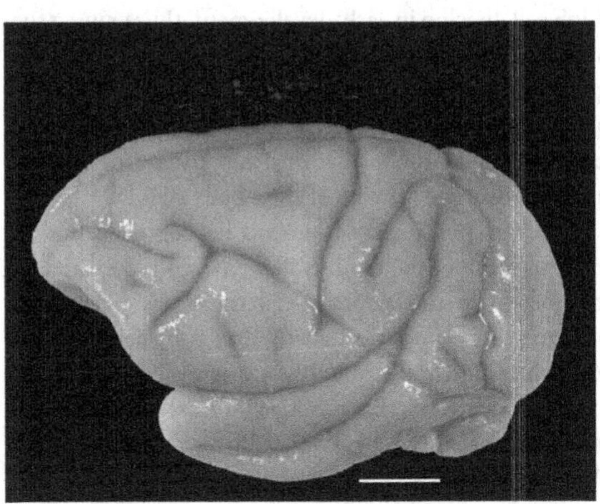

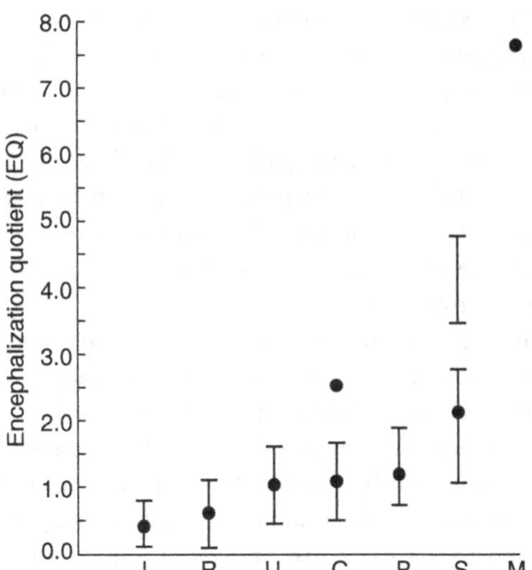

Figure 5.4. Left, brain of *Cebus apella*, lateral view of the left side. Scale is indicated by a one centimeter bar in the lower right. (Photograph courtesy of Peter Strick). Right, ratio of brain weight to body weight (Encephalization Quotient, EQ) across vertebrate classes and primate superfamilies (reprinted from Passingham 1981 with permission from the publisher). I = Insectivores, R = Reptiles, U = Ungulates, C = carnivores, P = Prosimian primates, S = Simian primates excluding humans, M = Humans. The higher values for simian primates (the upper bar) are the two values for one male *Cebus albifrons* and one male *C. apella*. The value for the third capuchin individual, a *C. capucinus*, was at the top of the range of the lower set of values. The range given for carnivores excludes the one high value given as a dot. See the text for further explanation.

follows we present one view of what is special about the capuchin brain; how these characters might have arisen in an evolutionary sense, and how they arise ontogenetically through evolutionarily conservative growth processes. As this research area currently enjoys great interest from evolutionary and developmental neurobiologists, our explanations in all certainty will be revised in the future as theories and data accumulate.

Brain tissue, along with the digestive tract, is the most metabolically costly tissue in the body. Across taxa, relative brain size (measured as weight) scales to body size (also measured as weight) and metabolic rate. Once body size is taken into account, the New World monkeys except the Callitrichidae (marmosets and tamarins), which have smaller brains per body weight in accord with their lower metabolic rate (Armstrong and Shea 1997), have the same relative brain size as Old World monkeys. Relative brain size scales closely to body size, but after body size has been removed as a source of variation, we still find some residual variation across anthropoid taxa in the relative size of the whole brain and in the proportions of the various parts of the brain to each other.

The theoretical debate about relative brain size focuses on explaining this residual variation in brain size across taxa. The primary data set used in this effort was produced by Bauchot, Stephan and colleagues (Bauchot 1979a, b, Bauchot and Stephan 1966, 1969, Stephan 1972, Stephan *et al.* 1970). These authors calculated volumes of brain structures by outlining the structures visible in enlarged photographs of 50 or more stained sections of the brain, where the brain had been sliced at regular intervals from front to back (like a loaf of bread). The outlines of the various structures, made visible by the staining, were then literally cut out of the photographs and the paper cutouts were weighed to calculate the volumes.

Working from Bauchot and Stephan's (1969) data set, Jerison (1973) calculated an "encephalization quotient," designated EQ, as the actual brain volume in relation to expected brain volume as predicted from body weight, where the mammalian average = 1.0. Using this measure, capuchins, with an EQ of 2.54–4.79, do indeed look anomalous. Their highest value falls outside the range of other nonhuman primates, even the apes (see Figure 5.4). But there are reasons to consider this result as misleading. In constructing his table, Jerison used the data from the single largest male specimen per species given by Bauchot and Stephan (1969). For the three capuchin species in Jerison's table, the weights and EQ values, respectively, for the males are 3.765 kg and 2.54 (*C. capucinus*), 1.640 kg and 4.79 (*C. albifrons*), and 2.400 kg and 3.49 (*C. apella*). It seems likely that the *C. albifrons* specimen, with a weight of less than 2 kg, was a juvenile; if that is so, the high EQ value of 4.79 for that individual (near double the value of all other nonhuman primate species except for the two other capuchin species) is problematic.

Modified formulas have been used on the same data set, such as the volume of the whole brain as a proportion of the volume of the medulla, a structure in the hindbrain (Passingham 1981), or the Encephalization Index used by Stephan (1972) that compares the brain volume of the taxon under study to the predicted brain volume of a "basal insectivore" of the same body size. In this latter formulation, capuchins rank among the top of the monkeys, along with talapoins (*Cercopithecus talapoin*, a genus of small African monkeys), another platyrrhine genus, the woolly monkeys (*Lagothrix*), and chimpanzees (*Pan*) (Gibson 1986), all with values between 10 and 12. To our knowledge these latter authors (unlike Jerison 1973) used data from all the individuals in each taxon, not just the largest male specimen. Using magnetic resonance images, Rilling and Insel (1999) confirm that capuchins are highly encephalized; they have proportionally as much extra brain volume for their body size as the apes.

The comparative studies of the relative volume of different brain structures by Stephan and his colleagues (Stephan 1972, Stephan *et al.* 1970, 1981, 1988) indicated particularly great enlargement of three structures (the cerebral cortex, the thalamus and the cerebellum) in the brain of *Cebus* relative to a hypothetical ancestral insectivore of the same body weight. We can explain why these parts of the brain are larger in capuchins more easily than we can explain why the whole brain is relatively large. Finlay and Darlington (1995) and Finlay *et al.* (2001) have shown that the uneven pattern of expansion across brain structures seen in mammals of varying body sizes, and thus varying brain sizes, is predictable from evolutionarily conservative patterns of ontogenetic growth. In essence, the argument is that the relative growth of parts of the brain, particularly the forebrain, covaries with the order of their neurogenesis. The order of neurogenesis derives from the basic axial structure of the developing brain (Finlay *et al.* 1998; Rubenstein *et al.* 1994). Areas of the brain that are at the dorsal and outer edges ("alar") from the center of the neuroaxis experience the most prolonged period of high cell division in early development, thus gaining the most neurons and becoming the largest. This is why the parts of brain that are enlarged in capuchins are the same parts of the brain that are enlarged in apes, humans and any primate taxon with brains that are, overall, larger than the average primate of their body size.

What is very clear from this conceptualization of the evolution of the brain is that any evolved change in the size of particular brain structures in a given taxonomic group is not necessarily driven by selection for particular functions. Initially, at least, they are a result of growth patterns that produce increases in the size of the whole brain (see Deacon 1990b and McKinney and McNamara 1991 for comprehensive treatment of this general issue). With reference to capuchins, this argument suggests that it is not (yet) fruitful to speculate that directional selection for a particular behavioral function (like extractive foraging) has produced enlarged cortices or cerebella (but see Barton 1999 and Barton and Harvey 2000 for a dissenting view). More likely, the enlarged neocortex, thalamus and cerebellum in capuchins reflect overall enlargement of the brain, which itself may reflect directional selection for behavioral complexity in one or more dimensions. Unfortunately, post hoc we cannot readily identify the originating source(s) of selection. The selective advantage conferred on an individual by enhanced motor control via the basal ganglia would have done just as well at driving encephalization as directional selection for enhanced memory of food resources, for example. Once neural resources are there, no matter from what evolutionary process, they can be recruited during development to produce redundant and multifunctional connections in the brain that can support efficient perception and action. Concerted detective work on a broad comparative scale must ensue before we can decipher the evolutionary causes of brain enlargement in *Cebus* relative to other primates. We know very little about growth and differentiation in the brain, and we know even less about the consequences of changes in brain size, in absolute or relative terms, on these processes.

Organization

Overall, the organization of the nervous system is highly conservative across and within vertebrate taxa. Thus, the organization of the capuchin's brain is largely like that of other primates. However, there are some differences in the organization of structures in the brains of capuchin monkeys and squirrel monkeys (*Cebinae*) in comparison with the brains of other platyrrhine and catarrhine monkeys. We do not yet understand the functional significance and evolutionary origins of these differences, but the differences seem to cluster on sensory functions and integrative functions with strong sensory involvement.

Visual cortex

Armstrong and Shea (1997), reviewing the literature on neural organization in New World monkeys, conclude that the organization of the visual cortex and of connections between visual and limbic systems differentiate *Cebinae* from other monkeys more clearly than differences within the association cortex. In the visual cortex, capuchin and squirrel monkeys do not have the strongly monocular organization into "columns" across the depth of the cortex that other monkeys do. Specifically, in layer IV of the primary visual cortex (also called the striate cortex), squirrel and capuchin monkeys have areas where inputs from each eye overlap. In contrast, Old World monkeys and *Ateles* (spider monkeys) among platyrrhines have segregated areas for each eye in that same layer (see Armstrong and Shea 1997 for details). At present, the functional significance of these differences is unknown, and we do not know whether other differences exist across taxa elsewhere in the visual system. Changed proportions in limbic areas suggest greater integration of sensory input in these regions in *Cebinae* than in other primates, as do more projections from higher order visual cortex to limbic areas. As for the differences across taxa in the columnar organization of the visual cortex, the functional significance of the taxonomic differences in limbic areas is unknown. The distribution of visual field representation across regions of the cortex does not distinguish capuchins from macaques, although many details of organization of the cortical visual areas vary across New and Old World monkeys (Rosa *et al.* 2000).

Somatosensory cortex

Capuchins, like all other monkeys except Callitrichids, have in the parietal cortex two areas where cutaneous sensory input from the body "maps" onto the brain, called somatosensory topographic mappings. The two mappings have complete and roughly mirror-image representations of the body, and appear in areas of the parietal cortex labeled "Area 1" and "Area 3b." In capuchin and squirrel monkeys, the two mappings are organized differently with respect to one another than in all other monkeys in that the representation of the trunk and the limbs appears in a different order, rostral to caudal and dorsal to ventral (Felleman *et al.* 1983). Capuchins also have a unique (among primates) representation of the forearm in Area 1 that is not matched in Area 3b. This pattern is common in nonprimates and prosimians, but not in other monkeys. Felleman *et al.* suggest that the pattern in capuchins is a primitive (evolutionarily conserved) feature of neural organization that other primate taxa have lost. The organization of somatosensory maps of other parts of the body is the same in all primate species examined (Carlson *et al.* 1986).

Somatosensory maps relate to how animals use their bodies in that a large area on the map devoted to a body part is associated with enhanced cutaneous sensitivity in that area compared with body parts with smaller areas of representation on the map. Capuchin monkeys, like all primates, have finely differentiated maps of the hands, particularly for the nonhairy (glabrous) skin on the palms and digits. The somatosensory map of the digits covers a larger area of cortical surface in capuchin monkeys than in squirrel monkeys; the map in capuchins looks much like that in macaques, and thus one might expect they would have equivalent abilities to discriminate objects by touch. However, Carlson and Nystrom (1994) found that both squirrel and capuchin monkeys learned to discriminate textures with more difficulty than macaques (*Macaca*). Two capuchin monkeys learned more quickly than five squirrel monkeys to discriminate textures by touch alone, but the genera were similar in how fine a discrimination they could make. Following lesioning of Area 1, one capuchin monkey did better, not worse, than before. Thus, it seems that the duplicate representation of the hand in Area 1 in capuchins may not aid discrimination of texture, although it does for macaques (Carlson 1984). However, in another study, capuchins were better than rhesus at discriminating between a grape and a stone by touch alone (Welles 1972, 1976).

These conflicting results are hard to interpret with respect to neural organization in capuchin monkeys, but more important, they highlight the fact that our understanding of structure/function relations in brains is very limited. We have at present no basis for predicting or interpreting a substantial difference in haptic sensitivity between New and Old World monkeys. If additional studies of haptic discrimination confirm Nystrom and Carlson's results that capuchins and squirrel monkeys reliably differ as strikingly from Old World monkeys in this domain of sensory function, the haptic system can be added to the visual system as distinguishing New and Old World monkeys.

One other distinctive feature of somatosensory mapping in capuchin monkeys is the large size of the representation of the tail, in keeping with the capuchin monkey's use of the tail as a prehensile organ (see above). The tail occupies as much tissue area in the somatosensory map as the foot, or approximately 10% of the mapped cortex. The comparable figure for spider monkeys (*Ateles*), whose tails are more prehensile than those of capuchin monkeys, is 16%. The tissue area devoted to the tail in species that do not have prehensile tails is an order of magnitude less (Felleman *et al.* 1983).

Overall, we are left with a picture of similarity in gross structure in the nervous system sprinkled with taxonomic variation in many fine details. The evolutionary origins and functional implications of most of the differences we now know about are not well understood. Current thinking is that the differences arise ontogenetically from relatively minor variations in early neural development that can have major ramifications for adult form, as suggested by Finlay *et al.* (2001) to be responsible for taxonomic variations in the size of neural structures.

SENSORY SYSTEMS

Vision

In many respects, capuchin monkey eyesight is like human eyesight. For example, capuchin monkeys have about the same ability to see fine detail (visual acuity) as humans do (De Valois 1971) and as macaques do (Weiskrantz and Cowey 1963). Similarly, capuchin monkeys have roughly equivalent sensitivity to brightness after adaptation to the dark (called scotopic sensitivity) as do humans and other New World genera. Their scotopic sensitivity is slightly greater than humans' at wavelengths above 530 nm (the yellow/orange/red end of the spectrum). Their temporal sensitivity (critical flicker-fusion) is about that of or slightly better than humans and two other diurnal New World monkeys (common marmosets, *C. jacchus*, and squirrel monkeys). Humans see flickers as a continuous light at 55 Hz; capuchin monkeys, at 58–60 Hz; De Valois 1971).

Color vision

The most interesting story about vision in primates, and one that has just recently been unraveled, concerns the extent and sources of variation in color vision among New World monkeys. To make this story easier to understand, we need to explain some basic features about color vision. Normal human color vision is decribed as "trichromatic" when the eye contains three distinct photopigments, each with a distinctive response curve to light of different wavelengths. Other kinds of color vision are known in humans. The most common variant is called "dichromatic" and occurs when the eye contains two photopigments rather than three. Individuals with dichromatic color vision have difficulty discriminating between certain hues (light of different wavelengths) that trichromats find easy to discriminate – for example, differentiating red from orange. Exactly which hues dichromats have trouble discriminating depends on which photopigment they lack.

An eminent vision researcher has labeled color vision in New World monkeys "a cornucopia of variation" (Jacobs 1994, p. 38). This is because six distinct forms of color vision have been identified in these monkeys: three forms of trichromatic color vision, including the form in humans with normal color vision, and three forms of dichromatic color vision. The distinct forms of color vision are associated with six combinations of photopigments (see Figure 5.5).

Strikingly, within *Cebinae*, individual variations in color vision are as great as taxonomic variations (Jacobs 1996). Male *Cebus* and *Saimiri* share a form of color vision that resembles forms of color blindness in humans. They see some wavelengths of light as pure yellow that a normal human trichromat sees as green or as red. They also have lesser ability to discriminate hues by brightness compared with trichromatic humans and trichromatic monkeys (all Old World monkeys and howling monkeys, *Alouatta*) (Amstrong and Shea 1997, Jacobs 1998). Some female squirrel and capuchin monkeys have trichromatic color vision, while others have dichromatic color vision.

The variations in photopigments observed in Cebid monkeys have a single sex-linked genetic component. The relevant single gene component, which is on the

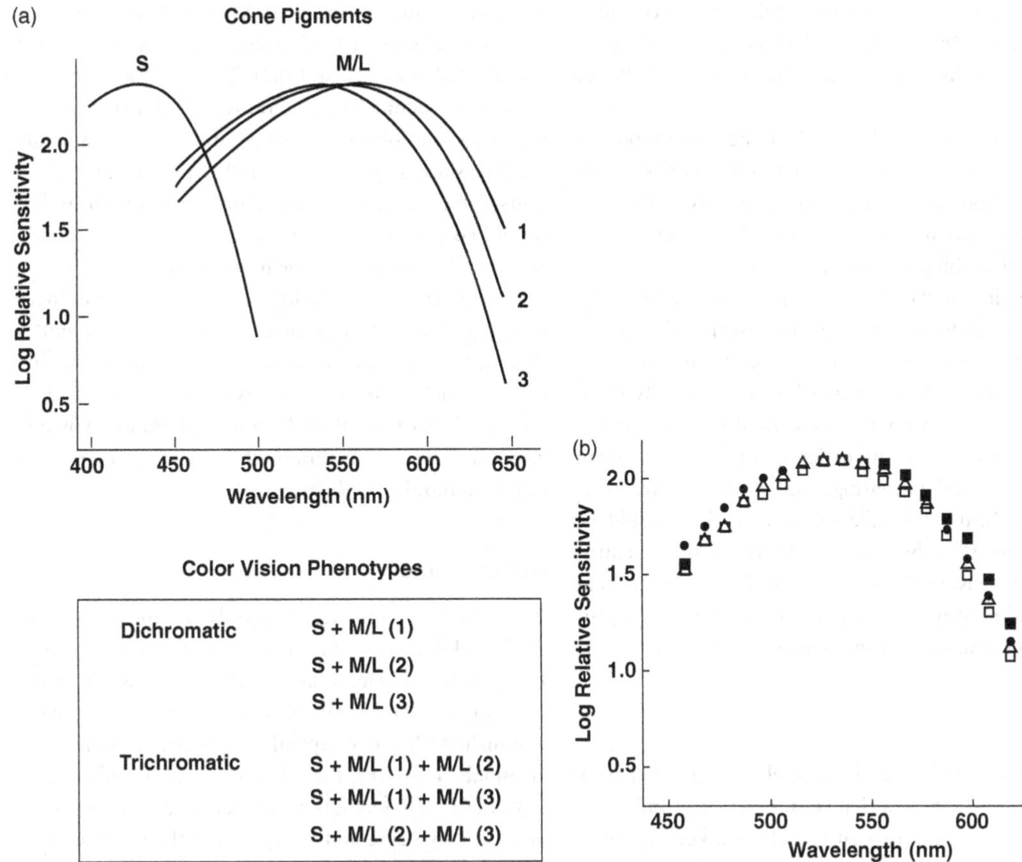

Figure 5.5. Color vision in capuchin monkeys. (a) The four varieties of absorption curves for platyrrhine monkeys, together with the six combinations of photopigments (box) that produce them. S = Short wavelength pigment, M/L = Medium and Long wavelength pigments. All individuals have the S pigment. Dichromats have a single type of M/L pigment. Trichromats have two different M/L pigments. (Reprinted from Jacobs 1998, reproduced with permission from Elsevier Science). (b) Cone spectral sensitivity in two male *C. apella* and two humans with dichromatic color vision. The monkeys' values are shown as solid symbols; the humans' values are shown as open symbols. Note that the curves are virtually identical and many values overlap. (Reprinted from Jacobs 1999, with permission from Elsevier Science.)

X chromosome, specifies the opsin proteins that comprise the photopigments. Three alleles of the "opsin" gene are thought to exist, each specifying one of the three known medium-long wavelength photopigments. Female squirrel monkeys homozygous for the "opsin" gene are dichromats, as are all males; female squirrel monkeys heterozygous for this gene are trichromats (Jacobs 1999). Similarly, female capuchin monkeys may have dichromatic or trichromatic vision, with at least two variations of photopigments in the dichromatic group (Saito *et al.* 2001). Saito's study used 13 female tufted capuchins, several of which are related to one another, so the breadth of the genetic variation in this sample was not large. We expect that the third combination of photopigments found in other New World monkeys will be found in female capuchins as well.

Very few amino acid substitutions are needed to explain the variations in photopigments, suggesting that evolution of different forms of color vision could arise rather easily. As Jacobs (1997, 1998) notes, the putative mechanism is now understood, but the selective regimes responsible for the evolution of different forms of color vision are not. Although male capuchin monkeys have dichromatic rather than trichromatic color vision, they can still distinguish colored surfaces quite sensitively (Gomes *et al.* 2002, Pessoa *et al.* 1997), as can humans with this form of color vision (Jacobs 1999; see Figure 5.5).

Cropp *et al.* (2002) analyzed the distribution of allelic variation in the X-linked opsin genes across species and populations of squirrel monkeys and found that no correlation exists between allele frequencies and behavioral or biogeographical differences. They conclude that it is unlikely that variations in color vision are selected to conform to a particular visual environment. Rather, they suggest that heterozygosity of the opsin gene supports trichromatism, which is a generally adaptive condition for primates (Osorio and Vorobyev 1996). With one locus having three alleles, assuming Mendelian inheritance, random mating, and equal allele frequencies, two-thirds of female squirrel monkeys could be heterozygotes for this allele and therefore have trichromatic color vision (Cropp *et al.* 2002). We expect that capuchin monkeys will be found to have the same proportion of female trichromats as squirrel monkeys.

Audition

As discussed in Chapter 11, capuchins, like other New World monkeys, have a rich vocal repertoire, and many other indicators tell us that they have a keen sense of hearing. For example, capuchins tap on hollow bamboo or dead limbs while foraging, apparently to locate a hollow place where a hidden prey item may be found. Recently, Visalberghi and Néel (2003) have confirmed that capuchin monkeys can select a filled nut rather than an empty nutshell merely by tapping the shells.

Laboratory studies have shown that all primates have similar abilities to hear low frequency sounds, but they differ in their sensitivity to sounds in low and middle frequencies, and in their ability to hear high frequency sounds (high-frequency cut-off). New World monkeys are the most sensitive to sounds in the frequency range of 7–10 kHz, within the range of best sensitivity of Old World primates (which varies across species from 1 to 16 Khz). Humans and apes hear sounds at lower frequencies (2–4 kHz) better than in this range, and prosimians hear sounds at higher frequencies (8–16 kHz) better. To put this into a human perspective, the lowest note on a piano is about 0.025 kHz, the highest note is about 4 kHz. New World monkeys can hear sounds at frequencies that are much higher than humans can detect, up to 45 kHz, and they can hear these frequencies at lower intensities than Old World species (Fobes and King 1982). The sound spectrum of most of capuchins' vocalizations includes frequencies within the audible range for humans. However, animals of all ages occasionally emit very high-pitched vocalizations, some elements of which humans cannot hear. We should keep in mind that we hear high-pitched sounds as softer (less loud) than the monkeys do.

Aside from their ability to hear and discriminate sounds at different frequencies (D'Amato and Colombo 1988a), we have little information about auditory abilities in capuchin monkeys. For example, we have no data on their ability to localize the source of sounds, which is important in the leafy, low-visibility environments that capuchin monkeys inhabit.

Olfaction and taste

The olfactory bulb occupies only a tiny proportion (0.001) of brain volume in *Cebidae* (Fobes and King 1982), a figure comparable with that of other monkeys. In general, simian primates are thought to have diminished olfactory capabilities compared with other mammals. However, they do have a Jacobson's organ, or vomeronasal organ, that is innervated by fibers from the olfactory nerve and is exposed to the nasal passage via a naso-palatine duct (Sarnat and Netsky 1981). This organ is thought to contribute to olfaction in other mammals and it may have a similar function in capuchins. Capuchins sometimes rub their external genital areas on substrates, an activity called "scent marking" because it may deposit an olfactory cue when they do this. They do this less often than do Callitrichid monkeys that in general appear to rely more on olfaction in social signalling than do other monkeys (Epple 1986). They also engage in urine washing and "anointing" behaviors, discussed below in the section on body care (see Figure 5.7). These behaviors, like scent marking, suggest that capuchin monkeys use olfaction in their daily lives. Ueno (1994a,c) has shown that capuchin monkeys can discriminate among genera of New World monkeys by odors in their urine. Although they rely on vision more than olfaction to locate food (in contrast to owl monkeys, *Aotus*; Bolen and Green 1997), they can discriminate among food odors, and more readily learn to discriminate fruity odors than fish or seaweed odors (Ueno 1994b).

Capuchin monkeys appear to have a primate-typical sense of taste, in accordance with the primate-typical density and types of receptors on their tongues (Fleagle 1999, Hill 1960, Sonntag 1921). Perhaps the most widely studied aspect of taste is sensitivity to sugars, especially fructose and sucrose (Simmen and Hladik 1998). All primates prefer soluble sugars. Capuchins can detect sucrose at concentrations of less than 10 mmol, a somewhat better sensitivity than predicted by their body size, but as expected for an animal with a frugivorous (rather than gumivorous or folivorous) emphasis in their diet (Hladik and Simmen 1996). Thus, capuchins can indeed taste when a fruit is ripe on the basis of sugar content. Hladik (1981) suggests a positive relationship across species between taste acuity for sugars (greater in frugivorous species than folivorous or gumivorous species) and foraging efficiency, given that frugivores seek out scattered high-calorie food resources, whereas folivores and gumivores rely on the long-term satiating effects of low-calorie foods.

Nutritional components of food (proteins, fats, starches, minerals, amino acids) and non-nutritional components, such as quinine and tannins, are tasted more variably by different primates than are sugars. Hladik and Simmen (1996) relate these variations to differences across habitats (e.g., African and Neotropical forests) in the nature and extent of toxic compounds in plant foods that animals would likely encounter as well as to differences in diet. We do not know much about capuchins' sensitivities or preferences for many aspects of foods and flavors, except that they are moderately tolerant of quinine, as are most omnivorous primates (Ueno 2001). Ueno (2001) points out that omnivorous primates accept even rather bitter foods if they contain sufficient sugars, and that they crave a variety of foods. A diverse diet enables these species to accommodate an unpredictable risk of ingesting toxins, such as those signalled by bitter substances, in any particular plant food. Together, accepting bitter tastes, seeking sweet foods, and seeking varied foods over short time periods constitute a foraging strategy Ueno suggests could be more accurately called "variovorous" than omnivorous.

Tactile/haptic sensitivity

In mammals, tactile sensitivity arises from an assortment of receptors in the skin and around hair follicles. Primates also have a special receptor in glabrous skin on palmar surface of hands and the soles of the feet, known in humans as Meissner's corpuscle, that function as additional friction and pressure receptors. Frictional and tactile sensitivity in the hands and feet of primates is also enhanced by the presence of epidermal ridges (dermatoglyphs). The ridge system plus Meissner's corpuscles provide primates with acute haptic sensitivity to pressure and friction in their hands and feet compared with other mammals (Martin 1990). The overlapping sensitivity of the various receptor types in glabrous skin results in a profile of sensitivity that may be likened to "feeling in color." Just as overlapping sensitivity to wavelengths of light of the several retinal pigments results in perception of hues and saturation in light, the multiple receptors in the hands result in a rich perception of touch (Bolanowski *et al.* 1988).

We have little information on species differences in tactile or haptic acuity (reviewed in Lacreuse and Fragaszy 2003). However, we have some information on capuchins' use of touch to search for objects. Capuchin monkeys can find grapes buried in shavings (Parr *et al.* 1997) and locate and prehend without vision sunflower seeds placed in crevices (Lacreuse and Fragaszy 1997). A young male capuchin monkey studied by Klüver (1933) detected threads (by which he could pull in a bit of food) down to 0.2 mm in diameter (the smallest threads used in the study) in total darkness. He did this by sweeping his hand across a glass plate on which the thread was placed. These findings fit well with what we know of how capuchins forage: they frequently reach into crevices and other places where vision is not possible, and their motive for doing so is to search for and retrieve food items.

Extended haptic perception (Burton 1993) is another aspect of active touch that might be particularly relevant to capuchins' way of life. When a blind person uses a cane to explore the ground surface over which he or she is walking, the information conveyed about the ground surface comes from the actions of the cane, and it is conveyed via haptic sensations at the hand surface. The tip of the cane is serving in this instance as an extension of the hand. Thus inert components are recruited for perception, allowing perception to go beyond the specialized perceptual organs (Burton 1993). Are capuchin monkeys acute at this kind of activity? This question has hardly been asked. We know that capuchin monkeys occasionally use an object to probe into enclosed spaces into which they cannot fit their hands, to wipe across a surface, or to touch an object out of reach, but we do not know what information the monkeys gain from these actions about the object or surface they contact.

Proprioception, kinesthesis and the vestibular sense

Proprioception refers to perception of one's posture, movement and changes in position, as well as perception of the position, weight and resistance to movement of objects in relation to the body. Kinesthesis refers to perception of the direction and extent of one's own movements through stretch receptors in the joints (e.g., sensing the length of one's stride). The vestibular sense allows monitoring of one's body position with respect to gravity. All of these senses are fundamental to posture and movement, and in primates they are all well developed. We do not now know of significant differences across species in their acuity or function with respect to control of normal bodily movements. However, differences in the availability of information from these senses for learning are implicated in differences across species in the ability to generate voluntary novel movements to solve a problem. Given the capuchin monkeys' greater abilities to generate novel actions with hands, feet and tail in problem-solving circumstances than are observed in other species of monkeys, we expect that capuchins are better able than other species to use perceptions of their own body movements in exploring the environment and in organizing new movements. For example, two capuchin monkeys reliably discriminated between nuts differing by as little as 2.1 g by hefting each of them (Visalberghi and Néel 2003). Hefting is the action of rhythmically lifting and lowering an object held in the hand, and it provides humans with dynamic information about heaviness and other properties of an object such as its shape and wieldability (Turvey 1996, Turvey *et al.* 1999). It seems likely that capuchins also seek information about weight through hefting objects.

Overall, the sensory equipment of different primates is quite similar. Species differences in the reliance on and sensitivity of the various senses are likely related to ecological specializations; in the case of capuchins, this is most likely relevant to active touch, haptic search, and senses of movement and body position (as opposed to vision, olfaction, or audition, for example). Additionally, and not independent of ecological specializations, differences in brain size probably impact the extent to which multiple sensory pathways exist and the integration of sensory information across modalities and across time. Relatively larger-brained species like capuchins may have an advantage in the acuity and/or the integration of sensory information from various modalities compared with their neurally less well-endowed close relatives, but this has yet to be demonstrated.

GENITALS

In most species of primates, including capuchins, the penis contains a bone (a baculum, or *os penis*). Capuchins have a relatively long penis that terminates in a distinctive disk-shaped glans. One common name for the capuchin in Brazil (*macaco prego*) translates in English as "nail monkey", because the shape of the penis resembles a carpenter's nail. Perhaps even more distinctive than the shape of the male's penis, however, is the size and the appearance of the clitoris in females. As in other genera of *Cebidae*, especially in the subfamily *Atelinae* (howling, spider and woolly monkeys), female capuchins have a very prominent clitoris (see Figure 5.6). Unlike these other genera, however, the clitoris in female capuchins is not pendulous; it is rather firm and, in younger animals, erectile (Carosi unpublished). The clitoris of capuchins also contains a bone-like structure (probably an *os clitoridis*), identified by palpation. The baculum in males and its counterpart structure in females are located in the distal part of the penis and clitoris, respectively. Radiographs confirm the presence of this tissue, although histological confirmation that it is bone is still needed. The putative *os clitoridis* ranged in length from 2.5 to 8 mm in 12 capuchins 2 years and older (Carosi and Haines 1999). Some other primate species, including prosimians, monkeys and apes also possess an *os clitoridis*, and in all of these species, males have a baculum. The functional significance of this feature of anatomy is obscure (Dixson 1998a). We discuss sexual behavior in Chapter 13.

At birth, the clitoris appears much like the penis, and it is common for infant female capuchins to be wrongly identified as males for several months after birth. Using data published on births in zoos, Debyser (1995) reports for *C. apella* and *C. capucinus* that out of more than 1000 births, only 160 females were identified and 469 males, and the rest were listed as "unknown." These statistics suggest that identifying females is more

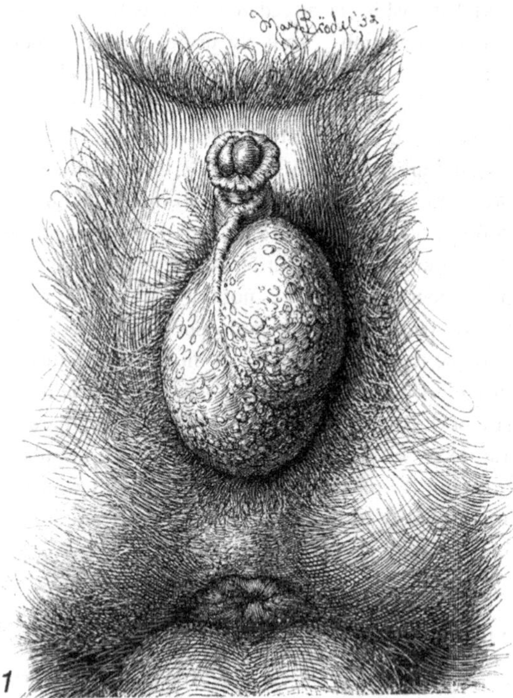

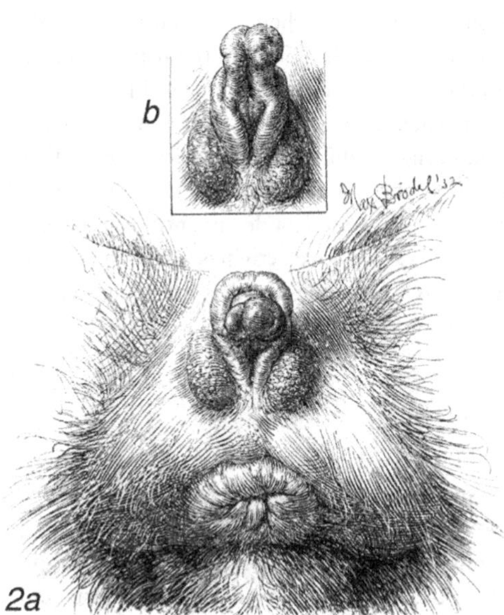

Figure 5.6. Genitalia of a male (1) and female (2a) *C. capucinus*. 2b shows the erect clitoris. Drawings by Max Brödel, published in 'The External Genitalia of the Simian Primates', by George B. Wislocki, from the *Journal Human Biology*, 8(3), Sept. 1936, published by the Johns Hopkins Press, Baltimore, MD. (Reprinted with permission.)

difficult than identifying males. The fleshy part of the clitoris appears to become thicker and shorter with age, although it is not clear whether the *os clitoridis* changes in size with age. In juvenile and infant tufted capuchins studied by Carosi *et al.* (2000), the clitoris was 2.0–2.8 cm long, whereas in adult females, it was reduced to 1.0–1.6 cm.

GUT

The gut of capuchins is notable for its exceptionally small caecum (Martin 1990). The caecum, a section linking the large and small intestines, is larger in animals more reliant on plant foods, and absent in carnivores, cetaceans, pinnipeds and certain insectivores such as moles. All primates have caecums, but this feature of the gut is exceptionally small in capuchins and humans (Martin *et al.* 1985). The small size of the caecum reflects the fact that capuchins, more so than any other primate except humans, specialize in eating foods that are easily digested and have relatively high energy content (Martin 1990).

Except for a small caecum, capuchins have a rather undifferentiated gut compared with other primates, which is consistent with a mixed diet of fruits, seeds and animal material (Chivers and Hladik 1980). Seeds that are rich in fats and protein (such as palm nuts, and nuts of the family Lecythidaceae, the Brazil nut family) are processed more like animal matter than the vegetative parts of plants, according to Chivers and Hladik. Perhaps the capuchin's reliance on nuts is as important as its reliance on animal material for the morphology of its gut.

SELF-CARE

Capuchins show a range of self-directed behaviors that function to care for the body, such as grooming. Virtually all primates groom themselves, and indeed, so do animals of many orders, but of particular interest in nonhuman primates is that often one individual grooms another. We treat this form of grooming in more detail in Chapter 11. An individual grooming itself uses the same actions as an individual grooming another (spreading the fur with one or both hands, then using one hand to pick, scrape, or prehend small objects in the hair or irregularities in the skin, or using the tongue or lips to touch the skin or hair). Two other self-care behaviors, fur-rubbing and urine-washing, are shared with some other New World genera, but are otherwise not common in primates. As we have no data documenting significant variations in these self-care activities across species in the genus *Cebus*, we assume that they are genus-typical in form, frequency, and function.

Urine washing

Urine washing is a distinctive behavioral pattern described in the literature most often for squirrel monkeys and capuchins, but seen also in howling monkeys, spider monkeys and other platyrrhine genera (Milton 1975, 1985, Roeder and Anderson 1991, Schwartz and Rosenblum 1985). Urine washing probably has multiple functions, among them hygiene, thermoregulation and response to irritation from biting ectoparasites (Robinson 1979). During an episode of urine washing, the monkey delivers a small quantity of urine into the palm of one hand, which is then rubbed on the sole of the ipsilateral foot. Sometimes the monkey performs the same sequence with the other hand and foot, or uses the hand into which it urinated, or less commonly, the foot that rubbed that hand, to scratch or rub another body part. Of 117 instances of urine washing observed in Fragaszy's colony of eight adult male tufted capuchins in a brief observational study of this phenomenon, 34 bouts were followed by rubbing or scratching another body part. The monkeys urine washed in many contexts; no particular precipitating event or context was detected. One monkey of eight was never observed to urine wash; one monkey urine washed each day these observations were made. These observations were made in an indoor setting where humidity and temperature were constant and no ectoparasites were present.

Urine washing has stronger correlates with environmental circumstances in more natural settings. Carosi and Rosofsky (1999) conducted more extensive observations of urine washing in two groups of tufted capuchins living in indoor/outdoor housing and found that washing occurred more frequently in warmer and drier conditions. Similarly, urine washing is more common in *C. olivaceus* in the dry season (Robinson 1979) and more frequent in captive *C. apella* on sunny days compared with cloudy days (Roeder and Anderson 1991). In Carosi and Rosofsky's study, urine washing was more frequent in the middle of the day in animals housed indoors (where temperature was relatively constant) as well as animals housed outdoors, suggesting that current ambient temperature is not the sole reason for this temporal pattern. The sexes do not seem to differ in the frequency of urine washing, although the frequency of urine washing increases in females during the luteal phase of the ovulatory cycle, a period when body temperature is elevated (Carosi and Visalberghi unpubl.). Carosi and Rosofsky (1999) conclude that thermoregulation is likely to be the most prominent function of urine washing.

Anointing or fur-rubbing

Capuchins in both captive and natural settings perform a very distinctive behavior called "anointing," "anting" (Longino 1984) or "fur-rubbing," in which an individual applies a substance, most often a plant material or tissues from a soft-bodied invertebrate, across large sections of the body using the hands and tail. Captive capuchins will even do this with peat, an activity that Ludes and Anderson (1995) called "peat-bathing." Capuchins salivate excessively (resulting in drooling) during bouts of fur-rubbing and they rub the excess saliva, along with the other material, all over the body including the flanks and back. Squirrel monkeys and titi monkeys (*Saimiri* and *Callicebus*) in captivity display similar behavior with substances such as ice cubes and onions (Fragaszy pers. obs.). However, capuchin monkeys engage in this activity with far greater enthusiasm than these other species, and apparently, from the dearth of reports from the field about this activity in these other genera, it is a more frequent and remarkable activity in capuchin monkeys.

Baker (1998) conducted an extensive study of anointing in a group of wild white-faced capuchins. The monkeys engaged in bouts of fur-rubbing on average once every other day in the dry season, and about once a day in the wet season. Sometimes a solitary animal rubbed its fur, but usually fur-rubbing involved groups of two to seven animals of all ages and sexes participating in a single bout (see Figure 5.7). Rubbing bouts could

Figure 5.7. White-faced capuchins engaged in fur-rubbing (anointing). This behavior involves enthusiastic rubbing of pungent plant or animal material onto the body accompanied by drooling. (Photograph courtesy of Katherine MacKinnon.)

last up to 15 minutes, but typically were about 3–4 minutes long. Baker describes the activity as "highly energetic, almost frantic in appearance" (p. 146). The monkeys repeatedly pounded and bit the plant materials, and smeared pulp, seeds and/or juice over the body. Many different plants were used for this purpose, including members of the genera *Clematis*, *Piper*, *Sloanea* and *Citrus*. Monkeys anointing together rubbed their bodies against one another. Baker (p. 156) describes the event as "a mass of wet, drooling monkeys with bits of . . . pulp or broken leaves sticking to their fur, squirming and rolling over and around each other." This is surely promising material for a movie production on the bizarre behavior of animals!

White-faced capuchins preferentially select pungent or astringent materials to use for fur-rubbing, although there seems to be intergroup variation in which materials are used. For example, groups in Lomas Barbudal, near Baker's site, use plant species that are not used by Baker's group, and vice versa, although the same plants are available at both sites. Baker (1996) suggests that a functional consequence of the behavior is that plant materials that may have bactericidal or insect-repellant functions are spread over the body. The insect-repellant function of anointing has been confirmed by Valderrama *et al.* (2000) for *Cebus olivaceus* at Masagural, in the Llanos of Venezuela. Wedge-capped capuchin monkeys anoint themselves with millipedes (*Orthoporus dorsovittatus*) in much the same way that white-faced capuchins use plant materials. Millipede secretion is so avidly sought by the monkeys that up to four of them will share a single millipede, and the manner in which the monkeys rub, drool and squirm is a replica of what Baker has described for white-faced capuchins.

The millipedes used by the wedge-capped capuchins secrete two benzoquinones, compounds known to be potently repellant to insects and topically irritating. They secrete these defensive chemicals in nearly pure form in abundant quantities when touched. Anointment with the secretion of these millipedes would likely provide protection to the monkeys against insects, particularly mosquitoes and the bot flies they vector, during the rainy season when these insects are prevalent. Bot fly larva develop under the skin for several weeks resulting in enlarged, painful and itchy lesions that the monkeys scratch and bite incessantly, leading to the possibility of infection. Thus, reducing the probability of being parasitized by bot flies would be beneficial to the monkeys. The monkeys anoint themselves exclusively during the rainy season, the precise time when they can be expected to be beleaguered by mosquitoes. The capuchins perhaps also benefit from the topical and oral disinfectant properties of these substances (Lauer *et al.* 1991, Stärk *et al.* 1991). However, according to Valderrama *et al.*, these compounds are potently irritating to the eyes, painful to inhale and noxious to the taste, and on top of that, they are both toxic and carcinogenic. Strong medicine indeed!

LOCOMOTION AND POSITIONAL BEHAVIOR

Capuchins use quadrupedal locomotion – they walk on four limbs and travel mainly on larger branches (6 cm in diameter or larger), occasionally using lianas and palm fronds as supports. They exhibit many patterns of locomotion while moving in a three-dimensional, richly varying habitat, using the ground and all layers of the forest (Fleagle and Mittermeier 1980, Freese and Oppenheimer 1981, Garber and Rehg 1999, Gebo 1992, Youlatos 1998). All capuchins share the same locomotor and positional behaviors. Although capuchins preferentially travel on larger substrates, they frequently use small branches (3 cm or less) while foraging, using the tail to support their weight as necessary (see above). Foraging activities involve clambering and climbing up and down on small substrates more frequently than locomotion. Locomotion is more likely than foraging to involve short leaps from 1–3 m, bridging, and dropping from a higher branch to a lower one. Capuchins will also occasionally leap gaps larger than 3 m.

The monkeys often use their tails in a prehensile manner when crossing a small gap, and to grasp something to steady the body after leaping. They use their tails more often and with more "tension" (strength) during feeding, especially when foraging below the supporting branch, as when in an inverted bipedal posture. In this case, the tail anchors the body and frees the hands. The tail also provides leverage while seated or standing for vigorous extractive foraging actions, such as pulling and pounding (Garber and Rehg 1999). In captive settings, capuchin monkeys frequently will move away from the perch or other support by which they descended to the floor only as far as their tail, still wrapped around the perch or pole, allows. It appears that maintaining contact by the tail with the support confers confidence to the monkeys in this somewhat risky situation.

Steudel (2000) argues arboreal locomotion is more costly than terrestrial locomotion for mammals. The locomotor style of capuchins, quadrupedal walking/running, combined with frequent leaping, clambering and climbing, fits the picture of a high-cost locomotor pattern. Capuchins travel considerable distances daily. No doubt expenditure of energy in locomotion and other rather costly postures, such as inverted bipedal positions, is part of the high energy budget of capuchins.

Capuchins adopt a variety of postures while resting and sleeping. During rest periods in hot weather, for example, the monkeys sprawl with all four limbs dangling below the branch on which they are lying (see Figure 2.1). While sleeping they may sit with the back curled and the head lowered to near the feet, nearly forming a ball. The tail is curled around the lower body in this position so that the head is virtually covered with the tail. They may also lie on their sides in sleep, slightly flexed and with the tail wrapped around a support or the body. Monkeys will huddle together, sides in contact, to rest and sleep. We discuss capuchins' selection of sleeping sites in natural settings in Chapter 2 (see Anderson 1998 for an overview of sleeping habits and their social and hygienic correlates in primates).

SUMMARY

The genus *Cebus* is physically distinctive from other platyrrhine genera in having a robust jaw and dental morphology, a large brain in relation to its body size, a moderately prehensile tail, and skeletal and neuromuscular features of the hands affording strong grips, some degree of opposition of the thumb to the index finger, and somewhat independent finger movements. Jointly, these characteristics support a wide range of locomotor and foraging actions. The short caecum in capuchins in comparison with all other nonhuman primates, a characteristic of the gut indicating a highly digestible diet, attests to the capuchin monkey's dietary specialization in high-quality, protein- and energy-rich foods.

The capuchin monkeys are typical platyrrhines in other aspects of the body. As far as we know, their sensory capacities are similar to those of other platyrrhine monkeys and with the exception of color vision, of catarrhine monkeys. Capuchin monkeys, like most other platyrrhine monkeys, vary in the forms of photopigments they possess, and thus in the details of their sensitivity to light of different wavelengths. The current consensus is that all male capuchin monkeys have dichromatic (rather than trichromatic) color vision, whereas females generally have trichromatic color vision.

Their self-care activities include one highly distinctive behavior, anointing, in which they rub pungent and sometimes topically irritating plant or animal materials on their fur. They also rub their bodies with urine, a behavior called urine washing. Both of these behaviors are shared with some other platyrrhine monkeys, but they appear to be more frequent and more elaborated in capuchin monkeys.

6 • Development

The female labors silently, repositioning herself on the broad tree limb in the sheltering darkness. She strains at brief intervals, sometimes touching her perineal area afterwards and then sniffing her fingers. Mostly, however, she is rather still, as are the others of her group who are resting around her. After half an hour of quiet labor a head emerges and the infant is born within moments. The mother bends forward, grasps the infant's body, and moves it toward her face, nuzzling and licking it. The newborn infant grasps the mother's hair with a strong and sure grip, flexing its limbs and bringing itself into contact with her body. The mother's attention shifts away from the infant after a few minutes, toward her own body again, as the afterbirth is delivered, inspected, and eventually discarded. By this time the infant has crawled onto the mother's shoulders and is resting with its head and shoulders on one side of her body and its hindquarters and tail draped down the other side. The rest of the group is still sleeping, and the mother settles down to rest for the few remaining hours before dawn . . . Sometime in the next few hours, the infant moves by gentle but persistent crawling, rooting for the nipple by sweeping its face from side to side across the mother's body. The mother lifts her arm accommodatingly as the infant moves off her shoulders towards her chest. She nuzzles its head briefly, but otherwise she is passive throughout the infant's travels across her body. Eventually the infant takes the nipple for its first meal. The mother is content to rest, expressing only fleeting interest in this new little body.

Developmental explanations encompass how the individual grows and differentiates over time. We can provide only a brief discussion of development in the space of this chapter. First we follow the individual through gestation and birth; then we examine development in three domains, physical, motor and social, from infancy to young adulthood. We consider in detail how individual monkeys assume the critical responsibility of feeding themselves. Finally, we mention what is known about reproductive maturation and the end of the lifespan as capuchins age in their fourth and fifth decades of life.

Although there are descriptive and normative reports about development of wild capuchins, most detailed information comes from studies of captive monkeys reared with their mothers in fairly species-typical social groups. Sometimes infants are orphaned or abandoned by the mothers; in captivity these babies are reared by humans. The conditions of development for hand reared infants are decidedly not species-typical. Nevertheless, looking at their development, and particularly how they differ from normally reared infants, provides some fascinating insights into how development proceeds in species-normal circumstances.

PRENATAL DEVELOPMENT AND BIRTH

Mammalian development begins at conception, when the sperm fertilizes the ovum, the zygote forms, and the developing embryo implants in the wall of the uterus. A delicately orchestrated differentiation of cells and the development of tissues and organs ensue. Although virtually nothing is known about the particulars of prenatal development in capuchins, we can safely assume that they are not dramatically different from other primate genera, especially given that prenatal development is a phylogenetically conservative process.

Corradini *et al.* (1998) provide normative data on fetal growth of tufted capuchins collected by ultrasound examination. Their work establishes a way to calculate gestational age from fetal growth parameters. Gestation length is closely correlated with maternal body size in placental mammals, including capuchins (Hartwig 1996), but the length of gestation may vary across individual pregnancies by as much as 10 days within a genus. Average gestations of 155–162 days have been reported for capuchins (Hartwig 1996); Corradini *et al.* (1998) report individual values from 154–162 days.

We have observed labor and delivery of healthy infants a few times in our captive groups of tufted capuchins, and these observations form the basis for the vignette that began this chapter. The mother begins licking the face of the new infant as it emerges from her body. The infant clings to the mother unaided from the moment of birth, although many females do provide support to the infant with one hand when it is on her front (see Figure 10.2). Other group members generally ignore the mother and new infant during and immediately after birth. Over the following days, however, youngsters as well as older monkeys approach the mother and smack their lips or teeth at, peer at or nuzzle the infant; adult females and subadults often sniff the genital area of the infant or touch its face. The infant is treated as an object of gentle interest, especially when it vocalizes or moves. Those approaching a newborn infant to inspect it often give a purring, low-amplitude vocalization (Di Bitetti 2001b). Mothers are more likely to allow siblings or other kin to come near the infant than unrelated animals. The social rank of the mother impacts how often she allows others to come near her very young infant; higher-ranking females are less tolerant of others approaching than are lower-ranking females (O'Brien and Robinson 1991).

THE NEONATE

Capuchins, like all primates, are born as relatively precocial mammals (Portmann 1990). Primates have hair and can open their eyes, locomote and regulate body temperature to varying degrees. In contrast, altricial mammals are born with little or no hair; eyes closed, cannot locomote, and cannot regulate their own body temperature effectively. Among primates, however, capuchins are relatively altricial in many ways. For example, neonatal capuchins are not able to maintain their body temperature on their own in ambient temperatures less than about 32 °C, and they have less postural control and locomotor capacity at birth and shortly afterward than Old World monkeys (macaques, baboons, talapoins, colobus monkeys etc.) and many other platyrrhine genera (Fragaszy 1990b).

Despite its apparent immaturity, the infant capuchin is well prepared to manage its most important affairs. The essential elements that the infant requires (nourishment, support and warmth) are provided by the mother. The neonate's abdomen is nearly naked, and the mother is a convenient source of radiant heat (roughly a cozy 38 °C). The mother's fur affords a tight grip; her body supports the infant and she permits the infant to suckle at will, provided she stops moving long enough for the rather wobbly infant to creep along her body to the right place.

Weight

Infant capuchins weigh approximately 9% as much as their mothers at birth (Fragaszy and Adams-Curtis 1998, Hartwig 1996; see Table 6.1). Average weights per species for newborn infant capuchins range from 228–238 g (Hartwig 1996; cf. average of 210 g and range 170–260 g, Fragaszy and Adams-Curtis 1998). Infant male *C. albifrons* are significantly heavier (251 g) than female *C. albifrons* (222 g) at birth, and at all ages thereafter (Fleagle and Samonds 1975). Sex differences in birth weights have not been reported for other species in the genus but this may be attributable to small sample sizes.

Appearance

Human observers first detect newborn infant capuchins as small lumps lying diagonally across the mother's neck, just in front of the shoulder blades (see Figure 4.2). They are camouflaged by their color and position; often the infant's tail alerts the observer to its presence: a telltale bit of brown "string" dangling below the mother's neck. Newborn capuchins are colored similarly to the adult, with perhaps less contrast between limbs and torso in *C. apella* and a grayer color in *C. capucinus*. Infants appear rather long and skinny, with thin, long limbs, long hands and feet and an elongated skull, reminiscent of a racing bicyclist's helmet. The external genitalia of males and females are superficially similar in appearance (see Chapter 5).

Behavior

Newborn infant capuchins cling tenaciously to their mothers. In neonatal capuchins, flexor activity

Table 6.1. *Body and brain proportions for newborn male (M) and female (F) infants in a variety of primate taxa. Sources are listed by number to the right of each column (in italics; see Note).*

Species	Neonatal brain weight (kg)	Adult brain weight (kg)	Neonatal body weight (kg)	Adult body weight (kg)	Neonatal/adult brain weight (%)	Neonatal/adult body weight (%)	Age at weaning (days)
Cebus albifrons	0.0326 (F) *3*	0.0668 (F) *3*	0.2277 (F) *3*	2.29 (F) *4*	49 (F) *3*	10[a] *3, 8*	270 *6, 8, 9*
	0.0344 (M) *3*		0.2375 (M) *3*	3.18 (M) *4*			
				2.54 (F) *4*			
Cebus capucinus	0.029[a] *6*	0.0792 (F) *6*	0.23[a] *2, 3*	3.68 (M) *4*	37[a] *6*	9[a] *6, 8*	365 *5, 9*
				2.52 (F)			
Cebus apella	NA	0.071[a] *6*	0.245[a] *10*	3.65 (M) *4*	49[a] *7*	10.9[a] *8*	416 *11*
Saimiri sciureus	0.0146 (F) *3*	0.0233 (F) *3*	0.110 (F) *3*	0.67 (F) *4*	63 (F) *3*	16.5 (F) *3*	330 *1, 9*
	0.0155 (M) *3*	0.0253 (M) *3*	0.1129 (M) *3*	0.78 (M) *4*	61 (M) *3*	11.4 (M) *3*	
				9.15 (F) *4*			
Ateles fusciceps	NA	0.1147 (F) *6*	NA	8.89 (M) *4*	NA	NA	365 *6*
				7.29 (F) *4*			
Ateles geoffroyi	0.064[a] *6*	0.1109[a] *6*	0.426[a] *6*	7.78 (M) *4*	58 (F) *6*	6 (F) *6*	870 *6*
Cercopithecus talapoin	NA	0.0377[a] *6*	0.18[a] *2*	1.12 (F) *4*	NA	16 (F) *6*	180 *6*
				1.38 (M) *4*			
Pan troglodytes	0.128[a] *6*	0.4103[a] *6*	1.756[a] *6*	39.5 (F) *4*	31 (F) *6*	5.6 (F) *6*	146 *6*
				50.0 (M) *4*			

[a] Sex unspecified or average of both sexes.

Sources: (1) Baldwin and Baldwin 1981; (2) Byrne and Suomi 1995; (3) Elias 1977; (4) Fleagle 1999; (5) Freese and Oppenheimer 1981; (6) Harvey and Clutton-Brock 1985; (7) Kirkwood and Stathatos 1992; (8) Ross 1991; (9) Rowe 1996; (10) Visalberghi and Anderson 1999; (11) Fragaszy and Bard 1997.

predominates, which aids the infant in maintaining firm contact with the mother. The fingers clench the mother's hair with a "double lock" grip (Napier 1980), in which the ends of the fingers (terminal phalanges) are flexed inward against middle and first phalanges to achieve a particularly strong grip on objects of narrow diameter (such as hair). The capuchin infant can open its eyes, turn its head, hiccup, cough, yawn, vocalize, lift the tail and hindquarters during elimination, and move by coordinated, if wobbly, movements of all four limbs, grasping and releasing with hands and feet. The tail is often wrapped tightly around the carrier's waist, providing a fifth point of contact. As the mother (or other carrier) often moves quickly, with erratic accelerations and decelerations and in varying planes (e.g., jumping up or down, twisting, and so on), a tight grasp is mandatory.

What else can a newborn infant do, beyond clinging to its mother and clambering back and forth from her nipples to her back? To find out, we have examined several infant tufted capuchins while they were separated from their mothers for a few minutes at a time for testing purposes (Fragaszy and Bard 1997). Capuchin infants can hold their heads erect for a few minutes, even on their day of birth, indicating that postural control of the neck and head is well developed. This makes sense in that the infant must hold itself on the mother's ventrum and support its head in a semi-vertical position while nursing or resting. They do not show stepping or crawling movements (as neonatal humans and chimpanzees do) when suspended above a surface and the hands or feet touch the surface. They can right themselves from a supine to a prone position (from back to stomach) with effort, and they will attempt to crawl on a flat surface, although without supporting the torso, from 4 days of age. In all these abilities, capuchins are motorically advanced relative to chimpanzees and humans, although they are delayed relative to rhesus macaques (cf. Fragaszy and Bard 1997 with Schneider 1988, Schneider and Suomi 1992). We have no specific information on sensory function in newborn capuchins except that visual and auditory orientation and tracking are present, as they are in chimpanzees, humans and probably most other primates (Antinucci 1989, Fragaszy and Bard 1997).

Infant capuchins cannot support the torso in a quadrupedal position until about 5 weeks after birth (see Figure 6.1). If placed prone on a flat surface before this age, the arms and legs splay sideways. In contrast, newborn monkeys of terrestrial Old World cercopithecine genera (baboons, macaques etc.) can toddle unsteadily with a quadrupedal gait within days after birth.

PHYSICAL DEVELOPMENT

Table 6.1 presents values of some important developmental parameters for capuchin monkeys and a few other genera chosen for phylogenetic closeness (*Saimiri*, squirrel monkeys), similar life history (*Ateles*, spider monkeys), similar size but membership in the Old World family *Cercopithecinae* (talapoins) and purported behavioral similarity (*Pan*, chimpanzees). Capuchin infants are intermediate to the other genera in the relative body size (neonatal/adult body weight); they are proportionally smaller than *Saimiri*, and proportionally larger than *Pan* (Table 6.1). However, infant capuchins have relatively large brains for their body weight (neonatal brain/body weight) and their brains grow proportionally more after birth than the brains of other monkeys (i.e., the ratio of neonatal brain weight to adult brain weight is small).

The most extensive quantitative data on growth of capuchin monkeys derives from studies conducted with captive *C. albifrons* and *C. apella* by K. Samonds, L. Ausman, J. Fleagle and their colleagues (e.g., Ausman *et al.* 1982, Elias and Samonds 1973, 1974, Fleagle and Samonds 1975, Fleagle and Schaffler 1982, Jungers and Fleagle 1980, Samonds and Hegsted 1973, 1978, Thurm *et al.* 1975). We make extensive reference to this data set below (see also footnote 1).

Weight gain

Capuchin infants gain weight at a rapid and linear rate for 6–8 weeks after birth, so that weight increases by more than 100% from birth to 60 days (*C. albifrons*; Fleagle and Samonds 1975). Thereafter the rate of weight gain slows, but at the end of the first year of life, young capuchins weigh more than 1 kg (*C. albifrons*: 1.17 kg, females; 1.32 kg, males; Fleagle and Samonds 1975; see Figure 6.2). Jungers and Fleagle (1980) report a rate of weight gain (as the slope of the least squares regression) as +0.38 in male *C. albifrons* and +0.42 in male *C. apella* over the first 1100 days of life (3 years).

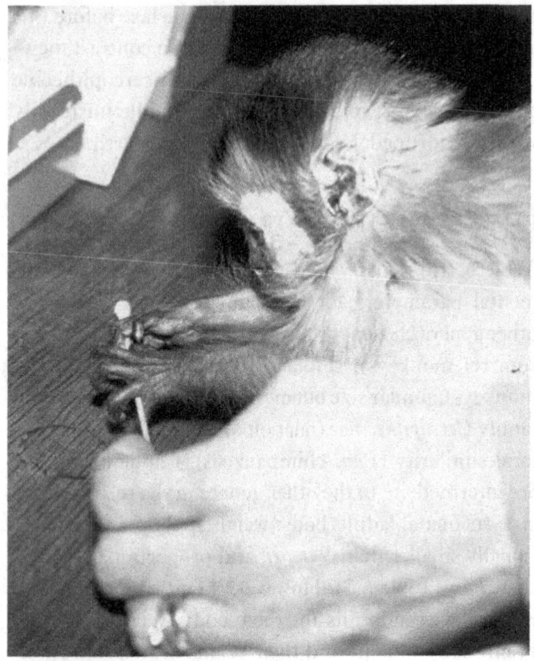

Figure 6.1. Top left, a 2 week-old hand-reared tufted capuchin resting on a stationary surface reaches out with both hands towards an interesting object. Top right, at 5 weeks, the same infant can push its torso off the surface by extending its arms. Bottom, at 8 weeks, the same infant, sitting upright on a stationary surface, reaches with both hands toward an object presented at its midline. Mother-reared infants begin to do these behaviors a few weeks later than hand-reared infants. (Photographs by Dorothy Fragaszy).

To put this in relation to its adult weight, in captive *C. apella*, by 1 year, the infant has grown to more than 50% of its mother's nonpregnant adult weight (Fragaszy and Adams-Curtis 1998). To reach adult weight takes more than another 4 years for females and more than another 6 years for males (*C. albifrons*, Fleagle and Samonds 1975; *C. apella*, Fragaszy and Adams-Curtis 1998; see Figure 6.2). A small prepubertal growth spurt in weight is evident in the third year of life in *C. albifrons* (Ausman *et al.* 1982, Wilen and Naftolin 1978).

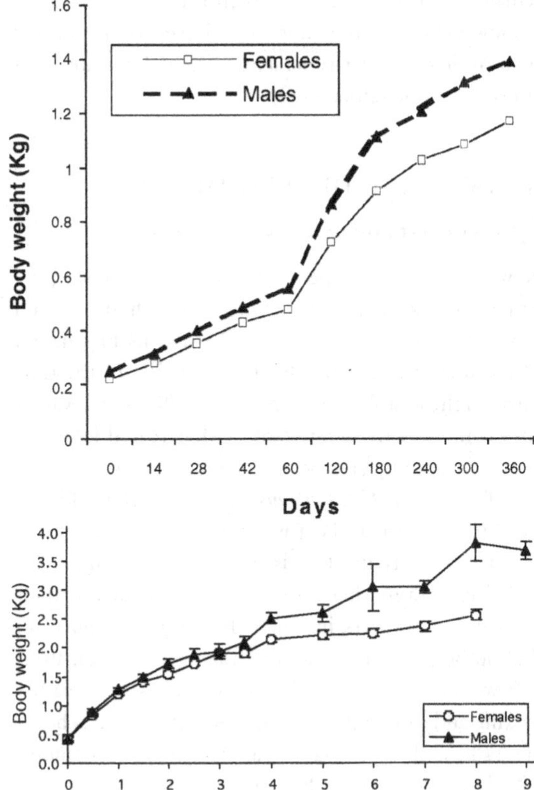

Figure 6.2. Body weights for young capuchins. Top, mean body weights for hand-fed white-fronted capuchins across the first year of life. Redrawn from Fleagle and Samonds 1975. Bottom, mean body weights for mother-reared captive tufted capuchins from birth to eight years. Redrawn from Fragaszy and Bard 1997.

Nervous system

The neonatal capuchin's brain is a smaller proportion of its adult weight (*c.* 50%) than is the brain of all other nonhuman primates except chimpanzees (Elias 1977, Hartwig 1996). Hartwig (1996) notes that although body growth and neural growth in platyrrhine genera are fundamentally conservative, two genera, *Saimiri* and *Cebus*, have unusually large adult brain sizes relative to their body weight. *Saimiri* have a long gestation and relatively large infants (17% of maternal weight at birth) followed by normal postnatal growth rates of brain and body. The brains of *Cebus*, on the other hand, undergo more postnatal development than in other platyrrhines, and postnatal growth of the brain (as of other parts of the body) occurs at its fastest rate immediately after birth. In all genera of primates with particularly large brains relative to their body size (apes, capuchins and humans; see Chapter 5), the period during which the brain grows at its fastest rate, as it does just before birth, is extended longer after birth compared with other primate genera with proportionally smaller brains.

An extended period of brain development after birth provides more possibilities for activity to influence neural development, cerebellar and cortico-thalamic tracts in particular, than if these fundamental brain structures were largely in place at birth (Finlay *et al.* 2001, McKinney 2000). Developmental neurobiologists have documented many ways, and hypothesize many more, in which extensive postnatal neural development enhances individual differences (Elman *et al.* 1996). This possibility fits comfortably with what we know of capuchins' activities compared with other genera.

Skeletal development

We can assess skeletal maturity by measuring the lengths and widths of various anatomical indicators, such as the skull and the long bones, and by identifying whether the growth plates at the ends of the bones have matured (as they do when growth ceases). Fleagle and colleagues made an extensive series of radiographs of 26 white-fronted and tufted capuchin monkeys from birth up to 3 years in most cases, and up to 5.5 years for a few animals.[1] Examination of these radiographs reveals that almost none of the capuchin monkeys' wrist bones are ossified at birth, a pattern shared most closely with *Pan* and distinctly different from *Macaca* (Watts 1990). Continuous skeletal development is evident throughout the first 5.5 years of life (Jungers and Fleagle 1980) and localized skeletal development may continue beyond this age (e.g., the digits, particularly the thumb, continue to grow in length into early adulthood in *C. apella*; Fragaszy *et al.* 1989). In the first 6–8 weeks after birth, growth in all infants was nearly linear for all measurements. Thereafter, a curve of the form

$$Y = a + c \log_e X$$

where Y = length (mm), a = intercept, c = regression coefficient and X = age (days) fit the data for most measurements through the first year of life (Fleagle and Samonds 1975). This growth pattern resembles that of other primates, including humans. Also, the parts of the body nearer the head grow more rapidly than the parts of the body nearer the tail (e.g., arms grow sooner than legs). The head completes 90% of the first year's growth in the first 180 days. Fleagle and Samonds (1975) suggest that those parts of the body that grow more over the course of the first year also spread out the growth more evenly. The jaw, or mandible, which is most robust in *C. apella*, grows at similar rates in both *C. apella* and *C. albifrons* (Cole 1992). Apparently species differences in shape and size of the jaw are already present at birth; they do not arise after birth from differential growth rates.

The flexibility of the joints changes in development; we know only a little bit about these changes in capuchins. Infants that are still predominantly being carried by others exhibit less dorsiflexion of the wrist and more flexible finger joints than older infants (Fragaszy *et al.* 1989). It may be that "stiffer" wrists and more flexible fingers (that afford a locked-knuckle grip on fine objects, such as hair) contribute to the younger infant's ability to hang on to a moving carrier.

Dental development

Dental development can be measured by documenting the ages at which the teeth erupt (are visible upon examination) from the gums (gingival eruption) or from the jaw bone (alveolar eruption), the latter occurring earlier. Galliari (1985) reports on gingival eruption of deciduous and permanent teeth in captive tufted capuchin monkeys to 30 months of age. Capuchins are born with deciduous incisors already visible above the gums. A full set of deciduous teeth appear in the next several months (by 30 weeks), accompanied by much chewing activity. The first permanent tooth, the first molar, appears at 13.5–14 months. The permanent incisors appear soon after, at 14–18 months. The second molar appears at 26–28 months. Fleagle and Schaffler (1982) report a similar schedule for *C. albifrons* for the first and second molars. They found that the third and fourth premolars appear by 3 years, and the third molar before 3.6 years. The last permanent teeth to emerge are the canines, which typically appear at the end of the fourth or during the fifth year. In both *C. apella* and *C. albifrons*, males get each of their permanent teeth 2 weeks to 2 months earlier than females. This difference between the sexes is unusual among primates; the more usual pattern is for the sexes to have similar timetables of dental development (Fleagle and Schaffler 1982).

BEHAVIORAL DEVELOPMENT

Cycles of rest and activity

Newborn capuchins spend most of their time sleeping and nursing (see Figure 6.3). One infant tufted capuchin was estimated to be awake only 14% of its first day of life, and alert and active for less than 5% of the time through the first 5 weeks (Fragaszy 1989). These values appear to be representative of mother-reared infant *C. apella* in captivity (Byrne and Suomi 1995, Fragaszy *et al.* 1991) and of *C. capucinus* (Mitchell 1989). The 8-week mark seems to be a watershed in how the infant organizes its activity. This is evident even in wild infant tufted capuchins that are precocious relative to their captive counterparts in when they begin to move off the mother, contact food etc. (Valenzuela 1992). Prior to 8 weeks of age, when the infant is alert, it is mainly looking and listening but not moving. It alternates bouts of waking and nursing lasting approximately 10 minutes with bouts of sleeping for 10–20 minutes (Byrne and Suomi 1995, Fragaszy 1989). After 8 weeks, infants are more often active while alert (reaching and moving), nurse for shorter periods and at less frequent intervals, and sleep for longer periods. Byrne and Suomi (1995) report that it was common for infants to be awake for a full hour by 11 weeks of age.

Postural and locomotor development

Soon after the infant can hold its torso off a surface with its arms and legs (by 5 weeks, see Figure 6.2), it will reach out into space, releasing its grip on the carrier's fur. Hand-reared infants will reach out with one or even both hands at much younger ages, but they do not have to cling to a moving carrier (see Figure 6.2). In this case, as in humans, postural security constrains early reaching. The hands and eyes work together from birth, but the very young infant must be on a stationary and stable surface before it will let go of the surface to reach out. Similarly, bimanual activity is affected by the infant's postural circumstances. Hand-reared infants sit upright and reach bimanually rather often by 8 weeks of age

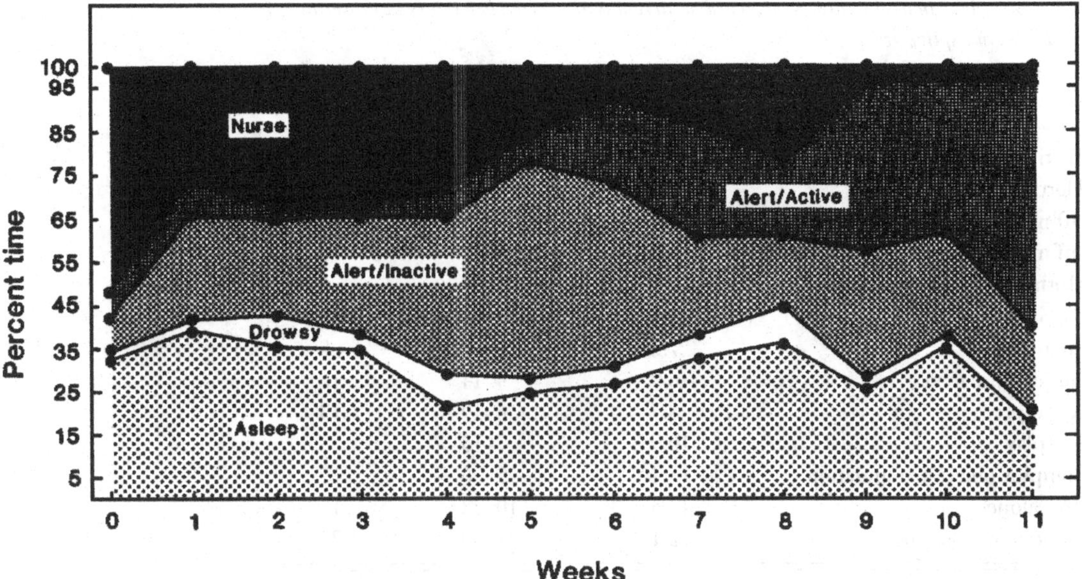

Figure 6.3. Proportion of time during daylight hours spent in different behavioral states across the first 11 weeks of life by a tufted capuchin. The infant and its mother lived in a captive group. (Reprinted with permission from Fragaszy 1989b).

(Fragaszy 1990c, 1995; see Figure 6.2). Mother-reared infants do so rarely until several weeks later (Adams-Curtis *et al.* 2001).

The infant capuchin's postural control and strength improve sharply over the first several months of life, and it begins to explore its surroundings more actively (see Table 6.2 and Figure 6.4). Eventually, sometime between 9 and 13 weeks in captivity, the infant makes its first foray off the mother, to sit next to her or to toddle precariously nearby. In La Macarena, Colombia, and Raleighvallen, Suriname, mother tufted capuchins may encourage infants to dismount from 5–6 weeks on, a few weeks earlier than in captivity (Valenzuela 1992, E. Ehmke and S. Boinski pers. comm.). At this age, Valenzuela notes, the infants are unable to do more than simply stand or move in place. Wedge-capped capuchins at Masaguaral, Venezuela, white-faced capuchins in Panama and tufted capuchins in Manu, Peru follow the timeline of captive tufted and wedge-capped capuchins in descending from the carrier and beginning to move independently some time in the second to third month (Byrne and Suomi 1995, Levy and Bodini 1986, Mitchell 1989, Oppenheimer 1968, Robinson 1986). In any case, infants improve their locomotor skills rapidly once they begin to spend time off the mother.

While this pattern of motor development is entirely predictable and is shared with all other arboreal monkeys, the precise timing of developmental milestones in capuchins is relatively delayed compared with other Cebid monkeys. For example, Fragaszy *et al.* (1991) found that capuchins developed virtually all behaviors one to several weeks later than squirrel monkeys reared under the same conditions in captivity and squirrel monkeys in the wild (Boinski and Fragaszy 1989). We know from comparing infant capuchins in natural and captive settings, and even captive infants among themselves (Byrne and Suomi 1995; see Table 6.2) that the precise timing of various developmental milestones can vary, and many could occur later in captivity than in the wild. The sources of the wide individual differences among captive monkeys are not easily discernible; they are not apparently related to sex, age of mother or conditions of housing, for example (Byrne and Suomi 1995). Time spent in an active, alert state in the first 11 weeks predicted amount of time spent alone or with others, in play or in exploration during months 2–6, but early state measures were not predictive of behavior after 8 months of age (Byrne and Suomi 1998). Variations in the mothers' behavior are partially attributable to their experience. More experienced mothers keep their infants on

Table 6.2. *Age (in weeks) of first observed occurrences of selected behaviors in infant tufted capuchins in captive groups.*

	Mean	Range	Difference
Tactile/oral exploration	4.2	3–7	4
Alert active > quiet[a]	5.6	4–7	3
Off mother, holding on	5.5	4–11	7
Off mother, in proximity	6.6	4–12	8
Mother and infant separate	7.4	4–12	8
Carried by others[b]	7.5	2–20	18
Infant alone	7.5	4–12	8
Eat/explore food	6.9	4–14	10
Social play	8.7	6–19	13
Self play	10.9	7–15	8
Complex object manipulation	15.4	9–26	17
50% alone[c]	17.5	10–28	18
Alone > with mother[d]	22.1	15–38	23

[a] Age at which alert active scores exceeded alert quiet scores in activity state scoring.
[b] One infant in Group 2 was never observed to be carried by another individual. The earliest case (at week 2) was due to kidnapping of an infant by a juvenile male, and is not counted as a voluntary separation of mother and infant. The next earliest incidence of carrying by others was at 4 weeks of age.
[c] Age at which infant was consistently alone for 50% of the observation time.
[d] Age at which infant was consistently alone more than with its mother. Taken from Byrne and Suomi 1995.

their ventrums (chest and abdomen) more than do first-time mothers for the first 9 weeks, and perhaps as a result groom and nurse their later infants more than they do first-born infants for the first few months of life (Byrne 1993). Obviously, the developmental story of capuchin monkeys (as of all life) has a more complicated plot than the unfolding of an intrinsic maturational timetable. This is nowhere more evident than in the development of manipulation, as we see next.

Development of manipulation

Given the prominence of manual activity in capuchins' life style, developmental scientists have been much interested in how young monkeys become proficient at handling objects. Adams-Curtis and Fragaszy (1994; see also Spinozzi 1989) describe the development of manipulation in young capuchins over the first 6 months of life (to the age at which infants in captivity are spending about half their time off a carrier; see Table 6.2). The first 8 weeks is a period of beginnings: the first tentative reaches; poorly aimed, gentle, slow, producing contact occasionally but not leading to grasping or retrieval. These early-appearing actions involve sustained visual orientation but not precise control of the hands; the fingers are widely splayed as the infant reaches, for example (D. Fragaszy, M. Busch and N. Wojda unpubl.). The infant is limited to contacting objects or animals that are a short reach from the mother's back. Manipulation is infrequent and tentative in nature. The second 8 weeks is a period of explosive increase, as the infant's improving postural control and stamina, among other concurrent changes, enable it to engage in much more activity. Manual activity increases approximately four-fold in frequency. Infants keep the fingers parallel during reaching more often than splayed. They begin to exhibit the main elements of the species-typical manipulative repertoire such as pounding objects on substrates, rubbing, tapping and so on, although these activities constitute a very small proportion (7%) of all

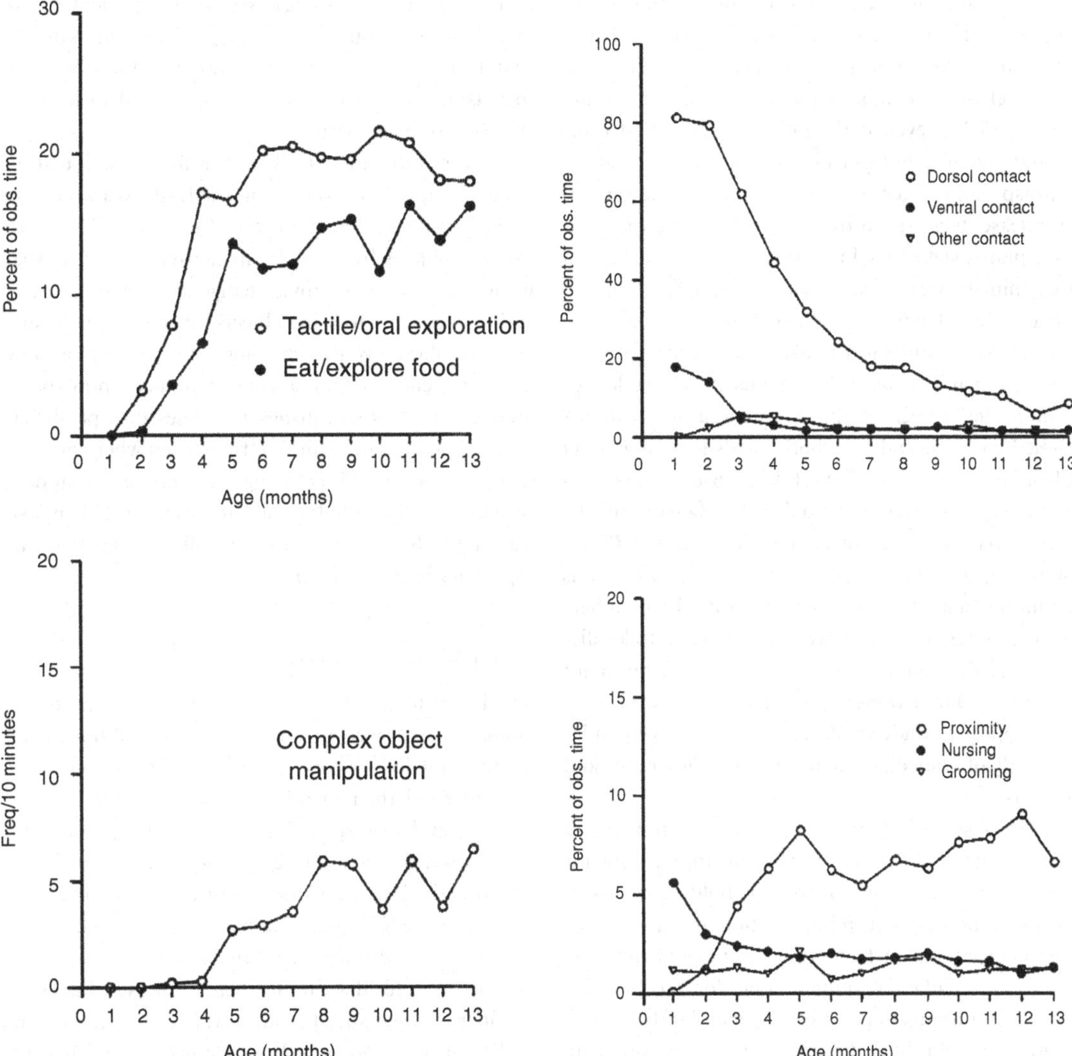

Figure 6.4. Left, exploratory behavior over the first year of life in captive group-housed tufted capuchins. Right, proportion of time spent in interaction with the mother by infant captive tufted capuchin monkeys over the first year of life. (Reprinted from Byrne and Suomi 1995 with permission of Wiley Liss).

manual activity – mostly they merely grasp and hold objects, or touch surfaces. In weeks 17 to 24 the repertoire of actions expands to include all adult activities, including those that require precisely controlled actions of the digits and precisely aimed small movements. Such actions become more frequent, and the rate of activity continues to rise (in fact, nearly doubles again, from 128 to more than 215 acts per hour).

Bimanual activity appears on the same timetable as unimanual activity, although capuchin monkeys at all ages predominantly use one hand to manipulate an object or surface (Adams-Curtis *et al.* 2001). In most bimanual activity, the two hands perform the same action, but complementary bimanual activity, such as one hand holding an object and the other hand picking at it, also occurs from the outset.

Postural changes that allow the infant capuchin to support itself on the unstable surface of the mother or carrier and to locomote independently are pivotal in the infant's changing manual actions. In the first month to 6 weeks of life, even if the infant is capable of coordinated reaching and grasping, it cannot often release its grasp on a carrier that moves much of the time. It can release its grasp on the carrier while the carrier is momentarily stationary, but surfaces and objects (except other animals) are not usually close enough for the infant to reach them. Under the right circumstances, however, even very young capuchin monkeys can reach effectively, although their control of their hands is still crude. For example, the 2-week-old infant pictured in Figure 6.1 reached with one and with both hands toward a target in front of him. He could do this because his torso was completely supported on a fixed surface (a paint roller). He had no need to cling to maintain his position. This is one reason that hand-reared monkeys seem precocious in some domains of activity compared with their mother-reared counterparts (see Fragaszy 1995 for a fuller discussion of the ramifications of hand-rearing on infant capuchins). Only actions requiring precise control of the digits appear to undergo development relatively independently of immediate environmental constraints and supports.

The onset of independent locomotion at about 12 weeks of age or earlier has important implications for the appearance of actions that require holding an object. The infant now can control its location in space in relation to objects and surfaces it may want to manipulate. However, the infant is quite wobbly for the first few weeks that it begins to locomote independently. It must sit down to achieve sufficient stability to manipulate objects in any sustained fashion. Once the infant has good quadrupedal balance and is spending more than 50% of its time moving independently (achieved by 6 months in captivity; earlier in natural settings), grasping and other more vigorous acts comprise the majority of its manipulative activity.

The first 6 months encompass the steepest slope in the young monkey's climb to manual proficiency. In the second 6 months of life, although all adult forms of manipulation have already appeared, virtually all other indices of manipulation continue to change gradually. A greater proportion of the young monkey's activity involves fine movements or vigorous movements, manipulating portable objects, and combining objects and surfaces (e.g., banging an object on a surface). Biting and chewing become even more predominant. After its first birthday, the young monkey continues to show increasing interest in portable objects and to increase the vigor of its activity.

Overall, the most striking feature of manipulation in young capuchins from a year to at least 3 years old is its sheer frequency. The high rate of manual activity with inedible objects by juveniles in captivity indicates that using their hands to move, change and explore objects and surfaces, and to bring objects into contact with surfaces, is inherently attractive for young capuchins. The high frequency of spontaneous manipulation in young monkeys in captive settings translates into persistent foraging activity in natural settings, even when foraging efficiency is low. Thus young monkeys are equipped, physically and psychologically, to meet the challenge of learning to feed themselves. We will come back to this topic later in this chapter.

SOCIAL DEVELOPMENT

The broad features of social development in capuchin monkeys are shared with other primates, although the infant capuchin's social experiences seem somewhat more relaxed than in other primate species, such as baboons and macaques. The prominent tolerance by adults towards infants in capuchin groups is in keeping with the largely tolerant nature of social relationships among adults (see Chapter 11). From the moment of birth, infant capuchins live in an intimate and largely benign social world including their mothers and all other members of their social groups. At first interactions are unilateral, initiated by others toward infants that are largely unable to reach out from the carrier. Within a few weeks, however, as infants become more active and eventually more mobile, they begin to initiate contacts with others, to play (most often with peers), to approach others as they feed or manipulate objects in the environment. They move freely among all members of the group during the first year and into the second. During the second year, their social lives become more complicated: the young monkeys face the challenge of weaning and towards their second birthday they increasingly begin to face adult rejections of their attempts to cling. As juveniles and young adults they assume increasingly adult patterns of social interaction, including sexual behavior, participation in intergroup encounters, and aggressive interactions within their group. In this section we review developmental patterns in some of the more important categories of social interaction.

Making and breaking contact with the mother

Valenzuela (1992) reports that mother tufted capuchins in La Macarena, Colombia initiated their infants' first times off their bodies, depositing them in areas with many small branches and twigs that the infants could grasp securely. However, once they have better motor control, infants are primarily responsible for leaving the mother. Infant capuchins studied by Fragaszy *et al.* (1991) were primarily responsible for breaking contact with the mother (80 to 100% of separations), and to a lesser degree, primarily responsible for reunions (about 70%) (see also Byrne and Suomi 1995).

Nursing and ventral contact

Mother–infant pairs vary greatly in how they manage their first few days together. Some newborn infants spend much of their first few days on the mother's ventrum, nursing and sleeping. Others spend much more time on the mother's back, coming to her ventrum only for nursing. Byrne and Suomi (1995) observed that infant tufted capuchins nursed during 9% of the time and spent 20% of the time in ventral contact in the first week of life (see Figure 6.4). Time spent in ventral contact declined rapidly after 2 weeks, stabilizing at just a few percent across the first 13 weeks, the same percentage of time devoted to nursing.

Carriage

Infants cling primarily to the mother, but they also routinely ride on other individuals (see Figure 6.5). Riding on individuals other than the mother is uncommon in infants under 4 weeks of age, although we have observed infants on others even on the day of birth. Very young infants are passive in such transfers: the new carrier grasps the infant so that it releases its grasp on the mother and recovers its grasp on the new carrier's body. In contrast, infants older than 4 weeks transfer from one carrier to another at least partially on their own initiative. Carriage by others peaks in weeks 6–15 in tufted capuchins in La Macarena (Valenzuela 1992), in months 3–6 in wild white-faced capuchins in Panama (Mitchell 1989) and in months 4–6 in wild wedge-capped capuchins (O'Brien and Robinson 1991). After 6 months, capuchin infants are more often moving on their own than carried by others.[2]

Who carries infants? The answer is simple: everyone. In addition to the mother, juveniles and particularly siblings (of both sexes) most often carry infants in natural settings, but adults, even unrelated females and males, also participate (MacKinnon 2002, Manson 1999, Mitchell 1989, O'Brien and Robinson 1991, Valenzuela 1992). Sometimes an adult devotes persistent attention to an infant and becomes a frequent carrier and/or associate of that particular infant (O'Brien and Robinson 1991). In one striking case that illustrates the generality of adults' attraction to infants, an orphaned infant apparently "adopted" the senior adult male of her captive group as her special friend and was carried more by him than anyone else in the group (Fragaszy, pers. obs.).

The duration of the mother's heavy involvement in carrying her infant apparently varies considerably with the setting. Mothers in natural settings carry their infants for a much shorter period after birth than do captive mothers. Wild infant wedge-capped capuchins interact more with others than with their own mothers in the months 3–6 (O'Brien and Robinson 1991), and wild infant tufted capuchins are carried more by others than by the mother by 6 weeks of age (Valenzuela 1992). In captive monkeys, however, the mother remains the primary carrier for longer periods (see Figure 6.5). For example, in Fragaszy *et al.*'s (1991) study of infant tufted capuchins in captive groups, infants spent on average 57% of their time in contact with the mother over the first 28 weeks. Byrne and Suomi (1995) provide similar data on infant tufted capuchins observed from birth through the first year (see Figure 6.4). Most likely the energetic constraints imposed by less abundant food, the greater distances to travel, the greater need to move quickly, and the varying requirements for jumping, climbing, etc. that capuchins face in natural environments make carrying a heavy infant a more demanding job for monkeys in the forest than in captivity (Steudel 2000). Thus the patterns of care we observe in captive monkeys represent the behaviour of the "leisure class," so to speak.

Allonursing

Unlike most other species of nonhuman primates, infant capuchins nurse from females that are not their mothers. This phenomenon, called "allonursing," appears to be

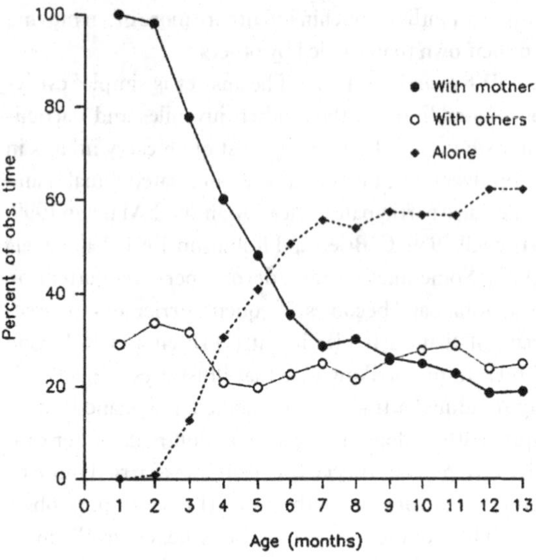

Figure 6.5. Proportion of time spent with the mother, with others, or alone over the first year of life in captive group-housed tufted capuchins. (Reprinted from Byrne and Suomi 1995 with permission of Wiley Liss).

more common in some groups than others, but we have enough reports across species and sites to conclude that it is a genus-typical phenomenon. For example, O'Brien and Robinson (1991) observed infant wedge-capped capuchins nursing from their older siblings and from unrelated females in just over 10% of observed instances of nursing; Perry (1996a) reports that all the lactating female white-faced capuchins she observed nursed infants other than their own at rates from 0.01 bouts/hour to 0.05 bouts/hour. In captive groups of tufted capuchins, allonursing is relatively frequent (Fragaszy and Visalberghi, pers. obs., Weaver 1999). Weaver (1999, Weaver and de Waal 2000) describes how infant tufted capuchins pester females other than their mothers for a chance to nurse, going into tantrums if denied. Infants seek to suckle from other females more frequently as their own mothers begin to reject their advances (Weaver 1999). Thus, the dynamics of allonursing, in captivity at least, are tied up with the weaning process, which we discuss below.

Grooming

As in other primates, grooming is a common form of intimate affiliative contact in capuchin monkeys; in fact, it is the most frequent form of affiliative interaction directed by the mother to the infant (Weaver and de Waal 2002). Capuchin mothers in captive settings groom their infants at a rate of 10–12 times per hour in the first 8 weeks of life; the rate declines steadily after that age until it reaches just 1–2 times per hour by 24 weeks of age (Fragaszy *et al.* 1991) (see Figure 6.4). Often the mother grooms the infant as it clings to her ventrum while nursing or sleeping. Other animals groom the infant while it is in any position on the carrier. All animals seem particularly attracted to the infant's anogenital area, often lifting up one of the infant's legs to gain better access to the most interesting places. Animals approaching the infant to nuzzle it or initiate grooming often make a soft guttural vocalization that sounds similar to a vocalization given by mothers inviting nursing. Young monkeys begin to groom others around their first birthday (tufted capuchins in Raleighvallen; E. Ehmke, pers. comm.; white-faced capuchins at Lomas Barbudal, S. Perry, pers. comm., and at Santa Rosa, MacKinnon 2002), but they do not become frequent groomers until much later, well after weaning (MacKinnon 2002 and S. Perry, pers. comm.).

Weaning

Assuming that re-conception indicates that the infant's demands on the mother are minimal, and thus using re-conception as a marker for completion of weaning, Fragaszy and Adams-Curtis (1998) estimated that capuchin infants in a captive population were completely weaned, on average, at 416 days (13.6 months). Weaver (1999) calculated the onset of weaning on the basis of the first substantial drop in hourly rates of nursing. She followed 11 infants from infancy into weaning, which began when the infants were from 8–12 months old (mean = 10.8 months). Eight individuals completed weaning (i.e., were not seen to nurse again from the mother) at ages ranging from 13–34 months. In the two cases where individuals nursed past 24 months, their mothers bore infants that did not survive when these individuals were 22 and 23 months old. Excluding these two cases, the mean age at completion of weaning was 20.5 months. It may be that weaning is completed earlier in natural environments than in captive environments because mothers face higher energy demands combined with more limited resources in natural environments than in captivity, but we have few data that bear on this issue.[3]

The weaning process can be relatively benign for infants whose mothers are reliable and generally affable social partners (i.e., infants with a secure attachment to their mothers; see below). The mother invites the infant to interact less often, and infants begin to monitor the mother from a distance and to beg to nurse, expressed by a particular posture seen in a variety of submissive/approach settings (Weaver 1999, Weaver and de Waal, unpublished). Other than begging to nurse, however, these infants show little disruption of their relationship with their mothers. On the other hand, some infants have a rougher time managing conflicts with the mother over nursing. They are characteristically provoked into tantrums, with loud screaming and accompanying postures (e.g., lying on their sides, in a curled position) and facial expressions of acute distress (see Figure 11.1). Infants may throw tantrums if they are denied nursing by females that are not their mothers, even if they do not have tantrums following denial of nursing by their own mothers (A. Weaver, pers. comm).

Emotional attachment to the mother

As suggested above by the different profiles of infant behavior during weaning, there are substantial differences evident across mother–infant pairs in the nature of their social relationship. Weaver and de Waal (Weaver 1999, Weaver and de Waal 2002 and unpublished) developed a classification index of relationship quality in capuchin monkeys based on 18 behaviors analogous to those used to evaluate attachment in human mother–child pairs (see Cassidy and Shaver 1999 for a general treatment). They also developed the Relationship Quality Index (RQI), a ratio of the relative frequencies of affiliation (grooming, sitting in contact) to agonism (most often avoidance and threatening postures, but occasionally including slapping, chasing or mock biting) within the mother–offspring pair. This index allowed them to identify youngsters as "securely attached" or "insecurely attached" on the basis of the youngster's behavior towards the mother over time.

Weaver observed 24 tufted capuchin monkeys from 3 months to 3 years old (at the start of observations) interacting with their mothers and others in their group over a 2-year period. Based on their RQI values, the young monkeys could be classed as securely attached (N = 22) or insecurely attached (N = 16) to their mothers. Although all mother–infant pairs exhibited disputes, the relative proportions of interactions that were scored as disputes varied widely across mother–offspring pairs. Youngsters classed as insecurely attached experienced nearly ten times the proportion of agonism by the mother as did securely attached youngsters. Insecure infants also exhibited the most agitation (characterized by nervous milling) around their mothers and "squabbled" at her seven times more often than securely attached individuals (Weaver and de Waal 2002). Insecurely attached youngsters experience unpredictable interactions with their mothers; Weaver and de Waal suggest that as a result, they develop greater behavioral arousal in the face of (potential and actual) conflicts.

The nature of the infant and young juvenile capuchin's relationship with its mother apparently presages the way it interacts with others in its social group: insecure and secure juveniles vary consistently in a suite of characteristics including how they cope with social conflict with others (Weaver 1999, Weaver and de Waal 2002). The interwoven development of social relationships, emotional attachments, regulation of internal (physiological) states of arousal and modes of coping with social conflict remains a rich subject for study in capuchin monkeys.

Social interactions outside the mother–infant relationship

Infants interact with other group members from birth, initially in benign affiliative ways as others investigate and groom them or play with them. Infants are particularly attractive to others when they begin to move and vocalize frequently. Once infants are able to move about on their own, social interactions with others become more frequent. As a general rule, infants and juveniles interact more with kin, particularly the mother, than with non-kin, except in play (O'Brien and Robinson 1991, Welker *et al.* 1990, Welker *et al.* 1992a, b). But this preference for the mother does not prevent the infant from engaging other members of its group in social interaction, especially in play. In Weaver's (1999) study, for example, although youngsters interacted more often with the mother than any other single individual, as infants they interacted with others as often (29 times/hour) as with the mother (25 times/hour), and with others more often than the mother (31 vs. 22 times/ hour) during weaning and after weaning (24 vs. 18 times/hour). They interacted with other adult females and peers at similar rates (5–10 times per hour) and with adult males 10–14 times per hour.

O'Brien and Robinson (1991) note that younger and lower-ranking wedge-capped capuchins carry the infant more; higher-ranking animals investigate more and interact with the infant in other ways that cost less energy, such as grooming. Infants approach adult males, sit with them, sometimes cling to them and play with them, but adult males approach, carry and groom infants less often than do females and juveniles (MacKinnon 2002, Fragaszy, pers. obs.). Infant and juvenile white-faced capuchin monkeys at Santa Rosa are particularly attentive toward the alpha male and/or their mother's favorite males, and infants direct some affiliative and investigatory behaviors exclusively to adult males, who are overwhelmingly tolerant of their attentions. Adult males even participated in boisterous play with larger juveniles (MacKinnon 2002).

The infant's relationship with its mother impacts how it manages affiliation and conflicts with others into the third year of life (Weaver 1999). For example, infants that are securely attached (in the sense used above) play more frequently with their peers and with unrelated adults than do infants with more difficult relationships with their mothers. Secure infants reconcile (approach and make affiliative overtures toward) opponents following conflicts in an adult-like manner, whereas insecure infants have an agitated conciliatory style that is distinctly different. Secure infants seek the mother following conflict with adult aggressors; insecure infants are more likely to contact the aggressor but not to seek the mother following conflicts with other adults. There are data to suggest that the infant capuchin's relationship with its mother is predictive of individual differences in social temperament later in life (Weaver and de Waal 2002). This is another important element of behavioral development where further inquiry with capuchins will be useful to developmental scientists.

VOCAL DEVELOPMENT

Infant capuchins vocalize from birth, producing a variety of high-pitched (2.7–6.1 kHz, in the case of one infant tufted capuchin recorded by Di Bitetti 2001b) ascending and descending pure tones. These sounds do not resemble adult calls, and Di Bitetti (2001b) characterized them as "babbling." Di Bitetti notes that because he was recording the vocalizations of infants in nature from farther than 4 meters away, he may not have recorded soft (low amplitude) vocalizations. In captive conditions, we hear very young infant tufted capuchins trilling softly, producing a cricket-like sound, during social interactions, as when another monkey approaches the infant closely. White-faced capuchins of all ages produce a trilling vocalization, but infants produce this vocalization most often (Gros-Louis 2002), and they produce it while approaching and interacting affiliatively with others. At about 7 weeks after birth, infant tufted capuchins begin to produce vocalizations that resemble the call types typically heard from older infants (that Di Bitetti labels trills, whistle series and squeals) but at first these calls are intermediate in structure between babbling and the calls of older infants. Trills (series of ascending pure notes) are given often by infants, but not adults, especially when they are moving independently at the rear of the group, or when they apparently seek to be carried or nursed. Infants' trills are initially structurally distinctive from adults' trills, and gradually become more adult-like as the infants mature. Squeals, very loud and high-pitched calls, are given when infants are distressed; whistle series (a series of two or more pure tone syllables that sound like short whistles) are given to attract their mothers or another carrier. At dusk, infants whistle at high rates, which may be related to the infant's efforts to find and get close to their mothers or others to sleep. Infants sleep at night with their mothers and close maternal kin (Di Bitetti *et al.* 2000).

PLAY

Play is difficult to define but easy to recognize. Here we consider play to be activity that has no apparent immediate functional consequences and that appears to be pleasurable to the participants. Capuchin monkeys of all ages play, but youngsters, whether in captive or natural settings, play far more often than adults. Young capuchin monkeys in captivity devote a good deal of their day to play – for example, 16 bouts per hour for nursing infants, to 28–30 bouts per hour for individuals during weaning and weaned juveniles (Weaver 1999; see also Visalberghi and Guidi 1998 and Table 6.3). In natural settings, play occurs most often when the group as a whole is stationary, but on the whole play is less common than in captivity. For example, Fontaine

Table 6.3. *Play in young capuchins. Values are percent of instantaneous scans taken during interval scan sampling of captive tufted capuchins living in stable groups with outdoor play space.*

	All[a]	Infants[a]	Juveniles[a]	Males[b]	Females[b]
Activity budget					
% Time play	23	21	24	24	20
% Time object play	2	3	2	2	1
% Time solitary play	6	6	6	6	7
% Time wrestle play	10	8	11	11	9
% Time group wrestle	4	4	4	4	3
% Time chase	1	<1	2	2	>1
% Time solitary play	8	9	8	8	8
% Time social play	15	13	16	17	12
Distribution of play type					
Object play	10	12	8	9	6
Solitary play	26	28	24	23	35
Wrestle play	41	37	44	43	43
Group wrestle	17	20	16	18	14
Chase	6	2	8	7	2
Sum	100	100	100	100	100
Solitary play	35	41	33	32	41
Social play	65	59	67	68	59
Sum	100	100	100	100	100

Note: Data from Landau and Fragaszy unpublished. [a]N=20:7 infants and 13 juveniles. [b]N=20:15 males, 3 females and 2 individuals of unidentified sex.

(1994) reports that he observed juvenile white-faced capuchin monkeys on Barro Colorado Island playing in a mere 3% of samples, far less than reported in captive studies. In captivity, play occurs most often when the group is relatively undisturbed. Very young infant capuchins (too young to move independently) participate in gentle play, reaching out to others while clinging to the mother, as others come to it. They initiate social play with others and join existing play groups nearly as soon as they are able to move about on their own. Social play prominently features chasing and wrestling in many varied forms (including swinging from another's tail, for example) and is accompanied by unmistakable facial expressions like those described for other species that signal play: mouth open wide, lips retracted at the corners but teeth covered by the lips (see Figure 11.1). In the wild social play consists of chasing, grappling, mock biting, tail-pulling, mounting and mock double-threats (where one monkey stands over the back of another and both make threats towards a target, often another monkey). Sometimes monkeys playing actively together make a low-amplitude staccato or pulsed vocalization, a sound somewhat like chimpanzee laughter (Provine 2000).

Young monkeys most often play socially with one other animal (Visalberghi and Guidi 1998, Weaver 1999, Landau and Fragaszy, unpublished; see Table 6.3). For example, in Landau and Fragaszy's study of 20 infants and juveniles up to 36 months old in two breeding groups, 71% of play wrestling bouts involved two monkeys. However, monkeys also routinely joined in larger groups of three (22% of wrestling bouts) four (4% of wrestling bouts) or even five (4% of bouts). Animals join and leave others in wrestling and chase play in rapid succession; wrestling bouts can go on for many minutes but the participants may change completely during the bout.

Infants older than a year engage in more wrestling play than infants younger than a year, but younger infants have the same propensity to play in groups of three or more animals, and devote a similar amount of their day to play as older infants. At Raleighvallen, Suriname, young tufted capuchins often include squirrel monkeys in their play groups (S. Boinski and E. Ehmke, pers. comm.)

Youngsters also play by themselves, engaging in repetitive or unusual locomotor motions (e.g., swinging around a bar; somersaulting on the ground, launching into space), and they manipulate objects in ways that appear playful. During play, including social play, monkeys avoid normal quadrupedal locomotion in favor of less common locomotor movements (Fontaine 1994). Fontaine suggests that the development of movement flexibility and dynamic stability, particularly in suspensory or other less common postures, is one outcome of the vigorous and varied patterns of locomotion during young monkeys' play. Solitary locomotor play and object play constitute a minority of play when monkeys have peers available to play with. For example, in Landau and Fragaszy's study, object play constituted 22% of all play, and solitary locomotor play a mere 6%.

In natural settings, infants' manipulation of objects is often categorized as early foraging activity (e.g., Fragaszy 1990b). The young infant's mouthing, feeling, biting and handling of objects leads naturally to ingestion when it is fiddling with edible objects. However, proficient foraging requires more than playful attention to nearby objects. We now turn our attention to how young monkeys make the transition from playful handling to proficient foraging.

BECOMING PROFICIENT FORAGERS

Becoming a proficient forager is one of the most important challenges facing the young capuchin. The nursing infant often mouths solid objects by the time that its deciduous dentition is complete (by 6 months), and these early explorations of objects lead naturally to some ingestion. However, it is unlikely that infants younger than eight months ingest very much during their frequent chewing; most of what very young infants in natural habitats put into their mouths is indigestible wood (twigs). Infants also routinely obtain small bits of food from others in the group, and they are uniformly tolerated when they approach group members who are eating (de Waal *et al.* 1993, Fragaszy *et al.* 1997a, b). Tolerance by others allows youngsters to eat at least small amounts of foods that they are unable to process or capture by themselves, such as husked fruits and nuts, palm pith and even vertebrate prey (Perry and Rose 1994). Nevertheless, infants must be strong enough and have sufficiently powerful jaws and teeth to eat the tough and hidden foods that capuchins rely upon before they can fully feed themselves. Perhaps they must also become skilled at processing specific foods in particular ways (Boinski *et al.* 2001).

Figure 6.6. A two-year old tufted capuchin pounds a pecan nut on the substrate. Although not visible in the drawing, the monkey had looped its tail around a support behind the one that its foot and hand are on, providing stability while striking forcefully. (Drawing by Stephen Nash from a photograph by Dorothy Fragaszy.)

Probably changes in body strength, dental equipment and developing skill all contribute to increasing efficiency of young foragers over time (Fragaszy and Adams-Curtis 1997). Perhaps because they are less efficient than adults, and coupled with their intrinsic interest in manipulating and mouthing objects, young foragers are extremely persistent (Figure 6.6). Young capuchin monkeys have more than just persistence on their side, however. The generativity with which capuchin monkeys in captivity produce varied actions with objects and surfaces, and combine objects with surfaces and with each other, translate in natural settings to varied foraging actions. Thus even inefficient and unskilled foragers are likely, eventually, to discover more efficient means of processing or finding the varied foods they eat.

Young and small monkeys try to eat the same foods as larger and older animals, but even after greater effort they are often less successful at exploiting palm pith, opening many kinds of nuts, and capturing vertebrate prey, for example (Boinski *et al.* 2001, Fedigan 1990, Fragaszy and Boinski 1995, Mitchell 1989, Perry and Rose 1994). We have little information on how youngsters learn to deal with specific foraging challenges such as these. However, in most cases it appears that adults do not use different actions than youngsters to try to open common food items. Rather, the species-typical actions that animals of all ages use work only when produced with enough force and with enough precision (e.g., Boinski *et al.* 2003, Fragaszy *et al.* 1997a). Sometimes young monkeys must also learn how to relate objects to surfaces in appropriate ways (e.g., selecting a hard surface rather than a flexible one on which to pound a nut). Young animals also become better at identifying where to look for hidden foods and how to handle problematic foods such as urticating fruits or stinging insects. Youngsters appear less proficient than adults at localizing "good" places to bite woody substrates, more often biting at sites that produce no edible items.

In summary, relatively inefficient foraging characterizes capuchins before, during and for a few years after weaning. Inefficient foraging is a serious problem facing the young capuchin, one so serious that some authors have suggested that the notably slow growth rates and the late onset of reproduction in this genus compared with other primate genera reflect the periodic risk of starvation (Janson and van Schaik 1993). Feeding itself is a daily challenge for a young capuchin monkey in a natural environment.

Young capuchins work at obtaining food in a socially supportive context. The most common aspect of social influence on youngsters' feeding is enhancement of interest in areas and objects that they see others exploring or eating (Drapier *et al.* 2003). Capuchin monkeys of all ages are motivated to look for food and to eat when they see others doing so (Visalberghi and Addessi 2003; see Chapter 13), although the monkeys' choices of what to eat apparently are not affected by what they see others eating (Galloway 1998, Visalberghi and Addessi 2001, 2003). In a typical situation, a young monkey will be inclined to forage at the same time and near the same place as another, leading to discovery of the same foods. Less often, it will manage to obtain some food from the other, or to smell the mouth of the other while or just after it has eaten something. Both of these processes can aid the young monkey to forage in the right places and at the right times and to select appropriate foods.

BECOMING ADULTS

Females begin to exhibit sexual interest in males in their fourth year although they typically do not conceive until they have nearly reached their full adult weight or even later (5 years or older; see Chapter 4). Males express sexual interest in the opposite sex from late juvenescence (about 3 years of age). Juveniles of both sexes engage in sexual play (mounting in all combinations of mounter and mountee). Females play less often in this way as they reach puberty but mounting remains an important element in males' play, merging into sexually motivated behavior. We discuss sexual behavior in more detail in Chapter 12.

Capuchin monkeys' transition through social adolescence into young adulthood has not been studied in detail. However, we have many years of experience watching pubescent (subadult) males and females in captive groups of tufted capuchins, and we can provide some casual comments about the most outstanding features of this age group. One can characterize subadult females (especially higher-ranking ones) as particularly active participants in disputes in their groups. They seem eager to "stir up trouble", sometimes initiating unprovoked conflicts with another and then soliciting support from others to join in the dispute against their chosen target. Arranging their social relationships seems a particularly important activity for them, as it is for many species of animals where females largely remain in their natal groups. This sort of activity subsides as females become mothers. Subadult males in captive groups may become the target of aggression by resident adults (females, primarily) but in general they are less likely to start disputes than are subadult females. They are equally likely as young females to join disputes started by others, however (Fragaszy, pers. obs.). We do not know how well this picture of social transition to adulthood carries over to other species of capuchins and to monkeys in natural environments. In particular, in white-faced capuchins, where males emigrate and immigrate into groups in cohorts and where aggression by males towards other males and towards infants is frequently lethal, subadults may behave very differently (see Chapters 3 and 11 for further discussion of differences across species in social dynamics within groups).

AGING

We have little information yet about aging in capuchin monkeys. We can say, however, that older capuchins remain active, socially and physically, into their forties as long as they remain in good health. For example, in both wild and captive populations, clearly elderly adult males can be socially "central" males and the preferred sexual partner for most of the females in their groups (see Figure 4.1). Although females may experience longer inter-birth intervals as they age, we have no evidence for menopause. In captivity, older females remain as socially integrated members of their groups as younger females. Maternal kin, both younger and older, remain important social partners, but females also maintain friendships with non-kin. In a study comparing older adult (16 years and older) to early adult (11–15) and young adult (6–10) females living in mixed age and mixed sex groups with their offspring, age did not significantly impact rates of grooming or sitting in contact, nor of exploring inedible objects or opening nuts (Fragaszy, Leighty and Branch, unpublished).

SUMMARY

Development in capuchins follows the familiar pattern seen in many other primates, but with some distinctive twists. Infants are attractive to others from birth, although as newborns they are incapable of responding to others with vigorous or differentiated actions. Capuchins are distinctive for the relatively small proportion of adult size of their brain at birth, indicating a greater proportion of brain development occurs postnatally in this genus than in others. Their mothers are important social partners for young, although they increasingly participate in the social life of their groups by clinging to others, nursing from others than their mothers, playing with others (including squirrel monkeys, in some places) in pairs or larger groupings, and eventually, grooming others, participating in sexual encounters, and other adult social activities. Differences across mother–infant pairs in the character of their relationship (ranging from secure to insecure) color many other aspects of the young monkeys' style of interacting with others throughout juvenescence and perhaps into adulthood. Capuchin monkeys remain vigorous members of their social groups until they near death.

Youngsters must become able to feed themselves before they can be fully weaned. To do so, they must become strong enough and large enough and develop strong teeth and jaws. For capuchins, these aspects of growth are in progress when the youngsters enter weaning, in their second year of life, but are not completed for several more years. Capuchins must also become skilled at selecting places to forage, at avoiding dangerous places and irritating prey, at opening hard surfaces, and at extracting food from surfaces and crevices. Developing these skills begins while infants are still nursing. Young capuchin monkeys, from the time that they can reach into space to contact a surface or grasp an object, are incessantly interested in manipulating objects and exploring surfaces with their hands and their teeth.

Sometime in the second year, the young monkey becomes fully nutritionally independent of the mother, although it is still rather inefficient at feeding itself for the next few years. The motor skills that support the capuchin's distinctive life style take years to develop in adult form. The process by which this happens, and the nature of the skills themselves, occupy us in the next chapter.

ENDNOTES

1 Radiographs of the extremities of capuchin monkeys (*C. apella* and *C. albifrons*) were taken at regular intervals from birth through 5.5 years by J. Fleagle and colleagues as part of a long-term project examining the effects of nutrition on growth conducted at the Harvard School of Public Health in the 1970s. The complete and annotated library of radiographs produced during the project is now available on five compact discs to researchers. They may be accessed by contacting Professor John Fleagle, Dept. of Anatomical Sciences, Health Sciences Center, State University of New York at Stony Brook, Stony Brook, New York 11794, USA. See Jungers and Fleagle (1980) for details of the x-ray procedure, the study populations and the sampling schedule, and Thurm *et al.* (1975) for printed examples of the radiographs.

2 The young infant (that cannot locomote well on its own) usually initiates the transfer from one carrier to another while

an individual sits near the infant and its current carrier. In most cases the infant is able to transfer (back to the mother, for example) when the current carrier stops moving and the mother is able to come close to the pair. Ready transfer back to the mother is not assured, however. Occasionally in a captive setting human intervention may be helpful (or necessary) in setting up a supportive circumstance for the mother and infant to be reunited.

3 Nagle and Denari (1982) list weaning in tufted capuchins as occurring from 16–20 weeks (4–5 months). We assume that these authors are reporting the minimal age at which infants can be removed from the mother and are able to feed themselves on semi-soft foods. Infants at 20 weeks of age would be unable to feed themselves in natural environments and we have never seen a mother initiate weaning until the infant is much older.

7 • Motor skills

It is early in the dry season in the llanos of Venezuela; the rains stopped 3 weeks ago. Daily, more of the forest floor reappears from the shallow waters that have covered it for months. The young wedge-capped capuchin darts to the base of a sapling, feeling in the muddy water for a snail the size of a small tangerine that she has spotted from her perch above. She retrieves it and carries it up to a large limb many meters above the ground, using one hand, her feet, and occasionally her tail for support as she moves. Once seated on the limb, she holds the snail in both hands. Raising it to the height of her shoulders, she bangs it repeatedly on the limb in front of her in a steady rhythm. After five strikes, she examines it carefully, then repeats the banging action. When a small crack appears in the shell, she bites there, levering her canine teeth into the opening, enlarging it by biting and pulling while holding the snail tightly in both hands. After a few more rounds of banging and biting, she has enlarged the hole to the diameter of a pencil. Finally she inserts her index finger into the exposed interior of the shell, and begins to scoop out the soft animal tissues hidden inside. It has taken her several minutes to prepare her meal.

The capuchin monkey in the vignette above uses a practiced set of smoothly coordinated actions with both hands to open the snail and scoop out the soft flesh inside. This is one of many commonly observed activities in this genus to which we accord the descriptor "skilled." What do we know about how capuchins master this skill, and about their motor skills in general? What range of actions do they perform skillfully? Discussing these matters requires a framework in which to consider skilled action, and a definition of action and of skill itself. Like many terms that are in common use in English, "action" and "skill" have multiple meanings, but we use these words in a particular sense. Following the definitions provided by N. Bernstein (1996), an action is an ordered sequence of movements that together solve a motor problem. The sequence is integral to the action, and adaptive variation in the performance of the elements is evident across repetitions of the action. In other words, no action is performed in exactly the same way twice, nor should it be. Think of sharpening a pencil: each time one sharpens a different pencil, one presses it with slightly different force into the aperture of the sharpener, turns the handle of the sharpener at a slightly different velocity, takes out the pencil and blows off the shavings after a varying number of rotations of the handle, and so on, in accordance with the sharpness of the blades, the hardness of the pencil wood, etc. Still, the action as a whole has a consistent sequence and outcome. According to Bernstein, an action is skilled when it effectively solves a particular motor problem across varying starting conditions (e.g., in the example of the pencil, when one achieves the desired outcome with different pencils and different sharpeners). For Bernstein, "motor skills" are a subset of "dexterity,"

> the ability to find a motor solution for any external situation, that is, to solve any emerging motor problem correctly (i.e., adequately and accurately), quickly (with respect to both decision making and achieving a correct result), rationally (i.e., expediently and economically), and resourcefully (i.e., quick-wittedly . . .).
>
> (Bernstein 1996, p. 228)

In this chapter, we focus particularly on motor skills associated with foraging in nature and with fine control of objects in captivity, with some attention to the use of other parts of the body (e.g., the tail) to solve unusual motor problems. We focus on these skills because they distinguish capuchin monkeys from other nonhuman primates.

In general, capuchins share locomotor patterns, sitting postures and general forms of reaching and grasping with other platyrrhine genera, particularly with other monkeys in their subfamily (*Cebidae*). They are shared most especially with squirrel monkeys that resemble capuchins morphologically and behaviorally, except capuchins are approximately three to five times heavier than squirrel monkeys (Janson and Boinski 1992). Comparing capuchins and squirrel monkeys is particularly instructive for understanding the role of motor skills in capuchins' life style. Like squirrel monkeys, capuchin monkeys derive a large portion of their protein from animal sources. However, as we saw in Chapter 5, capuchins have a suite of cranial characteristics and a degree of individuated control of the digits, characteristics that enable them to process tough plant materials and hard foods through skilled activity involving handling objects and surfaces. Squirrel monkeys, in turn, are better than capuchins at detecting small invertebrates on surfaces, and snatching them rapidly (Janson and Boinski 1992). Hence, the differences in motor activity between capuchins and squirrel monkeys, along with differences in strength and size, define their respective foraging ecologies.

In what follows we describe the many forms of manipulation seen in capuchin monkeys in natural settings, to provide a basis for understanding the nature of manual skills in these monkeys. Then we move to the laboratory to consider the features of perception and motor control that support the skilled actions seen in nature, and to consider how these skills are acquired and the range of conditions across which a skill is expressed as a fluid, rapid and economic solution of a problem. We note three hallmarks of manual action in capuchins: variety, generativity and combinatorial character.

VARIETIES OF ACTION

Capuchin monkeys spend about half of their daytime hours foraging (see Chapter 2). Much of this time involves manipulating substrates and ingesting prey found in substrates (e.g., dead branches, bases of bromeliad plants, termite nests, bamboo cane, bases of palm fronds; Boinski *et al.* 2001, Fragaszy 1986, Fragaszy and Boinski 1995, Izawa 1979, Janson and Boinski 1992, Robinson 1986, Terborgh 1983). For example, female *C. olivaceus* at Hato Masaguaral, in Venezuela, spend a quarter of their day sifting, pulling and biting apart dead wood material and sifting in the crowns and frond bases of palm trees; they obtain 39–45% of their prey in this way (Fragaszy and Boinski 1995).

We expect that capuchins use a variety of manual actions during foraging because of the amount of effort that capuchins devote to searching through and handling foods and food-containing substrates and objects, but in fact the full range of movements and actions they use has been difficult to document. The number of distinct manual actions that one can recognize depends upon how close one is to the monkey, how clear a view one has through foliage and one's angle of regard. All published descriptions of capuchins' manual actions in field conditions and also most in captivity directed at assessing the general form of capuchins' spontaneous activity have used real-time scoring, which does not allow detailed notation nor examination of quick movements. Thus, the descriptions that follow reflect the exigencies of recording data in "real" time.

Capuchins' manipulation ranges from extremely strenuous (as when they pound hard objects against a surface, or rip a branch off the trunk of a tree) to extremely delicate (as when they pick small invertebrates out of holes, or sift through loose debris on the forest floor). Capuchins routinely unroll or rip apart leaves; grab paper wasps' nests and, after escaping most of the angry wasps, delicately pull out the pupae and eat them one at a time; capture lizards, frogs, snakes, birds, coatis, squirrels or other vertebrate prey with their hands and then proceed to pound them on available surfaces or dismember them by biting and pulling; snatch flying insects out of the air; tap gently along a branch in the course of searching for prey to be found inside hollow branches; and so on. Capuchins make a lot of noise in the forest when they bang and break branches and other hard objects; field researchers (and hunters) searching for capuchins often find them more effectively with their ears than their eyes! The precise details of what materials are attacked or sought after by capuchins varies across sites, in accord with the available resources, but the general picture of an active, destructive, intensely manipulative monkey holds across all species and all sites.

This picture also holds for capuchins in captivity, where they have a reputation for undoing latches and locks that other species cannot, and for general cleverness with their hands in many ways. Studies comparing capuchins with other species of nonhuman primates have documented repeatedly that these monkeys are anomalously dexterous and manipulative (Chevalier-Skolnikoff 1989, Costello and Fragaszy 1988, Parker 1974a, b, Torigoe 1985, Welles 1972, 1976). In Chapter 5, we explained some of the anatomical features of capuchins' hands that support their dexterity; we will come back to the details of flexibility of hand movements later in this chapter. The important message here is that persistent and varied manipulation of objects and surfaces is characteristic of capuchin monkeys in all environments, particularly juveniles, as we discussed in Chapter 6.

GENERATIVITY OF ACTIONS

A second hallmark of manual activity in capuchins is its generativity, or production of different forms. Captive capuchin monkeys make comprehensive use of all available objects and surfaces, applying every movement to every surface, and involving nearly every portable object with every surface. For example, 31 tufted capuchins (infants to adults) housed socially in cages containing straw bedding, some hanging toys, some small branches, some small pieces of PVC and food objects, produced 211 distinct kinds of actions, and manipulated something in 51% of observation intervals over a 13-month period (Fragaszy and Adams-Curtis 1991). The monkeys most frequently held and bit an object (28% of all actions); the next most frequent set of actions (constituting 13–14%) involved contacting a surface or object directly and gently with the body (scratch, hold, lick). The vigorous acts of hit, push and rub constituted 4–5% of all actions scored. The monkeys manipulated food more often than any other object, but they performed all acts with a variety of objects, except "sift", which by our definition could only occur in the straw on the floor of the cage.

Overall, capuchin monkeys display considerable generativity in action with the same objects, the variety of objects to which the same act is applied, and the frequency and proportion of combinatorial behaviors (see below) even when manipulating familiar objects in the home cage (Byrne and Suomi 1996, Fragaszy and Adams-Curtis 1991; see Table 7.1). Captive adult capuchins express greater interest in objects, and greater generativity in manual action when they encounter novel objects or something that can be altered through action rather than familiar objects (Fragaszy and Adams-Curtis, unpubl.). Juveniles handle objects in generative ways even when the objects are very familiar and do not provide food. Apparently their interest in handling objects and the probability that they will generate novel manual actions become more contextually specific as capuchins reach adulthood.

Table 7.1. *Mean occcurences per hour for manipulative acts in group-housed capuchin monkeys.*

Act	Infants	Juveniles	Adults
Single object	72	202	108
Cage (substrate) only	65	61	13
Object with cage	11	33	9
Object with object[a]	0.5	3.0	< 0.1
Body with an object or the cage	6	27	6
All acts	155	326	136

[a] Values for the category "object with object" are presented to the nearest tenth; values for other categories are presented to the nearest whole number. Values represent the number of events scored during 10-s intervals (360 intervals per hour). Combinatorial actions include the last three categories. From Fragaszy and Adams-Curtis (1991).

COMBINATORIAL ACTIVITIES

The third hallmark of manual activity in capuchins is that they frequently combine objects and surfaces, or objects and other objects, which we label for convenience here "combinatorial actions." Such activities, even though they are a small proportion of all manual activity in both natural and captive settings (Byrne and Suomi 1996, Fragaszy and Adams-Curtis 1991, Fragaszy and Boinski 1995, Natale 1989, Panger 1998; see Table 7.1), still occur very reliably. For example, Fragaszy and Adams-Curtis (1991) report that each captive tufted capuchin monkey observed during routine conditions in its home cage acted with an object on a surface 16 times per hour, on average; the monkeys also combined objects with each other a few times per hour (see also Byrne and Suomi 1996). The monkeys spontaneously perform these actions even when there is no detectable reason to do so.

Table 7.2. *Relations embodied in common actions performed by capuchin monkeys.*

	Relational category	Examples
Zero order		
Simple	Act directly with the body on a surface or an object	Bite, hit, rub, scrape, pull, etc.
Specific	Act on a target zone of a surface	Bite at a certain location on a branch Insert a hand into an opening
First order		
Simple	Combine an object with another object Combine an object with a surface	Bang one block on another block Bang a block on a perch or a fruit on a branch
Specific	Combine an object with a surface	Rub or bang specific side of fruit against a surface
	Combine an object with another object, where the moved object is oriented or aligned to the other	Insert a stick into a hole Insert an object into a cup held in the hand
Second order		
Simple	Combine one object with two others	No common actions (but observed in tool use; see Chapter 10)

Combinatorial actions are particularly interesting to behavioral scientists because: (a) these actions allow the monkeys to gain access to foods they could otherwise not get through direct biting and pulling; (b) they require the coordination through action of objects and/or surfaces relative to each other, a feat not routinely accomplished by nonhuman primates, and (c) these actions are the precursors of using tools, another distinguishing characteristic of capuchins (see Chapter 10).

All species of capuchins perform combinatorial actions routinely during foraging for both plant and animal materials, and captive monkeys produce these actions during spontaneous activity with familiar objects during the first year of life, well before weaning. Older animals in captivity are more sedentary than juveniles in routine familiar conditions, but they combine objects and surfaces when it is useful and interesting to them to do so (i.e., while exploring novel objects; see above). They maintain a strong interest in performing this kind of activity even when it has little obvious utility. For example, tufted capuchin monkeys of all age–sex classes with few other objects to manipulate banged plain wooden blocks enthusiastically against all available surfaces, including other blocks, even after several encounters with them (Visalberghi 1987, 1988).

Relations embodied in actions combining object and substrate

To bring some conceptual order to the varieties of combinatorial actions produced by capuchins, we have classed them by the number of relations embodied in the actions (see Table 7.2). In zero-order actions, the actor manipulates an object or surface directly. As we have already noted, zero-order actions predominate capuchin's manual activity. These actions can be quite precise, as when a monkey inserts a digit or its whole hand into a crevice. First-order actions combine an object with a fixed substrate or another stationary object. Simple first-order combinations require only that the object and surface be brought together; specific alignments are not needed. The overwhelmingly most common combinatorial actions capuchins produce in captivity and in nature, rubbing and pounding an object against a substrate, are simple first-order combinations.

Specific first-order relations are more difficult for the actor to achieve because they require producing a particular spatial relation (such as alignment) between object and substrate. Izawa and Mizuno (1977) provide a striking illustration of specific first-order combination in their descriptions of tufted capuchin monkeys opening hard fruits by pounding them against

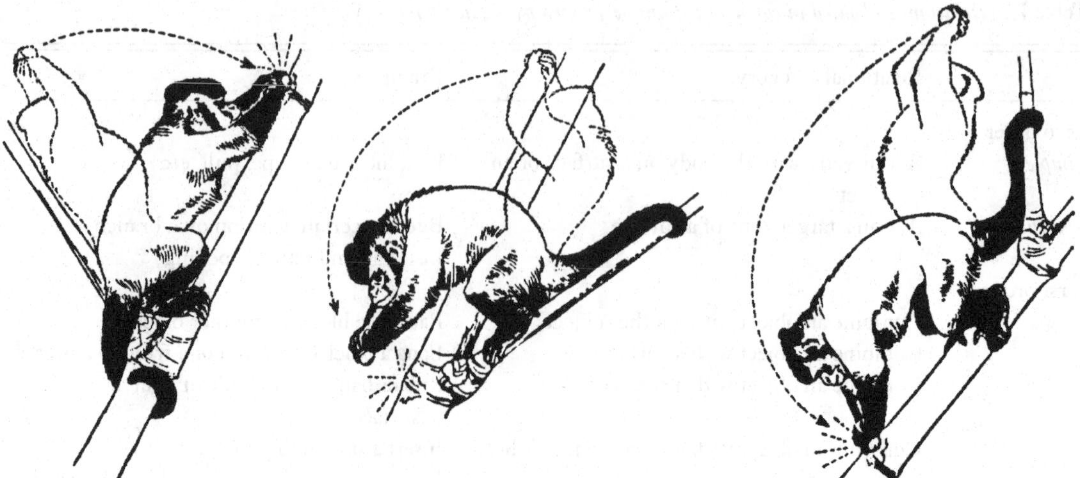

Figure 7.1. Three different positions in which a tufted capuchin monkey strikes a cumare fruit (*Astrocaryum chambira*) against a kind of a bamboo tree (*Bambusa guadua*) just above the joint. (Izawa and Mizuno 1977, reprinted with permission of the Publisher).

the protruding growth node of a bamboo trunk (see Figure 7.1). Sometimes monkeys consistently pound the longer axis of an elliptical or linear object in a perpendicular relation to the tree limb or other relatively straight edge, in essence using the substrate as a fulcrum (Panger 1998, Boinski *et al.* 2001). The border between simple and specific first-order actions is fuzzy, and likely to be crossed as individuals become skilled at a particular activity.

Combining loose objects with each other is also a specific first-order action. In this case both elements must be controlled for the proper relation between them to be maintained. We have a few examples of this kind of activity from monkeys in nature: white-fronted capuchins in Peru sometimes bang two hard nuts against each other, sometimes opening one of the nuts in this way (Terborgh 1983), and wedge-capped capuchins bang two snails against each other occasionally, but not to the point of cracking either snail (Fragaszy, pers. obs.). Banging two nuts together occurs rather commonly in captivity when the nuts are abundant but again the action is rarely effective at opening either object. An example of a specific first-order action in a captive capuchin monkey comes from Fragaszy's lab, where one monkey habitually holds one piece of pelleted chow in his teeth and a second piece in both cupped palms, rotating his head back and forth to grind the pellets against one another. At the end of a grinding sequence, the monkey licks up the powdered chow he has produced.

The only examples we have from natural settings of capuchin monkeys producing second-order relations while manipulating objects came from recent reports of monkeys pounding open nuts placed on stones by using a second stone (inferred by Langguth and Alonso 1997; observed by Ottoni and Mannu 2001 in semi-free monkeys and in wild monkeys by Oxford 2003 and Fragaszy *et al.*, submitted); we discuss these further in Chapter 10. As we shall see in later chapters, captive and wild capuchins alike most frequently discover how to achieve first-order relations, where one object is related to a second (fixed) object or substrate, to solve problems. The monkeys far less frequently discover how to solve second-order problems, where they must manage two relations simultaneously. We elaborate on this subject in Chapters 8, 9 and 10.

Development of combinatorial activity

Young capuchins begin to perform combinatorial actions almost as soon as they can hold an object securely. They start at about 4 months of age, holding something at about shoulder height with elbow bent at about 120 degrees, and directing the hand in a rather gentle downward swipe to the substrate on which they are sitting, sliding the object next to their feet in a backward direction. As their actions become stronger and more vigorous, these initial tentative swipes become more like hits, and eventually they bang the object directly in front of their bodies with one or two hands using the characteristic rapid downward stroke seen so often in foraging contexts. At the same time, the action of rubbing, wherein the object is drawn backward and then forward against the substrate, differentiates from swiping.

Sometimes juveniles in captivity incorporate objects into play sequences, and much of their manual activity has a rather playful character anyway. For example, juveniles will sometimes swing 360 degrees around a bar while holding an object, banging it on each swing past the wall. In natural settings, identifying object manipulation as playful is problematic; manual activity outside of grooming is more often recognized as foraging (perhaps because more field researchers study foraging than play). However, very young capuchins in natural settings also appear to manipulate objects in a playful way. Of course, in all settings, young capuchins that are teething spend a good deal of time chewing nearly any surface of the appropriate size, such as twigs in nature, or metal, cardboard or plastic in captivity.

SKILLED MANUAL ACTIVITY IN NATURAL SETTINGS

Researchers at different sites across capuchins' distribution turn up interesting variations of capuchins' generative manipulation and the skill with which they pursue foraging. For example, Brown and Zunino (1990) describe an unusual specialization of tufted capuchins living at the margins of their geographical range, in northern Argentina (El Rey National Park, Misiones). At this site capuchins eat the leaf bases of bromeliads, plants so well protected by thorns and fibrous leaves that other mammals, it is thought, do not harvest them. The monkeys rely heavily on this tough food source, as fruit is almost absent from this habitat during the dry season (Brown *et al.* 1986).

Other examples of unusual foraging actions come from wedge-capped capuchins in Venezuela that bang open snails with very hard shells by pounding them on tree trunks, as in the vignette that opened this chapter (Fragaszy 1986, Fragaszy and Boinski 1995), and tufted capuchins in Raleighvallen, Suriname, that spend many minutes at a time banging open large husked fruits using large branches as anvils and may use distinctive actions with various species of fruits (Boinski *et al.* 2001, 2003). The tufted capuchins at Raleighvallen can spend more than 30 minutes processing a single large husked fruit (Boinski, pers. comm.). Chapman and Fedigan (1990) and Rose (1998) describe effective predatory behavior in *C. capucinus* in Santa Rosa, where capuchins routinely capture coati pups and other vertebrate prey that both flee and, when captured or cornered, defend themselves fiercely. Terborgh (1983) describes the skilled way in which *C. albifrons* search through fallen palm nuts on the forest floor, inspecting many to find the few that are damaged enough to open, but intact enough to be worth opening:

> When an animal found one that seemed to possess the desired qualities, it usually went up onto a low branch for the opening operation. There it would often bash the nut vigorously against the branch, then begin to bite it with its premolars. After a bite or two the nuts would be rotated to a slightly different position and bitten again. If it failed to yield, it would be rejected, and the whole selection process would begin anew. If the nut cracked, the endosperm would be laboriously excavated from the shell, using canines or fingernails to pick at the firm material." (Terborgh 1983, p. 83).

These foraging actions, so characteristic of capuchins, require strength, stamina and persistent and careful manual action. The generativity of manipulation that is so striking in captive capuchins, along with an ability to modify actions quickly and sensitively in accord with variable conditions, are both evident. Capuchins' foraging actions are skilled *sensu* N. Bernstein (1996) to a greater degree than in other New World monkeys, and perhaps to a greater degree than in any other monkeys.

INDIVIDUAL DIFFERENCES IN MANIPULATIVE ACTIVITY

On close inspection, many dimensions of variation can be identified in the organization of any action. For example, individual capuchin monkeys open pumpkin seeds to extract the kernels in very idiosyncratic ways. Some hold the seeds perpendicular to the incisors and then pull them apart with their fingers; some hold the seeds parallel to the incisors and essentially slit them open with their teeth, some use yet other techniques (Visalberghi, unpublished). However, at a cruder (functional) level, there are stronger commonalities in capuchins' actions than there are differences (Fragaszy and Boinski 1995). The manual actions we have mentioned in this chapter can be observed, if the opportunities are present, in capuchins in captivity and in nature, and for the most part in young and old alike. In captivity, the sexes interact similarly with objects, using the same actions and expressing the same degree of interest (Byrne and Suomi 1996, Fragaszy and Adams-Curtis 1991), although adult males may have better access to interesting objects or sites than adult females (Visalberghi 1988).

Despite sharing the same repertoire of actions, however, animals of different sizes (and correspondingly, sexes or ages) do experience substantial differences in the effectiveness of the same actions. Infants up to weaning age are patently ineffective at many of the actions that are staples of the adult foraging repertoire (Fragaszy and Boinski 1995, Mitchell 1989). Sex- and age-related differences in foraging reflect more than just size and strength, most likely, as animals of different age/sex classes choose to forage in rather different places (Fragaszy 1990a, Fragaszy and Boinski 1995, Terborgh 1983, Rose 1994a; see Chapter 2). Both Fragaszy (1990a) and Terborgh (1983) note that sex differences are more evident than age differences in the details of foraging behavior (e.g., substrates selected).

Do certain individuals have a greater likelihood of manipulating objects? We have a few studies bearing on this issue. First, Fragaszy and Adams-Curtis (unpublished data) found that individual tufted capuchin monkeys' frequency and variety of activity in routine conditions did not predict which individuals would discover how to solve a task by using an object in a particular way (i.e., to use it as a tool). In other words, spontaneous activity did not serve as an index of "manipulative IQ." However, rates of activity with familiar objects did predict juveniles' activity with novel objects and juveniles were more likely to discover how to use an object as a tool. Byrne and Suomi (1996) looked at this question in a similar way. They report a positive correlation between the frequency with which individuals manipulated familiar and novel objects. Only about a quarter of their 43 subjects ever combined one loose object with another loose object, reminding us once again that the distinctive characteristics of capuchins' manipulation, in everyday terms, are its persistence, vigor and likelihood of combining object and substrate (in pounding and rubbing) in first-order relations.

EXPLORING MANIPULATIVE SKILLS

Developing manipulative skill requires controlling specific spatial and movement relations of the body and of objects that the body can contact. In captivity, we can engineer tasks to probe capuchins' abilities to control the movement (trajectories, velocities and accelerations) of single objects; to place an object in relation to another object or a location on the substrate; and to organize the position or movement of two objects relative to each other and the body. In this section we review individuated movements of the digits, control of objects using haptic perception alone, using the hands to apply force, placing an object precisely, and moving a cursor on a computer screen using a joystick.

Precision handling and individuated movements of the fingers

Those who observe capuchin monkeys carefully at close quarters have always appreciated the degree of precision in moving objects that they can achieve. We now know that, whatever the form of the thumb joint in New World monkeys, capuchins do manage functional precision grips, although not of the classic tip-to-tip form seen in humans, and they can move the fingers in individuated, or independent, ways (see Chapter 5 for a discussion of the anatomical and neural correlates of opposition of the thumb to the other digits and of individuated finger movements). Costello and Fragaszy (1988) described several varieties of precision grips achieved by capuchins picking up small objects, including thumb to index (the most common form, seen in more than half the cases), thumb to more than one other finger, a scissor grip in which the object was held between two fingers (either in the same hand or between two hands), and wrapped in a single digit. The principal requirement to achieve a precision grip is the ability to produce sufficient lateral or oppositional pressure between two digits. However, lateral opposition does not provide the precision of control that pad opposition does, because in a lateral grip the object can only be rotated in a frontal plane or moved towards or away from the palm. With pad opposition, the fingers can move the object in many more directions. Thus, an individual with the capacity to use a pad opposition grip is likely to do so in favor of a lateral opposition grip when precision is important to the outcome of the actions.

Christel and Fragaszy (2000) found that when tufted capuchins reached for small pieces of food placed in shallow depressions in an otherwise flat surface, or pieces of food placed on top of slender sticks, they most often moved all the fingers in synchrony and with the same joints flexed. Less often, the monkeys stretched and adducted the thumb and flexed all the other digits slightly. The monkeys were able to grasp the food items in the fingertips using the fleshy tip (volar aspect) of the index finger to contact the top (dorsal aspect) of the thumb while the thumb flexed in its distal joint. Thus, the objects were pressed against the thumbnail at its inner (ulnar) side. This type of grasping pattern was associated with fast reaching times (average of 1.4 seconds, from initiation of movement to placing the food in the mouth) and the monkeys completed many action cycles. In other words, prehending an object between thumb and index finger seemed relatively easy for the monkeys.

Using the thumb and index finger in opposition can be effortful for capuchins when the situation demands more individuated finger movements, however. In Christel and Fragaszy's study, when the monkeys prehended objects from narrow grooves (just wide enough for one finger), they completed fewer reaching cycles and each cycle took longer than taking food from a flat board, the sticks or the wider depressions. When the food was in a narrow groove, they used an extended index finger to pull the food toward the dorsal surface of the thumb. In this situation they pre-shaped the hand slightly during the approach to the object to approximate the posture used to oppose the digits after contact with the food. Capuchins achieve far less pre-shaping of the hand during reaching than do humans, but we nevertheless consider it functionally significant because this posture enabled the index finger to enter the groove by itself in an efficient manner.

Visual attention during reaching and grasping

When we reach for an object that we see, we use vision to support precise reaching and grasping, and there is a predictable positive relationship between the duration of visual attention throughout the action and the demands for precision in the task (Ballard *et al.* 1997). The organization of visual attention during action plays an important organizing role in human behavior, according to Ballard *et al.* (1997).

Busch, Christel and Fragaszy (unpublished) examined how capuchin monkeys organize manual reaching and grasping actions in conjunction with visual attention, using the simple food-retrieval situation described above. Monkeys were tested singly in this project in a quiet room away from their companions, in a familiar and safe context. The task involved reaching for small bits of preferred food placed directly in front of them – an easy task, and one which capuchin monkeys are highly motivated to complete. Thus, the conditions were as optimal as we could provide for uninterrupted attention to the task at hand. Capuchin monkeys, like humans, looked at an object as they reached and while achieving a secure grasp. However, as soon as they achieved a secure grasp, capuchins (unlike humans) stopped looking at the object (now in the hand) and instead directed their attention to their surroundings, looking around the test room (although nothing was moving in the test room). They devoted more time to this activity than to looking at the object during reaching and prehension. Moreover, the longer they spent looking at the object during reaching and grasping, the longer they monitored their surroundings after prehending the object.

It appears that visual monitoring of their surroundings is a constant priority for capuchin monkeys under most circumstances, and that a demand for precise action can supersede this priority only for a brief period. This implies that activities requiring successive visually monitored actions to complete will be interrupted many times during completion. Capuchins, in other words, are far more distractible than the typical adult human, and not just because they may be more sensitive to extraneous sounds, etc. This aspect of their behavior no doubt has implications for their ability to learn the relations between actions and consequences, especially if they must visually monitor the consequences of an action immediately after completing it. We know of important social circumstances in which capuchins largely suspend monitoring their surroundings for longer periods (as during courtship; see Chapter 12). However, we expect that they visually monitor their surroundings very frequently in the course of normal activity.

USING THE HAND TO LOCATE AND REPOSITION OBJECTS

Welles (1972, 1976) found that capuchin monkeys (*C. apella* and *C. albifrons*), compared with other arboreal genera of monkeys (*Saimiri, Miopithecus talapoin* and *Callicebus*) and terrestrial monkey species (*Erythrocebus patas*, the patas monkey and *Macaca mulatta*, the rhesus monkey), excelled at prehending an object without vision, discriminating a food item from an inedible item using touch alone, pulling knobs of varied shapes and reaching around transparent barriers. In overall manual proficiency at the set of tasks that Welles set for these individuals, the capuchins ranked below chimpanzees and humans but they were more dexterous in virtually all tasks than other arboreal species. The one exception is illuminating: Squirrel monkeys (*Saimiri*) were better than capuchins (and all other species except humans) at snatching an object on a rotating platform. As squirrel monkeys routinely forage for mobile invertebrate prey that they flush from vegetation, this exception makes sense.

Capuchins will search an unseen irregular surface using their hands to locate objects, an activity called "haptic search." Humans use six highly stereotyped movement patterns in haptic search that have been labeled "Exploratory Procedures" (EPs) by Lederman and Klatzky (1987), who describe these movements and when they are used. Each EP is optimal (in terms of accuracy and speed) for the detection of a specific property (shape, texture, size and so forth). To find out whether capuchins, like humans, would use EPs to search an unseen surface, Lacreuse and Fragaszy (1997) presented 12 clay shapes to four adult humans seated at a table, and to 21 tufted capuchin monkeys in their home groups. The clay shapes were intended to elicit all the EPs described in humans. Each shape held 6–12 sunflower seeds distributed over the surface. The shapes were presented individually, behind an opaque screen, and the subjects (human and monkey) had to reach one hand through an aperture in the screen, feel the clay object and pick out the seeds. Planar geometric shapes were designed to elicit contour-following movements, and were baited specifically on the sides. Other geometric shapes (cube, pyramid, sphere) were baited on the top surfaces and were expected to elicit more lateral movements, such as repetitive and lateral rubbing motions. Three shapes provided concave places that were likely to elicit probing activity, and three other shapes were designed to elicit a variety of movements, combining lateral, contour following and probing. The experimenters re-baited the objects with sunflower seeds several times in a session and collected video recordings of each monkey and each human exploring the shapes. Analyses for both humans and monkeys concerned the number of EPs per reach, and for each category of EP, the percentage of occurrence (in relation to all EPs), the mean duration and the efficiency.

The capuchins exhibited all the same forms of exploratory actions with the fingers that humans did in completing this task (Lacreuse and Fragaszy 1997). As in humans, the performance of different EPs depended on the shape being explored. For example, for both species, *lateral movements* were more frequent for the shapes baited on the top surfaces while *probes* were restricted to shapes with concave places. In humans, shape of the object affected the percentage, duration and efficiency of the EPs. In contrast, shape affected only the proportional occurrence of EPs for the capuchins. These discrepancies may reflect a functional difference in the way capuchins and humans gather haptic information – this topic clearly requires further systematic inquiry. The important point for our discussion here is that capuchins use their hands to extract information about objects of interest to them, as do humans. Haptic sensitivity is a primary contributor to our own dexterity, guiding manual activity in a fundamental way. It is important to recognize that the same is true for this other highly dexterous animal, the capuchin monkey.

Differences between humans and capuchin monkeys in exploration were also evident, however. Specifically, capuchins performed far less *contour following* than humans. This difference reflects the most obvious contrast in the way the two species performed the task: humans were scanning the shapes in an exhaustive manner, to collect all the seeds; capuchins limited their explorations to a very small portion of the shapes, usually the area surrounding the point where the hand initially contacted the object. Thus the definition of the task seemed to vary for humans and for monkeys.

Humans use dynamic actions of the fingers to move objects that are already prehended (Elliott and Connolly 1984). Elliott and Connolly divide the human repertoire of intrinsic movements of the hand into two categories, according to whether the movements require simultaneous or sequential coordination of several digits. In sequential movements, the fingers reposition an object repeatedly during the action, as in rotating the lid of a jar. In simultaneous movements, two or more digits oppose one another, called simple synergies (as while pinching or squeezing), or the digits each do something different, called reciprocal synergies (as when rolling a pencil between thumb and forefinger).

We know that capuchins use simple synergies (pinch, squeeze), but we have not yet observed reciprocal synergies or sequential digit movements, nor have we hit upon a task that requires these movements naturally. Reciprocal synergies clearly pose greater demands for inter-digital coordination (a kind of relational problem among elements within the body) than simple synergies, and for this reason we should expect to find capuchins using simple synergies more than the other forms of intrinsic movements. However, establishing the occurrence (or absence) of the other two forms is an important future task in our quest to understand how manual abilities differ across species of primates.

USING THE HANDS TO APPLY FORCE

Capuchin monkeys use several different positions of the hand to grasp objects in the course of applying force with them. How an individual holds an object impacts directly the precision and forcefulness with which it can be handled, and therefore it is of interest to know how capuchins hold objects while applying force. Westergaard and Suomi (1994d, 1995b) considered this issue in capuchin monkeys practiced at throwing stones into small containers a short distance away (after the monkey threw a stone into a container, the experimenter returned it to the monkey, coated with peanut butter). With their arms extended through wire mesh, capuchins threw stones (15–60 g) up to 60 cm into a container (15 cm diameter) with 80–100% accuracy at 20 cm, to less than 50% accuracy (by the best subject) at 35 cm. They also threw stones into a bucket moving slowly in a pendular motion in front of their cage. Capuchins used both overhand and underhand movements to throw the stones, most often with the stone resting in the palm and all the fingers closed around it symmetrically, but also occasionally with the object pressed between the lateral edges of the thumb and index finger. The story is the same with grips used to hold stones to pound and to cut: power grips predominate when the task requires the application of relatively strong forces; precision grips are used occasionally when accuracy is important.

Humans grip objects with strong precision grips and power grips, translate and rotate them using the fingers alone in many different ways, and hold long objects with a grip (termed the "forearm squeeze") that allows the object to serve effectively as an extension of the arm (Wilson 1998). Chimpanzees differ from humans in all these elements (Marzke and Wullstein 1996) and it appears that capuchin monkeys, along with baboons, do not use the same grips as humans either. Thus we should not expect that capuchins will master all the same means of using objects, or that they will master them as easily or with the same endpoint of skill, as humans. A monkey trying to move an object in a particular way faces a bigger challenge than a human attempting the same movement.

FINE PLACEMENT OF AN OBJECT

Capuchins appear to rely preferentially on touch (haptic sense) rather than vision to achieve fine placement of an object. For example, monkeys quickly became skilled at passing objects through same-shaped cutouts (Staton 1995). The monkeys could slide a cylinder through a cut-out aperture more quickly than the human experimenter, and could manage passing symmetrical objects of many shapes. However, they had great difficulty with an irregularly shaped cross, where successful alignment required placing the object in a specific orientation relative to the cutout. The monkeys monitored the position of all the objects largely without vision, feeling rather than seeing when the object caught the edge of the opening. A capuchin monkey that skillfully opened a variety of catches closing the door of a box likewise preferentially used touch to align the handle with the surface of the catch, moving the catch along with one hand until it rested in a certain position against the other hand. However, when prevented from using both hands concurrently, the monkey solved the (now-familiar) task quite well, relying on vision to guide one hand's movement of the latch (Simons and Holtkotter 1986). Relying on touch rather than vision to guide manual action permits the individual to look around at its surroundings as it searches or modifies a substrate. Capuchins normally make frequent brief looks around at their surroundings, even during manual action, as discussed above.

Visually guided placement of an object can challenge capuchin monkeys. For example, when a monkey picks up a stick to sweep in a distant object, positioning the end of the stick skillfully at the right place relative to the target object requires much practice. Cummins-Sebree and Fragaszy (unpublished) observed that adult monkeys picked up a hoe and used it to pull in the food when it was first presented (with the food in the center of the tray, directly in front of them). We expected this outcome; capuchins readily discover and repeat this kind of relation among action, object and second object. However, we wanted the monkeys to be able to use the hoe to sweep in food from any position on the tray, not just the center. This, to our surprise, was initially a very difficult problem for the monkeys. The monkeys would sweep the hoe far beyond the food or short of the food, or they did not correctly modulate the force of the sweep and would fling the food off the platform. These errors diminished with practice and eventually the monkeys could maneuver the hoe to any position on the platform skillfully (i.e., with fluid, economical and accurate movements).

MOVING A CURSOR IN TWO-DIMENSIONAL SPACE: TILTING FOR SUCCESS

Capuchin monkeys can master using one object to control another in a two-dimensional context as well as in the normal three-dimensional world. The context in question is using a joystick to move a cursor on a computer monitor. Since Richardson *et al.* (1990) produced a self-paced interactive computerized training system for use with nonhuman primates, individuals of a number of primate species have mastered using a joystick to control the movements of a cursor on a computer screen. How do the monkeys master this highly artificial problem? Using a joystick involves a physical separation between the locus of action (the joystick handle) and the locus of effect (on the monitor). Thus, to control the cursor, the monkey must learn that moving the joystick produces an effect somewhere else (not at the end of the joystick), and that some events somewhere else have consequences (i.e., that moving the cursor into a "goal box" produces a food treat). Thereafter, it must learn the directional relationship between moving the joystick and moving the cursor, and finally it must learn how to move the joystick to produce the desired movement of the cursor.

Leighty and Fragaszy (2003) studied four capuchin monkeys mastering the joystick system using the self-paced task developed by Richardson *et al.* (1990) (see Figure 7.2). The task involves moving the cursor from a central position on the monitor to a visually distinctive area (the goal area) on one of the margins of the display. The location of the goal varies randomly over trials. When the cursor reaches the goal area, visual and auditory cues signal success, and the monkey receives a favored food treat. The size of the goal decreases systematically as the subject improves its efficiency at reaching it. At the outset, the goal is any point on all four sides of the monitor; at the final stage of the task (mastery), the goal area is barely larger than the cursor itself.

Two of the monkeys encountered the normal isomorphic relationship between joystick and cursor: pushing the joystick to the left moved the cursor to the left, and pushing the joystick to the right moved the cursor to the right. The other two monkeys encountered the reverse arrangement: pushing the joystick to the *right* moved the cursor to the *left*. The four monkeys mastered the task in an average of 2820 trials. As they improved at the task, all four capuchins increased the proportion of each trial in which they visually tracked the movement of the cursor on the monitor, confirming that one important feature they learned early on was that the display provided useful information. Fine control of the cursor's movement was the last feature of the task to be mastered.

The two monkeys mastering the reversed relationship between joystick and cursor learned the task as quickly as the monkeys mastering the normal relationship (see Figure 7.2). When, after having mastered the reversed relationship, these same two monkeys encountered the normal relationship between cursor and joystick they quickly mastered the new relationship. Apparently they learned in the first series the orderly directional relationship between joystick movement and cursor movement, and they learned to use the display to monitor whether the cursor was moving in the correct direction. When they encountered a new relationship, they had only one element of the set to relearn (the specific directional relationship between cursor and joystick movements).

Surprisingly, perhaps, to those who have not watched capuchins using joysticks, all four monkeys displayed a characteristic tilt of the whole torso to the side towards the goal as they moved the joystick

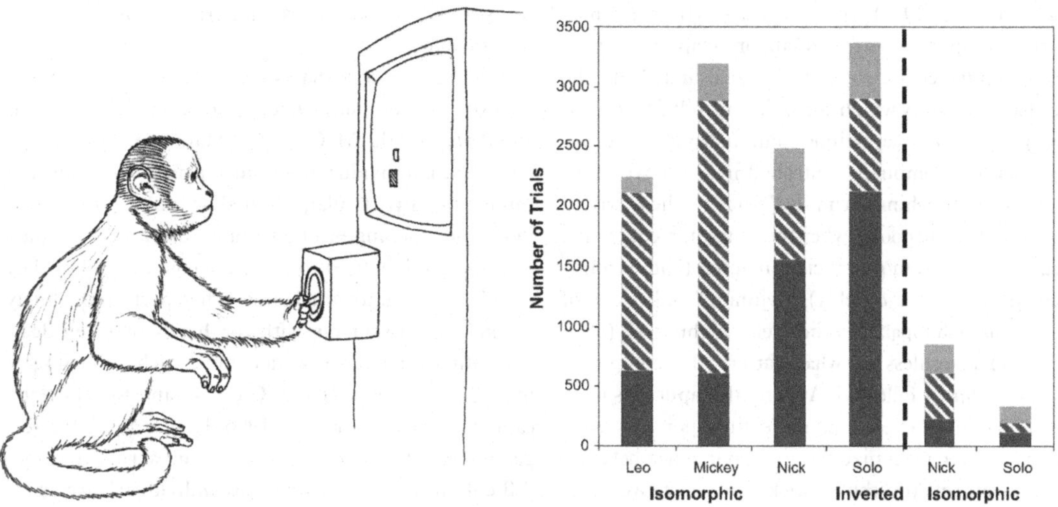

Figure 7.2. Left, a tufted capuchin monkey using a joystick to move a white cursor towards a dark target on a computer screen. This setting illustrates the task in the Mastery phase (see below) (drawing by Stephen Nash). Right, trials to criterion of joystick mastery. The acquisition task required subjects to direct a cursor to a highlighted area on the margin of the monitor. The highlighted region began as all four margins of the monitor and automatically reduced one margin at a time, or eventually, to a smaller portion of one margin (a titration), as soon as the subject reached criterion. The black region represents number of trials to criterion (successful completion of nine out of 10 consecutive trials from four margins, three margins, two margins, to one margin). The lined region represents number of trials to criterion for titrations from approximately 2/3 of a margin to an area slightly larger than the cursor itself (the smallest titration). The gray area (Mastery) depicts the number of trials to criterion for joystick acquisition which was defined as successful completion of 18 out of 20 consecutive trials at the smallest titration. In these 18 trials, the goal region was present on each margin at least two times and the subject could not bring the cursor in contact with the margin of the monitor outside of the highlighted goal region on more than one occasion per trial. (Redrawn from Leighty and Fragaszy, 2003). Nick and Solo first mastered an Inverted relationship, then an Isomorphic relationship.

laterally, as do other capuchins that are proficient at using a joystick. They did this only when they had mastered this task. Many interpretations of this phenomenon come to mind. Perhaps it has to do with bringing the face closer to the goal area, for example. Our preferred hypothesis at this time is that capuchins' tilting is similar in perceptual origins to the body tilting that humans exhibit when they are watching an object that they cannot touch directly in a context where they desire to control the object (as when watching a bowling ball headed for the gutter, a tennis ball headed for the boundary line, or a golf ball headed for the cup). Casual observations and inquiries indicate that humans also tilt when playing video games in roughly the same circumstances as the capuchins tilt: when moving an icon in a two-dimensional display using an interactive device (joystick, controller box, mouse, etc). To our knowledge, no other nonhuman species from the many that have used the same training system tilts while using the joystick. We are most curious to know if tilting is one more feature of motor innovation that is more evident in capuchins than in other nonhuman primates, or if other investigators, not knowing what to make of this phenomenon, have neglected to discuss it. At the moment we are "tilting" toward the first explanation.

Capuchin monkeys that had learned to use a joystick in the same fashion as the monkeys studied by Leighty and Fragaszy had more trouble transferring to a new version of the task where the horizontal axis or the vertical axis, or both, of the joystick controller reversed randomly across trials. In this situation, the actor must determine on each trial what rule governs cursor movement. Two capuchins out of three tested (like two chimpanzees out of three tested with the same paradigm) took longer to complete the trials where a reversal appeared, particularly during the first 40 trials in which they encountered reversals (Jorgensen 1994). In a rather different form of the joystick problem,

Jorgensen, *et al.* (1995) presented a task to the capuchins where a computer-controlled cursor "competed" with the capuchin's cursor to reach the goal first. Here the actor had to identify which cursor it controlled to move it appropriately. The capuchins could manage this problem when the computer-controlled image moved randomly (the capuchins "winning" 76% of these trials). However, they did poorly when the computer-controlled image moved in a more efficient manner (and encountered the goal more quickly), "winning" only 30% of these trials. Chimpanzees did well on this task (65% or better) regardless of what pattern the computer-controlled cursor followed. Apparently capuchins need more experience overall or more time per trial than chimpanzees to recognize the correspondence between their movements (on the joystick) and the movements of one out of two cursors.

MANUAL ASYMMETRY

Most humans use the right hand more frequently than the left hand for many activities. Asymmetries in hand use are matched by asymmetries in the organization of functions and morphology in the human brain, most notably the localization of language functions in the left hemisphere in most individuals (matching the left-hemisphere control of movement on the right side of the body). In addition to these well-known asymmetries, humans have many others (e.g, preferentially using one foot to kick or to lead onto a stair and one eye to look through a peephole). The puzzle of how such a broad spectrum of lateral asymmetries in structure, function and behavior evolved in humans has been the subject of much discussion among behavioral scientists. The human case is not unique, of course; asymmetry in behavior, neural organization and morphology is the normal condition for other species as well as for humans (Bradshaw and Rogers 1993). However, fascination with our own manual asymmetries, and with their possible link with language, has fed interest in manual asymmetries in nonhuman primates. Much effort has gone into determining if any other primate species has, as humans do, a population-level asymmetry in how they use their hands (Hopkins 1996, McGrew and Marchant 1993, 1997, Ward and Hopkins 1993, Wesley *et al.* 2002). Capuchins, with their anomalous manual skills and their status as users of tools, have been popular subjects in the search for manual asymmetries in preference or performance.

The evidence to date suggests that, with the possible exception of chimpanzees (Hopkins 1996, Hopkins and Pearson 2000, McGrew and Marchant 1997, Wesley *et al.* 2002), individual nonhuman primates overall, and capuchins in particular, often show no particular bias in simple reaching or other ubiquitous activities (such as which limb is used to initiate locomotion), but they readily exhibit individual biases when performing finely modulated movements with the hands or when dealing with strong postural demands, such as maintaining a bipedal stance (Hook-Costigan and Rogers 1996, capuchins: Anderson *et al.* 1996, Panger 1998, Westergaard *et al.* 1997, 1998a, b, c). As a general rule, the more difficult the task, the stronger are individual biases, suggesting that the bias can reflect practice with the task and developing skill with one hand (e.g., Fragaszy and Mitchell 1990, Limongelli *et al.* 1994, Spinozzi and Cacchiarelli 2000, Spinozzi and Truppa 1999, 2002). However, there is no consistent direction of asymmetry across the (small) populations. For example, Westergaard and Suomi (1993) report that eight of 14 capuchin monkeys that used a stone to crack open a nut preferentially used the left hand; four preferentially used the right hand, and the other two exhibited no bias. This mixed outcome (in terms of direction of bias across individuals) is typical of studies assessing manual preferences in nonhuman primates.

A group-wide bias in the same direction has been found in more than one group of capuchins only in tasks with a high demand for fine spatial positioning and repositioning of the fingers and incorporating a strong haptic component. The tasks in question involved locating and prehending seeds placed in crevices of irregularly shaped objects, discriminating seeds from similarly shaped pieces of tinfoil, or searching for grapes buried in wood shavings or under water and discriminating them from stones of similar size and shape. A left-hand bias was evident in three groups of capuchin monkeys in these conditions (64 monkeys, three studies combined; Lacreuse and Fragaszy 1999, Parr *et al.* 1997, Spinozzi and Cacchiarelli 2000). Forty-two of the monkeys preferred left or right hands equally often when they merely picked up small pieces of food from a tray whether they could see their hands or not, suggesting that reliance on touch alone is not sufficient to induce consistent use of the left hand (Lacreuse and Fragaszy 1999, Spinozzi and Cacchiarelli 2000).

Perhaps capuchins have a right-hemisphere superiority for processing spatial information during haptic exploration, as has been suggested for some other species of monkeys (e.g, spider monkeys, *Ateles*; Laska 1996). The monkeys in Spinozzi and Cacchiarelli's study also were more accurate at discriminating seeds from tinfoil on their first choice when they used their left hand rather than the right. Thus we are beginning to see some patterns in what produces a consistent bias for the use of one hand, although we have a long way to go before we can specify precisely what aspect(s) of the tasks contribute to the use of the left hand.

Although we can show that the monkeys exhibit a preference for the use of one hand in some circumstances, rarely is any monkey as faithful in the use of one hand for any single task of daily living as humans usually are. We are very far at present from understanding the origins or functional significance of motor asymmetries in nonhuman primates, or whether the asymmetries that we have documented are similar in any substantive way to the asymmetries we humans exhibit.

USING OTHER PARTS OF THE BODY IN INSTRUMENTAL ACTION

Capuchins use their mouths extensively in their actions on substrates and objects during foraging. This happens in the most straightforward way when biting and chewing, but they also use their mouths and tongues in less obvious ways. The tongue, for example, seems particularly long and extendable. In captivity it is common to see a *very* long pink tongue delicately reaching to explore or retrieve something beyond where fingers can go, as into a pipe with jelly on the inside surface. The teeth are used not just to crush and tear, but also to puncture and to hold fast when some other area is torn. The teeth and head together can be used to hook and pry, with joint action of the head and neck (upward) and the arms (downward). The feet sometimes hold items that cannot be held conveniently in the hands while seated or during locomotion, although they are almost never used to prehend a loose object or to handle objects without the hands.

Capuchins can use the tail in ways that are quite unrelated to the primary locomotor/postural uses of this appendage described in Chapter 5. For example, adult male white-faced capuchins will hold a branch in the tail during vigorous agonistic displays when the branch has broken off during the course of a display (Fedigan, pers. obs.) In captivity, capuchins can, with much practice, use the tail in precise ways. Tails can be used to sweep in objects that are out of reach on a flat surface, to move objects along from beneath a floor grid, and even to pick up objects. For example, a capuchin monkey observed by E. Visalberghi at the Primate Center of the University of Brasilia routinely used its tail to pick up grapes placed out of reach of its hands. The caretakers encouraged the monkey to practice this behavior. Normally the tail curls tightly around a solid, more or less rigid object (a branch or vine), but in this case it delicately curled around an object that would slide away if not surrounded. Rather simpler but functional uses of the tail include a report by Phillips (1998) that wild juvenile *C. albifrons* at Bush Bush Wildlife Reserve, Trinidad, can use the tail as well as hands and feet in obtaining water from a tree cavity. Tails, like feet, can be used to hold extra objects or to keep objects together if there are too many to be held in the hands, or even to carry them if the hands are full. As will be discussed in Chapter 8, using the tail and feet as aids in keeping objects together gives capuchins an opportunity to master sequential actions with loose objects that would be more difficult if the objects were scattered around.

SUMMARY

The motor skills that set capuchins apart and are critical elements in their distinctive life style all involve acting on the material world to produce a change in objects, surfaces or the relations between objects. Most actions involve direct contact of the body with a surface or object. A smaller proportion of actions involve first-order relations, where an object is brought into contact with a surface. An even smaller proportion of actions involve two relations, bringing two objects into a particular relation with each other (a second-order problem). As we shall see in Chapter 10, these latter two kinds of actions can lead to the spontaneous discovery of how to use an object as a tool. The monkeys use all appendages of the body in this effort: hands alone, but also tail, feet and mouth in concert with the hands. The ability to move the index finger somewhat independently opens up opportunities for capuchins to extract foods from a matrix, but we do not yet know if capuchins perform the intrinsic movements of the fingers that are so important to human manipulation.

Capuchins do not display marked asymmetry in manual activity, although there is a consensus at present that capuchins exhibit a measurable, albeit modest bias to use the left hand when the task involves precise control of the hand in space.

Returning to the themes we set out at the beginning of this chapter, let us consider whether capuchins can be accorded the accolade "dexterous" for any domain of skilled activity. We believe that wild capuchins earn it for those diverse direct actions on substrates and first-order relational actions combining an object with a substrate that monkeys in natural circumstances use on a daily basis in extractive foraging. Extractive foraging, as capuchins practice it everywhere in their natural range, involves the generative use of a broad repertoire of vigorous actions and finely controlled manual actions, combining the use of hands and teeth, and diverse grip patterns. Capuchins can earn the accolade of "dexterous" in relational tasks in captivity where they can encounter them often enough and in diverse enough circumstances to master many variations (as some monkeys have with joystick tasks and others have with some tasks involving using objects to push or probe in a container – see Chapter 10). We do not know if they can become skilled in second-order tasks where they must fully control two loose objects. Their difficulty in mastering second-order relations seems to limit capuchins' use of objects as tools.

Part III
Behavioral psychology

8 • Perceiving the world. Memory and perception

The player sits looking attentively at a computer screen, his hand resting comfortably on a joystick placed at the lower edge of the display. What he sees on the screen might discourage a less masterful player: a tangle of narrow alleys, interconnecting with one another at right angles at several points. He has never seen this particular pattern of alleys before, although he has solved other mazes like this, each new to him. The player's task? He must use his joystick to move a small cursor, a cross-shaped light area standing out from a dark background, through the alleys to reach the goal. At a T intersection, should he go toward or away from the goal? When the path he is travelling passes an intersection, should he take it, or should he keep going? This player has a general strategy: if a new path presents itself as he travels along, he usually takes it. But he manages to attend sufficiently either to the location of the goal or the location of the next choice, and perhaps every now and then to both things at once, to solve many mazes without making any wrong choices at all. This time, he makes an error on the third choice, takes the cursor half way to the end of the alley before he notices the error, then retraces his path to the choice point and travels the other way. Moving the cursor toward the goal is easier to manage than looking ahead to the next choice!

Yes, this vignette is about a capuchin monkey, a skilled player of this video game (see Figure 9.2). The description could be about a human, a chimpanzee or rhesus monkey – all of these species have also mastered moving a cursor through novel mazes on a computer display. How does an individual of any species solve problems that require anticipating the outcomes of choices still in the future? This is a question about perception, memory, and eventually, about prediction and planning. In this chapter we consider the achievements of memory and perception of capuchins as we now understand them.

MEMORY

Psychologists recognize memory as an essential part of the cognitive tool kit. As Tulving (1995, p. 285) put it, "Memory is a trick that evolution has invented to allow creatures to compress physical time. Owners of biological memory systems are capable of behaving more appropriately at a later time because of their experiences at an earlier time, a feat not possible for organisms without memory." How well can capuchin monkeys remember events that they witness and contingencies that they experience? Anecdotal evidence suggests that they can recognize individual people and other monkeys with whom they have interacted after years of separation. What about less emotionally significant objects or events? What about events encountered just once? What about remembering which item of two, both presented very briefly, is correct? These kinds of situations (and more) have been the subject of experimental laboratory research with capuchin monkeys, as with many other species (for a general review see Shettleworth 1998).

Memory for single events and over long delays

D'Amato and Buckiewicz (1980) studied capuchin monkeys' memory of where they had been prior to finding food using a paradigm popular to study memory in rats. They placed four tufted capuchin monkeys for five minutes, one at a time, in a T-shaped enclosure (a "T-maze"), one arm of which was painted black and the other arm painted in black and white stripes. The next day, they placed each monkey in the same maze, this time allowing it to stay only in the arm that it spent the less time in the day before. Then they moved the monkey to a waiting area for 30 minutes. Finally, they returned the monkey to the start point of the maze, where the monkey found ten raisins. After eating the raisins, the monkey was removed once again for a few minutes, and then returned to the start box. The variable of interest was where the monkey would go next – to the arm preferred on Day 1, or the arm inhabited on Day 2, 30 min before finding raisins in the start area. Two other groups of four monkeys each experienced variations of this procedure as controls for the effects of exposure to the maze and for exposure to food. All 12 monkeys were tested three times. The experimental monkeys switched their preference to the side they inhabited 30 minutes prior to finding the raisins on Day 2. Moreover, 4 months later, with no other experience in this maze, they still strongly preferred this side, spending nearly 80% of their time there. The two control groups of monkeys did not show this preference. In subsequent studies the monkeys learned in as few as 20 trials to choose a side of the maze by its paint markings to acquire food, even when delays of 30 minutes intervened between making a choice and learning the outcome of their choice (food or no food; D'Amato *et al.* 1981a, b). Although these studies used quite arbitrary arrangements, nonetheless the capacities that the monkeys displayed to remember significant features of their surroundings can be understood in relation to their abilities to make decisions and to evaluate alternatives on the basis of prior experience.

As suggested by D'Amato's studies with captive capuchins, wild capuchin monkeys at La Suerte Biological Reserve in Costa Rica can learn the "place" where food is located in one experience (Garber and Paciulli 1997). In Garber and Paciulli's study the monkeys encountered six platforms in one cluster and seven platforms in a second cluster about 25 m apart. The platforms in each cluster were 1.5 m above the ground and 2.8–3.2 m apart from each other, simulating food patches the monkeys might encounter naturally. Twice daily the experimenters baited five of the 13 platforms with bananas and put plastic bananas on the other platforms. The real bananas were in the same location for five days in a row, moved to a second set of locations for another five days, and then a third set of five locations. In the third series, the bananas were hidden. Next, the bananas were hidden in a fourth set of novel locations, and a large yellow block indicated the location of the real bananas. Following a shift in the bananas' location, the capuchins searched to find the new location, but a single experience finding the banana at the new location was sufficient for them to shift their initial choice of platform on the next visit. Some capuchins also learned within six trials to use the yellow block, an arbitrary "landmark" cue, to find hidden food. In a similar experiment, tamarin monkeys did not learn to use arbitrary landmarks to find hidden food (Garber and Dolins 1997). Capuchins, who forage regularly on hidden foods, seem to be more attentive to indirect indicators of food than are tamarins.

Short-term memory

Memory has been conceptualized as short-term (lasting a few seconds) or long-term (lasting more than a few seconds) (Baddely 1995, Rolls 2000). Short-term memory (also called "temporary" or "working memory") is accessible to decision processes. For example, it allows an individual to remember that a relevant event just happened, and to act accordingly. A good deal of experimental work with nonhuman animals has investigated memory in relation to this conception (Shettleworth 1998). Capuchin monkeys can remember a visual stimulus (two-dimensional outlines) they have just seen and to select one that matches it (or if the rule requires, the one that does not match it) after delays of up to 10 minutes (Colombo and D'Amato 1986, Tavares and Tomaz 2002). Although the monkeys make proportionally fewer correct choices as the delays increase, at 32 seconds delay they still made better than 90% correct choices (compared with nearly 100% correct at 0.5 second delay) in Colombo and D'Amato's (1986) study, and 67% at 10 minutes in Tavares and Tomaz's (2002) study, in which unique (novel) stimuli were presented on every trial. The monkeys' ability to match stimuli in this way was stable over the day, although reaction times on the task were faster earlier in the day than in later afternoon, as in humans performing the same tasks (Tavares 2002).

The monkeys were nearly as good at remembering an auditory stimulus and picking a match to it after seconds of delay. With auditory cues, they selected the correct match nearly 100% of the time at 0.5 second delay, and nearly 80% of the time at 32 seconds. This is a significant decline, and a greater decline than seen with visual stimuli, but still a very respectable level of retention. It is particularly respectable because these monkeys had previously experienced much difficulty in learning to discriminate between two auditory stimuli (D'Amato and Colombo 1985). They appeared not to remember patterned auditory stimuli (tunes), but rather to rely on local features (such as the last note or the first note) to discriminate such sounds (D'Amato and Salmon 1984). Because monkeys learned auditory discriminations so much more slowly than visual discriminations, it had been thought that perhaps they would also forget auditory stimuli more quickly. However, this was not so: Capuchin monkeys can remember cues in both modalities equally well although they cannot discriminate cues in the two modalities equally well. Perhaps these results reflect better discrimination of cues that last longer; a stationary cue persists as long as the animal looks at the test array, whereas sounds typically are available only for moments at a time.

Capuchin monkeys can recall objects or two-dimensional images that they have seen many minutes earlier and select a matching object, and they can remember where they were in relation to an important event (even if the event happens afterward, and happens somewhere else). Memory capacities like these are rather widely shared among nonhuman primates and other nonhuman animals (Shettleworth 1998, Tomasello and Call 1997).[1]

Spatial memory

Travelling in larger spaces

Capuchin monkeys, like most primates, move throughout their home range on a daily basis. Field observers invariably have the impression that the animals they are following are moving in an orderly, goal-directed manner from one location to another and they can predict quite accurately where the animals will stop next when they set off in a particular direction. Careful comparison of the direction of travel in relation to various models indicate the human observer's impression of directed travel is correct: capuchin monkeys travel in a forward direction towards food sources and do not retrace their steps on the same day (Robinson 1986).

Experimental evidence that capuchin monkeys remember the location of food resources in their home range comes from a series of studies by Janson and his colleagues with a group of tufted capuchin monkeys living in Iguazú National Park, Argentina (Janson 1998b, 2000, Janson and di Bitetti 1997). Janson (1998b) set out feeding platforms at 17 locations throughout the monkeys' home range and studied the monkeys' travel paths in relation to the platforms' locations (see Figure 8.1). The platforms (0.8 × 0.8 m and 1.5 m or more above the ground) were at least 180 m apart. For 50 consecutive days, when the monkeys traveled toward a platform, the research team provisioned that platform with 10–80 tangerines (a fruit that grows wild in their home range). When the provisioned platforms first appeared in their home range, the monkeys detected them at least 50% of the time only if they came within 41 m of a platform (Janson and Di Bitetti 1997). However, after 2 weeks the monkeys traveled in direct routes between platforms.

Locating objects in near space

Consistent with their ability to travel efficiently among multiple food sites in natural spaces, capuchin monkeys can learn to search smaller spaces systematically. De Lillo *et al.* (1997) suspended 16 small cups containing peanuts from the ceiling of a capuchin monkey's home cage in four clusters of four cups, simulating a clumped resource. After the monkey took the food out of a cup, the cup reverted to its original position and the monkey could not see whether it was empty. The monkeys could search the cups in any pattern. After some experience with this arrangement, the monkeys emptied the cups in each clump before moving on to the next, and did not return to clumps they had already emptied. Thus they learned to organize a systematic search in a defined space with clumped resources. When nine cups were distributed the monkeys searched more effectively when the cups were arranged as a line or a circle, rather than as a 3 × 3 matrix or a cross (De Lillo *et al.* 1998). With the linear arrangment, eventually some monkeys adopted the strategy of always starting from the same location and traveling systematically the length of the array, an effective way of organizing search in a predictable space

Potì (2000) has extended this line of work by investigating what forms of spatial reference capuchins are using in search situations. Researchers recognize two basic frames of spatial reference that humans use to locate objects in space: an egocentric frame (e.g., in front, behind, above the perceiver), and an allocentric or external frame (e.g., object next to a wall, in the tree, under a bush). A side bias in action reflects a powerful reliance on an egocentric frame of reference; a cue-bias in action reflects a powerful reliance on an allocentric frame of reference. To determine which frame(s) of reference capuchins used, Potì presented monkeys with two identical boxes placed on a tray (see Figure 8.2). A black

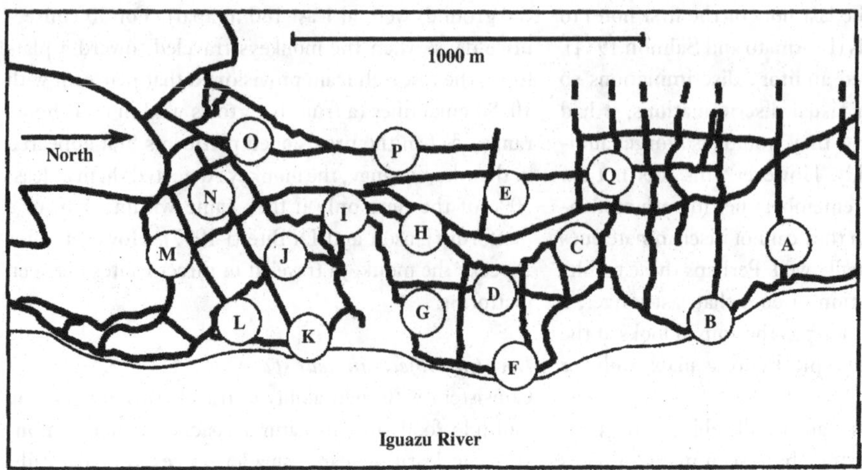

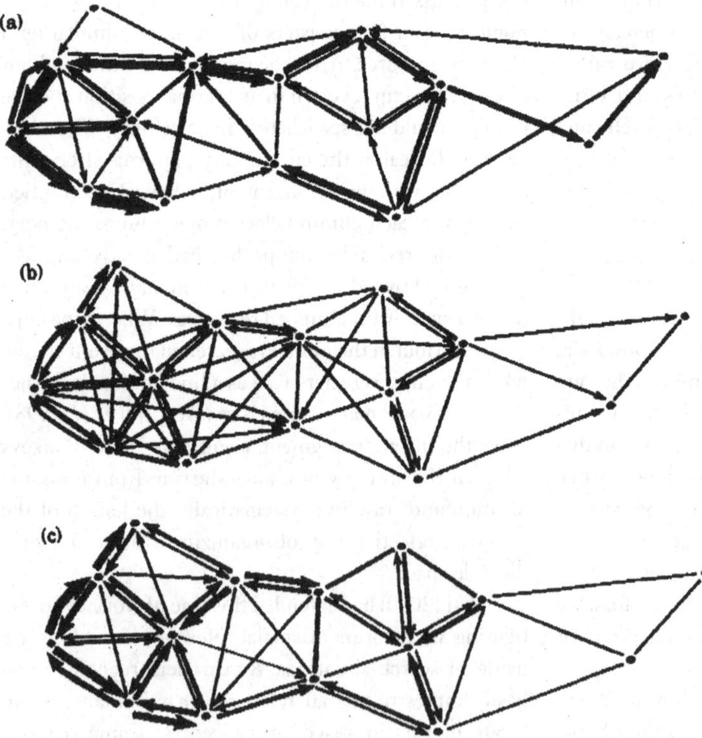

Figure 8.1. Top, the distribution of 17 platforms baited with abundant fruit in the home range of a group of tufted capuchins in Iguazú National Park, Argentina used by Janson (1998) to study spatial memory in these monkeys. Bottom, the travel paths of the group of tufted capuchins in their home range on three typical days (a, b, c) when the 17 platforms were baited with fruit. Note that the monkeys frequently traveled directly from one platform to the next no matter where they started from, indicating that they remembered the locations of the platforms in relation to their own position. (Reproduced from Janson 1998b with permission of Academic Press).

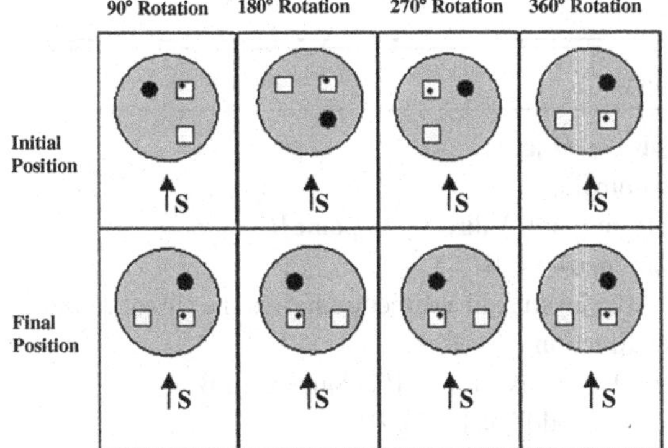

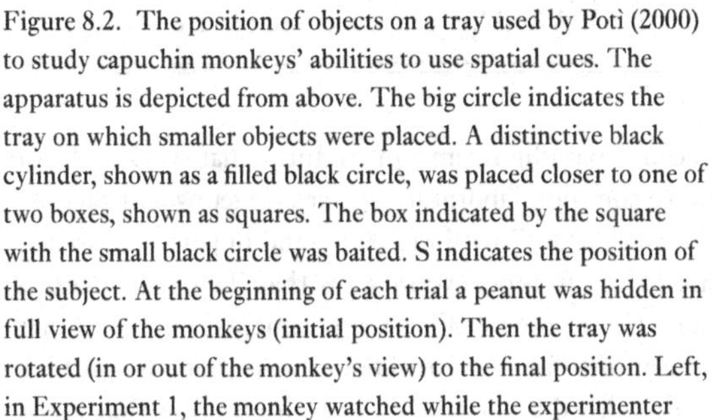

Figure 8.2. The position of objects on a tray used by Potì (2000) to study capuchin monkeys' abilities to use spatial cues. The apparatus is depicted from above. The big circle indicates the tray on which smaller objects were placed. A distinctive black cylinder, shown as a filled black circle, was placed closer to one of two boxes, shown as squares. The box indicated by the square with the small black circle was baited. S indicates the position of the subject. At the beginning of each trial a peanut was hidden in full view of the monkeys (initial position). Then the tray was rotated (in or out of the monkey's view) to the final position. Left, in Experiment 1, the monkey watched while the experimenter baited the container closer to the cylinder with a small piece of food; then a panel blocked the monkey's view while the tray rotated 90, 180, 270, or 360 degrees. Right, in Experiment 2, the experimenter randomly baited the container closer to, or farther from, the cylinder, and then rotated the tray either behind a panel (as in the first experiment) or in full view of the monkey. The monkeys reliably chose the baited container in the first experiment, but had difficulty doing so in the second experiment when they could not see the tray rotate. (Redrawn from Potì 2000)

cylinder, visually distinctively different from the boxes, was closer to one box than the other (and thus served as a landmark, or allocentric cue of the identity of the boxes). After the monkeys saw the experimenter hide a food treat under one of the two boxes on the circular tray, a panel blocked their view, and the experimenter rotated the tray. In this way the boxes moved in relation to the monkey's body axes and in relation to external cues (i.e., the room where she tested the monkeys). However, the boxes remained in the same relation to the black cylinder on the tray. Then the experimenter raised the panel and the monkey could choose to lift one box (and retrieve the food, if it made the correct choice).

The monkeys initially relied upon an egocentric frame of reference to choose a box (i.e., a side bias), but they eventually learned to use the black cylinder as a landmark cue. Thus, Potì replicated Garber and Paciulli's (1997) findings that capuchins could use landmarks to find hidden food. Potì then conducted a second experiment with new subjects, to evaluate whether the monkeys were learning a specific association between the landmark and the baited box, or learning a more general relational rule. In the second experiment, the baited box alternated randomly between the closer and the farther position in relation to the cylinder landmark (see Figure 8.2). On one-third of the trials they could see the platform as it turned; on the other two-thirds they could not. The monkeys preferentially selected the correct box when they could see the tray rotate, and when the objects rotated out of view a full 360 degrees (and thus reappeared at the original locations). However, they did not preferentially choose the correct box if the tray rotated less than 360 degrees while they could not see it. Usually in these trials they selected a box on the basis of an egocentric frame of reference (i.e., on their left side, or on their right side). Thus they could remember a specific spatial association (Experiment 1) and track the target visually (Experiment 2) but they could not remember a spatial relation following an unseen rotation. These results suggest that capuchins do not ordinarily use remembered relations among external objects or landmarks to guide their actions. In larger spaces also, they may know where they are in relation to other places, and thus travel efficiently from place to place, without forming a detached "cognitive map" of the environment. The same may be true of many animals that navigate efficiently, including humans (Gibson 2001).

Table 8.1. *Relations that capuchins can recognize and remember.*

Relational category	Definition	Examples
Zero order	X is present	Recognize an object
First order	X in relation to Y	Spatial domain:
		Travel from point A directly to point B
		A is a landmark for B
		Pushing the joystick in a direction moves the cursor in the same direction
		Identity: A belongs with A; B belongs with B
		Oddity: A is "odd" of {A, B, B}
Second order	X in relation to Y and Y in relation to Z	Items in a set belong in a fixed order
		Elements in Set A and Set B are ordered by size within set

Dubois *et al.* (2000, 2001) studied a different kind of spatial reference in capuchins: how monkeys respond to an object encountered in different places. Captive wedge-capped capuchins habitually performed activities differentially in specific locations in their home cage: they manipulated objects more often in one site than in other sites, they looked outside the cage more often in a different place, and so on. When the experimenters placed several apparatus containing food in the monkeys' home cage, the monkeys worked more persistently on the apparatus in the area where they habitually manipulated objects than in the area where they habitually visually monitored the area outside the cage (Dubois *et al.* 2001). These findings remind us that space is not a uniform quantity: it is woven unevenly into the monkeys' everyday activities and rhythms.

LEARNING ORDERS, PATTERNS, CATEGORIES, AND RELATIONAL RULES

> During 17 years of research with *Cebus* monkeys we have alternated between marvelling at their cognitive accomplishment and being plunged to the depths of despair over their inability or reluctance to learn a variety of apparently simple tasks.
>
> (D'Amato and Salmon 1984, p. 164).

As D'Amato and Salmon's comment poignantly suggests, the researcher trying to understand how and what a capuchin learns about relations and patterns faces a formidable challenge. The challenge arises in part because paradigms and procedures that work well for more common laboratory species (pigeons and rats) do not work as well with monkeys, and in part because the monkeys often behave in ways that the experimenters have not foreseen. These experiments are oftentimes educational for the researcher in ways that have nothing to do with the particular research question that motivated the experiment. One lesson we have learned is that how well capuchin monkeys can solve a problem involving patterns or categories depends strongly on the number of relations that the actor must recognize or use. Adult humans handle relational problems very easily, but monkeys, like very young human children, apparently do not (cf. Klahr 1994).

In the next section we examine how well capuchin monkeys detect relations among events or conditions (e.g., that B is dependent in some way on a preceding event or condition A). For example, picking out an object that is unlike two other objects is a relational problem. No single property of an object *except* its relation to the others determines if it is the odd one. Keeping track of relations allows the actor far greater precision in perceiving, predicting and controlling events than does remembering specific associations (for example, that the food is hidden under the block nearest the landmark block, in Poti's experiment 1). Keeping track of two relations concurrently affords greater precision than keeping track of just one, or of two sequentially. Throughout the remainder of this chapter, we classify problems as zero order, first order and second order, in accordance with how many relations must be remembered or managed at one time (see Table 8.1).

Learning orders and patterns

Capuchins have participated in many studies about learning to choose one object, visual display or tone from among a set of two or more according to an arbitrary rule assigned by the experimenter. The monkey must discover the relation between its actions and an outcome (usually, obtaining or not obtaining a piece of food). We know that capuchin monkeys, like squirrel monkeys and other primates, can learn to discriminate among objects on the basis of an abstract rule such as "choose the odd object of the set" (Thomas and Boyd 1973). To do this they must perceive one relation: whether an object is like others (or not).

Perhaps of greater interest is capuchin monkeys' ability to use second-order relations. For example, they can learn to choose pictures in a predetermined arbitrary order. D'Amato and Salmon (1984) showed that the monkeys remember the pictures in the order they appear in a five-item series (ABCDE) and can place a picture into its correct position in the series at any point (see also D'Amato and Colombo 1988b). Capuchins act more quickly to order pairs that are farther apart in the series, and more slowly to order pairs that are closer together in the series (a pattern called the "symbolic distance effect"). This pattern is shared with humans, suggesting that the monkeys organize the series in memory in a similar way to humans (Colombo and Frost 2001, D'Amato and Colombo 1990). Furthermore, capuchins can put a "wild card" (a neutral placeholder) into a series to preserve the order (for example, AXCDE, or ABCXE) (D'Amato and Colombo 1989; see Figure 8.1). Capuchins, in other words, work in accord with the concept of transitivity (D'Amato 1991). Transitivity is a second-order relation because the relation of the parts to the whole determines the relation of each part to all other parts; additionally, each element has a set relation to the ones immediately adjacent to it.

Capuchins failed to demonstrate a similar ability to recognize the relation among auditory elements that they heard as a pattern of increasing or decreasing tones; apparently, monkeys do not hear melodies (D'Amato 1988, D'Amato and Colombo 1988a). Recognizing a pattern in a temporal sequence of sounds may be a peculiarly human proclivity, or one shared with song birds and other species that use patterned sounds in species-typical communication, such as gibbons (*Hylobates* spp.) and titi monkeys (*Callicebus* spp.), in which mated pairs sing duets. In any case, as D'Amato (1988) put it, capuchin monkeys cannot hum a tune because they do not hear one!

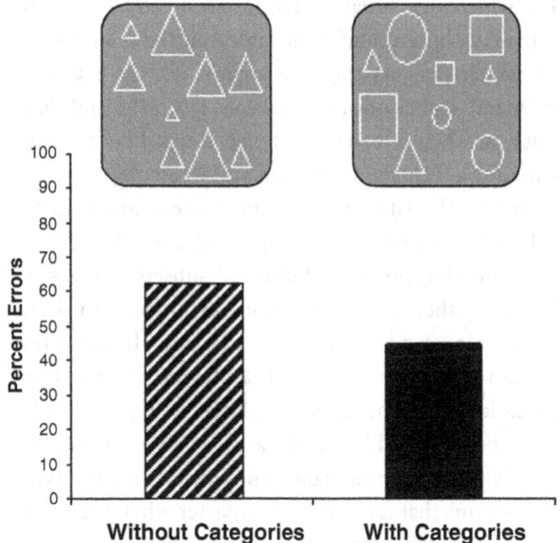

Figure 8.3. Capuchin monkeys using a touch-screen display can sort nine items by size (upper left). They can also sort nine items within shape by size (upper right), for example, touching in order the large, medium and small circles, then the large, medium and small squares, etc. As shown in the graph below, capuchins make fewer errors when sorting within categories by size than when sorting without categories. Redrawn from McGonigle and Chalmers (2001).

Working with two relations

More recently, using touch-screen interactive displays, capuchins in McGonigle's laboratory have shown impressive abilities to work with relationally ordered sets with a large number of elements, selecting up to nine images in a fixed order (e.g., by ascending size) (McGonigle and Chalmers 2001, McGonigle *et al.* 2003). They can do this even though the items are positioned randomly on the computer monitor from one trial to the next so that spatial memory does not contribute. Even more impressively, capuchin monkeys can sort, for example, three squares in ascending order of size, then three circles in ascending order by size, then three triangles by size, when all the squares, circles and triangles are presented on the screen simultaneously (see Figure 8.3). The capuchins' ability to do this indicates that they can effectively organize several elements in working memory (i.e., which objects have been chosen already; which still remain unchosen), and at the same time they can remember which category they are sorting. This is a second-order activity because two relations, size order and shape order, are being managed at one time. The monkeys gained an advantage in performance when they could use a categorical relation (shape) and a size relation, just as do humans (see Figure 8.3). McGonigle *et al.* (2003) suggest that conceptually "chunking" subsets of the same shape together reduces the number of individual elements that must be managed at one time. It seems that monkeys, like humans, adopt strategies that make the job easier (McGonigle and Chalmers 2001; see also De Lillo et al. 1997). This result, as we shall see in the next two chapters, has important implications for the types of problems that capuchins can master when they must control or produce more than one relation at one time.

Learning to use relations among stimuli

Acting in accord with a conceptual rule affords a powerful way to deal with new situations. Rather than learning the particular rules that apply to each new situation, one can apply perceptual strategies already honed by previous experience. Comparative psychologists have been interested for a long time in how well animals of different species can apply a previously acquired relational strategy toward a novel problem. A classic way to compare species' abilities to adopt rule-guided behavior is to study an individual's performance on a series of similar but not identical problems. The aim of such studies is to determine if, after substantial experience with similar problems, correct performance on new problems is evident from the outset or is acquired more quickly than in initial problems. If either outcome occurs, the individual is said to have exhibited "transfer," or to have mastered the relational rule(s) embodied in the task.[2] Capuchin monkeys show transfer, in this sense, in various identity problems (e.g., choose the object that matches the sample; do not choose the one that is different from the sample; D'Amato and Salmon 1984a, D'Amato *et al.* 1985, 1986) and in oddity problems (Thomas and Boyd 1973). In this respect they are similar to other species of primates that have been tested so far in these kinds of first-order relational tasks (Adams-Curtis 1990). They also accurately identified the stimulus that did not match a novel sample stimulus (the "non-match to sample" procedure, or NMTS) after experience with similar tasks, providing evidence of generalized identity matching (Barros *et al.* 2002). This faculty is not well documented in nonhuman primates, although it is thought to be fundamental for many aspects of behavior, such as complex social relationships (Schusterman *et al.* 2003).

To compare conceptual abilities across primate species in a more discriminating manner than afforded by measuring performance in first-order relational tasks, Rumbaugh and Pate (1984) proposed assessing "mediational learning" with testing procedures that draw on the logic of relational learning tasks but that add more relational demands (as described below). Rumbaugh and Pate (1984) summarize a body of data on several species of primates using the mediational learning paradigm showing that monkeys did more poorly on these tasks than did apes. Capuchin monkeys, however, were not represented in their data set.

De Lillo and Visalberghi (1994) filled this gap by testing four capuchin monkeys with the "mediational learning" paradigm used by Rumbaugh and Pate (1984). The monkey first learned to choose the correct object in a pair (e.g., A+ [A is correct] paired with B– [B is incorrect]). Next, the monkey learned to select the correct object of the same pair after the assignment of "correct" changed from A to B (A– paired with B+). This switch is called a reversal. Each subject learned to reverse from A+ to B+, and back to A+ again with 10 pairs of objects until they chose correctly on 67% of trials in each phase (where a phase is defined by which object is correct: A+, B+, A+) with the 10 pairs. Then they repeated the process with 10 new pairs to a criterion of 84% correct in each phase for each pair. The monkeys did reasonably well in the reversal tasks, reaching the criterional number of correct trials in about 17 trials per pair per phase for the lower criterion (67% correct), and 19 trials per pair for the higher criterion (87% correct).

After completing this part of the experiment, the monkeys received a new task: they learned to choose A+ in the same way as before. Then they encountered one of three conditions: (1) the familiar reversal, to A– B+; (2) a new object "C" replaced A, and it was incorrect (B+, C–); or (3) A remained, and was incorrect, and a new object "D" replaced B (A–, D+). Substituting the new stimulus for A or B, and determining whether the monkeys could maintain correct selection of the original B or the B substitute, is one way to evaluate "mediational learning" according to Rumbaugh and Pate (1984). Each monkey worked on 12 problems of each type.

The capuchins' performances on the reversal tasks in De Lillo and Visalberghi's (1994) study are similar to those reported by Rumbaugh and Pate (1984) for gibbons (*Hylobates* spp.) and for talapoin monkeys (*Miopithecus talapoin*). Their performances on the mediational tasks were better than those reported for most species of monkeys tested in this paradigm, but not as good as those of great apes or of rhesus monkeys tested using an interactive video format (Washburn and Rumbaugh 1991). The capuchins' biggest source of errors in the mediational trials, it appears, was a tendency to select the new object regardless of whether it substituted for the incorrect or the correct object. D'Amato's work (D'Amato and Salmon 1984, D'Amato *et al.* 1985, D'Amato and Colombo 1988b) using a different testing format (where the monkeys had to work with three or more images at once) demonstrates that capuchins use mediational abilities (although of a different sort) in other kinds of learning tasks, where the monkeys' preferences for novelty do not impede correct performance. Overall, we learned something fundamentally important about capuchins in this work: when given a choice, even in the face of probable short-term failure, they often choose the new thing. As we shall see, this tendency of capuchins resurfaces in virtually every format of testing used in experimental studies of problem-solving. Accordingly, we must be very careful in interpreting these monkeys' performance on tasks in which the monkey's choice is supposed to reflect evaluation of the "correct" status of one object *versus* another. The capuchin monkey's interest in manipulating or inspecting an object that is so evident in their foraging activity (see Chapters 3 and 7) seems to impact how they approach the opportunities we give them to select and sometimes to manipulate objects.

PERCEPTION

> Perception forms the portal between reality and knowledge . . . In the theatre of the mind, it is the opening scene.
>
> (Kellman and Atterberry 1998, p. 1).

Animals actively seek information with their senses (Gibson 1966); we call this process "perception." Perception can be studied at several levels, from the ecological (the nature of perceptual information available in the environment and the actions taken to collect it) to the biological (receptors, neural mechanisms, often considered the province of sensation rather than perception) and to the psychological (what information is sought, how specific information is extracted from the rich incoming array; how perceptual information is remembered and is used to guide action). We focus here on what information is sought and extracted, and how that information impacts action. Most of the studies reviewed below concern visual perception. It is important to note that we do not usually know whether an individual is explicitly aware of the outputs of perception – we know in some cases for humans that we are *not* consciously, or reflectively, aware of all perceptions. For example, we are usually unaware of swaying while standing, although we make continuous postural adjustments to maintain balance. We make no claims about capuchins' reflective awareness of any of their percepts – there is no empirical basis for making any claims in this domain.

Seeing objects as wholes

Some early experimental work in problem-solving by monkeys used a task that required the subject to trace a string connected to a piece of food; pulling the one correct string of many overlapping strings brought the food within reach (the "patterned string" problem). The number of strings and the degree to which the strings overlap and change direction affect the difficulty of the task. Harlow and Settlage (1934) tested two white-faced capuchin monkeys, 28 individuals from eight catarrhine species, and one spider monkey on patterned string problems using from two to four strings. Capuchins did as well on these tasks as the other species, making errors on as few as 8% of trials with the easiest pattern to 82% errors on the most difficult one and leaving one problem unsolved. The best subject in this study was the spider monkey, in the sense that it solved a pattern that no other monkey solved. The monkeys behaved as if they expected the strings always to run in straight lines (whereas in fact they often zig-zagged). This might reflect inattentiveness to the finer details of the string's position or limited visual acuity to discriminate one string from another where they intersect. However, it seems as plausible that the monkeys expected that a string that is straight along part of its length would remain straight! In this case, they apparently falsely perceived one (straight) string where in fact there were two or more separate strings. This suggests that the monkeys perceived a "whole" even when the objects (strings, in this case) were actually separate.

Do monkeys perceive objects as "whole", even when parts of them are occluded? To find out, Fujita

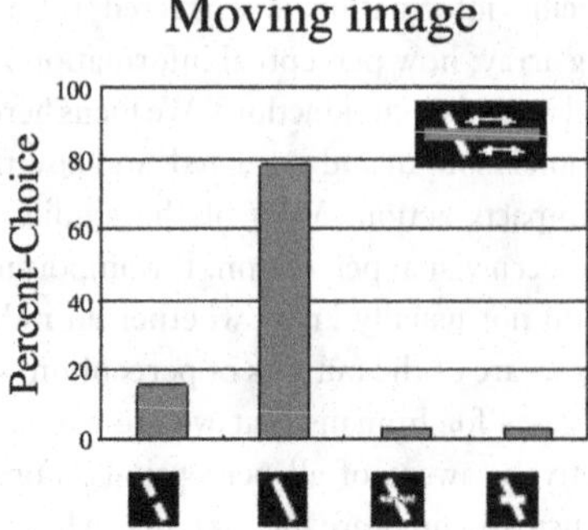

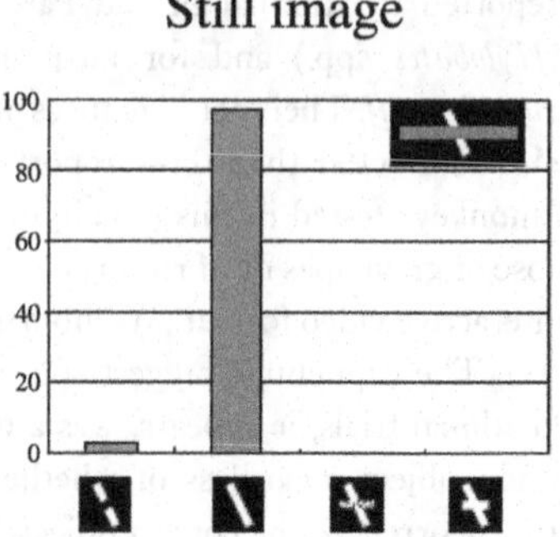

Figure 8.4. Capuchin monkeys perceive partially occluded objects as wholes. The monkey looked at a moving two-dimensional image, or at a still image of a split rectangle (shown in the black boxes at the top of each graph). Then it had to select from among the four images shown (at the bottom) the one that matched the sample. Regardless of whether the sample image moved in unison behind the occluder as a unit (left panel), or remained stationary (right panel) the monkey consistently chose the full rectangle (second from left) to match the split rectangle rather than choosing the two unconnected rectangles (far left), or two other variants (the two images on the right). When the two parts of the split rectangle sample moved independently of one another (not shown in this figure), the monkey matched the sample to the two unconnected rectangles. (Redrawn from material provided by Kazuo Fujita).

and collagues (Fujita 2001, Kuroshima et al. 2002) presented a capuchin monkey with two-dimensional line drawings of geometric shapes (hereafter, "objects") on a computer monitor (see Figure 8.4). The monkey first viewed a screen depicting one of four sample objects at the center of the monitor, together with a long rectangular shape (a "belt"). The belt appeared at the top, middle or bottom of the monitor, and the sample object was partially occluded by the belt. One sample object was a complete rectangle, another was two smaller rectangles with a gap between them (a "split" rectangle). When the belt occluded the center portion of the sample, the split rectangle and the complete rectangle appeared identical; thus, this image was ambiguous with regard to the full form of the "occluded" sample. On some trials the stimulus object moved horizontally back and forth, with the visible portions above and below the belt moving in unison. After viewing the display for 2.3 seconds, the monkey immediately saw another display showing the four possible sample objects. The monkey's task was to select the one sample object that matched the one just seen in the previous display. The monkey matched the ambiguous sample with the whole rectangle, whether or not the sample had moved, on nearly 80% of the trials, as do humans from a young age.

Fujita's study shows that adult capuchin monkeys use visual information about object contours as adult humans do. They see the world as composed of whole objects, rather than as unconnected segments, even when the objects are partially occluded. In the natural world, this makes good sense. For example, the monkey can perceive a tree limb as one segment, even if its view of the limb is partially occluded by the trunk of another tree.

There is a crucial difference between the laboratory situation described above and how individuals encounter objects in the real world. Perceivers (monkey or human) gather, through their own movements, a great deal of information about object unity and many other properties of objects and surfaces (Gibson 1966, 1979). Movement changes the properties of the visual array, auditory array, or other sensory arrays in orderly ways. Simultaneous changes in proprioception permit the perceiver to identify many kinds of invariant relations among elements in the array that signify, for example, solidity of an external object, or the perceiver's position with respect to the surface of the earth (Stofregen and Bardy 2001). Much of this self-produced information is not available to a subject passively viewing two-dimensional "objects" moving in front of him or her.

Klüver (1933) presented a capuchin monkey with a problem where the monkey's actions clearly contributed to its perception of the relation among elements in the situation. The monkey encountered a long length of string strung between two hooks and draped within reach of the monkey. If the monkey pulled both ends at once, she could pull down a lever and bring a piece of food within reach. If she pulled one end, the string slipped through the hooks and fell to the ground. Klüver describes how the monkey failed at this task several times, but then pulled both strings at once, apparently by accident. The monkey pulled both ends of the string in subsequent trials while watching the string and the lever, with one end of the string in each hand. Thereafter she proficiently pulled both ends of the string.

Using visual information under altered relations between vision and location

One can ask whether monkeys can perceive the relations between movement and action when the normal relation between vision and location of an object is altered. We have seen with the joystick tasks mentioned in Chapter 7 that capuchin monkeys, like several other species of nonhuman primates, can do this: they can learn to use a joystick to maneuver a cursor on a computer screen, even though the locus of action is separated from the locus of viewed object movement on the screen by a substantial distance. They can learn to move the cursor to a predetermined point, to track a moving icon, to intercept the path of motion of an icon, and to move around "barriers" and through narrow "alleys."

Another situation where the normal relation between vision and location of an object is altered occurs when the monkey uses a mirror to guide its hand to an object seen only in the mirrored reflection (Marchal and Anderson 1993). When honey-dipped raisins (a truly decadent delight for capuchin monkeys) were stuck onto a surface below the front edge of the monkeys' cage (and thus out of their view), two monkeys out of three learned to locate them efficiently with their hands when they could look in the mirror, but searched randomly when they saw only the non-reflective side of the mirror. One monkey became so proficient that he got the raisin in a single reach per trial after the eighth session with this task, compared with two to six attempts per raisin without the mirror. However, even after exposures ranging from hours to days, capuchins do not use their image in a mirror to direct actions to body parts they cannot see, in the way that apes can (Anderson 1994, 1996, 2001, Riviello *et al.* 1993; see Chapter 13).

PREDICTING MOTIONS OF OBJECTS

An adult human knows that unsupported objects fall until they reach a solid horizontal surface, that objects moving in a straight line continue to move in a straight line unless other forces (e.g., gravity) intervene, and that solid objects cannot pass through other solid objects. Do capuchin monkeys recognize these same properties of moving objects? Fragaszy and Cummins-Sebree (unpublished) have examined these questions by presenting to six adult male tufted capuchin monkeys a "puppet" display of a metal ball rolling across various surfaces (see Figure 8.5). First the monkeys learned that retrieving a metal ball and returning it to the experimenter produced a food treat. They did this after the ball rolled to a stop at one of several designated locations in the display. Subsequently, they learned to indicate the location where the ball would stop rolling, rather than actually to prehend the ball. At this point, their behavior indicated where they expected a rolling ball to come to rest. The experimenters then presented a series of displays to them where the ball moved in front of them but stopped short of the possible learned "end" locations. After they indicated their expectation for where the ball would come to rest, the ball continued on a linear path to its (physically logical) end point. A correct choice earned the monkey a food treat. Essentially, the experimenters asked the monkeys to predict where an object that they saw moving across a particular substrate would come to rest.

After they became proficient with a few familiar layouts of surfaces and motions, we conducted experimental sessions in which novel paths and arrangements were included among the familiar ("training") patterns. We asked the monkeys to predict the ball's resting point in three situations: (a) when it rolled horizontally toward a gap in the supporting horizontal surface that was twice the diameter of the ball (Figure 8.5a and c; the "Gravity" experiment); (b) when it could roll horizontally in either direction, and a solid panel obstructed the path in one direction (Figure 8.5b, the "Solidity" experiment); and (c) when it rolled in a linear motion, in the absence of an intervening obstacle or force (Figure 8.5d, the "Continuity" experiment). In all these cases, the ball appeared to the monkeys to move autonomously, although in fact the experimenters controlled its motion from behind a thin wood-fiber panel by means of a strong

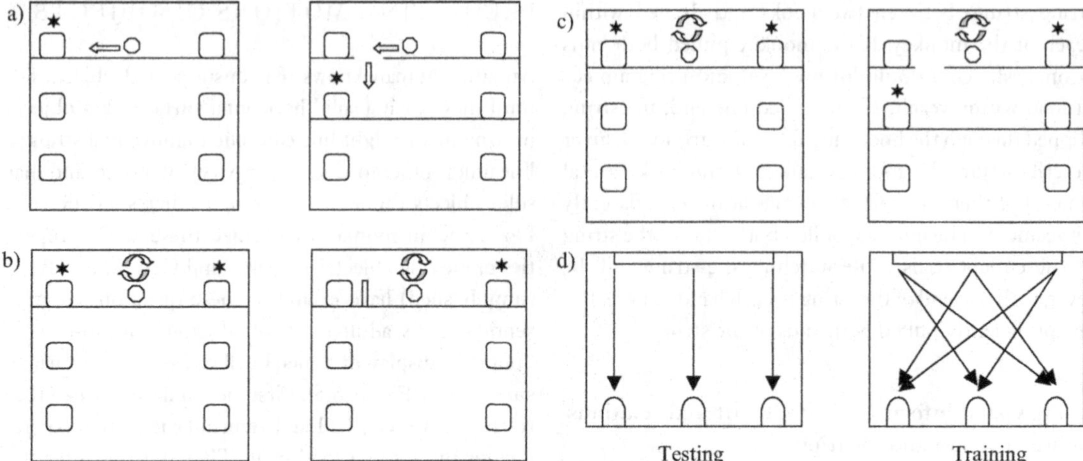

Figure 8.5. Experimental arrangements used to evaluate capuchin monkeys' ability to predict the future position of a rolling ball. By placing its hand on the window, the monkey indicated the window (shown by an outline in the figure) where the ball would appear if it continued to move to the left or right along the shelf. For all of the diagrams, the arrows indicate the movement of the ball, and the asterisks indicate which choices were correct (and resulted in the monkey receiving a food treat). (a) Gravity Experiment – Training (left): Half of the time the ball rolled toward the left window, the other half to the right window. The ball stopped rolling at a point about two-thirds of the way toward one window. For any training trial, the correct (rewarded) choice was the left window when the ball moved to the left and the right window when the ball moved to the right. Gravity Experiment – Testing (right): the ball first moved about two-thirds of the way toward one window and then stopped (just before reaching a gap). Once the monkey chose a window, the ball was moved over the gap so that it fell to the next platform. On half the trials, the gap was toward the right window and the ball rolled to the right; on the other half, the gap was on the left and the ball rolled to the left. (b) Solidity experiment – Training (left): the ball moved in a circle and came to rest in the center of the shelf; the monkey then chose a window. Choice of either window was rewarded. Testing (right): a barrier was present on one side of the shelf. The correct choice was the window opposite the barrier. (c) New Gravity, conducted after the Solidity experiment – Training (left): the ball moved in a circle and came to rest in the center of the shelf; the monkey then chose a window. Either window was correct. Testing (right): a gap was present on one side of the shelf; the correct choice was either the window opposite the gap or immediately below the gap. (d) Continuity experiment – Training (left): the ball rolled on a linear path toward the lower end of the tray. It stopped several centimeters above the windows. The monkey then chose one of the windows (indicated by outlines in the figure). Testing (right): the ball rolled along a novel (diagonal) path toward the lower end of the tray, stopping a few centimeters above the windows. The monkeys became proficient at predicting the ball's future position in the Solidity and Continuity experiments, but had more difficulty with the Gravity problems. (Drawings by Sarah Cummins-Sebree.)

magnet that they moved in the desired manner. Each monkey completed 12 to 18 trials with novel patterns in each type of task, and 24 to 54 familiar patterns. The monkeys maintained nearly perfect performance on the familiar paths during the test sessions, so we will discuss only their performance on the trials where the monkeys encountered novel conditions.

Fragaszy and Cummins-Sebree found that three of six subjects correctly predicted the end point of the ball rolling in a linear manner (thus solving the "continuity" problem) at frequencies significantly above chance. One monkey correctly predicted the end point when a solid barrier appeared to block movement toward one of two possible end points on 14 out of 18 trials (nine correct choices are expected by chance); two other monkeys were nearly as good (13 out of 18 trials). However, none of the monkeys correctly predicted the ball's resting point when it rolled toward a gap. In fact, all of the monkeys chose the incorrect end point (that had been correct during training) more often than the correct choice (nine or more trials out of 12). We reasoned that our training condition might have induced a preference for the goal closest to the ball when it stopped moving. Consequently, we repeated this experiment using the neutral starting condition that we had used with the solid barrier, in which the trial began after the ball oscillated back and forth but eventually stopped in the center of the display, where it had appeared at the start. A "correct" choice in this situation could be either the goal window opposite the gap (as was the correct goal when the barrier blocked access to the goal beyond it) or the window on the shelf below the gap. With this procedure, all six monkeys selected a correct goal more often than an incorrect goal, but no monkey had a perfect performance (12 correct trials) needed to meet the statistical criterion of a less than 5% probability of a chance occurrence. The best performance was 10 correct trials out of 12. Throughout this testing some monkeys retained very strong side preferences, a characteristic of capuchins in laboratory choice tasks that is mentioned by many other authors.

These studies demonstrate that the monkeys attend to some features of substrates more than others, and that they are able to predict some aspects of an object's movement in relation to the features of the substrate it will cross. They are not uniformly proficient, however, and it seems likely that they require experience at moving objects themselves (and experiencing proprioceptively the consequences of an object encountering particular features of substrates, such as holes) to anticipate the movement of objects across some kinds of surfaces. The answer to the question "Do capuchin monkeys recognize some of the same properties about moving objects that humans do?" is a qualified "yes." One significant feature of these studies is the passive role of the subject: the monkeys did not move the objects, they merely observed movement (as in many studies of visual perception, with humans and nonhumans alike). Passively viewing a scene is not likely to be the best context for monkeys to learn about the properties of objects and their movements across varying substrates. In subsequent studies we have explored what experiences are most informative for the monkeys in relation to their ability to move objects across substrates that vary in their properties of support. We report these studies in Chapter 10.

SUMMARY

Capuchin monkeys' memory and visual perception are similar to those of other nonhuman primates. They travel efficiently through familiar spaces from one location to another and they can use landmarks to locate objects they cannot see. They have good memories for significant events and they perceive objects, object movements, and surfaces in much the same way as humans do, although they are less attentive to some features of surfaces (such as holes) than adult humans are. There are many gaps in our knowledge of perception and memory in capuchins, reflecting the state of knowledge about these domains in all species. For example, we know almost nothing about perception in haptic, auditory or olfactory modalities, or about prospective memory (keeping a goal in mind).

Capuchins are like other monkeys in that they can recognize an abstract relation among objects (e.g., sorting objects by size). Capuchins can also recognize two relations at the same time, and use them in a nested or hierarchical fashion, such as sorting pictures by shape, and sorting pictures of a given shape by size. Capuchins do less well when they must infer the role of a new element in a relation (as in the mediational discrimination learning arrangement used by De Lillo and Visalberghi 1994). Overall, there have been only a few clear demonstrations of recognition of two nested or integrated relations (second-order relations) in capuchins, and they have all dealt with classifying and ordering visual stimuli. Recognizing, predicting and producing second-order and higher relations among objects and events confers great power and precision in human behavior, and thus there has been strong interest in determining how other species compare with humans in this respect (Langer 2000, Tomasello and Call 1997). As we discussed in Chapter 7 and revisit in the next two chapters, capuchins appear to use first-order relations in several situations involving controlling objects. They may use second-order relations in some circumstances, but this is less clear.

ENDNOTES

1 An alternative categorization of memory concerns underlying neural systems supporting different functions. Two systems are recognized: one mediated by corticostriatal circuits, supporting stimulus-response associations and habit learning, and another mediated by corticolimbic circuits, supporting both recognition and recall of sensory representations (Overman 1990). The latter system is thought to be required for concept

learning. In human children and in rhesus monkeys, the first system functions in a mature manner at younger ages than the latter system (Bachevalier 1990, Overman 1990). Resende *et al.* (2003) reported that capuchin monkeys (*C. apella*) estimated to be 1–3.5 years old learned concurrent visual discriminations as proficiently as adults, but not delayed non-match-to-sample (DNMS) discriminations, replicating the developmental dissociation between "habit" and "concept" learning observed in rhesus monkeys and humans. This is the first report of a developmental dissociation of these memory systems in platyrrhine monkeys.

2 Note that demonstrating that an animal is using a concept, rather than associating specific objects or configurations, requires careful experimental design (Burdyn and Thomas 1984, Thomas 1996). The attribution of conditional concepts is particularly problematic; Thomas (pers. comm.) proposes there is as yet no clear evidence of conditional concepts in nonhuman animals.

9 • Engaging the world. Exploration and problem-solving

The young white-faced capuchin spots the swollen thorn tree along the familiar path between the big fig tree and the stream. The monkey is attracted to the swollen thorns, that contain delicious ant larvae, but it does not climb the tree directly to break off a thorn. A previous attempt to do this led to painful bites from the ants living in this tree, and the monkey had quickly retreated when the ants attacked. This time the monkey sees two small branches hanging a meter above the thorn tree. Quickly it moves to the larger of the two branches, walks out to a point where the branch arcs above the tree, suspends itself by its feet and tail, and snatches a thorn. Three ants come with the thorn and immediately attack the monkey's hand, but the monkey eats them quickly. He licks the bitten areas for a moment, and then climbs back up to a larger, more stable limb, sits down to crack open the thorn, and delicately picks out the ant larvae from inside.

This vignette captures something very ordinary and very impressive: goal-directed activity, with some degree of recognition of risk and opportunity, planning and judgment occurring in the course of achieving the desired result. In the example, the monkey (a) recognizes the relation between the thorn and a previous pleasant experience of eating the ant larvae and the unpleasant experience of having been bitten by the adult ants; (b) uses a sequence of activities to obtain a thorn and the (hidden) resource within it; (c) adopts a mode of approach quite different than that used with other food resources not protected by biting ants; and (d) chooses a particular branch, from among those available, to reach the thorn. The monkey must perceive several things about the immediate environment and monitor the outcome of a sequence of actions as it carries them out. Doing all this is so commonplace that we do not recognize the significance of the activity until we have to explain how an individual ordinarily does it, and why, in some circumstances, it is not able to do it.

In this chapter we consider how capuchin monkeys organize their actions in the world. We start with some provocative anecdotal observations of capuchins doing innovative and clever actions with their bodies and with objects, to further convince the reader that capuchins are really quite extraordinary in this aspect of their behavior. Thereafter we turn to the scientific literature to review their problem-solving behavior in a variety of circumstances, both natural (entering novel spaces, manipulating substrates containing food) and patently artificial (searching small cups for hidden food items; seriating cups, placing objects together, pulling a handle synchronously with another monkey, undoing latch puzzles). Our interest in this review is in what sorts of information the monkeys are seeking and producing through action, what they are learning from their activity, and what aspects of the various problems challenge them. We close the chapter by considering the kinds of tasks that capuchins find easy or difficult to master, framed in relation to how they know the world through action. This gives us a foundation for thinking about tool use, the subject of the next chapter.

GENERAL CHARACTERISTICS OF CAPUCHINS AS ACTORS AND PERCEIVERS

> The world can only be grasped by action, not by contemplation . . . The hand is the cutting edge of the mind.
>
> Jacob Bronowski (1973)

It is fair to say that capuchins epitomize the popular notion of the active monkey, and if we take Bronowski at his word, they have very active minds! Capuchins act on their world, more than other monkeys, through coordinated actions of hands, mouth, feet and tail. The destructive tendencies of capuchins with the things they manipulate are well known. They manage to undo latches, untie ropes, unsnap hooks and unscrew bolts. They are also notorious for what they do to objects that come into their control. They touch things, handle them, rub, bite, pull, push and pound them, exploring their properties as they destroy them. Other monkeys are not like this. For example, squirrel and titi monkeys, two other platyrrhine genera, touch and may bite or pull on objects, but neither one attempts to smash anything (Fragaszy and Mason 1983). Macaques, while destructive, are less interested in the properties of objects (Klüver 1933).

Capuchins are primarily interested in opening up, getting inside, putting inside, taking apart and otherwise altering objects and surfaces. In captive situations in particular, incessant fiddling with objects can take an exploratory or entertaining tone, and in such cases it appears to the human observer that the monkey is trying out new actions for the pleasure of mastering something new or exploring a property of an object or surface, or their combinations. A random list of such observations observed in our captive groups includes:

(1) Many capuchins use a cup as a container for solids or liquids, an inventive and effective behavior. But cups afford more inventive activity than this. For example, we have watched a young capuchin (about 3 years old) become skilled at filling a cup with water, and then balancing it in the hand so as not to spill water as he moved all across the cage. He could even move across the mesh ceiling upside down, holding the cup out to one side with one hand and using both feet, tail, and a hand to move. He practiced this skill intermittently over a period of months and eventually managed to keep most of the water in the cup no matter where he moved in the cage.

(2) A young monkey balanced a pellet of chow (a rectangular, flat object about 2 cm × 3 cm) on her nose, then walked bipedally sideways on the perch a few steps to the left, then a few steps to the right, like a trapeze artist. This action was observed only a few times.

(3) Monkeys put a box, bowl or paper bag over their heads with their hands, and then moved tentatively around quadrupedally for a few moments before taking it off, as if exploring movement without vision (see also Chevalier-Skolnikoff 1989).

(4) Capuchin monkeys of all ages typically splashed water out of a pan with their hands. They often followed this action by throwing small bits of straw, food, etc. into the pan, then scooping them out (with splashing again), repeating the cycle of throwing in and then taking or splashing out several times. Crab-eating macaques with the same materials played enthusiastically, even wildly, in the water, but did not put objects into the water in repetitive playful sequences. However, they often carried objects along with them as they entered the water (Visalberghi and Fragaszy 1990a).

(5) A young hand-reared capuchin (about 6 months old) held a tube at an angle against the floor, then dropped whole pecans (nuts that roll nicely) through it one at a time. She watched each one travel through the tube and roll along the floor. She did this many times over several days (Fragaszy, pers. obs.).

(6) This same monkey routinely took the cap off a sturdy plastic bottle, took out bits of food placed in there by her human caretakers, then put something else in the bottle, and put the cap back on! (Fragaszy pers. obs.)

(7) A juvenile capuchin encountering a mirror for the first time hit the mirror with her hand, with her foot, and then again with her hand, all the while continuing to look closely at the surface of the mirror. It seemed as though she were interested in the "invisible" surface that felt solid but could not be seen, as well as in the image on the surface, to which she directed social signals (Riviello et al. 1993; see Figure 9.1)

All these behaviors are self-discovered; they reflect the strong intrinsic interest of these monkeys in acting on objects and surfaces with the body, and their creativity in doing so (Fragaszy and Adams-Curtis 1991). We see these kinds of behaviors most often in young monkeys, but data on response to novel objects indicate that adults will also perform "rare" acts when novel opportunities arise (Fragaszy and Adams-Curtis, unpublished). It is not that old animals lose the ability to innovate; they merely require a specific motivation to do so.

Although the appearance and longevity of an unusual behavior are unpredictable, the occurrence of

Figure 9.1. A young tufted capuchin monkey encountering a mirror for the first time expresses interest in the image and the surface of the mirror. She directs social behavior at her own image. (Photograph by Elisabetta Visalberghi.)

some innovative behaviors involving objects and surfaces is predictable if captive capuchins are given the opportunity to generate them. This is a remarkable characteristic of these monkeys, shared with apes but not with other nonhuman primate genera (Klüver 1933, Parker 1974a,b, Torigoe 1985). This characteristic of capuchin monkeys is clearly relevant to their propensity to discover how to incorporate objects into action on the environment to attain some goal (i.e., to use tools), as we discuss in Chapter 10.

The observations listed above and many more like them, while inspiring and entertaining, must not be taken for compelling evidence of any particular motivation or skill on the part of the monkeys. They suggest what questions we might want to ask the monkeys but anecdotal observations do not provide any answers. We do not know if the monkey would perceive the same relations between actions and outcomes that an adult human would, nor do we know what they are seeking when they perform these actions. To begin to answer these questions, we turn to the scientific literature on capuchins' behavior while solving a variety of movement problems.

ORGANIZING SEQUENTIAL ACTIONS

Capuchin monkeys can learn, as can many other kinds of animals, to perform several actions in succession to achieve a single objective. Psychologists have shown that animals can use one action or outcome as a cue to perform the next one; this enables the animal to learn "chains" of associated actions without necessarily keeping the whole sequence in mind from the outset. Organizing sequences of actions where decisions are made at each step is a psychologically more complicated form of sequential activity. This is because the actor must keep the end point of the activity in mind as the sequence is performed for an effective or efficient performance, thus placing additional demands on working memory. For example, selecting a particular object from a set to use in a certain way requires that the eventual use of the object guide the selection. Westergaard *et al.* (1998a) gave a task of this format to a capuchin monkey. The monkey first learned that when he selected a yellow chip (and not red) or a black chip (and not red) and gave that chip to the experimenter, the experimenter would give him an object that he could use to obtain food. Next the experimenter gave him two different tasks, one at a time in random sequence, each task requiring the use of a different object (hereafter, "tool") to get food. Now the monkey received one tool if he chose the black chip and a different tool if he chose a yellow chip. The monkey learned to choose the chip of the correct color for a particular task when the same task appeared for five or ten consecutive trials. Finally, the monkey encountered the two tasks in a random order, and on more than 90% of trials he selected the chip that produced the tool matching the task he confronted. This monkey's behavior became strategic, in the sense that the second part of the action sequence (using the tool) guided action at the first part of the sequence (selecting the chip). Klüver's (1933) capuchin monkey also chose and used objects in sequence. In the longest sequences Klüver provided to his experimental subject, the monkey acted with one object to retrieve another to retrieve a third, which it used to retrieve a bit of food placed beyond her reach. Sometimes the first object was out of view of the others and the monkey had to retrieve it to begin the sequence. Klüver provides wonderfully detailed descriptions of the capuchin monkey's behavior as it faced varying challenges to coordinate actions with objects to retrieve food.

As most of the problems he presented to this monkey are classified as "tool" problems, they are treated in more detail in Chapter 10.

Planning incorporates memory in that the goal of the activity must be kept in mind (i.e., in working memory) as the activity is carried out. Planning also includes monitoring actions to determine if the outcome leads one closer to the goal, and if not, restructuring activity. This aspect of planning is also recognized by the term "problem solving" (Klahr 1994). Contemporary psychological theories treat planning and problem-solving as a challenge in relating multiple elements to each other sequentially or simultaneously (e.g., Bidell and Fischer 1994, Case 1992, Fischer and Bidell 1998, Klahr 1994, Case and Okamoto 1996, Siegler 1998).

How well can capuchins plan sequential actions? We have studied this question by presenting three capuchins with 192 mazes, each maze only once, on a computer screen, and asking them to navigate a cursor to a goal point (Fragaszy *et al.* 2003). We presented the same mazes to four chimpanzees, allowing us to compare capuchins with another species performing the same task. The mazes each contained from one to five binary choice points, and zero to three of the choice points required selecting a non-obvious choice (such as the path leading away from the goal) as the correct path. Two representative mazes are illustrated in Figure 9.2. We presented the mazes in what we considered an ascending order of difficulty, from few to many choices and from few to many non-obvious choices. The most difficult mazes, presented last, each contained five choice points, of which three presented non-obvious choices.

Although overall the chimpanzees made fewer errors than did the capuchin monkeys, two of three capuchins solved more mazes without error than expected by chance and made fewer errors than expected by chance. Subjects of both species frequently corrected their path after making an error, moving back through the choice point to the other path before they encountered the end of the alley. Animals of both species self-corrected in this fashion following about 40% of errors. Three subjects (one capuchin monkey and two chimpanzees) made proportionally fewer errors than expected at choices closer in sequence to the goal, and proportionally more errors than expected at choices farther in sequence from the goal. This pattern suggests the subjects planned a set order of choices from the goal back to the start, and carried them out in reverse order (from start to goal). All of these findings indicate that both capuchins and chimpanzees employed some aspects of planning as they navigated the mazes. The general similarity in the pattern of choices and reactions to errors by capuchins and chimpanzees indicates that these two species perceived the mazes and organized their navigation in similar fashion.

At first sight, a three-latch problem might seem simpler than a 5-choice maze. But consider the problems in this way: in solving a maze, as in solving a problem requiring the use of multiple tools, such as those presented to chimpanzees by Köhler (1925) and to a capuchin monkey by Klüver (1933), the subject uses at each point in the problem an action it has already mastered (moving something through the choice points) although the sequence is new. A sequential latch puzzle presents the additional task of learning the various actions at the same time as learning the sequence in which to perform them. It is a task with two kinds of learning – action and sequence – happening simultaneously.

Adams-Curtis (1988, Adams-Curtis and Fragaszy 1995) presented two groups of monkeys a latch-puzzle containing three components that had to be moved in a fixed sequence to uncover a raisin. Only one juvenile female mastered the puzzle. She solved the puzzle first during the third experimental session, and solved it again 13 more times in that session. Subsequently, she solved the puzzle more than 300 times in the remaining 16 sessions. Adams-Curtis (1988) closely analyzed all of the monkey's contacts with each part of the puzzle to determine how her solutions changed with practice. At first, she made many incorrect contacts to parts out of order, ineffective efforts to open all parts, and correct solutions to one part, only to shift her activities and not pursue opening the next part that was now available for solution. By the end of the fifth session, she had mastered opening all the latches. The very last aspect of the puzzle that she mastered was, perhaps surprisingly, where to begin. She did not begin to contact the first latch of the puzzle consistently at the start of each presentation until the sixth session, although she had solved the puzzle many times before. Prior to the fifth session, she was apparently following the rule "Contact any part and try to solve it. If that part does not open, try a different part." The analysis of the one solver's progress from haphazard but persistent actions with a set of objects,

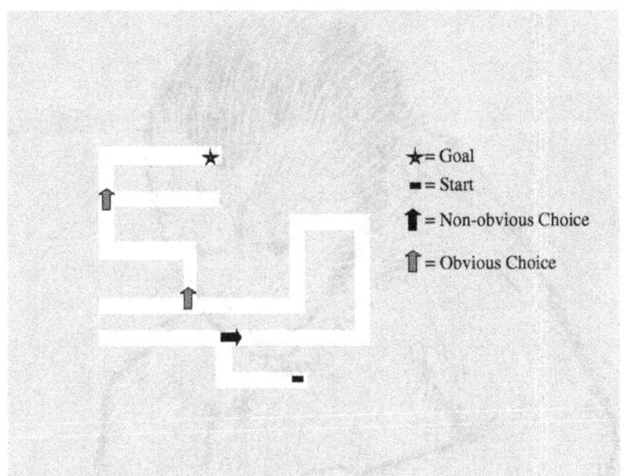

Figure 9.2. Two sample mazes presented to capuchin monkeys and chimpanzees by Fragaszy *et al.* (2003). The actor used a joystick to move a cursor through the white alleys of the maze to a goal region (shown as a star) on a computer monitor. Black arrows indicate choice points in which the incorrect choice appears to lead more directly to the goal than the correct one (a non-obvious choice); gray arrows indicate all other choice points. The maze on the left contains three choice-points, one of which is a non-obvious choice. The maze on the right contains five choices, three of which are non-obvious. Capuchin monkeys and chimpanzees could solve mazes like those shown here, although they made errors while doing so. Adapted from drawings by Julie Johnson-Pynn and Sarah Cummins-Sebree.

to a sequentially ordered set of differentiated actions, illustrates that the double task of simultaneously discovering a sequence and a set of different actions for different objects is not easy for capuchin monkeys.

ORGANIZING MULTIPLE OBJECTS

Spontaneous constructions

As highlighted in Chapter 7, one of the most striking features of manual activity in capuchin monkeys is that they frequently act to combine objects in various ways. Can they organize multiple objects into sets or structures? Do they do so spontaneously? The answer to both questions is "yes, sometimes." Spinozzi and Natale (1989) studied how capuchins organize activity with multiple objects by providing a monkey with a set of six small objects (cups, crosses, rings and sticks) shaped from four different materials. Each set of objects belonged to one of three conditions: (a) two sets of three identical objects that differed in one property only (e.g., three wood cups and three wood crosses); (b) three identical objects that differed in both form and material (e.g., three wood cups and three acrylic rings); and (c) six objects of three different forms and two different materials, or vice versa. The individuals were free to do as they liked with the objects, without reward or interference, for a period of 5 minutes. The monkeys encountered each condition once per session, over eight test sessions.

One of the aspects of the monkeys' activity that interested the experimenters was the way they moved objects into proximity (closer than 10 cm) with one another (i.e., more than one object within arm's reach) or moved them apart from each other. When young children are given objects like this, they routinely construct and disassemble sets and even assemble sets of sets in a hierarchical organization (such as placing all metal objects in one place, ordering them by shape, then placing all wooden objects together, ordering them also by shape). These actions presage the logical operations used in addition, subtraction, multiplication and so on (Langer 1980, 1986). If object placements were random there would be smaller numbers of multi-object sets than two-object sets. However, Spinozzi and Natale (1989) found that two 4-year-old capuchin monkeys placed three or more objects in 40% of their sets, roughly equivalent to what human children of 15 months do in similar circumstances. Similarly, if objects were placed randomly, sets would not likely overlap in time or follow one another in close temporal sequence. Capuchin monkeys made temporally sequential sets in 101 instances out of 129 sets (82% of their constructions), but rarely made simultaneous sets (14 of 129 sets; 11% of constructions). Human children show a different pattern; they increasingly compose simultaneous sets. Producing simultaneous sets (as humans do) enables exploratory composing of related sets. The monkeys' limitations (compared with young humans not yet 2 years old) in managing more than one relation of one object to another at a time is evident in these findings. Nevertheless, the propensity to place objects into spatial correspondence with one another differentiates capuchin monkeys from the macaques that participated in Spinozzi and Natale's study.

Potì and Antinucci (1989) analyzed the same data set as Spinozzi and Natale with a focus on the nature of the movements of objects into and out of successive sets. They found that two capuchin monkeys achieved first-order constructions (in which each successive construction is independent of the previous one), but very rarely produced second-order constructions (in which one set is linked to another set by the order or nature of the actions imposed upon them both). In this aspect of behavior the capuchins were similar to macaques studied with the same method.

Figure 9.3. Capuchin monkey in the process of nesting five cups into a seriated structure, as studied by Johnson-Pynn *et al.* (1999). The monkeys were reliably able to seriate five cups and to insert an additional sixth cup into the middle of the seriated set they had just created. (Drawing by Stephen Nash from a photograph by Carrie Rosengart.)

Guided constructions

Other methods of probing individuals' organizational skills involve providing subjects with a goal, as opposed to the unguided (spontaneous) manipulative actions analyzed by Spinozzi and colleagues. A classic test of children's developing organizational skills is whether they can seriate sets of nesting cups (Greenfield *et al.* 1972), an activity that children do spontaneously in play, but can also easily be prompted to perform. Initially, young children (about a year old) usually simply pair two cups. Later on, they make multi-cup structures, eventually managing frequently (in their third year) to order the cups correctly. Later, in their third or even fourth year, they can routinely insert a middle cup held out from the rest into its proper place in an already-seriated set.

Fragaszy and colleagues presented nesting cups to adult capuchin monkeys and chimpanzees and to children between 11 and 21 months of age using the same experimental design as Greenfield *et al.* (1972) and Johnson-Pynn *et al.* (1999, Fragaszy *et al.* 2002; see Figure 9.3). The results were surprising. Capuchins produced fully-seriated sets of five cups on half of the trials in which they got the cups; apes (*Pan troglodytes* and *P. paniscus*) did so on a similar percentage of trials (55%). When given a sixth cup to insert into a seriated set of five cups that they had just constructed, capuchin monkeys succeed at inserting the extra cup on 56% of these trials; apes, on 36%. Moreover, both capuchin monkeys and apes frequently combined objects using what Greenfield *et al.* (1972) identified as a hierarchical method. That is, they put a small cup into a larger cup, picked up the set, and then placed the set into a third cup or set of cups. Greenfield *et al.* (1972) labeled this way of combining cups "subassembly." Capuchin monkeys and apes used subassembly on 17% of five-cup trials, and 43% of sixth-cup trials, where the sixth cup could be any of those between the bottom and the top. Apes in similar conditions used subassembly on 28% and 60% of trials, values not significantly different from the monkeys' values. In all these features, capuchin monkeys and chimpanzees did not differ. The proportional use of subassembly to combine cups was positively correlated with monkeys' and apes' success at inserting a sixth cup, as it was in Greenfield *et al.*'s (1972) study with children. The monkeys and apes were more successful at seriating five cups and inserting a sixth middle cup in the seriated set than were all of the very young children (11–21 months old) in Fragaszy *et al.*'s (2002) sample, although they used subassembly far less than the older (up to 36 months old) children in Greenfield *et al.*'s study that were proficient at seriation.

It is highly unusual to find that chimpanzees and monkeys perform as well as children older than 2 years on a task purported to tap emerging attentional and planning skills. Our interpretation of these findings is that the specific conditions present in this task helped our nonhuman subjects master what is properly appreciated as a rather complicated task (Johnson-Pynn and Fragaszy 2001). This task provides immediate proprioceptive feedback to the actor about the success of each placement when he or she tries to place too large a cup into another (i.e., the cups do not fit together). This is what we mean when we characterize constructions of seriated sets of cups as "guided" (as in the heading of this section); the properties of the cups themselves "guide" the actor's actions.

After much experience with blocked cups, the actors adopt various strategic behaviors that improve their efficiency. In the case of capuchins and chimpanzees, strategic reactions to errors include taking one or more cups out of the existing structure and inserting a different cup or set of cups (Johnson-Pynn and Fragaszy 2001). Even if the actor does not select the next cup very accurately, as long as one structure is modified but not fully disassembled, eventually the strategy of sequential placement will produce a seriated set. A second aspect of behavior also helps achieve success: capuchin monkeys become quite good at containing the cups in a small working area, using the tail and feet to keep unruly cups from rolling away. Because they have one working stack, persistent action with the one stack permits them to seriate the cups. Thus, although this task appears to require mastery of an ordered sequence, a simpler strategy is also effective: keep working on one stack, and replace a blocked cup with some other cup. This simpler strategy is sufficient to get the capuchin monkey and the chimpanzee (and the young child!) through this task.

Could the monkeys, with experience, act more efficiently than the "persistence" strategy described above? Rosengart (2001) examined this question by replicating the original procedure, eight trials with five cups, with two monkeys that had not participated in the previous study. Then she studied how they became "experts" at combining cups. The monkeys practiced combining two cups, then three cups, then four and then five, to criterional levels of success with each number of cups. They were given a food treat for producing a seriated structure, no matter how that was achieved. Finally, Rosengart presented them once again with the same testing regime with which they had started. Both monkeys shifted the manner in which they combined cups, adopting a preference for the subassembly method in the second round of testing that they did not have in the first round. This finding indicates that experience plays a critical role in how the monkeys organize their combinatorial activity; they learn how to be more efficient at the task. We expect the same holds for apes and children.

Fragaszy and Rosengart (unpublished) also examined whether the use of different means of combining the cups reflected something about the cups themselves. We did this by comparing how the monkeys combined five same-sized cups into one structure with how they combined five seriating cups into one structure. All the monkeys used the three possible methods of combining in different proportions with the nesting cups than with the same-sized cups, and all expressed a stronger preference for one method of combining cups with the more permissive same-sized cups. Evidently the properties of the cups influenced how the monkeys go about combining them; the same-sized cups permitted the monkeys to use an individually preferred method.

Overall, capuchin monkeys, like chimpanzees and young children, can produce impressive multi-cup structures, and can manage to re-assemble sets to include additional elements. Capuchin monkeys do this without apparent reliance on a systematic strategy; rather, they develop preferred but relatively simple ways of managing multiple objects (keeping them together, replacing a blocked cup with a different cup). How they combine cups depends both on their prior experience at assembling cups, and on the properties of the cups themselves. They become more effective with relatively little practice. We think that this way of conceptualizing the monkeys' approach to the problem of seriating cups works equally well for chimpanzees' and very young children's behavior in the same tasks. One need not have a reflective understanding of relative size to work effectively with objects of different sizes, so long as errors can be detected immediately, and so long as simple variations of combinations can resolve errors (Johnson-Pynn and Fragaszy 2001). A simple procedural strategy suffices. How one passes from a procedural strategy to a full-fledged abstract conception of seriation, as an adult human possesses, is not yet clear, but it is clear that neither monkeys nor humans begin with the abstract conception. Rather, they begin with action routines, and from their routines, they develop effective procedural strategies (see Lockman 2000 for a similar view of how children acquire skills at acting with objects).

INNOVATION: MASTERING NEW ACTIONS AND NEW RELATIONS

> Making variations on a theme is really the crux of creativity.
>
> Douglas Hofstadter (1982)

Animals that are able to use their bodies in novel ways are able to innovate when they are faced with a new problem, as Heinrich (2000) reminds us in explaining how ravens are so good at solving novel problems. Using the body in a novel way requires flexibility in movements. Anatomy provides some advantages to primates in this domain over other orders that lack, for example, the range of movements of digits, hands and arms that primates can produce. In addition to anatomy, however, variable behavior also rests upon initiative of the actor, and for this characeristic we look to the brain (Reader 2003).

Capuchins fall into the more privileged group of taxa with relatively larger brains than other primates (see Chapter 5). In some way, the larger brain available to capuchins underlies their generative motor activity. The generativity in manipulation that capuchin monkeys produce, and their generativity in using other parts of the body such as feet and tail to explore and interact with their surroundings, support their ability to discover new ways to act with objects that more behaviorally constrained animals cannot discover. Coupled with persistence and an adequate ability to produce and monitor fine movements, these characteristics permit capuchins to interact with the world in a different way than do other monkeys.

SOLVING PROBLEMS WITH FIRST-ORDER AND SECOND-ORDER RELATIONS THROUGH ACTION

Are there any features that characterize the challenges in manipulating objects that capuchins can master? We suggest that one prominent feature of capuchins' problem-solving strategy is that they master zero-order and first-order relations far more easily than second-order relations (see Table 7.2). Recall that a zero-order relation is produced by direct contact of the body with the object or substrate, as in pounding a substrate with the hand. A first-order relation is produced by contacting one object with another object or substrate, as in pounding a nut on a branch. A second-order relation is produced by simultaneously contacting one object with two others. Table 7.2 provides illustrations of spontaneous activities with nesting cups embodying zero-order, first-order and second-order relations in action.

Mastering second- and higher-order relations, such as composing sets of objects that have some relation to one another, getting through sequences of choice points in a maze, or synchronizing the timing of action with another individual's actions pose a greater challenge than bringing an object into combination with a fixed surface. The less one can rely on proprioception and kinesthesis to confirm success, the more difficult a task appears to be for capuchin monkeys.

Regardless of the number of relations in the task, precise positioning of objects or ongoing monitoring of force is far more difficult for capuchin monkeys than simply acting with vigor. Capuchins require more practice to become skilled at a given force relation or positioning problem than adult humans; what is easy for us is not easy for them. We do not yet know what differences in kinesthetic sensibilities, anatomy, attention, memory or other factors underlie these differences.

COORDINATING ACTION WITH ANOTHER INDIVIDUAL

We have not yet talked about a special kind of problem facing social animals: coordinating their activity with each other. Determining if and how nonhuman animals coordinate action with another individual holds our interest for several reasons. First, coordinating activity in time or in space with another individual is a central supporting element in social learning (Coussi-Korbel and Fragaszy 1995), and thus an important contributor to the generation of shared practices within groups (Fragaszy and Perry 2003b; see Chapter 13). Second, certain kinds of coordination with another may indicate something important about the complexity of social cognition possessed by an individual (see Tomasello and Call 1997 for a review of this argument). Experimental evidence suggests that capuchins cannot spontaneously mislead one another in a competitive situation (Fujita *et al.* 2002). Capuchins are not sensitive to what another individual sees or does not see (Hare *et al.* 2003), nor can they readily make use of human gaze or pointing to direct their attention to a hidden object (Itakura and Anderson 1996, Mitchell and Anderson 1997, Vick and Anderson 2000), although they can learn to do so (Kuroshima *et al.* 2002). In this dimension of behavior, capuchins are like other monkeys, but they seem to differ from the great apes, that show in a variety of ways a nascent ability to use another's directional gaze, to direct a human's attention by pointing, and to anticipate what another may do (see Anderson 1996 for review).

When the action of two or more individuals contributes to achieving a joint goal, we call coordination "cooperation." Increasingly diverse forms of cooperation have appeared across human evolution, and behavioral scientists consider evaluating the distribution of cooperative tendencies across extant primates an important means of understanding the evolutionary foundations of this aspect of human behavior. From a psychological perspective, one wants to know how an individual comes to organize its behavior with respect to another's behavior in space, time and form, and the complexity of joint organization that different species can achieve. For example, cooperation may be scaled according to the degree and nature of coordination between the parties, from two parties performing the same action synchronously to multiple parties performing complementary actions while monitoring each other's progress (Boesch and Boesch 1989).

Capuchin monkeys coordinate their activity in time and space so that it converges with that of other capuchin monkeys in many settings, ranging from group travel to mobbing snakes to inspection of foods (Boinski 1988,1993; Boinski and Campbell 1995, van Schaik and van Noordwijk 1989, Fragaszy *et al.* 1997a). Coordination in this manner is a common feature of group life in many animals. Do capuchins also cooperate? Answering this question is not easy, as it requires identifying that the individuals share a goal and determining that joint action contributes to achieving it. The most promising candidate activity during which cooperation may occur in natural settings is hunting. Capuchin monkeys might cooperate in the sense of division of labor during hunting (see Chapter 3) but we require more data before any strong conclusions can be drawn about this phenomenon. In any case, the data from observational studies of spontaneous behavior do not support strong conclusions about the nature of cooperation; that is why researchers have embarked on experimental studies of cooperation with monkeys in captive conditions.

Experimental studies concerning cooperation in captive capuchin monkeys have used apparatus that requires synchronous pulling by two animals. This form of joint action has not been reported for wild capuchin monkeys, but it is one that can easily be engineered in captivity and it has a long history in comparative studies, as it was first presented to chimpanzees in the 1930s by Crawford (1937). In the first study of cooperation with capuchins, Chalmeau *et al.* (1997) showed that individual capuchin monkeys easily learned to use two handles (both at once) to operate a dispenser that delivered a single piece of food. Gradually the handles were moved apart until they were so far apart that two individuals had to pull the handles at the same time to make the mechanism work. The groups of monkeys encountered the apparatus in their home cage, and they could come and go from the apparatus at will. The monkeys continued to pull at high rates when the handles were far apart and they succeeded occasionally at getting food in this way. However, they did not coordinate their pulling attempts in relation to another individual's presence or actions. In other words, they did not transfer the rule "use two handles at the same time" to the new situation, where the rule would have been "use my handle when the other actor is using the second handle".

Subsequently, Visalberghi *et al.* (2000) gave capuchin monkeys a variation of the joint pulling task. This time they tested pairs of monkeys, and the monkeys had to stand on an elevated platform at the front of the apparatus to reach the handles (see Figure 9.4). The new apparatus allowed the monkeys to see the food on a panel and to see the food rolling toward them following joint pulling. An individual monkey could pull its handle at any time but the panel in front of it tipped (allowing the food to roll to it) only when another animal was pulling at the same time 60 cm away. The feedback

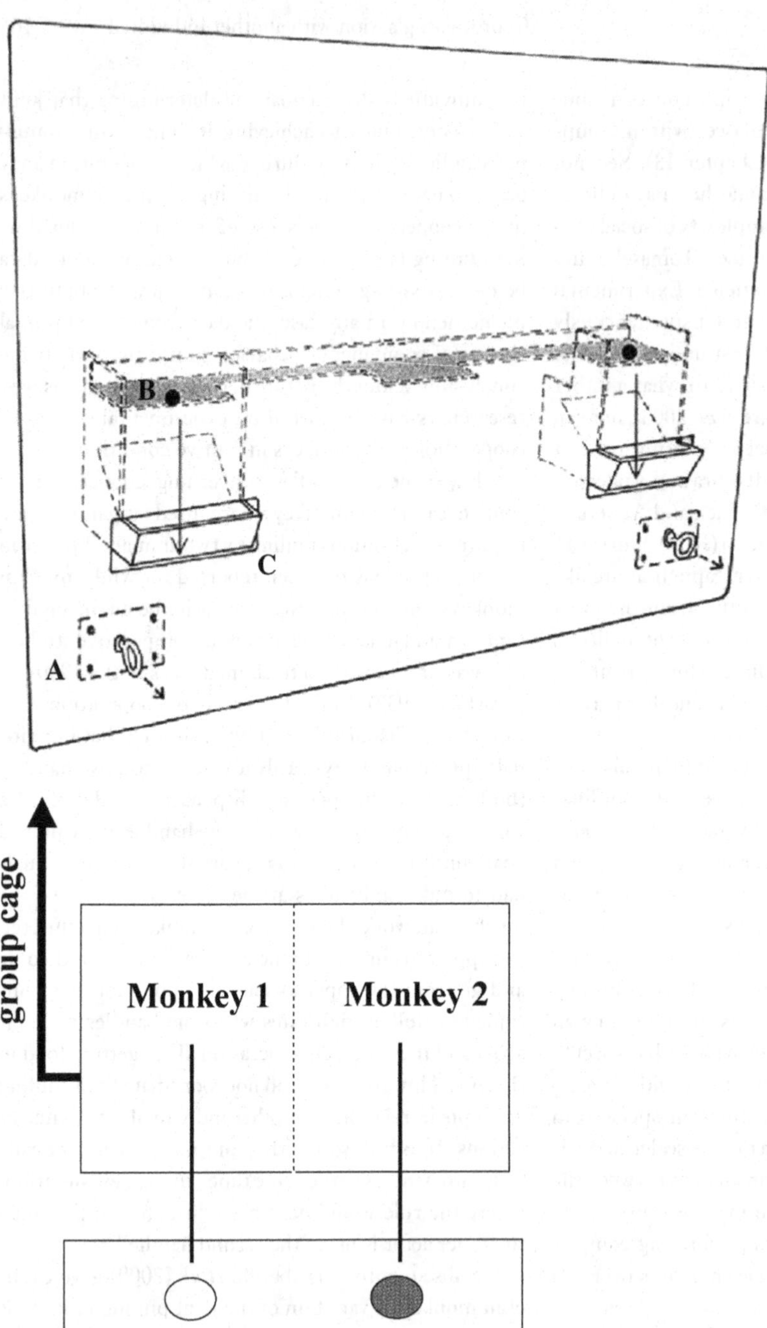

Figure 9.4. Top, apparatus used by Visalberghi *et al.* (2000) to study cooperation in capuchin monkeys. When two monkeys pulled the handles (A) simultaneously, a dispenser (B) was triggered and delivered a piece of food within the monkeys' reach in both troughs (C). The monkeys could move freely in the large room (3 × 1.90 × 2.60 m) in which the apparatus was present. Reprinted with permission of the American Psychological Association. Bottom, plan view of the setting used by Mendres and de Waal (2000) to study cooperation in capuchins. Each monkey stayed in one side of the partitioned test chamber (each side 0.70 × 0.60 × 0.60 m). The tray was counter-weighted and fitted with one or two pull bars. One or two transparent cups (ovals) rested on the tray. Each monkey could pull one bar (the black lines). The number of pull bars, which cup was rewarded, visibility through the mesh partition separating the monkeys, whether the partner on the left was confined to the test area or allowed to enter from and return to the group cage, and counter-weight of the tray were manipulated in a series of experimental conditions. (Redrawn by Elsa Addessi from Mendres and de Waal 2000.)

to the puller from the handle's movement was the same whether or not the other animal pulled. This time the monkeys learned part of the contingencies: they pulled more often when they could see food on the panel than when there was no food on the panel but they pulled equally often whether a partner was on the platform or not. When a partner was on the platform, the monkey pulled most often when the partner was within reach of the handle (although not selectively when the partner was also pulling). The monkeys apparently learned that the partner's presence close to the handle was more predictive that pulling would produce food than when the partner was elsewhere on the platform, but they were still willing to pull even when there was no partner.

Mendres and de Waal (2000) used another variation of the joint pulling task with pairs of capuchin monkeys, with two important differences in the arrangement of the task from that used by Visalberghi *et al.* (see Figure 9.4). First, the two monkeys stayed in mesh boxes, side by side, rather than moving about freely in a larger space. Second, the task (pulling in a weighted tray) permitted a monkey to feel a difference when it pulled alone (a heavier weight) *vs.* when another monkey was also pulling. Pairs of monkeys pulled a handle to bring toward them a tray on which two transparent cups were placed. The tray was weighted so that the monkeys were pulling against a fixed weight to move it. Each monkey could see if its cup contained food before and during pulling. In this situation, the monkeys pulled more often when they could see that a partner was in the other box than when they could see that the partner's box was empty. They pulled equally often when they could see their partner and when the other box was blocked from view (but the partner was there), but they were less successful when they could not see the partner than when they could see the partner. They also glanced at the partner more often when the partner's assistance was needed *vs.* when the weight was light enough that one monkey could pull it alone. Thus it seems that in this arrangement the monkeys were able to learn more about the significance of the partner for solution of the task than in the previous studies.

Why did Mendres and de Waal's (2000) monkeys learn to work together better than the monkeys in the previous studies? It seems likely that the addition of kinesthetic information allowed each monkey to appreciate the significance of the partner's position with respect to the handle and perhaps also the significance of the partner's actions. Learning to notice either of these contingencies could account for better success when the actor could see the partner than when it could not see the partner. In this task, each actor could feel the pull of the weight and see (if it glanced at the partner) over many seconds when the other monkey was facing the handle or also pulling, aiding in solution. Moreover, one individual could pull for many seconds and the second join in after some period; then both would be successful. In the tasks used by Chalmeau *et al.* (1997) and Visalberghi *et al.* (2000) the synchrony of action could last for a much briefer time but still be effective. An individual is less likely to notice brief synchrony, particularly when it can be detected only through vision. In short, our interpretation of the different outcomes across these three studies is that relevant cues in multiple sensory modalities over a longer period of time improved capuchins' learning of what conditions lead to success, compared with cues in one modality for a shorter period of time.

It seems likely that kinesthesis arising from manual activity is particularly potent in this regard, given the signficance of skilled and vigorous manual activity in the daily lives of these monkeys. It may be that when the actor uses the same sensory modality for monitoring its own actions and detecting the consequences of the partner's actions, the partner's influence is easier to detect and, possibly, to associate with the sight of the partner acting. We can ask whether cues in multiple modalities, especially kinesthesis, lasting over a period of seconds could enable capuchin monkeys to learn to perform complementary behaviors in a cooperative task, and how well the coordination of activity learned in the task used by Mendres and de Waal (2000) is sustained when the context is broadened to allow freer movement of the participants.

de Waal and Berger (2000) examined how capuchin monkeys, experienced at the synchronous pulling task described above and illustrated in Figure 9.4, coped with the situation in which both monkeys had to pull but only one monkey (R) was rewarded with food in its cup; the other monkey (no-R) had an empty cup; that is, it obtained no reward. The authors report that both capuchins pulled even when only one was rewarded. The monkey that had no food in its cup often reached through the mesh to pick up bits of food from the other side, and it did so more often when it had pulled a handle than when it had no handle. The monkey with food tolerated its partner retrieving bits of food from its side of the dividing mesh in all conditions. de Waal and Bergher described the R monkey, because it tolerated the no-R monkey reaching through to its side of the mesh to retrieve bits of food, as "paying" the other monkey for its help.

An alternative interpretation focuses on the immediate motivational sources of each animal's activity. For both monkeys, pulling in the experimental setting was associated with obtaining food. As we know from much work on the effects of varying schedules of reinforcement in operant tasks, an act performed to produce an outcome is especially likely to be performed persistently when it is reinforced occasionally. Thus, both monkeys should pull, even if only one receives food after doing so on any given trial. Similarly, both monkeys are motivated to find/collect the fruit when they pull the tray in. Thus, the no-R monkey is likely as highly motivated to seek food after pulling as the R monkey. It may simply reach more often into the R monkey's side of the testing cage to collect small bits of food under these conditions than when it had not pulled. Captive tufted capuchins routinely tolerate others taking food from their near vicinity or even directly from their hands or mouths (de Waal 1997, 2000, de Waal *et al.* 1993, Fragaszy *et al.* 1997b), and very occasionally actively give food to another (de Waal *et al.* 1993). Thus finding that capuchins allow another to take food through a mesh is not, by itself, sufficient evidence to claim that a special relationship allows this transfer of food.

Altogether, the data presented by de Waal and Berger (2000) show that capuchin monkeys persist in pulling even when they do not see food in their own cup, and they permit others to reach into their side of the cage to retrieve bits of food. They may be able to learn to cooperate in the sense that de Waal and Berger suggest is occurring in this setting (i.e., to "pay" a helper), but we think that drawing this strong conclusion must await further study.

de Waal (2000) reports that the social "attitude" of partner capuchins influences tolerance towards the partner taking food through a mesh partition, outside of a particular shared task while in the testing area. In the weighted pulling task both monkeys are working toward retrieving food. This situation may support positive attitudinal reciprocity as de Waal (2000) conceives capuchins can develop (see also de Waal 2002). We look forward to continuing work with this paradigm to clarify the character of shared access to food that capuchins display in this setting.

Work with the cooperative pulling task developed by de Waal and colleagues continues. The same monkeys that learned to pull in a weighted tray with another monkey in the confined circumstances described above later worked the apparatus together in a more variable arrangement (de Waal and Davis 2003). Individual monkeys' rate of pulling on a handle varied in accord with several variables, including how the food was distributed on the tray and who their potential partner was. For example, if the partner was dominant and the food was clumped, the subordinate individual faced the probability that the dominant partner could monopolize most or all of the food, and the subordinate's rate of pulling declined on trials with those particular conditions. This paradigm has proven its value as a means to study how capuchin monkeys make decisions about action in relation to current resource conditions and social circumstances. The monkeys' behavior indicates a remarkable sensitivity to social considerations, as is often proposed to characterize primates and to have been the result of distinct selection for social complexity (Byrne and Whiten 1988, Dunbar 2001).

Capuchin monkeys display a manner of coordinating activity with another individual in another sense that is relevant to this discussion. The coordination we speak of here is the prominent tendency of capuchin monkeys to place objects they are holding into the hands of a human, an act that humans invariably interpret as "giving." Indeed, we routinely take advantage of this propensity when we train our monkeys to "give" objects to us upon verbal request (a very handy habit when we need to retrieve our pens, keys, or glasses that have inadvertently gotten into the hands of the monkeys!). We suggest, however, that the natural human interpretation that the monkey is "giving" something to the other inaccurately attributes a social intention to the monkey. A capuchin monkey that is comfortable in the presence of a human will spontaneously approach the person and place any small object it happens to be holding into the person's palm, often rolling or rubbing the object on the person's palm, and picking it up and replacing it repeatedly. It seems that the person's palm serves as an interesting surface for the monkey to explore through touch. There need not be an intention to "share" motivating this behavior.

S. Brosnan and F. de Waal (pers. comm.) took advantage of capuchins' propensity to place objects in a human's hand to study how they valued objects when they could barter what they had for something the human experimenter offered "in exchange." The monkeys first learned that one token could be exchanged for a piece of sweetened cereal (highly desired), and a different token for a piece of bell pepper (a green vegetable; less desirable). When both foods were offered, the monkeys preferentially gave the experimenter the token associated with the preferred food. However, in subsequent testing, if the monkey could give the experimenter either token, but received food only if the token given matched the food offered, the monkeys did not selectively offer the token that matched the less-preferred food. Females consistently gave the experimenter the token associated with the preferred food, whether or not it was available on that trial. Males, on the other hand, simply gave the experimenter either token. Thus, the sexes differed in how they acted when the available food was less preferred, females acting more selectively (albeit ineffectively). Perhaps females are more attentive to the value of foods, the authors suggest, an interpretation that fits with the sex differences evident in other situations in which females are more selective than males about social exchanges that can be seen as holding differential value to the participants (e.g., grooming; see Chapter 11).

It is important to keep in mind that capuchin monkeys very rarely deliberately place objects, particularly objects of any value, in another monkey's hand (or mouth). Over many years of watching capuchins we have very rarely seen monkeys place food or any other object in another's hand. For example, de Waal *et al.* (1993) report 17 instances where a monkey was judged to have dropped food directly in front of another or placed food into the hand of another animal out of 3389 interactions over food that they observed. The rarity of this action contrasts with the routine tolerance capuchins display when they allow others to take food directly from their hands or mouths or to pick up nearby bits. Those very rare instances in which one animal actively places food into the hands of another may be interpreted as "sharing" or "giving" (see Chapter 13), or they may be interpreted as the exploratory placement of an object on a new surface, as seems to occur with humans when the monkeys place food or other objects on a human's hand. For this reason, documenting the contexts in which capuchins reliably give food to another by active, direct placement into the other's hand is an important task. We encourage field observers to be particularly careful about documenting food transfers among capuchins in natural settings.

COPING WITH RISKS

Risk assessment and coping with risk

In addition to generating innovative behaviors and to coordinating action with others, solving problems involves balancing opportunities against risks. Capuchin monkeys in natural settings must balance their use of resources (food, water, resting areas, travel routes etc.) against the risks of running afoul of more dominant animals, predation, poisoning, falling, parasitic infection and so on (see Chapters 2 and 3). Behavioral ecologists have conceptualized risk in the abstract terms of the cost accruing from the loss of a resource or the potential for bodily harm (e.g., from starvation; Janson and van Schaik 1993). Psychologists, in contrast, conceptualize risk in terms of the affective value of a situation, event, or object, assessed as the power of the situation to produce withdrawal, inhibition or defense. It is this psychological conception that we address here. This conception of risk is related to stress. Stress is a factor that disturbs the organism's equilibrium in a fashion that elicits processes which result in regaining equilibrium (see Barrows 2001 for elaboration).

Assessing risks in nature

Capuchin monkeys in natural settings indicate what features of the environment they monitor for risk by looking around, including upward (for aerial predators) and downward (for terrestrial predators). Such vigilance varies with age, sex and setting. Obvious risk factors for all ages include being low in the canopy or on the ground (exposed to cats and snakes) and being in tree crowns exposed to aerial predators. Monkeys are more vigilant when in these places, and also when farther from other animals (van Schaik and van Noordwijk 1989).

Capuchin monkeys' assessment of what constitutes a predatory threat may sometimes surprise a human observer. In Fragaszy's study group of *C. olivaceus*, all animals reacted strongly to the sound of a dog barking by becoming completely silent, remaining perfectly still for a few seconds (a common reaction to an observed mild threat, such as a man on a horse), but then fleeing rapidly and silently like ghosts through the forest in the opposite direction from where the sound came. In contrast, they did not react to the sound of a gunshot that was, to the human ear, about the same distance away as the barking dog. Apparently unnatural noises, although perhaps startling, do not have an inherently aversive quality for these monkeys. The human observer, on the other hand, had reactions opposite to those of the monkeys to these same sounds.

In general, it seems that male capuchins are less averse to risk than are female capuchins. Males are more likely than females to travel and forage on the ground and apparently more willing to endure biting and stinging prey. For example, Fragaszy (1986) observed individual male *C. olivaceus* breaking paper-wasp nests off the branches to which they were attached, then running off with the nest with the adult wasps in fierce pursuit. Once the monkey had left most of the wasps behind, it settled down to enjoy the larvae, extracting them one by one from the nest, while more or less ignoring the occasional wasp still attacking his hands and face. Females did not seek out wasp nests, foraging instead in the quieter realms of the crowns and trunks of palm trees and on diverse small branches and vines (Fragaszy and Boinski 1995). Similarly, among *C. capucinus*, males are more likely to participate in hunting mammalian prey, which (like the wasps) often defends itself vigorously against the monkeys' marauding attacks (Fedigan 1990, Rose 1997). Adult males are more likely to stay near a predator after the rest of the group has fled, and males often push large dead branches until they fall crashing to the ground, making a satisfyingly loud noise, after encountering a predator or another disturbance (including human observers). This is not to say that females are uniformly uninvolved in hunting or in dealing assertively with prey or with potential predators, only that a sex difference is evident in the relative frequency with which they engage in such activities in natural settings. Indeed, in Carlos Botelho State Park, Brazil, female *C. apella* take wasp nests (Patrícia Izar, pers. comm.), although Fragaszy saw only male *C. olivaceus* taking wasp nests at Masaguaral. As we shall see next, female capuchin monkeys are capable of the same array of assertive defensive behaviors as males.

Responding to risks in captivity

In a study with two groups of captive tufted capuchin monkeys, Vitale and Fragaszy (unpublished) presented two potentially fear-invoking objects (the glove used by the animal technicians during infrequent capture procedures, and a battery-operated toy penguin, 25 cm tall, that flapped its wings and made a clacking sound when remotely activated). The objects appeared inside a box in the home cage when the experimenters opened a door on the box by remote control. The same box was presented many times with nothing inside it, and the monkeys did not respond to the empty box or the closed box as potentially threatening. To the experimental objects, however, they directed full-scale mobbing behaviors. Each group responded with intense mobbing only to one of the objects, and it was a different object in each group. In one group, an adult male responded the most quickly and for the longest time; he also got closest to the apparatus (i.e., he "led the attack"). In the other group, a young adult female took on this role (although there was an adult male in this group as well).

In a similar study, Vitale *et al.* (1991) presented a model rubber snake (that could be moved by the experimenter like a puppet in a life-like, sinuous manner by pulling on strings) individually to six tufted capuchin monkeys and four crab-eating macaques (*Macaca fascicularis*). The experimenter moved the head of the model snake in and out of an opaque container along a wall in the monkey's room for four to five seconds at irregular intervals over a 5-minute period. The capuchin monkeys responded intensely to the snake, as did the crab-eating macaques. An interesting difference emerged, however, in how individuals of each species behaved towards the model snake. The capuchin monkeys rather quickly changed their behavior toward the snake from full alert and defense to include assertive challenges. They moved close to the snake while directing threatening facial expressions and vocalizations in its direction, lunged at it, and occasionally tossed small objects toward it that were at hand on the floor of the cage. This pattern sounds like the captive equivalent to the branch-dropping behaviors of wild capuchin monkeys that cause field observers to watch out for disturbed monkeys overhead (see Chapter 3). One monkey broke off twigs from a tree branch and pushed these into the area where the snake appeared. Other monkeys touched the snake directly. Eventually, one capuchin attacked and dismantled the snake and the experiment came to a natural and satisfying end from the monkey's point of view. In contrast, the crab-eating macaques remained as far away from the place where the snake appeared as the layout of the room allowed, and directed threatening and fearful facial expressions and vocalizations in the snake's direction when it appeared.

For captive monkeys, threat of capture is a diffuse, repeated risk that is not under their control to avert. The appearance of a capture glove, net or the person most predictably associated with these objects and capture events (such as the veterinarian) produces profound changes in the monkeys' behavior. They may give threatening or fearful vocalizations, they stop eating or manipulating objects in a routine fashion, and they move around the upper parts of their enclosure, apparently attempting to move away from the threat. Even after the veterinarian or other disturbing person leaves the housing area and all the capture paraphernalia has been stowed out of sight, the monkeys remain vigilant and they continue to display disturbed behavior for several minutes to hours. Studies in Fragaszy's laboratory (Landau and Fragaszy, unpublished) have shown that having a food-searching activity to perform after the veterinarian leaves allows the monkeys to return to normal behavior almost immediately, rather than to continue pacing and giving fear vocalizations. In this case, a diversionary activity that affords them control, occupies their hands and provides food is an effective coping strategy. Manipulating objects and searching for food is, of course, virtually always possible in natural settings.

Novel space

Moving into a strange area is a challenge for many animals, including capuchin monkeys. Boinski *et al.* (2000) document the ways in which animals can behave cautiously when entering a strange area. In natural settings, individuals control in what manner and when they enter strange areas. In captive conditions, they have little control over their location. Nevertheless, their behavior upon entering strange spaces can illuminate species-normal characteristics of spatial exploration and adjustment. Fragaszy and colleagues watched two groups of tufted capuchin monkeys adjust to life in spacious outdoor enclosures (30 m × 30 m, with grassy substrate and a multi-level perch system throughout the cage interior) after having lived for all their lives (in the case of younger animals) or for many years (in the case of a few older animals) in much smaller indoor quarters (Fragaszy, Matheson and Johnson-Pynn, unpublished).

To our surprise, the two groups responded to the challenge of the novel space in very different ways. One group contained a resident elderly, arthritic male that spent a lot of time on the floor of the cage in the indoor quarters. This group moved out quickly (following the male's lead) and immediately began to move around the cage on the ground as well as the perches. In the first 15 minutes after the door was opened, the male moved immediately on the ground along the periphery of the cage and then into the interior. His group mates followed suit, although they moved along the perches as well as on the ground, and they did not follow him but instead traveled in diverse directions around the cage. The other group, which had no resident male at that time, was so hesitant to leave the indoor area that we finally forced them from the interior and closed the door so that they could not re-enter the indoor area for a few hours. This group did not travel as far into the interior nor as far along the perimeter of the cage as the other group, and they did not travel on the ground as much as the other group. Over the next 2 weeks, adult females in the group with the resident male continued to spend more time on the ground than females in the other group. We interpret the differences between adults in the two groups as an indication that the behavior of a central member of the group can powerfully influence how others assess the risk of moving into new areas. We take up this aspect of social influence again in Chapter 13.

How do captive monkeys adjust to a more natural space? Do they immediately move into the trees and act like wild monkeys? Unfortunately, not at all. Four adult monkeys (two males, 5 and 7 years old, two females, both 17 years old) released for the first time into a fenced evergreen forest (1.5 hectares) at Monkey Jungle, near Miami, Florida, displayed wildly varying reactions to their new-found freedom in this rich environment (Oetting *et al.* 1994). After spending eight days in a small mesh enclosure inside the forest, all four monkeys exited the enclosure after it was opened, but they did not go far. With the exception of one female's circumnavigation of the entire perimeter of the fenced forest area on the ground, a trip that took her about 15 minutes, the monkeys remained near the enclosure. The female that had traveled around the forest on the ground re-entered the enclosure on her return and did not leave it again voluntarily for days. Her three comrades were braver during this period, venturing out a short distance into the forest, but generally staying nearby and returning often to the small enclosure as if reluctant to leave her and/or to move too far from the enclosure. One monkey even became "lost" on one of the first days following their release after he moved out of sight of the small enclosure that appeared to serve as psychological home base for the introduced monkeys. He gave species-typical "lost" calls and did not find his way back to the enclosure and his companions for several hours.

Over 3 months during which these monkeys were tracked carefully, the released monkeys used the space and substrates differently from two resident tufted capuchins. The released monkeys showed the same postures and locomotor behaviors as the residents, but they travelled less widely and spent far more time on the wire mesh walkways than did the resident monkeys. As the wire mesh walkways were gradually removed from the forest for repairs during the period following the monkeys' release, the introduced monkeys shifted their use of space to those areas where wire walkways remained. Even at the end of 3 months, they spent less time feeding and more time inactive than the resident animals, and they spent more time on artificial substrates.

Altogether, although wonderfully flexible when faced with diverse challenges in captive settings, captive capuchin monkeys are inept at many aspects of living in natural settings. The skilled travel patterns, attention to potential food sources, and even patterns of locomotion on varying substrates that seem so effortless to the capuchins that we watch in natural settings – all require experience for their normal practice (see Menzel and Beck 2000 and Beck *et al.* 1991 for a similar evaluation of captive lion tamarins, *Leontopithecus*, released into a forest).

SUMMARY

Capuchins are powerfully motivated to engage the world through direct physical contact with objects and substrates, initiated primarily with their hands. They are capable of using the body (hands, tail and feet) in innovative ways (e.g., surrounding or holding multiple cups with their tail and feet while working with them). Innovative use of the body is an essential aspect of human behavior as well, although we do not often recognize this aspect of ourselves. Also like humans, capuchin monkeys are endowed with unusual proclivities to combine objects and to string together sequences of actions with objects. For example, they can seriate nesting cups effectively and they become more strategic at doing so with minimal practice.

According to several theories of human cognition, knowing the world through action can proceed without or before developing abstract reasoning (Case 1992, Johnson 1989, Klahr 1994). These theories posit that action enables the development of abstract relational mental activity. The extent to which capuchins develop abstract knowing, and how action might contribute to this development, is a subject for continuing research.

In routine conditions capuchins act synchronously (e.g., traveling and feeding together), and researchers have wondered whether they might coordinate activity intentionally (i.e., cooperate) to solve a shared problem. Studies with captive monkeys have used joint pulling tasks to assess how well capuchins work together. To date, capuchins have learned to work together most effectively to achieve a shared goal when each monkey can feel as well as see the partner acting over a period of seconds to minutes. We suggest that the duration and the multimodal nature of the feedback each actor gets from participating in this task aids the monkeys' learning.

Capuchins assertively threaten intruders or potential predators, chase down prey, and in some situations readily explore novel spaces. On the other hand, the assertive demeanor so evident in capuchins when they are in familiar circumstances does not necessarily help the monkeys cope with completely novel situations (e.g., large spaces with limited visibility etc.). Captive monkeys cannot solve everyday problems that wild monkeys can without extensive practice, and vice versa. The lesson here is that individual experience and immediate context powerfully affect capuchins' exploration and problem-solving.

10 • Fancy manipulators. Capuchins use objects as tools

Toko, an adult male tufted capuchin, faces the problem of cracking open a nut that seems as hard as iron when he bites it. Recently, when he and other members of his group encountered piles of wooden blocks in their enclosure, they produced outstanding pounding ovations with them. Their enthusiastic activity was so noisy that I wondered whether producing loud sounds by pounding was as rewarding for them as it is for children. When I gave them nuts and wooden blocks, at first Toko pounded a nut on the hard surfaces of the cage, but he did not manage to open it this way. As time went by, Toko sometimes pounded the nut on and by the wooden block. He also ardently pounded the block on the floor or wall while the nut was in his mouth and afterwards, took the nut out of his mouth as if to check if it had broken (see Figure 10.1a). Toko tried every possible spatial combination between nut and block. At last he produced the correct combination of objects and the right action, pounding the nut with the block as the nut rested on the hard floor. This time the nut broke! He paused, staring at the broken nut as if he hardly believed his success. He proceeded to place another nut on the floor and began pounding the block nearby. He seemed to realize that pounding is the "key" but he still had to learn how to get this key into the "keyhole". Eventually he mastered this task, becoming skillful at pounding open nuts using a wooden block (see Figure 10.1b).

Notwithstanding Toko's initial problems, scientists have known since the dawn of European exploration of the New World that capuchins occasionally use diverse objects as tools for many purposes. They can become quite skilled in doing so and, indeed, much of their reputation as "clever" monkeys rests upon the fact that they use tools. Our fascination with tool use by members of nonhuman species reflects a profound appreciation of the importance of tools to our own species, and our curiosity about the mentality of these other species that use tools, like us.

Toko's problems illustrate an extremely important point about how capuchins arrive at using an object as a tool. The reader might think that Toko's initial ineffective behavior with the blocks and nuts reflects his lack of experience with nuts, with hard objects to pound, and with both things together, and that wild capuchins would not behave so naively. However, all capuchins, wild or captive, exhibit inefficiencies in coordinating objects, surfaces and actions (like Toko did) upon encountering a new problem. Capuchins are not unusual in this regard: apes and humans (especially young individuals) also have problems achieving effective coordination among actions and objects when they start out in a new tool-using situation (Inoue-Nakamura and Matsuzawa 1997, Lockman 2000, Smitsman 1997). We expect early attempts to be inefficient when the individual must manage a new set of spatial relations, particularly when the task involves mastering a dynamic relation, or combining two or more relations in the same task. In this chapter, we use a scheme that is an initial attempt to classify tool use within the perception-action perspective adopted in Chapters 7–9 (see Table 10.1). We propose this scheme as a starting point, acknowledging that it does not, in its current formulation, permit an unambiguous classification of all variants of tool-using behaviors. In the example with Toko provided above, to crack the nut he had to position the block with respect to the nut, after placing the nut on a hard horizontal substrate. According to our classification scheme, this problem involves a single dynamic relation; we call this a first-order relation.

We start by defining tools and tool use, provide some historical background to our interest in and knowledge

(a)

Figure 10.1. (a) Captive capuchins spontaneously learn to use objects to crack open nuts. As described in Visalberghi (1987), the naïve monkey explores nuts and blocks with a repertoire of actions, many of which are ineffective; gradually more effective actions come to predominate. The monkey pounds the nut on the tool (top left); the monkey pounds the nut on the floor next to the tool (top right); the monkey pounds the tool with both hands on the floor and keeps the nut in its mouth (bottom left); the monkey places the nut on the tool and then pounds tool and nut on the floor (bottom right). (Drawing by Stephen Nash from Elisabetta Visalberghi's photographs.) (b) An adult male effectively cracks open a nut by striking it with a log, demonstrating skillful use of a tool. This is an example of a dynamic first-order relation; see Table 10.1. (Drawing by Stephen Nash from Elisabetta Visalberghi's video.)

of tool use in nonhuman primates, and describe the behavioral and ecological factors that foster the occurrence of tool use in capuchins. Next, we summarize the research findings pertaining to tool use in wild and captive capuchins. We organize these findings loosely in accordance with the relations that must be mastered to use tools in a particular context. Finally, we discuss the extent to which capuchins' behavior in tool-using contexts reflects abstract or generalized recognition of the causal relations between their actions and the movements of objects, and how their mastery of the abstract elements in using a tool compares with the other famous nonhuman users of tools, the chimpanzees.

DEFINING TOOLS AND TOOL USE

"Tools" is not a natural category of object (such as "plant" or "animal"), and "tool use" is not a natural category of behavior (such as "locomotion" or "feeding"). An animal uses a tool when it uses an object as a functional extension of its body (mouth, beak, hand, claw etc.) to act on another object or a surface to attain an immediate goal. Thus the behavioral expressions of "tool use" are potentially infinitely variable and appear in all functional categories of behavior (e.g., feeding, communication, self-care, exploration). A tool acquires its function by its use and the goal it helps achieve defines

(b)

Figure 10.1. (*cont.*)

what "kind" of a tool it is. For example, one can use the heel of a shoe to pound a nail, a toothbrush to sweep something off the table, and a cane to explore the sidewalk in front of one's feet. In these cases, the shoe is "a hammer", the toothbrush is "a broom" and the cane is "a sensory substitute for the hand" (see Burton 1993). This definition, like others before it, leaves uncertain the status of some activities for which it is difficult to specify the goal of the action (e.g., use of objects in the service of exploration, or activities with objects directed to the body such as anointing; see Chapter 5). Given their ambiguity, these latter actions (exploration, anointing) are not usually classified as tool use.

To be conservative, we focus on cases in which capuchins use objects to achieve a tangible goal. Moreover, we add to the usual definition of tool use given above (using an object as a functional extension of the body to act on another object or surface) the requirement that the actor *produce* a relation between the tool and another object or surface, and not simply use an already-existing relation (see Table 10.1). This definition excludes some situations that others commonly include as examples of tool-use, such as pulling in a stick or other object that is already in contact with a target (say, a piece of food) when the actor arrives on the scene. In our scheme the monkey has to place the stick in relation to the food to classify the action as using the stick as a tool.

TOOL USE IN NATURE AND IN CAPTIVITY BY CAPUCHIN MONKEYS

More than 100 years ago, keen observers reported that captive chimpanzees used tools. The phenomenon was at first dismissed by the scientific community as an unnatural behavior prompted by humans and, until recently, tool use was viewed as one hallmark of human distinction. Nowadays, we know that wild chimpanzees use a large variety of tools, and they do so in different ways across Africa and within local communities (Boesch and Boesch-Achermann 2000, Humle and Matsuzawa 2002, Matsuzawa and Yamakoshi 1996, McGrew 1992, Sugiyama 1993, Whiten *et al.* 1999, 2001, Yamakoshi and Myowa-Yamakoshi in press).

Like chimpanzees, capuchins readily use tools in captivity (Figure 10.2). Tool use by captive capuchins had already been reported hundreds of years ago in chronicles of the first European scientific expeditions in the New World (de Oviedo 1526/1996). However, as was the case for chimpanzees until 40 years ago and for orangutans until very recently (van Schaik 2003, van Schaik *et al.* 1996, 2003), we have little evidence of *systematic* use of tools by capuchins in the wild. Despite occasional reports of tool use by wild and semi-free ranging capuchin monkeys (summarized in this chapter) the scarcity of tool use by wild capuchins still contrasts with their impressive achievements in captive settings.

Visalberghi (1993a) commented on this puzzling contrast by noting that capuchins have not been studied in the wild as extensively as apes and that their more arboreal lifestyle limits their opportunities to use tools compared with apes. In the trees, their hands are more often needed for support; moreover, loose objects that could be used as tools are less available and less easily set aside and retrieved, and stable, strong and appropriately shaped supporting substrates are less available

Table 10.1. *Relations produced through action with an object that are evident in capuchins' use of tools. In our view, an action involving a zero-order relation is not tool use; tool use requires producing a first-order relation. Order refers to the number of relations between objects and surfaces that are required to reach the goal, and not to the number of actions in a sequence.*

Relational category	Definition	Examples
Zero order	Act on one object; action on second object occurs by default	Pull in a cane positioned with food inside the hook and the straight part of the cane within reach Pull in cloth with food on the cloth
First order		
Static first order relations	Acting with an object on a fixed surface (or on a fixed object) to reach the goal	Probe into an opening with a stick ("dip") Pound a stone on a nut fixed on a surface
Dynamic first order relations	Acting with an object A in relation to an object B that moves. Since action with A alters the state of B, B must be monitored as action progresses	Push food out of a tube with a stick Pull in an object with a stick when they are not already positioned so that pulling is effective Pound a loose nut with a stone
Second order		
Sequential second order relations	Acting with an object A in relation to object B following placement of object B in relation to a third object C (surface or object). In this case, one static relation between B and C and then one dynamic relation between A and B are produced	Pound a stone against a nut placed on a second stone
Simultaneous second order relations	Acting with an object A in relation to object B while maintaining B in relation to C (surface or object). In this case two dynamic relations (between A and B, and between B and C) are coordinated simultaneously	Push food through tube with a stick while avoiding a hole Pull food with a rake across a surface with a hole Pound a stone against a nut on an anvil surface while holding the nut (to prevent the nut from falling off the anvil)

in the trees than on the ground. Imagine pounding a round nut on a log or stone that rests solidly on the ground. Then imagine the same activity while sitting in a tree and pounding the nut on a sloping tree branch! Finally, activities carried out high in the forest canopy are more difficult for terrestrial humans to see than activities occurring on the ground.

All of these are plausible explanations for the rarity of observations of tool use in wild capuchins. However, although arboreality may limit opportunities for capuchins to use tools or for us to observe such activity, we know that chimpanzees and orangutans do sometimes use tools in trees (Boesch and Boesch-Achermann 2000, van Schaik 2003, van Schaik *et al.* 1996). Thus arboreality alone does not preclude tool use. Instead, we must consider what aspects of capuchins' behavior and ecology might support the discovery of how to use an object as a tool in the wild. We will discuss this point

Figure 10.2. An adult female tufted capuchin (Roberta) dips for apple sauce while holding her newborn infant in one arm. She holds the stick with a power grip. This is an example of a static first-order relation; see Table 10.1. (Photo by Elisabetta Visalberghi.)

later in the chapter. For the moment, this consideration might suggest other ways we can look for tool use in wild capuchins, and help us to understand why we observe it more often in captive monkeys.

One can turn the question around and ask why we see tool use at all in capuchin monkeys. Like other primates, capuchins possess the necessary sensory and anatomical characteristics for using objects as tools. As described in Chapters 5 and 7, they have a well-articulated hand with anatomical adaptations that favor the fine manipulation or precise positioning of objects, and they have sufficiently long limbs, postural control, and strength to generate considerable forces (when pounding, for example). This, however, does not distinguish them from most other monkeys, especially Old World monkeys, although all other monkeys use tools less often than capuchins.

Capuchins possess two behavioral characteristics that are less widely shared with other primates and that are particularly relevant to using objects as tools. First, although using a tool is an individual endeavor, it is acquired more readily in socially supportive contexts where experts tolerate novices nearby (Coussi-Korbel and Fragaszy 1995, Fragaszy and Perry 2003a, van Schaik 2003). Capuchins are relatively tolerant of one another, particularly adults of youngsters (see Chapters 6, 11 and 13). Second, and fundamental for the discovery of tool use, capuchins generate a great variety of explorative and manipulative behaviors that involve acting with objects and on surfaces (see Chapters 7 and 9). Capuchins reliably spontaneously combine objects with substrates and with each other by pounding and rubbing; they also insert their hands and objects in holes and crevices (Byrne and Suomi 1996, Fragaszy and Adams-Curtis 1991). When captive capuchins encounter objects they consider benign, whether novel or familiar, they quickly approach, explore and manipulate them with enthusiastic interest (e.g., Jalles-Filho 1995). Their interest towards objects persists over time, even towards familiar objects (Visalberghi 1988, Westergaard and Fragaszy 1985). Although wild capuchins initially often avoid novel objects (Visalberghi *et al.* 2003), they explore and manipulate familiar objects and substrates persistently and routinely engage in many actions. This can allow them to discover the consequences of actions combining objects and surfaces (Terborgh 1983, Fragaszy 1986, Chevalier-Skolnikoff 1990, Panger 1998, Boinski *et al.* 2001).

All of these behavioral characteristics make it likely that a capuchin monkey upon encountering an interesting set of objects or an interesting substrate with loose objects available, and with the motivation, time and security to investigate, will produce actions with objects on surfaces. Tool use relies upon perception/action routines (e.g., pounding, inserting) that are applied to virtually any set of objects and surfaces they encounter. As a routine behaviour the monkey may occasionally combine one object or surface with another object, and so discover that using an object helps it achieve some goal. This scenario is sufficient to support the frequent discovery of tool use by captive capuchins, but one can see that it might not occur as often in natural settings. When might these conditions apply in nature? If we knew that, we could predict when and where we would find capuchins using tools in nature. Let us start the "story" from the beginning, that is, from the early reports of tool use.

EARLY REPORTS OF TOOL USE BY CAPUCHIN MONKEYS

The earliest report of tool use in *Cebus* dates back about 500 years, to the Spanish naturalist Gonzalo Fernández de Oviedo y Valdés. Urbani (1998), who recently brought attention to this report, found in Chapter XXV of the *Sumario de la Natural Historia de las Indias* (de Oviedo 1526/1996) the following passage [translated from Spanish by Urbani 1998]: "Some of those cats (monkeys) are so astute that many things they see men do, they imitate and also do. In particular, there are many that when they see how to smash a nut or a pine nut with a stone, they do it in the same way and, when leaving a stone where the cat (monkey) can take it, smash all that are given to them. They also throw a small stone, of the size and weight of their strength, as would be thrown by a man". Urbani suggests that since de Oviedo traveled mainly in the Darien region (today Panama and northwestern Colombia), he probably referred to *Cebus capucinus*.

Erasmus Darwin, the grandfather of the more famous Charles, was also impressed by a captive capuchin monkey without teeth cracking open a nut using a stone (Darwin 1794). A century later the British psychologist George Romanes (1883/1977) reported several observations of a capuchin using tools, including nut cracking, aimed throwing and the use of sticks for raking. From that time on, many naturalists and psychologists have periodically reported serendipitous observations and studies of capuchins using tools (Belt 1874, Bierens de Haan 1931, Cooper and Harlow 1961, Garner 1892, Nolte 1958, Vevers and Weiner 1963, Warden *et al.* 1940, Watson 1908; for further details, see Beck 1980).

EXPERIMENTAL STUDIES OF CAPUCHINS USING TOOLS

The first major "modern" contribution to the study of tool use by capuchins was made by Heinrich Klüver (1933, 1937). Klüver systematically studied a few captive capuchin monkeys with no prior experience that specifically prepared them to use tools. One of his subjects, PY, participated in more than 300 experiments. These experiments, which are reminiscent of those carried out by Hobhouse (1915), Yerkes (1927a, b), and Köhler (1925/1976) with great apes, involved presenting food in various locations (lying beyond reach on the floor, suspended from the ceiling, etc.). Klüver also varied the objects available to the monkey (stick, rope, sack, brush etc.), where the objects were presented (near or at a considerable distance from food, within or beyond reach), the number of different actions necessary to obtain the food, and whether or not the monkey had to construct the tool. Overall, PY was very proficient, although less successful in those tasks in which capuchins' most common activities with objects (pounding, rubbing, biting etc.) were not likely to result in a solution (e.g., straightening a wire to use it as a stick, or stacking boxes in a particular [stable] combination). Despite Klüver's successful debut, the study of capuchins' tool-using abilities was short-lived (with the brief exception of studies by Harlow reviewed in the Prologue; see Fig. 10.3); systematic investigations of capuchins' propensities for using objects as tools were not resumed until 40 years later.

Table 10.2 lists the articles on tool use in capuchins published in the two decades since Beck's (1980) exhaustive review. Some of these studies have also been described extensively in recent reviews (Anderson 2002, Tomasello and Call 1997, Visalberghi 1990a). A glance at Table 10.2 is enough to see that in the 1980s researchers started to undertake systematic investigations with impressive diligence! Parker and Gibson (1977, 1979) helped spark renewed interest in capuchins' tool use behavior. They argued that higher forms of sensorimotor intelligence (*sensu* Piaget) evolved as an adaptation for extracting embedded food resources, and Gibson (1986) further argued that relative forebrain size was larger in species that were extractive foragers than in species that relied on other foraging strategies. Extractive foraging includes searching for food without tools (e.g., unrolling leaves, peeling/stripping bark, or turning

Figure 10.3. Harry Harlow conducted experiments about tool use in capuchin monkeys in the 1930s. Left a food reward hanging from a rope is too high to reach directly and the capuchin uses a stick to knock it down. Middle now the reward is even higher and the capuchin places a wooden box underneath, and (right) uses the stick from this elevated position to reach the reward. (Courtesy of Harlow Primate Laboratory, University of Wisconsin-Madison.)

over rocks) as well as using objects to process foods (e.g., cracking open nuts with a hard object). According to the extractive foraging hypothesis, the common occurrence of tool use in capuchins and chimpanzees is the result of convergent evolution related to their shared exploitation of embedded foods. Some researchers inspired by this view provided descriptions of instances (or possible instances) of spontaneous use of tools in the absence of specific efforts to elicit it (Chevalier-Skolnikoff 1989, 1990, Gibson 1990). Others undertook controlled testing within the Piagetian framework (Natale 1989, Parker and Potì 1990) with the aim of understanding the development of tool-using behaviors in young individuals as well as the achievements of adults. In the latter studies, the experimenters placed a desired food out of the monkey's reach and provided one object (a stick) that could be used to rake in the food. Inspired by Köhler's (1925/1976) studies, Natale (1989) and Parker and Potì (1990) presented the capuchins with many different conditions. For example, they varied the position of the food in relation to the stick, or they made it necessary to use one stick to get another. Capuchins initially solved the problem inefficiently but later became skilled at manipulating the stick as required by the variations of the task. They succeeded in retrieving the reward from various locations even when they needed to reposition the stick.

In one of the few developmental studies of tool using, Parker & Potì (1990) followed how one capuchin began to use a stick as a rake. They tested the capuchin once every week until 1 year of age and thereafter once a month. At 9 months of age, when testing began, their capuchin pulled a stick straight in to obtain food (a problem involving a zero-order relation; see Table 10.1). At 18 months, it was able to reach the food with the stick to rake it in, even when it had to reposition the stick to do so (a problem involving producing a first-order relation). However, by the time testing ended at 24 months of age, the capuchin had not yet used a short stick to get a longer one (a task involving two first-order relations in sequence). Sequential problems of this sort are readily solved by adult capuchins (Klüver 1933, Westergaard and Suomi 1993, Westergaard *et al.* 1997), although positioning a rake accurately requires practice.

At present the theory proposed by Parker and Gibson (1977, 1979) linking intelligence (expressed in tool use) to extractive foraging is considered problematic because most extractive foragers do not routinely, or even rarely, use tools. Moreover, Tomasello and Call (1997) present three additional flaws in the theory: (a) it

Table 10.2. *Recent studies on tool use in capuchins, chronologically ordered.*

(a) Studies in captivity[a]

Task	Relational category[b]	Specific aim(s) of the study	N tool users/N total[c]	Species	Source
Nut cracking	First dynamic	Selection among differently effective tools	1/6	*C. apella*	Antinucci and Visalberghi 1986
Nut cracking	First dynamic	Acquisition of the behavior and social learning	2/42	*C. apella*	Visalberghi 1987
Dipping Sponging	First static First static	Acquisition of the behavior and social learning	6/9 9/9	*C. apella*	Westergaard and Fragaszy 1987a
Stick directed to a wound	First static	Serendipitous observation	n.a.	*C. apella*	Westergaard and Fragaszy 1987b
Stick directed to a wound	First static	Serendipitous observation	n.a.	*C. apella*	Ritchie and Fragaszy 1988
Raking/digging/probing	Ambiguous description	Observational study	n.a./12	*C. apella* *C. albifrons*	Chevalier-Skolnikoff 1989
Nut cracking Stick to push	First static First static	Social influences on tool use acquisition	5/20 5/20	*C. apella*	Fragaszy and Visalberghi 1989
Stick to rake	First dynamic	Sensorimotor intelligence	3	*C. apella*	Natale 1989
Stick to push a reward out of a tube	First dynamic	Appreciation of how the tool should be modified	3/4	*C. apella*	Visalberghi and Trinca 1989
Nut cracking	First dynamic	Benefits in terms of time and success due to the use of tools	5/6	*C. apella*	Anderson 1990
Stick to rake	First dynamic	Development	3/5 1	*C. apella*	Parker and Potì 1990
Sticks to push a reward out of a tube	First dynamic	Selection of the appropriate tool	4	*C. apella*	Visalberghi 1993a
Nut cracking and probing	First static First static	Sequential use of tools (tool-set)	3/9	*C. apella*	Westergaard and Suomi 1993
Probing	First static	Selection of the appropriate tool	2	*C. apella*	Anderson and Henneman 1994

Dipping	First static	Tool acquisition in juveniles and influence of the context	3juv/9juv	*C. apella*	Fragaszy *et al.* 1994b
Nut cracking	First static		3juv/9juv		
Stick to push a reward out of a tube	Second simult.	Understanding of cause-effect relations	4	*C. apella*	Visalberghi and Limongelli 1994
Stone flaking	First dynamic	Production of flakes	6/11	*C. apella*	Westergaard and Suomi 1994a
Stones as cutting tools	First static	Stones as cutting tools	3/15		
Nut cracking	First static	The use and modification of bone tools	3/9	*C. apella*	Westergaard and Suomi 1994b
Bone fragments as cutting tools	First static		3/9		
Bone modification due to the use of tools	First static	Modeling early hominid technology	5/10	*C. apella*	Westergaard and Suomi 1994c
Aimed throwing	First static	Modeling early hominid technology	4	*C. apella*	Westergaard and Suomi 1994d
Stick to displace a reward out of a tube	First dynamic	Comparison with apes	6	*C. apella*	Visalberghi *et al.* 1995
Digging tools	First static	Modeling hominid subsistence technology	4/10	*C. apella*	Westergaard and Suomi 1995a
Stone throwing	First static	Modeling hominid throwing capabilities	4	*C. apella*	Westergaard and Suomi 1995b
Dipping	First static	Modeling hominid bamboo technology	5/18	*C. apella*	Westergaard and Suomi 1995c
Cutting	First static		6/18		
Pestle use	First static	Use of different objects as pestle	10/18	*C. apella*	Westergaard *et al.* 1995
Nut cracking	First static	Modeling hominid metal-tool technology	5/14	*C. apella*	Westergaard *et al.* 1996
Cutting	First static		5/14		
Stones as cutting tools	First static	Transfer of tools and food	3/11	*C. apella*	Westergaard and Suomi 1997

(*cont.*)

Table 10.2. *(cont.)*

Task	Relational category[b]	Specific aim(s) of the study	N tool users/N total[c]	Species	Source
Ant gathering	First static	Use of sticks to extract ants	7/14	*C. apella*	Westergaard *et al.* 1997
Stones as cutting tools	First static	Use of a tool-set	3/14		
Nut cracking	First static	Use of tokens to request tools	1	*C. apella*	Westergaard *et al.* 1998a
Dipping	First static	Role of sex and age on tool use acquisition	21/36	*C. apella*	Westergaard *et al.* 1998b
Dipping	First static	Factors associated with tool use and modification	31/61		
Container for water	First static zero	Serendipitous observation	1/11	*C. olivaceus*	Urbani 1999b
Sponging stick as cane	n.a.		1/11		
			1/11		
Rake with a hoe	Second dynamic	Role of surface irregularities	4/4	*C. apella*	Cummins 1999
Rake with a cane	First static and first dynamic	Role of tool position	6/6	*C. apella*	Cummins-Sebree *et al.* 2000
Bait for fishing	First static	Observational study	4/6	*C. apella*	Mendes *et al.* 2000
Cracking open a baited box	First static	Modeling hominid behavioral evolution and the transport of tools	8/13	*C. apella*	Jalles-Filho *et al.* 2001
Transport tools to the box	First dynamic		1/13		
Transport tools to the nuts and nut cracking			7/8		
Dipping	First static	Influence of task location of tool use	2/4	*C. olivaceus*	Dubois *et al.* 2001
Rake with a cane	First static	Casual comprehension	4/4	*C. apella*	Fujita *et al.* 2003
Rake with a cane	Second dynamic	Causal comprehension	40/4	*C. apella*	Fujita *et al.* 2003

Table 10.2. *(cont.)*

(b) Studies in semi-free and wild conditions[d].

Task	Relational category	Specific aim(s) of the study	Cond.	N tool users/N total[c]	Species	Source
Strike snake with a branch	First static	Serendipitous observation		n.a.	*C. capucinus*	Boinski 1988
Throwing, probing	n.a.[e]	Observational study	w	n.a./21	*C. capucinus*	Chevalier-Skolnikoff 1990
Pounding to open oysters	First static	Serendipitous observation	w	n.a.	*C. apella*	Fernandes 1991
Exploratory probing	First static	Serendipitous observation	w	n.a.	*C. capucinus*	Garber and Paciulli 1997
Nut cracking	Second sequential	Serendipitous	w	n.a.	*C. apella*	Langguth and Alonso 1997
Nut cracking for inside larvae	Second sequential	Use of suitable pounding tools and anvils	s-f	n.a./44	*C. apella*	Rocha *et al.* 1998
Dipping	First static	Selection and modification of tools	s-f	3/11	*C. apella*	Lavallee 1999
Nut cracking	Second sequential	Use stones and anvils and pounding tools	s-f	15/18	*C. apella*	Ottoni and Mannu 2001
Leaves to absorb liquid	Zero	Serendipitous observation	w	n.a.	*C. albifrons*	Phillips 1998
Nut cracking[f]	Second sequential	Observational study	w	n.a.	*C. apella*	Boinski *et al.* 2001
Nut cracking	Second sequential	Use of stones and anvils	s-f		*C. apella*	Mannu 2002
Stick to push	First dynamic	Acquisition of tool use by providing a tool task	w	0/15	*C. capucinus*	Garber and Brown 2002
Nut cracking[g]	Second sequential	Use of anvil and pounding tools	w	4/10	*C. apella*	Fragaszy *et al.* unpubl.

[a] Captivity includes cages, outdoor enclosures and small islands.

[b] Relational categories are defined in Table 10.1 according to the number (zero, 1st or 2nd order) and type of relations (static and dynamic) embodied in the task. We include a few cases reported in the literature as tool use involving a zero order relation that do not fit our criterion of tool use.

[c] Number of individuals using tools and total number of individuals tested. When there is only one value, it means that the study focused only on the subjects indicated. n.a. = not applicable, meaning that the information is not provided by the Author(s).

[d] In semi-free ranging (s-f) conditions the animals have access to large areas from which they obtain a substantial part of their food.

[e] In our view, the instances described are not cases of tool use. Most of them refer to explorative behaviors and to dropping branches.

[f] Boinski *et al.* 2001 did not actually see the capsule of the *Couratari oblongifolia* open or the capuchin bring its content to the mouth. No complete instances of tool use have been observed yet.

[g] This study took place as this book was going to press.

assumes that intelligence is a unidimensional construct so that different species have "more" or "less" (an assumption no longer widely held); (b) extractive foraging is more widespread across species than Parker and Gibson realized; and (c) extractive foraging does not predict strong performance on Piagetian scales of intelligence.

A more recent effort to identify broad correlates of tool use in nonhuman primates uses the statistical methods of comparative biology. Reader (2003) found positive correlations in various genera of nonhuman primates between the frequency of reports of tool use in the scientific literature on the one hand and various measures of brain size in these genera on the other. The relationships among brain size, foraging behavior and the use of tools in human evolution and in extant primates is an active domain of scientific inquiry, and likely to remain so for some time to come. For now, the precise relationship between capuchins' relatively large brains and their propensity to use tools is not clear. However, how their tool-using behaviors relate to foraging activity is becoming more apparent, and will be made clear in our review of experimental studies with captive monkeys.

REPORTS ABOUT TOOL USE AT THE CLOSE OF THE TWENTIETH CENTURY

Although there is still no consensus about which theory should guide studies on tool use in nonhuman animals, studies with captive capuchins appeared regularly in the last two decades of the twentieth century. Because most of these studies were not theoretically driven, they are methodologically heterogeneous and thus difficult to compare. Nevertheless, we can discern some general features from the experimental studies listed in Table 10.2. Experimenters have usually given capuchins sticks to insert in holes (e.g., to dip for liquid foods), to push objects out of tubes and to pull objects within reach; blocks or stones to crack open nuts or closed containers; and sharp objects to penetrate surfaces. Capuchins have used a large variety of objects as tools to achieve several different aims. Overall, their use of objects as tools relies upon perception/action routines (e.g., pounding, inserting, throwing objects against surfaces) that are applied to virtually any set of objects and surfaces they encounter.

As Klüver (1933) showed years ago, capuchins can use two different objects in succession, such as a stone to penetrate a barrier covering a container and then a stick to extract ants from the container (Westergaard *et al.* 1997) or a stone to crack open a nut followed by a stick to extract the inner meat (Westergaard and Suomi 1993). In these studies, capuchins that solve two-step tasks are usually already proficient in each step of the task, or in very similar tasks; the new aspect they master is doing both actions in sequence.

As we saw in Chapter 9, learning a sequence is easier when the order is immutable. In the case of cutting a barrier and then probing, the actions cannot be performed in the wrong order. As a consequence, the features of the task guide, in some sense, the monkey's discovery of the correct sequence. We predict that capuchins can master set sequences of virtually any length, given sufficient ingenuity on the part of humans to engineer tasks with a set order of options. Learning to perform action/positioning combinations in a set sequence where the order of actions is important but the objects and surfaces permit any order would, of course, be much more difficult to master. We have no reports to date of capuchin monkeys using objects as tools in such tasks.

A general finding, evident from Table 10.2, is that not all individuals are successful at any particular task. Some individuals ignore the task while others, although they explore the situation, do not ever reach a solution, even if they have many opportunities to watch others using an object and gaining a reward. No single factor predicts which individuals will discover how to use an object as a tool in any given situation, and studies searching for correlates have produced mixed findings. Fragaszy and Adams-Curtis (unpublished) found no clear relationship between individuals' spontaneous rate of combining objects in routine conditions and the discovery of how to use a tool when a "task" is presented. In another study, Fragaszy and Visalberghi (1989) found that in dipping and pounding tasks involving a first-order relation, exploratory behaviors involving the prospective tools and objects in relation to the tools and the apparatus were more frequent in solvers than nonsolvers. However, the animals that solved one task did not necessarily solve the other task, given equivalent access to both tasks. In this study, sex and age were not predictive of discovery of how to use an object as a tool.

Westergaard *et al.* (1998b) found that sex did not predict which animals would use a probing tool, whereas age did. In particular, there were proportionally more solvers below 10 years of age than above, and juveniles were more likely to use tools than infants. Moreover, juveniles whose mothers died before they were 3 years old were less likely to use tools than juveniles whose mother survived (0 tool users out of 5 juveniles whose mother died versus 9 tool users out of 11 juveniles whose mother did not die). Interest in the apparatus without tool-objects available or in the objects without apparatus available did not differ among capuchins that used tools and those that did not in Westergaard *et al.*'s study. Similarly, Jalles-Filho (1995, Jalles-Filho *et al.* 2001) reports no correlation between the incidence of other manipulatory goal-directed behaviors (digging in the soil to acquire a hidden piece of food) and using sticks or stones to break open the cover of a box filled with maize, or between the exploration and manipulation of the box and use of tools to crack open the cover of the box.

Overall, tool use is more likely to be acquired by individuals who have the opportunity to explore objects and surfaces for extended periods of time and that, while acting with the objects, are in social contexts conducive to learning (i.e., with other tolerant adults, or with their mother, routinely using tools nearby; see Chapter 13). However, these optimal conditions are not very likely for wild capuchins; in nature, easy access to appropriate objects, close proximity to expert tool users at work, etc. are seldom the case unless they are on the ground. In an instructive case, a group of semi free ranging *C. apella* living in a regenerating forest (Tietê Ecological Park, São Paulo, Brazil) feed seasonally on palm nuts (*Syagrus romanzoffiana*). In this group all individuals except infants cracked open palm nuts (Mann 2002; Ottoni and Mannu 2001). As we further discuss below, one can identify several characteristics of this group's living conditions that might support the discovery of tool use by all individuals in the group. We predict that in other sites with conditions like those present in Tietê Ecological Park, most or all the individuals will acquire tool use.

Although most of the reports in Table 10.2 concern *C. apella* studied in captive settings, all other traditional species in the genus have at least one entry. At present we consider all species in the genus to be equally likely to discover how to use an object as a tool. Very few reports about tool use exist for wild or semi-wild populations; those that have appeared mostly concern cracking open enclosed foods. We next turn to the question of when this form of tool use is likely to be discovered by wild capuchins.

TOOL USE IN NATURAL AND SEMI-FREE RANGING SETTINGS

Ecological correlates of tool use: reliance on enclosed foods and feeding on the ground

Capuchins feed on many foods that require extraction from substrates (such as nuts, snails, husked fruits, invertebrates hidden in wood, etc.). Because they live in habitats with pronounced seasonal changes in the abundance and varieties of potential foods, their diet shifts accordingly. When fleshy fruits (such as figs) are scarce, capuchins shift their diet to feed on cryptic or embedded foods that take more effort to find and/or process. Moreover, even when other foods are abundant, embedded foods may be highly desirable due to their nutritional value (such as the brazil nut, *Bertholletia excelsa*, palm nuts of *Syagrus romanzoffiana* parasitized by insect larvae, and oysters, *Crassostrea rhizophorae*) or because they contain water, such as the fruits of many species of palms. Sometimes the monkey cannot break items by direct pounding or even by its powerful bite. In such cases, using a tool is more likely than when direct action suffices (Visalberghi and Vitale 1990).

The general disposition to act with objects is more likely to lead capuchins to tool use for exploiting embedded food resources when alternative foods are scarce or undesirable (Visalberghi 1987). Thus we should expect to find capuchins using tools when (a) seasonal reductions in availability of fruit are particularly harsh, (b) an embedded food (that is difficult to open) is abundant, and/or when the diversity of foods is consistently low and an embedded food is an important staple item in the diet, (c) they spend time on the ground, (d) stones and anvil sites are relatively abundant and (e) risk of predation is acceptably low. As we will see below, evidence supporting this picture is accumulating.

Tool use during feeding

The idea that the discovery of how to use objects to crack open foods is more likely when capuchins spend a good

Figure 10.4. An adult tufted capuchin cracks palm nuts (*Syagrus romanzoffiana*) using a stone after placing the nuts on a stone surface. A second individual watches at close range. Capuchins often tolerate others nearby while engaging in nut cracking, and this tolerance is thought to promote acquisition of the skill by younger animals. This picture was taken in Tietê Ecological Park, São Paulo, Brazil by Briseida Resende. (Printed with permission.)

deal of time on the ground (Visalberghi 1987) receives support from five recent reports. In the first, Fernandes (1991) saw an adult male capuchin using a piece of oyster shell to pound repeatedly on oysters (*Crassostrea rhizophorae*) fixed to the roots or branches of red mangroves, much as Dampier was thought to have reported long ago[1]. Note that the oyster was in a fixed location; the actor positioned the pounding object correctly and applied the appropriate force. This is a first-order relational problem (see Tables 10.1 and 10.2).

Second, Rocha *et al.* (1998) reported that several individuals of a group of *C. apella* living in Parque Municipal Arthur Thomas, Londrine, Brazil, use stone and wood objects to pound open palm seeds on stone anvils (see Figure 10.4). Typically, capuchins climb into the palm and eat the pulp of the fruit (*Syagrus romanzoffiana*) in the tree and discard the seeds, which drop to the ground. These seeds are often parasitized by larvae of Coleoptera, and capuchins are fond of these larvae. Several days later, capuchins collect the discarded seeds on the ground and transport them in their hands and mouth until they find a stone large enough to use as an anvil and a smaller one or a branch to use as a pounding tool.[2] The stones or branches used to pound the seeds weigh between 200 and 1000 g, or roughly between 10 to 30% of an adult monkey's weight. Capuchins place a single seed on the anvil and, holding the pounding tool with two hands, they strike the seed until it cracks open. Then, using their teeth and fingers, they extract the larvae. If they do not find any larvae they may discard the seed or eat its inner part. Rocha *et al.* (1998) estimated that 46% of the seeds contained larvae. Males and females estimated to be older than 2 years pound nuts open in this way. This problem embodies a sequential second-order relation: the nuts must be placed in a specific location, and the pounding object must be brought into proper relation to the nut with sufficient force to crack it (see Table 10.1).

Third, Ottoni and Mannu (2001; Mannu 2002) observed a group of semi-free ranging *C. apella* living in a reforested area close to São Paulo (Brazil) cracking nuts. All individuals in this group except infants cracked open palm nuts (*Syagrus romanzoffiana*). Because the ground was rather soft, they had to place the nut on a harder surface (usually a stone) and hit it with another hard object, such as a stone (see Figure 10.4). In another study, Jalles-Filho *et al.* (2001) showed that capuchins will transport nuts to a pile of rocks to crack them open (see also below *Tool use in capuchins and in ancestral hominids*).

Fourth, on the basis of strong but indirect evidence, Langguth and Alonso (1997) inferred that during a period of severe drought, members of a group of wild *C. apella* used a stone to pound open *Syagrus* nuts placed on an anvil stone. These authors found the remains of cracked nuts on and near stones on the ground, with other stones nearby, in areas where tufted capuchin monkeys lived. Given the reports from Rocha *et al.* (1998), Ottoni and Mannu (2001) and Oxford (2003), Langguth and Alonso's inference seems reasonable. These capuchins apparently responded to a period of severe food restriction by searching for viable food alternatives on the ground, including nuts. Most likely pounding the nuts is the only, or the most efficient, method by which they could open the nuts. Strong motivation imposed by harsh ecological conditions, a certain degree of terrestriality, and the ready availability of stones and hard surfaces on which to place the nut to be pounded, all favor the emergence of pounding by capuchins as a means to open hard foods.

Finally, wild *C. apella* in Piauí, Brazil routinely eat the hard fruits (also called nuts) of several palm species (Fragaszy, Izar, Visalberghi, Ottoni, and Oliveira, unpubl.). The monkeys open the hard fruits by first placing them on a large stone ("anvil") on the ground and then pounding them with a stone; the pounding stones may weigh more than a kg (see color plate 4). Since July 2000, these capuchins have been provisioned with palm nuts in one area with several anvils to encourage their nut-cracking activities in this site (see Oxford 2003 for a photo report). At least two groups of monkeys use this site often, although not every day. The monkeys in this area also crack nuts in many sites in addition to the provisioned site, including on the tops of sandstone mesas rising 20 meters or more above the valley floor, where the palms grow. The monkeys transport both nuts and pounding stones to these sites. The largest loose stone we found at an anvil site at the top of a mesa weighed 825 g – an impressively heavy stone for a monkey to carry.

In general, wild monkeys are more likely to use tools to obtain foods that they could not get in any other way, and we can expect that tool-use in feeding will be most evident when more easily obtained food is scarce. Of course, other factors will impact this behavior as well. Spending time on the ground declines as the risk of predation by terrestrial predators increases, for example. In any case, the capability to use objects to open enclosed foods may give capuchins a chance to inhabit otherwise inhospitable areas. For example, Fernandes (1991, p. 530) suggests that the ability to open oysters allowed capuchins to be "the only permanent primate resident" in the mangrove swamp he surveyed. Wild chimpanzees may also deal with food scarcity by increasing tool-using activities. Yamakoshi (1998) has recently demonstrated that the exploitation of the nuts and the pith of oil-palms (*Elais guineensis*) by wild chimpanzees in Bossou, Guinea, in West Africa, is strongly related to fruit scarcity. Tool-using supports an alternative feeding strategy when other keystone resources become diminished.

Use of objects for defense

Wild capuchins flail branches or drop objects on intruders (Chevalier-Skolnikoff 1990); these behaviors are seen in many monkey species that do not otherwise use tools (e.g., guenons). Dropping branches can be either the fortuitous by-product of the animal's excited noise-making displays (which include branch-shaking) or be aimed at disturbing/attacking the intruder; we cannot specify which explanation applies better to capuchins dropping branches. Other actions by capuchin monkeys with objects are more clearly aimed at a specific other animal. Boinski (1988) observed a male *C. capucinus* killing a snake by hitting it with a branch obtained from nearby vegetation. Captive capuchins tossed objects towards a model snake and, when the snake retreated into a hole, one capuchin probed the hole with a stick (Vitale *et al.* 1991). Captive capuchins can throw objects with a reasonably accurate aim (Westergaard and Suomi 1994d, 1995b), but only Vitale *et al.* have reported observing capuchins throwing something toward a frightening object. Captive capuchins have been seen to position a loose piece of wire mesh (the only loose object available in their cage) as a "shield" between themselves and a disturbing stimulus (a snake; King 1982).

STUDIES OF TOOL-USING IN CAPTIVE CAPUCHIN MONKEYS

Using sticks to extend reach

Many studies have investigated capuchins' ability to use a stick to probe or dip for food, or to insert a stick into a tube to push food out (Figure 10.2). In the dipping/probing task, a container is filled with a viscous food (e.g., syrup, apple sauce, yogurt) that can be retrieved through openings too small for a capuchin's hand. The apparatus is fixed to a rigid surface and suitable objects (stick, straw, dowels, bolts and branches from which smaller pieces can be used) are presented. Inserting a stick into an opening is a static single-relation task. Capuchins can master this task even before their first birthday (Westergaard and Fragaszy 1987a, Westergaard *et al.* 1998b), or shortly thereafter (Fragaszy *et al.* 1994b). Westergaard *et al.* (1997b) presented a version of the dipping task in which capuchins dipped sticks into a container holding ants. Five adults (out of six) and two younger capuchins (out of the seven whose age ranged between 11 months and 5 years) succeeded in the task. The juveniles that gathered ants with a stick were 3 and 4 years old.

In most studies where the experimenters gave the monkeys a dipping apparatus, they provided one or more sticks with a narrow diameter to insert into an opening of a much larger diameter, so the alignment of stick to opening was relatively permissive. When objects of various sizes and shapes are presented, the monkeys first attempt to insert them into the opening at random; they select other objects or modify the selected object if their initial attempts fail (e.g., Visalberghi and Trinca 1989, Visalberghi *et al.* 1995). This suggests that they perceive the fit between stick and opening through action, not through vision. We come back to the topic of how the monkeys select objects to use as tools later in this chapter.

The behaviors observed in this task provide us with information about several aspects of capuchin's long-term and short-term memory. Capuchins that have learned to dip for food do not forget how to do this, even after several years. For example, two capuchins that learned to dip for syrup in captivity (Westergaard and Fragaszy 1987a) dipped years later in semi-free ranging conditions when given similar opportunities (Lavallee 1999). However, applying a strategy successfully adopted in the past is not necessarily efficient in the task at hand: Figure 10.5 shows a female tufted capuchin that years before had used sticks to dip for syrup. Now she has an opportunity to work for a new food item: a walnut. She is holding a straw and touching the shell of the walnut with it. Clearly, "dipping" will not work in this situation. In this case, the monkey will have to abandon the remembered action-object combination and learn to apply new ones. It is evident from this example that capuchins do not appreciate the elements of a situation in the same way as the adult human observer.

Figure 10.5. An adult female tufted capuchin (Erna), proficient in using straw for dipping syrup, touches a nut glued on a wooden board with a straw, using the same action that she used previously to retrieve syrup from a closed container (see Figures 4.1 and 10.2 for illustrations of dipping tasks). She ignores the adjacent hard objects (one shown in the foreground) that could be used effectively to crack open nuts. Striking a nut glued in place would be a first-order static problem. (Photograph by Elisabetta Visalberghi).

In Lavallee's (1999) study, when the dipping apparatus was moved out of view away from the vegetation used by the capuchins as the source for tools, the monkeys left the apparatus to search for branches and twigs and returned to the site to use them. This observation replicates those of Visalberghi (1987) and Fragaszy and Visalberghi (1989) where capuchins collected tools from another room and brought them to the work site. Planning is implied by the selection of tools from a distance away from the work site. In one study, four capuchins, competent at using a stick to push food out of a tube, were given a choice between appropriate tools (sticks placed in an artificial tree) and inappropriate tools (stones placed inside a container), both of which were in an adjoining room out of view of the baited tube. In the first trial, two subjects struggled to get the reward with an unexpected technique: they resorted to pounding the tube with a stone until the vibrations moved the food close enough to one of the two openings for them to reach it. In the second trial these same capuchins solved the task very efficiently with the appropriate object (Visalberghi 1993a).

As mentioned above, capuchin monkeys readily discover that they can push food out of a tube using a stick, just as they discover that they can use a stick to collect viscous food through a narrow opening. Pushing something with a stick, like dipping with a stick, is a first-order spatial relation problem. However, it involves a dynamic relation between the held object (the stick) and the food, which moves through the tube, and thus presents a greater challenge than dipping. Visalberghi and Trinca (1989) reported that four tufted capuchin monkeys (*Cebus apella*) mastered this problem, a finding since replicated with other capuchins and other primate species (chimpanzees, bonobos, orangutans, Visalberghi *et al.* 1995; children, Troise 1991; Japanese macaques, Tokida *et al.* 1994; for a review see Visalberghi 2000). Visalberghi and colleagues used a transparent tube, which allowed the experimenter and subject alike to view the food inside and to see the tool entering the tube. We shall return to this set of studies later in this chapter to learn how capuchins dealt with the challenges of modifying tools to fit the tube, selecting tools from among several presented, and pushing the food when the interior of the tube contained a hole.

Using one object to break another

Categorizing pounding tasks

Cracking open a nut (or any husked fruit or a shell; we shall use the generic label "nut" for all such foods) can qualify as either first-order or second-order relations in our classification scheme (see Table 10.1). As in dipping in the simplest circumstance, the task involves producing a single spatial relation between a held object and a static target location. The nut is firmly attached to a substrate and at least one object to be used as a pounding tool is provided. Once the monkey holds the object in its hand, striking the nut with this object is a static first-order relational act. When the nut and the potential tool are loose and there is a cement floor, the problem embodies a dynamic first-order relation (pounding the nut with the object and monitoring whether the nut moves when it is struck). When the nut is loose, most of the ground is relatively soft, but hard objects that might serve as pounding implements and anvils are present, the monkey must produce a second-order relation by first positioning the nut on the anvil surface (first relation) and then pounding the nut with the hard object (second relation), meanwhile monitoring that the nut stays on the anvil as it is struck. The monkey must organize and monitor successively more relations to succeed going from static first-order to second-order relational tasks. This way of looking at the demands inherent in pounding open nuts with a stone is similar to that presented by Matsuzawa (1996), Inoue-Nakamura and Matsuzawa (1997), Westergaard (1999) and Boesch and Boesch-Achermann (2000) in distinguishing single and multiple relations. However, it specifies more explicitly the form of the relations (static *vs.* dynamic). Whereas to date the studies of nut-cracking in captive situations fall into the category of first-order problems, those in more natural settings include second-order relations. We will focus our attention on studies concerning dynamic first-order and second-order relations.

When and why capuchins use a pounding tool to crack nuts

Using a hard object to crack open a nut is clearly related to capuchins' tendency to pound objects directly on a substrate, a behavior that emerges before 6 months of age. Thus we can expect monkeys older than 6 months to be capable of discovering the utility of pounding one object with another, although a young animal's small size may constrain its success at pounding something open. In any case, using an object as a pounding tool significantly reduces the time it takes a capuchin monkey to open hard nuts versus banging them open by hand (Anderson 1990, Antinucci and Visalberghi 1986, Jalles-Filho, pers. comm.). For some nuts, cracking with a tool is the only means to achieve success[3]. Anderson (1990) noted that capuchins used pounding tools on approximately 75% of the trials with hazelnuts (which are hard and small), and less frequently to open walnuts (which they could pound open directly with moderate effort) or almonds (which they could bite open). Opening a hazelnut with a tool could take a mere 5 seconds, but without a tool it could take up to 30 minutes (360 times more!). Moreover, capuchins preferentially use objects which are more effective tools (a stone) more often than less effective ones (a plastic box or a block of wood) (Antinucci and Visalberghi 1986).

A nut-cracking sequence embodying a dynamic first-order relation typically consists of a capuchin picking up a nut and carrying it to the stone that it will use to crack the nut, placing the nut on the ground beside the stone, then lifting the stone with one or both hands and bringing it down on the nut. As illustrated in the vignette that opened this chapter, when naïve capuchins encounter nuts together with other hard objects, they combine the objects and nuts in all possible combinations of actions and spatial orientations (e.g., holding the nut in the mouth while pounding the tool on the floor, or placing the nut on top of the tool and pounding both of them – in that spatial configuration – on the floor; Figure 10.1) (Visalberghi 1987). Occasionally, the monkey first places the nut on the ground and pounds it with the other object. The occurrence of this effective spatial and action combination becomes more frequent with time (see also Anderson 1990). Individuals as young as 2 years old have been seen to use a hard object to crack open loose nuts (Anderson 1990) or nuts attached to a board (Fragaszy *et al.* 1994a). As with other kinds of tool use, often not all capuchins in a captive group (not even all adults) succeed at using an object as a tool to crack open nuts.

DEALING WITH VARIABLE SUBSTRATES AND OBJECTS

We are all familiar with the flexible ways in which humans use tools, including anticipation of which object-action combination is going to work, the ability to imagine possible obstacles that prevent our actions from being effective, and our ability to modify actions or objects in response to the potential reasons for failure. In other words, we can reflect on our actions with objects in time and space, and the relations among actions, objects and outcomes. When we see a child learning how to use objects as tools we can appreciate that humans are not born with this reflective knowledge (Lockman 2000). The child needs extended experience of his or her own actions in many variations of a given circumstance to become a skilled tool user. However, mastering skill at a given task does not necessarily involve reflection on what the task requires for its solution. When we see a capuchin monkey facing a problem that it can solve by using an object as a tool, such as using a stone to pound open a nut, we realize that it also has to learn what spatial relations to achieve, and what actions to use to achieve them. In this section we describe how flexibly capuchins can adjust to variations of objects and surfaces when they use one object to move another. This information provides us with insight into how differently capuchin monkeys perceive these problems compared to humans.

Tool selection and modification

Two sets of studies have investigated whether capuchins differentiate appropriate objects from inappropriate objects for a particular purpose, how they behave when the objects require modifications to be used as tools in a particular context, and how they use various objects to solve the same problem.

Selecting and using varied objects to push food out of a tube

As mentioned above, four capuchins became proficient at pushing food out of a transparent tube using a stick. These four subjects subsequently encountered the same apparatus in three new conditions in which the tools had to be modified or combined (Visalberghi and Trinca 1989, see Figure 10.6). In one condition the sticks were so short that two of them had to be inserted one behind the other inside the tube to move the food far enough for the monkey to reach it. In another condition the object (a bundle of thin canes held together by tape) was too large in diameter to fit into the tube. In the third condition the stick had thin pieces of dowel, which could be pulled out or broken off, inserted transversely at each end so

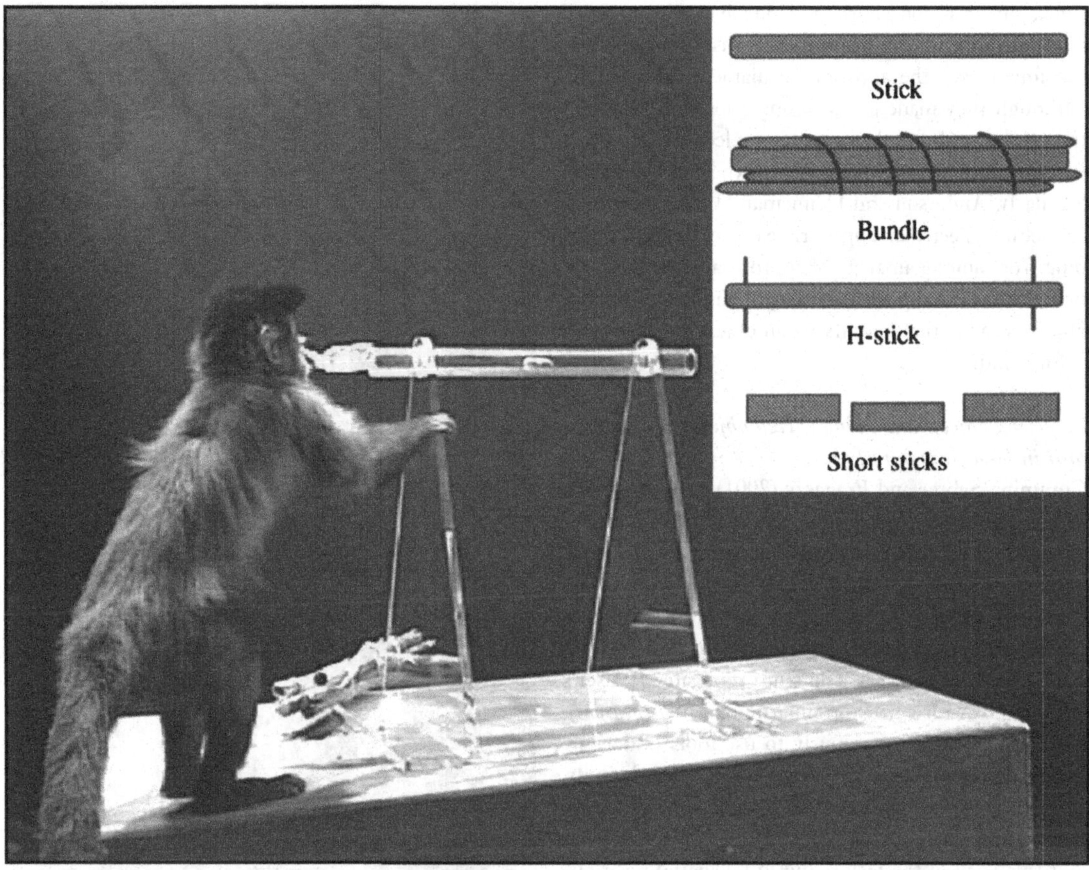

Figure 10.6. The tube task consists of a transparent horizontal tube baited in the center with a food treat. The objects provided to the subject that can be used as tools are shown in the top right of the figure: from top to bottom, a stick that can be used to push the treat out of the tube. A bundle consisting of several reeds firmly held together by tape; the diameter of the intact bundle is too large to fit into the tube. The H-stick consisting of a dowel with two smaller sticks placed transversally near the end; the transverse sticks block the insertion of the dowel into the tube. Short sticks at least two of which must be inserted into the tube one behind the other to displace the reward. (Drawing by Stefano Marta). The capuchin in the figure has dismantled the bundle of reeds (visible at her feet) and inserts the tape, not a reed, into the tube. Errors of this type (selecting inappropriate objects as tools) are common in capuchin monkeys. (Photograph by Elisabetta Visalberghi).

that the ends of the stick could not enter the tube. The capuchins succeeded in all conditions within a few minutes. Despite their success, they made many attempts to use the original object without modifying it and to use parts of the object (e.g., the tape, a splinter) that did not have the necessary properties (e.g., rigidity, length) to displace the food from the tube. Over the 10 trials in each condition the number of errors produced by each monkey and the nature of their errors decreased only slightly. After an interval of 5 years, when these same capuchins encountered these objects and the tube again for a filming session, they made exactly the same kinds of errors (Visalberghi and Limongelli 1992). These findings indicate that the monkeys did not quickly learn what properties of the objects, surfaces and actions were most important for success.

The same four capuchins were also given a choice among four different tools in another experiment (Visalberghi 1993a). Three of the objects did not afford insertion or would not reach the food (one was too thick, one was too short, and one had a transversal block at one end that prevented its insertion) whereas the fourth was the appropriate diameter and length. Although they made a few wrong choices throughout the 16 test trials, all the capuchins selected the correct tool far more often than would be expected by chance. Similarly, Anderson and Henneman (1994) found that capuchins selected an appropriate object to use for dipping from among an array of appropriate and inappropriate objects. It appears that recognizing an appropriate object is easier than modifying an object appropriately beforehand.

Selecting objects and using varied objects to pull in food from out of reach

Cummins-Sebree and Fragaszy (2001) examined what properties of an object influenced capuchins' choice when they prepared to use the object to pull in a piece of food. The objects they could choose from varied from each other in functional ways (with or without a curved segment, for example) or irrelevant ways (diameter, shape, color, texture). This study replicated Hauser's (1997) study with cotton top tamarins (*Saguinus oedipus*), a species not noted previously to use tools under any conditions. In the situation presented in Hauser's study, when they merely had to pull in a hooked stick to bring food within reach, the tamarins readily complied. Note that this form of the task required no spatial positioning of one object to another or to a surface by the actor. It is thus a zero-order relational problem and does not meet our criterion for tool use provided at the beginning of this chapter (see Table 10.1). The actor in this task does not produce a relation but rather makes use of an already-existing relation when it chooses which object to pull.

Six capuchin monkeys (*Cebus apella*) first demonstrated proficiency pulling food towards themselves across a flat surface using a hooked blue stick. Then, in a series of experimental sessions, they encountered two objects on every trial. After they selected one, they could use it to try to bring in a piece of food (see Figure 10.7 for an example). When the food was positioned outside the curve of one cane and within the curve of the other cane, capuchins tended to choose the one containing the food within the curve. Thus they recognized the importance of the position of the food with respect to the curve of the cane. When the choice was between two objects differing in color, size, shape or texture (i.e., irrelevant features) capuchins did not show significant preferences for the property that matched the familiar training cane. When they had a choice between an object already positioned appropriately (so that it just required pulling in), they preferred it to another

Figure 10.7. Tufted capuchins use tools of variable shapes to retrieve a piece of food. Top, the tool has a C-like shape; bottom, the tool has a bumpy surface and a V-like shape. Once the monkey has placed the hook around the food, the difficulty of the task depends upon the properties of the tool. Although both tasks concern a dynamic first-order relation (see Table 10.1), the task in the lower picture is more difficult because the monkey must continuously monitor whether the food slips underneath the bumpy tool. (Photographs by S. Cummins-Sebree).

that they had to reposition before pulling. Furthermore, when they were given a choice between novel or familiar objects (either of which could already be positioned or required positioning), capuchins not only continued to prefer the already-positioned object, but also preferred novel objects to familiar ones. Fujita *et al.* (2003), using a similar experimental design and evaluating choice of tools, replicated several of these results with four tufted capuchins. These authors suggest that the capuchins chose tools largely on the basis of the spatial relation between the tool and the food.

In Cummins-Sebree and Fragaszy's (2001) study, capuchins succeeded more often than tamarins at getting the food partly because they occasionally managed to position objects accurately with respect to the food. Tamarins did not reposition objects (Hauser, pers. comm.). Thus the capuchins managed the first-order relational version of the cane problem; the tamarins did not.

Cummins-Sebree and Fragaszy's and Fujita *et al.*'s results show that, when given a choice, capuchins select objects according to features that are relevant for their intended use (as did the tamarins – Hauser 1997) and not to irrelevant ones . A similar selection for the more functional objects was also evident in the nut-cracking studies we discussed above (see also Visalberghi 1993a). It is very likely that recognition of functional properties of objects develops in parallel with use and feedback from their use and, thus, interactions with different objects allow capuchins to learn what each of them affords in a given context. An affordance is here defined as the relationship between one's ability to exhibit a particular behavior and the properties of the immediate environment that allow the occurrence of that behavior (Adolph *et al.* 1993).

Coping with varied surfaces

Visalberghi and Limongelli (1994) presented a third variation of the transparent tube problem to four monkeys already proficient in the basic task. In this third variation, the subject encountered a tube with a hole in the middle (a "trap"; see Figure 10.8). The reward was placed nearer one of the two sides of the trap. To get the food, the capuchin had to insert the stick into the side of the tube from which it could push the food away from the trap. This second-order problem embodies two dynamic relations, one that the monkey must produce (between the stick and the food) and one that it must recognize (between the movement of the food over the trap and the food falling into the trap). Would capuchins recognize the second dynamic relation? In other words, would they take into account the outcome of their action with the stick on the movement of the food (toward or away from the trap), and avoid moving the food over the trap?

In this task an actor can succeed on half of its attempts by systematically inserting the stick into the same end of the tube or by inserting it into one of the two ends randomly. Rates of success higher than chance can be obtained by avoiding the trap through a rule of action based on associative processes (such as insert the stick from the side farther from the reward) or by recognizing the outcome of pushing the food into the trap. The four capuchin monkeys were tested in 140 trials. The rate of success was at chance level for three monkeys, whereas that of the fourth subject, Roberta (3 years old), was significantly higher than chance (86%) in the second half of the experiment.

Careful analysis of Roberta's behavior revealed that she adopted a distance-based rule of inserting the stick in the opening farthest from the reward. Looking inside the tube from either end, Roberta saw the spatial configuration among stick, tube and reward. Covering the tube, except for the trap, with opaque plastic (see Figure 10.9 top left) did not diminish her success. However, when the tube was modified so that the trap was not centered (one "arm" was longer than the other; see Figure 10.10) and the distance rule became counterproductive, Roberta's rate of success fell significantly below chance level (Limongelli *et al.* 1995).

Roberta encountered additional situations designed to explore her comprehension of the trap, such as a condition in which the trap was directly above the tube (Visalberghi and Limongelli 1994; see Figure 10.9 bottom left). In this circumstance she continued to use the distance rule to insert the stick (and in the process, she pushed the food away from the ineffective trap). Roberta also encountered an opaque tube with no trap in which the reward was placed slightly to the right or left of the middle (as was also the case in trials with the trap tube, see Figure 10.9 top right) and, in a separate set of trials, a straight transparent tube with no trap alternating with a trap tube (see Figure 10.9 bottom right). In these tests

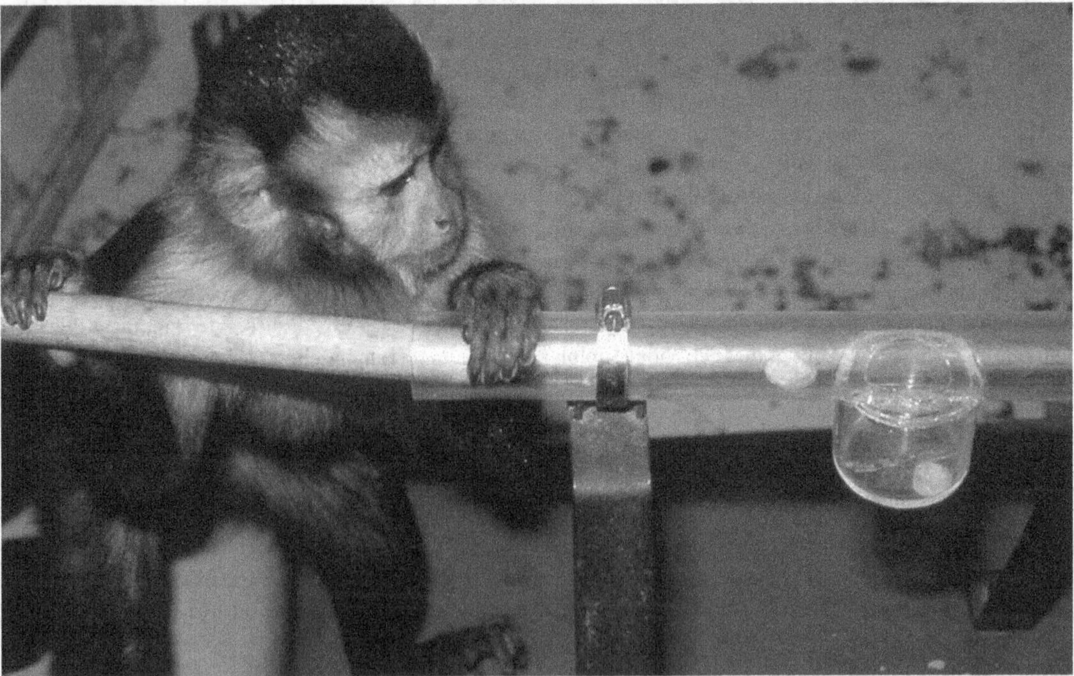

Figure 10.8. The trap-tube is a transparent tube with a hole in the center and a trap underneath it. The top panel shows an example of correct insertion of the stick; the reward is on the right side of the trap. Note how delicately the monkey (Roberta) moves the stick with the fingertips of her right hand while at the same time monitoring the slow movement of the reward. The bottom panel shows an example of insertion of the stick in the wrong side of the tube. Note that a reward lost in a previous trial by Roberta is already inside the trap. (Photographs by E. Visalberghi)

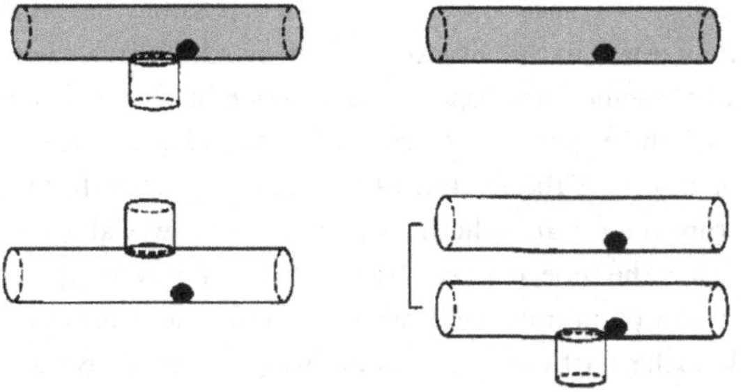

Figure 10.9. Control tests presented to Roberta, the only capuchin that solved the trap-tube task consistently. The figure exemplifies the cases in which the initial location of the reward is to the right of the center of the tube; the opposite location was also used but is not illustrated. Top left: The trap tube is covered with opaque plastic; for the purpose of illustration we show the position of the reward, but it was not visible to the monkey except by looking inside the tube from either end. Top right: The straight tube is covered with opaque plastic and again we show the position of the reward. Bottom left: The trap is rotated 180 degrees above the tube and therefore is ineffective; Bottom right: One tube and one trap tube are presented alternately. (Redrawn by Stefano Marta from Fig. 5 in Visalberghi and Limongelli 1994).

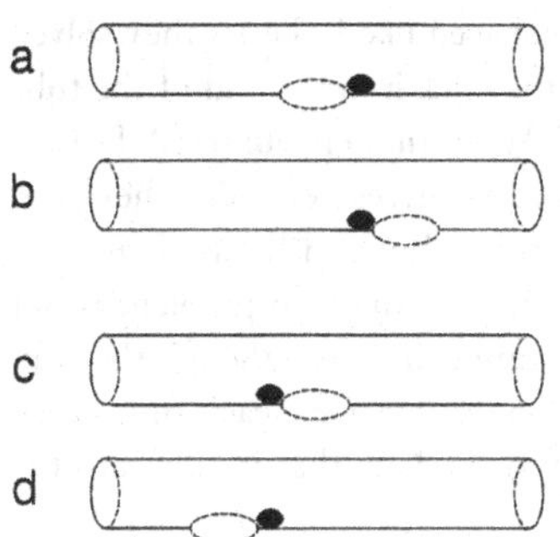

Figure 10.10. Tubes used to test whether the capuchin Roberta was using a distance strategy to solve the trap-tube task. When the reward falls into the hole, it is lost to the monkey. The figure shows the possible locations of the reward at the beginning of each trial. In a and c, the hole is centered; in b and d the food is in the same position as in a and c but the hole is displaced from the middle part of the tube. Note that in a and b the reward is in the same position along the length of the tube and offset from the center; the same holds true for c and d. Roberta did not solve the problem when it was presented as in b and d. (Redrawn by Stefano Marta from Limongelli *et al.* 1995).

Roberta continued to apply the rule concerning distance of the food from the end of the tube and pushed the food away from the center of the tube (where the trap had been in previous trials with the trap tube). These are examples of inappropriate (but not penalized) generalization. This might suggest that Roberta did not detect the hole in the original trap tube. However, perseverance of a particular spatial rule is also evident in other tasks where failure to detect surface properties seems an unlikely explanation. For example, an adult capuchin in Fragaszy's laboratory successfully used straw to dip for syrup through holes in the vertical sides of an opaque box. More than 5 years later, when she encountered a new dipping task (a transparent box with holes in the top), she repeatedly tried to dip into the vertical sides of the box although there were no openings there (see Figure 10.5). The tendency to generalize both actions and spatial relations provides an indication of how the monkeys perceive the task. It is clearly different from how an adult human perceives the same task.

Adult humans anticipate or can imagine the effect of pushing the food with the stick and (simultaneously) the fate of the food when it moves above a hole. Thus the position of the food with respect to the trap is integral to how we decide to push the food. The four capuchins probably anticipated that pushing with the stick causes the food to move but they did not (simultaneously) recognize that the food will fall into the hole when they push it toward the hole (when the hole is below the food). This appears clear from the behavior of the three monkeys that never scored better than chance with the trap tube, and it is also supported by thorough analysis of the behavior of the fourth monkey that developed an effective strategy based on a spatial relation. Visalberghi and Limongelli (1994) argue that their results show that capuchins, even when expert at using sticks to push food from a tube, fail to appreciate the causal relation between their behavior and the food falling into the trap.

When the trap tube was presented to five chimpanzees, two solved it above chance level (Limongelli *et al.* 1995) but their strategy was not based on the same distance rule as the successful capuchin (see experiment 2, in Limongelli *et al.* 1995). It is possible that these two apes might have understood the relevant relation between the food and the hole, as do children above 3 years of age (Visalberghi and Limongelli 1996, Want and Harris 2001). Reaux and Povinelli (2000) found that several chimpanzees behaved like Roberta; they solved the task by inserting the stick into the end of the tube farthest from the food. When they encountered the tube with the trap rotated 180 degrees vertically, they continued to use the same rule as Roberta. Thus identifying the second spatial relation in the trap problem is not easy for either apes or capuchins, even though they can see the hole and the food and can see each time when they push the food into the hole that it falls into the trap.

To better understand what makes the trap tube difficult for capuchins and for chimpanzees, we should consider how they perceive the situation. Do they perceive the hole in the tube as easily as we do? Do they anticipate the path of motion of the food when it enters the hole as we do? How do they learn to perceive these properties of surfaces and objects moving across them? Experiments on these issues have begun with capuchins; we turn to these next.

Managing action on irregular surfaces

Cummins (1999) investigated the ability of four capuchin monkeys to deal with two kinds of aberrations of a surface (a hole and a barrier) while using a hoe to retrieve a piece of food (Figure 10.11). When the hoe struck the hole, the monkey could see and feel the blade of the hoe falling into it. When the hoe struck the barrier, the monkey could see that the hoe was partly occluded and could feel the impediment to movement. Her results show that capuchins detect barriers on surfaces more readily than holes and that they move an object past a barrier more successfully than past a hole. Three capuchins learnt to avoid the barrier; and two learnt to avoid the hole. The monkeys managed to move the food past the barrier (9 times out of 10 trials) in the first 10 trials (two subjects) and after 40 trials (one subject), whereas the two monkeys reaching the same level of competence at moving food past the hole required 68 trials and 126 trials, respectively. When given the choice to retrieve food by moving it past a hole or past a barrier, one capuchin chose the barrier significantly more often than expected by chance, two chose the hole (despite losing the food into the hole proportionally more often than losing the food behind the barrier) and one did not show a preference for either condition.

Because in Cummins' (1999) study the monkeys practiced using the hoe to move an object around a hole in just two places on the surface, it is possible that what they learnt in this situation was to avoid the area where a hole sometimes appears. This possibility is consistent with the behavior of Roberta when she adopted the rule of inserting the stick on the side of the tube farther from the reward whether or not the trap was above or below the tube, or even absent. However, in subsequent testing (Cummins and Fragaszy, unpublished), the monkeys that participated in Cummins' (1999) study avoided moving a reward toward a hole placed anywhere on a surface, and they readily moved the reward across the location where the hole had been on their previous testing. Additional testing is needed to confirm how generalized the monkeys' behavioral adjustments would be in other circumstances to avoid moving an object toward holes or other surface irregularities.

Overall, these data show that capuchin monkeys do not perceive the affordances of a surface with a hole as readily as a surface with a barrier. However, capuchin monkeys can learn to use one object to move another object past a hole if they have enough of the right kinds of experience at doing so, just as they can learn to move an object past a barrier. Capuchins also had more difficulty with holes than with barriers when the task involved prediction of object movement, rather than action to move the object. As we described in Chapter 8, capuchins observing a ball rolling along a solid surface could predict its linear trajectory and they correctly predicted that a ball would not pass through a solid barrier. In contrast, when the path of a ball traversed a surface with a hole they did not predict that the ball would fall into the hole (Fragaszy and Cummins 1999).

These findings do not address whether the monkeys acquired a reflective understanding that objects fall when unsupported (that is, an abstract understanding of why food falls into the hole). Roberta, the monkey that used a distance rule to solve the trap-tube problem, apparently did not; she avoided the trap even when it was irrelevant to her actions. The monkeys using the hoes may have: they only avoided the hole when it was in the way of their movement. However, in all these problems the monkeys can develop procedural rules that may or may not be accompanied by an abstract conception of the invariant physical law (such as gravity) governing the movement of the tool or of the target object. Determining unambiguously that an individual's behavior reflects an abstract conception of a problem is difficult or impossible except by asking for a judgment prior to action.

Figure 10.11. Xenon chooses to retrieve one of the two pieces of food with a hoe according to the type of surface the food must traverse. Top, the monkey moves the food around the barrier (the raised block on the right) rather than toward the hole on the left. Coping with the hole posed a greater challenge to the monkeys than coping with the barrier; Bottom, the monkey chooses to move the food across a solid surface (on the right) instead of the surface with a hole (on the left). (Photographs by Sarah Cummins-Sebree.)

TWO WAYS TO THINK ABOUT TOOL USE

As will have been evident in our discussion, we have drawn on two different ways of thinking about tool use: a representational view, and one that we have labeled the action-perception view that is most often associated with the Gibsons (J. Gibson 1966, 1979, E. Gibson and Pick 2000) and Thelen (Thelen and Smith 1994). These two points of view draw on disparate philosophical traditions and notions of cognition; they emphasize different sources of competence and developmental processes, and they suggest different interpretations of the same data (for a spirited comparison that favors the action-perception view, see Reed 1996).

The action-perception perspective emphasizes the actor's search for information and the significance of learning to perceive relevant features of the situation (i.e., what surfaces should be related to other surfaces in what way), to perceive the relation between actions with objects, and to control the body to achieve the desired forces and positions of objects. This approach emphasizes knowledge as embodied in action and emphasizes the basis of learning to use tools (as to do any skilled action) in the discovery of actions, perceptual learning and practice in a particular context (Bernstein 1996, Smitsman 1997).

From this point of view, to understand the basis and limits of capuchins' abilities to use objects as tools, we need to look at how quickly and how precisely capuchins master object relations and how many relational elements they can manage at one time. We also need to look at the physical aspects of moving objects and applying force with them; for example, how the monkeys improve their control of placement, force, tempo, etc. These are not trivial aspects of learning to use an object as a tool, as anyone who has worked to master a new skill can verify. We can also look at the contributions of different forms of perceptual information (kinesthetic, visual, auditory, etc.) to precise placement and alignment of objects and their movements across surfaces. Experimental work on these issues with capuchins is just beginning, so we can as yet draw few conclusions. One prediction that this perspective brings is that capuchin monkeys will master most readily tasks in which the relations between action and outcome are immediately perceptible to them, and as a corollary, that proprioceptive (as well as visual) information about the outcomes of actions with objects will be especially helpful for them.

This perspective also suggests that the number of relational elements in a task (whether in the body or in the objects and surfaces that must be moved or related) and the precision with which these elements must be controlled will influence the difficulty of mastering the task. Increasing the requirements for precise placement and movement control (of force, torque, direction, etc.) will make mastering the use of an object to move another object increasingly difficult. This perspective suggests that the limitations exhibited by the monkeys in Visalberghi's trap-tube tasks might be overcome by monkeys that master moving objects across irregular surfaces in simpler contexts than the trap-tube task presented. Perhaps capuchins can improve their perception and management of relevant elements with appropriate experience, although they may improve more slowly than young children, for example. In other words, with this perspective we ask how, and how well, capuchins can master tasks with increasingly complex demands for relational control.

A representational perspective emphasizes mental reflection on the abstract relation between action and outcome. In this view, generalization emerges from an abstract recognition of the properties of similar tasks (i.e., causal understanding). Adult humans can express relevant elements of a tool-using situation in abstract terms; being able to manipulate these elements mentally is thought to underlie effective action in varying circumstances. According to this view, an appreciation of causal relationships among events and actions allows an organism to engage with its environment in more creative, flexible and foresightful ways than if it lacked this appreciation (Visalberghi and Tomasello 1998). For instance, causal understanding enables an individual to infer the cause by looking at the effect, and to intervene in situations to produce or block a cause so that the effect is under the actor's control. Causal understanding provides a means of moving (mentally) forward and backward in time.

From the representational perspective, capuchins do not appear to understand *why* or *how* certain familiar conditions (such as a hole in the surface) influence the way an object moves across a surface. They do not readily (if ever) come to an abstract causal understanding of the relations involved in a particular set of movement conditions. How do capuchins compare to the other champion tool users, the chimpanzees, from this perspective? Are chimpanzees more likely than capuchins to arrive at an abstract understanding of how and why an action with an object produces a specific movement of another object? This is still unclear. Povinelli (2000) presented a long series of problems involving rakes, holes, tubes and so forth, that are similar in many ways to the tasks we have presented to capuchins. Like the capuchins, the chimpanzees adopted a variety of strategies, but more than capuchins they learned to solve the problems effectively. However, in Povinelli's view, none of the chimpanzees, whether effective at using the tool or not, unambiguously anticipated cause-effect relationships to organize their actions. On the other hand, Boesch and Boesch-Achermann (2000) argue that wild chimpanzees anticipate the requirements of future action by choosing and transporting tools over long distances before using them. They also argue that when a chimpanzee hunts cooperatively with other chimpanzees it comprehends the causal relations among the other chimpanzees' actions, and its own actions, and the prey's behavior. In accord with this view, the finding that two chimpanzees solved the same trap-tube tasks given to the capuchins, apparently by taking into account the relative position of the food and the trap in the tube, suggests a level of sophistication that no capuchin has yet achieved (Limongelli *et al.* 1995).

We await further information about the behavior of capuchins and chimpanzees in identical contexts to make a more informed comparison between these two genera. At this time, we should recall Anderson's (1996) conservative conclusion following a comprehensive review of the literature: there is little basis at present to make strong claims of differential ability to use objects as tools in chimpanzees *vs.* capuchins.

TOOL USE IN CAPUCHINS AND ANCESTRAL HOMINIDS

Some studies have considered whether capuchins might illuminate the proximate and functional processes surrounding the emergence of tool use in early hominids. Oldowan hominids (human ancestors living in the Plio-Pleistocene, 1.5–2.0 million years ago) modified stones and used flakes as tools (McGrew 1992). Anthropologists have studied whether and how apes produce flakes from striking two stones together, a behavior that can only be inferred for hominids. One orangutan (*Pongo pygmaeus*) and one bonobo (*Pan paniscus*) learned to produce stone flakes and to use them to cut a cord to open a box. However, their flakes were judged to be primitive compared with those produced by Oldowan hominids (Toth *et al.* 1993, Wright 1972).

Westergaard and coworkers tested capuchins in several different conditions fostering uses of objects reminiscent of the uses of early hominids (see Table 10.2). Capuchins pounded stones, producing flakes; they then used the flakes to cut the acetate cover of a baited box (Westergaard and Suomi 1994a). In other studies capuchins were presented with sticks and peanuts hidden in 5 cm of packed earth, to simulate digging roots (Westergaard and Suomi 1995a), and with stone tools and copper nuggets to penetrate acetate barriers to explore the possible origins of metal tool technology (Westergaard *et al.* 1996). In addition to assessing whether and how the capuchins solved the tasks, one other aim of these studies was to evaluate what kind of "material culture" their activities would produce (i.e., what kind of flakes, what kind of modifications to the stone-tool or to the copper nuggets, etc.).

The products of the monkeys' tool-making activities proved on the whole very different from human artifacts. For example, the flakes produced by the capuchins are not as sharp as those produced by Oldowan hominids. Capuchins "do not produce large flakes through hard-hammer percussion, use precise, highly controlled striking actions, actively search for acute core angles, use flake scars as striking platforms or consistently produce invasive flake scars" as required for producing flakes similar to those of *Homo habilis* (Westergaard 1995, p. 3). Moreover, the tasks presented to capuchins (to cut, to probe, to pound) usually embodied first-order static relations, unlike most situations faced in natural settings. For example, capuchins often worked with target objects fixed to substrates. Thus these problems do not simulate the multi-relational situations our ancestors must have encountered. The act of cutting or digging or throwing is not by itself the challenge for the user of a tool, whether hominid, ape, or monkey; the challenge is achieving multiple relations, especially simultaneous and dynamic relations.

Early hominids also transported modified objects over long distances. Jalles-Filho *et al.* (2001) investigated whether capuchins in a small island would transport a stone to where it was needed (e.g., to a closed box containing food). Several capuchins (out of a group of 13) used stones to crack open the box when the stones were nearby but they failed (with the exception of one female) to bring them to the box when the stones were out of view (15 meters away). Thus they did not transport the potential tool to the site of use, although capuchins in other situations did collect branches, sticks, and other materials from places out of view of the site of use, and brought them to the site to use them (see above in this chapter). In another situation, the capuchins studied by Jalles-Filho transported nuts (fruit of *Terminalia* sp.) to distant stones that could be used to pound them open, but not the stones to the nuts. This behavior may reflect the relative efficiency of searching for the stone with the nut *vs.* carrying the stone or leaving the nuts unattended to search for the stone. In any case, on the basis of the rarity of tool use in the wild and the absence of tool-making in capuchins beyond breaking off branches or twigs to use as probes, capuchins' use of objects as tools "is not functionally analogous to the technological behavior (as inferred from archeological records) of the Plio-Pleistocene hominids" (Jalles-Filho *et al.* 2001, p. 375).

Studying how capuchins use objects as tools does not enlighten us about the origins of specific forms of tool use in hominid evolution. Instead, capuchins help us to appreciate the challenges that acting to achieve the correct relations pose to the naïve individual, and the ecological constraints affecting such activity. These are problems faced by individuals of any species; their study in capuchins may help us to understand the origins of skilled actions with objects in our own species and to understand why species differ in these abilities.

SUMMARY

European explorers of the New World reported 500 years ago that capuchins used objects as tools. We now have many systematic studies of this phenomenon, mostly of capuchins in captivity, and mostly concerning *C. apella*. However, all species in the genus *Cebus* have been seen to use objects as tools in captivity, and some species also in nature.

In the wild, a greater degree of terrestriality and a relative shortage of readily obtainable foods seems to promote the emergence of tool use. The most frequently observed instances of tool use in the wild or in semi-free ranging conditions are pounding a stone on a nut that the monkey has positioned on a hard surface (an "anvil"). This activity involves relating three objects to one another, a problem embodying a second-order relation.

The variety of ways in which capuchins use objects as tools, and the flexibility with which they do so, are remarkable in comparison with what we know of tool use in all other nonhuman primate species except chimpanzees and orangutans. Capuchins often generalize how to use an object from one context to another, and they can solve the same task using different objects. They can also modify an object to accommodate it to the task at hand, although the extent to which they will modify objects beforehand is still unclear. In general, they choose objects to use as tools that are suited for solving the task at hand and they learn to avoid irregularities in a surface that make it difficult for them to succeed. They are better at detecting some irregularities than others. The presence of a hole on the surface along which they have to move an object gives them a particular challenge.

Capuchins learn to perceive spatial relations among objects and to achieve specific outcomes through action with objects. Whether and when this knowledge can lead capuchins to an abstract appreciation of causal relations among actions, objects and surfaces, and outcomes needs further investigation, but it is clear that they do not develop this appreciation as readily as humans. Consequently there are fundamental differences in the way that capuchins and humans use objects as tools. We do not yet know exactly where to place great apes in this comparison. Comparing "shopping lists" of what sorts of problems members of each species can solve by using an object will not provide a satisfactory view of what is distinctive about any of the species compared. A deeper understanding of tool use requires one or more theoretical frameworks that would allow us to probe how such behaviors develop and how they are altered in the face of changing conditions. We have presented two frameworks that are guiding work now: an abstract representational view, and an embodied action-perception view. This is a vigorous field of scientific inquiry, and work from many different disciplines, particularly developmental psychology and behavioral neuroscience, will undoubtedly continue to shape how we understand tool use in nonhuman species.

ENDNOTES

1 Hill (1960, p. 427) writes that "Buffon quotes Dampier (1697) (*Voyage*, iv, p. 225) as authority for the statement that monkeys on the island of Gorgonia" [capuchins on Gorgona Island, Colombia] "would come to the shore when the tide was low in order to eat oysters, the shells of which they opened by banging upon the rocks, or by taking a stone to use as a tool for smashing the shells." However, when we read Dampier's book we found that Dampier merely wrote that "the monkeys come down by the sea-tide, and catch them [periwinkles and mussels], digging them out of the shells with their claws" (Dampier 1697, p. 173). Unless different editions of the book contain different information, the earlier references to Dampier as reporting tool use were apparently in error. In other words, Dampier did not report that he saw capuchins using tools.

More recently, Hernández-Camacho and Cooper (1976) mention unpublished observations by Carpenter of capuchins using pounding tools. We cannot verify if this was indeed the case; although there is widespread evidence that wild capuchins pound encased foods directly on a substrate, without the use of tools (e.g., Boinski *et al.* 2001, Fragaszy 1986, Izawa and Mizuno 1977, Struhsaker and Leland 1977, Terborgh 1983).

2 Note that we purposefully avoid labeling the object used as a tool as a "hammer," or hammering to indicate the pounding action. We consider these words potentially misleading because a hammer, according to Webster's Dictionary, is "a hand tool consisting of a solid head set crosswise on a handle . . ." and "to hammer" is "to strike blows, especially repeatedly, with, or as if with, a hammer." Stones are not hammers; using a stone to pound is not hammering. True hammering, with a head piece mounted on a shaft, requires less energy to apply the same force than pounding with that same stone held in the hand.

3 Nuts vary in size, hardness and elasticity (or brittleness) depending on the species. Hardness and brittleness are related, among other things, to ripeness and size. Size affects the way in which capuchins try to open a nut. On the one hand, they cannot crack large nuts with their teeth but they can pound them without hitting their fingers. On the other hand, they can crack small nuts with their teeth but cannot pound them without hitting their fingers. Hardness and brittleness vary tremendously even within the same species of nuts (e.g., walnuts; Visalberghi and Vitale 1990). Above a certain level of hardness, tool use becomes increasingly more advantageous than biting and/or pounding the nut on a substrate.

11 • Living together. Social interactions, relationships and social structure

Glancing at my watch I realize it is nearly noon and that the capuchins and I have been on the move without a substantial rest period for more than 6 hours. My breakfast is a distant memory and I long to sit down, even on the wet forest floor. Just then, the alpha male crosses over two trees to reach the foraging female I am watching and flops down in front of her on a wide branch. She is the top-ranking female, the social facilitator of the group and is nearly certain to groom the alpha male whenever he presents like this. Sure enough, she hesitates for only a moment before starting to groom his head and shoulders. Almost as if the noon hour whistle has sounded, the tenor of the group changes. The alpha female's yearling, whom I haven't seen for 2 hours, suddenly appears at her side and pushes his head under her arm from behind, seeking the nipple. Her adult daughter approaches and begins to groom the alpha male's hip and leg. The beta male takes up a lofty position in a nearby tree and looks out into the distance. Another infant appears and lures away the alpha female's offspring – their romping game leads them to scamper back and forth over the bodies of the grooming cluster. In the next tree over, some juveniles are wrestling and mounting one another. Only the low ranking adults continue to forage, using the rest period as an opportunity to eat from the highly desired fig tree that the high-ranking core has temporarily abandoned. Suddenly a noisy quarrel breaks out among the juveniles. The alpha female stands up and threatens in their direction – her son is one of the squabblers. The alpha male hops into the next tree and puts first his arm and then his body over the high-ranking juvenile. With their heads moving in unison, mounted one on top of the other, they produce a double threat face that causes the youthful mob to scatter. The alpha male tires of this effective two-headed display and looks back at the two females, who are now grooming each other. Instead of returning to them, he moves into the tree with the beta male and sits down beside him. These two males have been close companions since they grew up together in a neighboring group. They emigrated by force into this group as a team, driving out the former resident males. The alpha male picks up the beta male's arm, but instead of grooming him, he places the other male's hand over his face, closes his eyes and visibly inhales. Holding the beta male's hand in place over his face, he appears to fall into a trance. Looking around, I notice that everyone except the infants and the low ranking adults are settling down to nap. I find a large rock to sit on, and pull my lunch out of my pack.

When we collect data on the social behavior of monkeys, we essentially record "who does what to whom, when, and for how long." Fundamentally we are asking "why is A doing X to B; why at this time and for this amount of time?". Hinde (1983) has pointed out that the description of social behavior involves phenomena at a number of levels of complexity. The most basic level (the "what, who, when and for how long" questions) refers to the study of *interactions* among individuals. A series of interactions between two individuals in which one interaction affects the next over a period of time to build a pattern of interactions can be referred to as a *relationship*. This constitutes the second level of description of social behavior. Each participant in a given relationship is also involved in many other relationships. Such a network of relationships forms a *social structure*, which is a third level of complexity. In this chapter, we will use Hinde's conceptual model of social behavior at these three levels – interactions, relationships and social structure – to organize the presentation of information on capuchin social behavior.

Figure 11.1. Examples of facial expressions in *C. apella*. Top, (left and middle) Silent bared teeth display in an adult and in a young individual, respectively; (right) Open mouth silent bared teeth display. Bottom, (left) Open mouth threat face; (middle) Relaxed open mouth display; (right) Scalp lifting. (Drawings reproduced with the permission of Bertrand Deputte) (see Table 11.1 for more details).

INTERACTIONS

Communication signals

Social interactions usually occur through the mechanism of communication "signals" that are exchanged between two or more individuals. Such signals help the recipient to predict the subsequent behavior of the signaler and to respond appropriately. Communication signals take many forms and may be expressed through the visual, vocal, tactile and olfactory modes. Capuchins have well-developed mobility of the facial musculature, which allows considerable expressive variability, and they have excellent visual acuity for discerning the signals of others (Weigel 1978; also see Chapter 5). Thus, capuchins exhibit a rich repertoire of facial expressions and body postures that convey an array of messages to conspecifics about their internal state (see Figure 11.1 and Table 11.1). Weigel (1978) used sequence analysis of communication signals in captive *C. apella* to document that each unique facial expression is followed by specific predictable responses from other group members and argued that facial expressions have "meaning" to recipients. Although there is some minor variation in the visual signals of different capuchin species (e.g., *C. apella* raise their brows and "grin" during courtship, whereas *C. capucinus* protrude their lips in a "duck face"; see Chapter 12), Weigel concluded that all four capuchin species are quite similar in their visual repertoire of signals. Furthermore, Oppenheimer (1968, 1973), Weigel (1978) and Valenzano and Visalberghi

Table 11.1. *Facial expressions in* C. apella. *The descriptions are based on more than 200 hours of observation of captive* C. apella *(Valenzano and Visalberghi 2002 and unpubl. data). Previous descriptions by Oppenheimer (1973), Weigel (1978) and Weaver (1999) are taken into account.*

Facial expression	Also called	Brief description	Accompanying vocalizations
silent bared teeth display (SBT)	grin, smile	The jaws are closed. The baring of the upper and lower teeth row is produced by the retraction of the mouth-corners	During courtship the female performs an acute squeal ("warble"). Both sexes "warble" when the male mounts the female
open mouth silent bared teeth display (OSBT)	open mouth smile	Similar to the SBT, but the jaw is somewhat open	A pulsed call given in bouts while head cocking
open mouth threat face (OMTF)	threat face	The mouth is wide open, baring the canines and usually the incisors, with a complete retraction of the corners of the lips	Bark-like staccato call, sometimes a pulsed breath
lip smacking (LPS)		Rhythmic lowering and raising of the lower jaw, producing an audible smacking sound. The tongue is sometimes protruded	Smacking sound
relaxed open mouth display (ROM)	play face	The mouth is opened in an oval shape by retracting the corners of the lips and opening the mouth; teeth largely remain covered by the lips	Low staccato vowel sound
scalp lift (SCL)		Eyebrows and forehead are raised, no change in the configuration of the mouth	—
head cock (HCK)	head tilting	The head is rhythmically tilted from one shoulder to the other while gazing at recipient	—
protruded lip face (PLF)	duck face	Lower jaw protrudes anteriorly, lips are tensed together and protruded	—

(2002) argue that many of capuchins' visual signals and their associated context are remarkably similar to those found in Old World cercopithecines. For example, both capuchins and macaques exhibit broadly similar threat faces and fear grimaces. This could be because the original prosimian ancestor of Neotropical and Paleotropical monkeys exhibited baseline similarities, or because there are functional reasons for the facial expressions exhibited (e.g., advertisement of the canines during an open-mouth threat face and protective shutting-down of the eyes and ears when frightened). In any case, it is obvious that capuchins display and accurately interpret a wide range of visual messages that facilitate their social interactions.

Capuchins also communicate in the tactile mode: they touch, sit in contact with, groom and inflict various degrees of physical pain on one another (e.g., pinch, push, pull, bite). Primatologists have mainly studied tactile signals in the form of grooming interactions, which will be discussed below ("Affiliative" interactions). Olfactory signals (called pheromones) are the most difficult for us to study and there is not yet any research focused directly and exclusively on olfactory communication in capuchins (but see Ueno 1994a for experiments on odor detection in capuchins). However, pheromones are clearly implicated in some of the more interesting and unusual behavioral patterns of capuchins, for example, handsniffing, urine washing and genital inspections of new infants and adult males. In addition, the plants that capuchins employ for fur-rubbing are highly pungent (see Chapter 5). *Cebus capucinus* and *C. olivaceus* males also sniff the substrate where females have been sitting and sometimes rub branches with their chests or throats; all capuchins regularly sniff multiple fruits hanging from a tree before plucking one – apparently checking for the odor indicating ripeness. Obviously then, capuchins use pheromones and olfactory clues from the environment to help them communicate and find food. We simply need to find better ways to study their olfactory capacities and communication signals.

Capuchins have a rich repertoire of signals in the vocal-auditory mode. This makes good sense in a monkey that spends much of its day in leafy trees out of direct sight of its fellow group members. Through their calls, capuchins constantly maintain contact with one another and appear to transmit information about their internal status (fearful, contented, aggressive) as well as information about the environment (e.g., food locations, directional movement cues and predator sightings). Indeed, a traveling, foraging group of capuchins in nature vocalizes almost constantly, and their trilling, twittering, peeping and cooing sounds are often the easiest way for us to find them, as well as the mechanism *they* use to monitor the location of their group mates. At times, this pattern of intermittent short-range and middle-range calls turns into a veritable din of noise, for example when capuchins spot a predator that they mob with screams and the breaking of branches. At other times, a capuchin group may fall entirely silent, such as when they spot unfamiliar people or another group of capuchins approaching, or immediately after sighting an eagle.

Although early researchers such as Oppenheimer (1968; see also Dobroruka 1972, Garner 1892, 1900; Nolte 1958) made careful verbal lists and some sonograms of the vocalizations given by capuchins, no one has yet documented and published the full vocal repertoire of any species in this genus. However, Di Bitetti (2001b) took a large step in this direction by including in his dissertation spectrograms and context-descriptions of 30 acoustically distinguishable calls of *C. apella* in Iguazú, Argentina. A few studies have addressed a specific capuchin vocal signal. Norris (1990a, b) for example, concluded that *C. olivaceus* give acoustically distinct alarm calls to different kinds of predators – something that all capuchin researchers notice by ear and that Di Bitetti (2001b) documented through acoustic analysis. Norris also argued that the snake alarms of *C. olivaceus* have different acoustical features when the snake is on the ground or in trees.

Di Bitetti (2001b, 2003) carried out extensive analyses of food-associated calls in *C. apella* at Iguazú through both acoustic analysis and the use of feeding experiments. He concluded that food-associated calls (which he terms "grgrs" and "whistle series") are functionally referential signals that provide information about food, such as amounts, preferences and divisibility. Furthermore, capuchins make flexible (and possibly deceptive) use of these signals by taking into account the number, distance and rank of neighbors ("audience effect") before they produce food-associated calls. These calls appear to indicate the willingness of the caller to be approached and to share resources. In some contrast, Gros-Louis (2002) argued that in *C. capucinus*, food-associated calls do not fit an information-sharing model. Instead, she concluded that these calls function to announce food ownership (rather than willingness to share) and that callers are approached less often than are silent discoverers.

Robinson (1982a, 1984b) suggested that *C. olivaceus* combine calls into compounds that express internal states on a continuum from contact-seeking to contact avoiding. And he argued that capuchins use vocalizations, such as the "huh" call, to regulate the spatial distances between themselves and conspecifics. Later Boinski and Campbell (1996) analyzed the context of the "huh" calls given by *C. capucinus* and concluded that these vocalizations function to maintain spatial separation among individuals that are feeding, thereby increasing foraging efficiency.

Boinski (1993, 1996, Boinski and Campbell 1995) also argued that another *C. capucinus* call, which she refers to as the "trill," is used by these monkeys at Santa Rosa and La Selva to regulate group movement, in other words, as a travel coordination call. In Santa Rosa, it is usually an adult female at the edge of a group who produces this vocalization, and she then begins to move out a short distance looking behind her. Often the group will start to move in the direction initiated by the female, but if not, she returns to the group. Sometimes more than one individual will "try to lead" the group in conflicting directions with this vocalization, but usually adult females of the group reinforce each other's efforts to coordinate group movement patterns through trilling.

Gros-Louis (2002) has recently documented that white-faced capuchins at Lomas Barbudal use the trill in close-range affiliative social interactions. At Lomas, infants in particular employ trills as a friendly gesture when approaching others in their group, and the effect of this vocalization on the recipient is to increase the likelihood of an affiliative response.

There are many more fascinating capuchin calls for which we have yet to analyze the full range of acoustical properties, such as the capuchins' variable-sounding agonistic, contact, estrous and lost calls (see Figure 11.2). We have inferential evidence that capuchins recognize each other's individual voices – e.g., they only respond to lost calls from members of their own groups and they will peer through the brush toward a fight that includes vocalizations from a close relative. But we know little yet about the range of information that may be carried in the acoustical variability of these calls. We assume that capuchins accomplish a lot of their interactions through the vocal-auditory mode and that rich new insights into their social lives will be revealed through further research on their vocalizations.

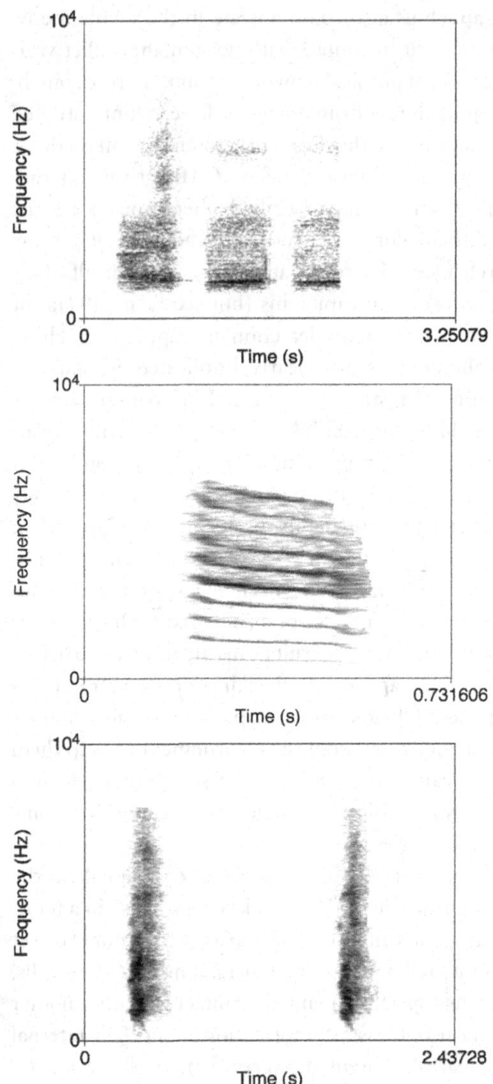

Figure 11.2. Top, spectrogram of a *Cebus capucinus* lost call (Santa Rosa). Middle, spectrogram of a *Cebus capucinus* food call (Santa Rosa). Bottom, spectrogram of a *Cebus capucinus* alarm call directed at a snake (Santa Rosa). Time in seconds is represented on the x-axis. (Courtesy of Shannon Digweed).

Agonistic interactions – conflict and reconciliation

Researchers find consistent patterns in the expression and resolution of conflict interactions in all primate societies, and capuchins are no exception. Most conflicts within the group take the form of dominance interactions, during which one individual supplants, threatens, chases or physically aggresses against another and the recipient responds with submission, avoidance (including fleeing the scene) or counterattack. In some primate species, all the members of a group can be easily ranked according to which of their conspecifics they consistently direct aggression and submission toward – that is, they can be placed in a linear dominance hierarchy. In other types of primates, the direction of agonistic signals during conflict interactions is too inconsistent for researchers to determine dominance ranks. Years ago, Bernstein (1966) conducted experiments on captive *C. albifrons* to show that male capuchins, unlike macaques, do not form linear dominance hierarchies, and field workers today experience some difficulties in determining dominance rank order below the alpha (top-ranking) male. In capuchins, it is almost always possible (easy in fact) to determine the alpha male and alpha female of each group in all the species. But there is diversity in whether or not linear hierarchies can be determined (Buckley 1983, Fedigan 1993, Izawa 1980, O'Brien 1991, Perry 1995; Robinson 1988a, J. Lynch, pers. comm.). That is, researchers commonly report that they can easily tell a subordinate male or female from the alpha animal, but it is not always possible to distinguish the 3rd from the 4th ranking individuals and so on. So, dominance ranks play a role in agonistic interactions of capuchins, but dominance hierarchies may not be quite as fixed, obvious or rigorous in this genus as in Old World cercopithecines such as macaques and baboons, for whom models of primate dominance hierarchies were first developed.

In captivity, capuchins manage competition and conflict with minimal fuss. The relaxed nature of their social relations is evident in spacing during feeding. One animal will commonly approach another to sit next to it while feeding. Rogers (1996) found that in captive *C. apella*, the type and abundance of food influences social dynamics. In particular, contest competition occurs in the presence of scarce highly preferred foods but is also present when highly preferred foods are abundant because capuchins try to get access to the most food items. Scramble competition is likely when low preference foods are scarce whereas abundant low preference foods result in no detectable competition. Even in small spaces with high monkey density relative to natural settings, highly desired resources promote attempted monopolization and theft but not overt agonism.

Agonism in capuchins can be thought of as occurring at three basic levels of severity: supplantation (when one individual approaches and forces another from its spatial location); threats and submission; and physical contact. Rose (1998) found that in *C. capucinus*, supplantation, the mildest form of agonism, made up 48% of all agonistic encounters she recorded, whereas threats and submission constituted 41% and physical contact only 11%. Indeed, it is common for researchers studying white-faced capuchins to report that levels of agonism in the wild are very low (Baker 1998, Fedigan 1993, Phillips 1995, Rose 1994b), often occurring less than once per hour among the members of a group. Rose also found that although chases and physical aggression are more likely to be directed down the dominance hierarchy, dominant individuals do not exhibit higher overall rates of agonism than do subordinates. Other species in the genus exhibit similarly low rates of aggression in nature (*Cebus olivaceus*; Fragaszy, pers. obs.) and in captivity (*Cebus apella*; Fragaszy *et al.* 1994b). Weaver (1999) found that in captive *C. apella*, conflict tends to occur more often between adult males and juveniles, but is infrequent among young animals and peers at all ages. Fragaszy *et al.* (1994b) and Cooper *et al.* (2001) found low rates of agonism towards newly introduced animals in captive *C. apella* even when new males were introduced into all-female groups.

Nonetheless, altercations are a notable aspect of capuchin social life, perhaps because participants vocalize loudly and attempt to rouse coalitionary support from bystanders (see Figure 11.3). Females tend to squabble more, employing threats and counter-threats whereas males have lower rates of agonism but resort to chases and physical contact more often. And on occasion, males inflict lethal violence on one another as well as on infants and females (see "Social structure" below).

This has quite naturally turned the interest of some researchers to the question of whether capuchins have mechanisms for reconciling after an agonistic encounter, as do many Old World catarrhines (Aureli and de Waal 2000, de Waal 1989). Perry (1995) argued that *C. capucinus* have a good behavioral candidate for a reconciliation mechanism in the "wheeze dance" or reunion display exhibited mainly between males (but also between the sexes) who are reuniting after a separation or a fight (see also Matheson *et al.* 1996). In spite of Perry's (1995) careful protocol in the field to distinguish whether or not affiliation is more likely to occur after an agonistic bout in white-faced capuchins, she was unable to decide conclusively that reconciliation occurs in this species, perhaps due to a small sample size. She suggested that if successful reconciliation does occur, it does so at very low rates and may occur long after the fight, or indirectly, which would be almost impossible to document. This could be because capuchins forage in a very dispersed manner, spending a good part of the day well out of each other's sight and may not "get around" to reconciling until the subsequent rest period. Verbeek and de Waal (1997) proposed a similar delay in reconciliation

Figure 11.3. Two female (top) and two male *C. capucinus* (bottom) perform a joint threat. Santa Rosa National Park, Costa Rica. (Photographs courtesy of Katharine Jack).

in captive *C. apella*. However, Leca *et al.* (2002) found that in captivity, *C. capucinus* do sometimes reconcile within a few minutes of a conflict, particularly if the antagonists are related to one another (kin male/female dyads = 48.1%; non-kin male/female dyads = 21.2%). In the Leca *et al.* (2002) study, capuchins reconciled at similar moderate rates to some species of Old World monkeys (e.g., long-tailed macaques), and the capuchins used intense affiliative gestures such as clasping and mounting to express reconciliation.

Verbeek and de Waal (1997) found that reconciliation occurred among captive *C. apella* following only a small proportion of agonistic episodes (e.g., former opponents exchanged affiliative signals in 19.2% of first contacts following conflicts), and only in the absence of highly attractive foods. Weaver's (1999, Weaver and de Waal 2000) careful documentation of the development of reconciliation in the same captive colony of *C. apella* (using several different measures of reconciliation) found that infant capuchins readily reconcile with adults following conflicts. However, reconciliation wanes as they age and becomes uncommon and indirect by the time they reach juvenescence. Why? Perhaps because her sample was composed mainly of immature *male* capuchins. In natural settings, immature male capuchins increasingly respond to social tension with evasion and (in the wild) with emigration, rather than reconciliation. Given that much of capuchin agonism in the wild occurs in a foraging context and that capuchins deal with resource constraints by increasingly dispersing themselves in space, this avoidance mechanism may have lessened the need for the active, direct face-to-face reconciliation interactions commonly found in terrestrial, open-country species like macaques and chimpanzees. However, the book is by no means closed on whether reconciliation occurs in capuchins – there are several more species and many more settings in which to ask the question and we have yet to find completely satisfactory ways to address the issue.

Affiliative interactions – proximity and association, grooming, play, and sexual behavior

At its simplest, individuals in a group may express their affinity for each other by sitting, sleeping, traveling and foraging in close proximity. This is similar to the way we choose to sit close to our friends and relatives in an auditorium or stay near them at any large gathering. Although one might argue that such proximity is not strictly speaking "an interaction," consistent close association between two individuals is often the best clue available to the ethologist that such individuals are in fact kin-related, or friends, or sexual partners or otherwise affiliatively connected. Of course an individual may be constrained from freely locating itself close to the others of its choice. For example, Janson (1990b) documented that the alpha male *C. apella* in his study group directed aggression differentially to other group members and thus played a major role in determining their spatial positions in the group.

Rose's (1998) study of proximity and association patterns in *C. capucinus* found that females are more likely than males to have other group members in proximity ($\leq$10 m), especially females with infants and during rest or social periods. She also found that females are more likely to be close to other females and males close to males. This bias toward same-sex association also occurs at the level of "nearest neighbor" (closest individual at any distance). Perry (1995) found that female–female dyads are more likely to spend time in proximity followed by male–female dyads and male–male dyads were the least common. Furthermore, Rose showed that dyads that tend to be nearest neighbors or to rest in contact also groom together more frequently.

Clearly, proximity and association patterns are not random with respect to social affiliation. In a study of *C. apella* in captivity, Byrne *et al.* (1996) argued that changing proximity patterns were the primary clue (indeed the only indication) that the beta male was about to challenge the alpha male and reverse ranks with him. Similarly, Perry (1998a) found that changing proximity patterns between the adult females and alpha male of the group preceded a male rank reversal. These observations demonstrate why primatologists use association patterns to help understand the changing nature of relationships within the group.

Grooming interactions are the most commonly studied affiliative pattern in primates (see Figure 1.7). Studies of grooming in capuchins have mainly focused on clarifying three related issues: (1) the social *vs.* hygienic functions of grooming in this genus; (2) determining which sex is more likely to direct and receive grooming; and (3) establishing whether grooming is more likely to be directed up or down the dominance hierarchy. Some capuchin researchers (e.g., Perry 1996a, Di Bitetti 1997) have been at pains to disprove Dunbar's

(1991) assertion that platyrrhines groom mainly for hygienic reasons (e.g., to remove ectoparasites and detritus from the fur) whereas catarrhines groom primarily to establish and maintain social bonds. O'Brien (1993a, b), Perry (1996a), Di Bitetti (1997), Parr *et al.* (1997) and Rose (1998) showed that grooming rates in all four well-known capuchin species are related to agonistic, coalitionary and dominance patterns as well as to estrus behavior, kin relations and presence and age of infants. Although the details differ among studies and species as to exactly how grooming relates to these other social factors, it is obvious that grooming by capuchins is far in excess of that needed to clean the skin and fur and that grooming interactions in this genus serve a variety of social functions.

In terms of which sex is more likely to direct and receive grooming, it is apparent that female capuchins are the groomers par excellence, at least in field conditions. In general, adult females are more likely to direct their grooming to other females (especially those with new infants), although the second most common form of grooming is directed from females to males, and the alpha male is a favorite recipient of grooming. Males, on the other hand, groom less than females and are unlikely to groom one another. There have been several attempts to determine the exact form of the social benefits females obtain by directing their grooming efforts to certain individuals – such hypothesized benefits include alliance formation/maintenance, coalitionary aid, access to new infants and to high-ranking females at the center of the group, and a reduction in aggression received. That is, monkeys who groom together generally get along better together, so that in effect individuals may "exchange" grooming for other forms of social affiliation. The results of different studies are mixed. For example, Perry (1996a) found that *C. capucinus* female dyads that groomed more frequently also participated in more coalitions, but Rose (1998) was not able to replicate this finding at a neighboring field site. O'Brien (1993a) also reported a strong correlation between grooming and alliance support. However, Rose found evidence for a more straightforward reciprocal exchange of grooming between partners rather than an exchange of grooming for agonistic support.

The most striking case of divergent grooming patterns found in capuchins concerns the relationship of grooming to dominance hierarchies. Seyfarth's (1977) model of grooming in Old World monkeys predicts that because individuals compete to interact with high-ranking females (who are the most valuable coalition partners), the amount of grooming a female receives should increase with her rank. Although there are confounding variables (e.g., kin relations, presence of attractive new infants), it has been argued for most catarrhines that grooming is largely directed from subordinates to dominants. Therefore, it was surprising when O'Brien (1993a) reported that in three *C. olivaceus* groups, most grooming is directed *down* the dominance hierarchy – he suggested that dominants groom subordinates as a form of appeasement. Later studies of *C. apella* both in the wild (Di Bitetti 1997) and in captivity (Linn *et al.* 1995, Parr *et al.* 1997) showed that this species also tends to groom down the dominance hierarchy – alpha females in particular direct more grooming than they receive. Parr *et al.* (1997) also found that adjacently ranked individuals are more likely to groom each other. Therefore, these authors suggested that capuchin grooming interactions may serve a different function than that reported for cercopithecine primates. Namely, capuchin grooming patterns substantiate the assumption of de Waal and Luttrell's (1986) "similarity principle" (like attracts like) rather than Seyfarth's assumption that dominants are more attractive to subordinates than vice versa. On the other hand, both Perry (1995) and Rose (1998) found that *C. capucinus* monkeys mainly groom *up* the dominance hierarchy, which is consistent with Seyfarth's model and with the Old World monkey pattern. This "up-versus-down the hierarchy" difference may well represent species variation within the genus *Cebus*. Thus, it would be helpful to know more (i.e., anything) about the relation between grooming patterns and dominance in *C. albifrons*. And more studies of grooming in the three better known capuchin species will also help to clarify the relevance of grooming to dominance hierarchies in this genus. What we do know is that grooming in capuchins is not an exclusively hygienic activity because it is interrelated with other social patterns such as agonistic ranks and affiliative networks. Recently Henzi and Barrett (1999) argued that Seyfarth's model of social grooming has not been well substantiated by empirical evidence from either New or Old World monkeys. They conclude that grooming is not exchanged for coalitionary aid, rather it may occasionally be employed by subordinates to engender greater tolerance from dominants, and that grooming is mainly exhibited for the

Figure 11.4. Playful wrestling interaction between two juveniles *Cebus apella*. (Photograph by Elisabetta Visalberghi).

direct benefits the behavior itself provides – removal of ectoparasites and release of pleasurable beta endorphins.

Playful interactions are another easily recognized form of affiliative expression in monkeys, although there are few reports directly and exclusively focused on capuchin play (see Figure 11.4). As is true of most primates, capuchin play is exhibited mainly but not exclusively by immatures and can take the form of solitary or social play (see Chapter 6). Immatures often engage in social play while the adults are resting from their foraging activities and play is more common when resources are more abundant (Freese and Oppenheimer 1981). Adults, especially males, will also join in rough and tumble games, but in the wild are far less likely to play than are the juveniles. Adult males seem more playful in captivity (Visalberghi and Guidi 1998).

Sexual interactions are important and complex enough in capuchins that an entire chapter (see Chapter 12) is devoted to their description. Here we will just note that capuchins, like other primates, engage in sexual interactions far in excess of what is needed for procreation. Manson *et al.* (1997) showed that the majority of heterosexual *C. capucinus* copulations occur during pregnancy and early lactation and are thus "non-conceptive" in nature. Furthermore, these researchers found that mounting with thrusts between two males commonly occurs during socially tense situations (e.g., reunions) and during play. So it is obvious that sexual interactions in capuchins, like grooming patterns, serve a number of social functions related to affiliation and agonism.

RELATIONSHIPS

Kinship

We do not yet have sufficient long-term data from the field to know the extent to which kinship is a vital factor structuring social relationships in capuchins. Although capuchins are characterized by male emigration and female philopatry, it seems likely that capuchins in the wild do not structure their social relations so tightly around multi-generation matrilines, as is characteristic of societies of Old World catarrhines such as macaques, vervets and baboons. However, Welker *et al.* (1990, 1992a, b) argued that their captive *C. apella* colony was very similar to the cercopithecines in that matrilines formed the structural foundation of social relationships. Martin (1994) also found that matrilineal kinship has a strong influence on affiliative and agonistic aiding patterns in captive *C. apella*, especially mother–daughter and sister–sister relationships. And we have tantalizing clues from the field that kinship structures at least some adult relationships in *C. olivaceus* and *C. capucinus*. O'Brien and Robinson (1991) found that relationships among female *C. olivaceus* are stable over many years and continue from the juvenile to the adult stage of life. At least some of these lasting relationships occur between female siblings (or half siblings) and between mothers and adult daughters. The main benefit of kinship in this species is a reduced rate of agonism rather than an increased rate of affiliation, therefore female kin are not always found to be close grooming partners. Rose (1998) also had a small number of known female kin in her study sample of *C. capucinus* at Santa Rosa and found that mother–daughter pairs had high rates of grooming and association, although not necessarily the highest in their groups. At Santa Rosa, male *C. capucinus* often transfer groups in pairs or small parties of three to four individuals, and these males are occasionally half-siblings or hypothesized to be cousins. At other times, a young male and an older adult male who have formed a close relationship in one group disperse from their home group and travel together before joining other groups, either together or separately. These males may be father–son pairs or simply "friends." We also have evidence of males emigrating into groups containing older, familiar males that dispersed from the formers' natal group years earlier.

Friendship

When two or more male *C. capucinus* transfer between groups together, they sometimes continue to associate closely in their new group. We have observed several pairs of males who were born to what we assume are only distantly related mothers to form close affiliative bonds as juveniles, such that they are almost always found in proximity and frequently play together. Or a juvenile will form a friendship with a subadult male who has emigrated into his group. The pair will then disperse from the natal group and enter a new group together, where they maintain close association. One such pair at Santa Rosa transferred groups together twice and sought each other's company preferentially over that of females, even though they held the alpha and beta positions in their new groups (Jack 2001). This pair rested in contact constantly and engaged in the rarely seen but apparently soothing interactions of mutual finger-sucking and handsniffing. In such interactions one individual places the other's fingers in its mouth, or over its nose and mouth (like a gas mask) or even up its nostrils, and inhales deeply with a trance-like expression on its face (see Perry *et al.* 2003a, b and Chapter 13). Since adult male capuchins are capable of inflicting terrible wounds on one another with their canine teeth, it is remarkable to see them willingly resting their fingers in another male's mouth.

Pairs of female *C. capucinus* may also form tight and exclusive affiliations, but in the absence of complete matrilineal information, it is hard to know whether these close associates are friends or relatives. Perry (1997) reports a long-term friendship/alliance between the alpha and beta females of her study group. Although matrilineal relations were unknown among the adults of her group, this alliance might have been based on some type of kinship bond. Whatever its basis, the relationship between these two females was stable and powerful within the group's dynamics and was arguably the determining factor in the downfall of one of the group's alpha males. Males sometimes tried to break up the alliance between these two females, but they always reunited.

Female–female relationships

At least two (*C. olivaceus* and *C. capucinus*) of the four traditionally recognized capuchin species appear to be female-bonded in the sense that females maintain long-term relationships with one another based on consistent partner choices for grooming, affiliative, coalitionary, proximity and dominance interactions (O'Brien and Robinson 1991, Perry 1996a; see below for more on "female-bonded social systems"). In captivity, we know that female bonds are not restricted to kin. Leighty and Fragaszy (unpublished data) found that older (>16 years) captive female *C. apella* living in social groups with stable female membership spent as much time in proximity to other adult females as to younger animals, including their own offspring. There was no decline in the degree of social integration across younger and older age groups in these females. Moreover, the frequency of social proximity and grooming among adult females was unrelated to whether or not the females were kin to each other.

In the wild, all the capuchin species are known to be female philopatric and a good inferential case can be made that at least some of the adult females of any given group in the wild are kin-related. That being said, there are several qualifiers. Adult capuchin females do sometimes transfer between groups, reducing the relatedness of the females of a given group. At some sites (e.g., Santa Rosa), relationships among adult females are not very stable over the years, whereas in others (e.g., Lomas Barbudal, Masaguaral) they are (O'Brien 1993b, Manson *et al.* 1999). Grooming relations between female dyads are not always reciprocal or correlated with patterns of aid and kinship (see above). On the whole, it would be fair to say that the pattern of female-female bonds in capuchins is not as stereotypically fixed and as strictly correlated with kin and dominance relations as it is in Old World cercopithecines, such as the macaque and baboon societies on which the model of female-bonding was originally based. However, female capuchins do form long-term affiliative bonds that knit them more closely together than do the bonds among male capuchins or between the sexes.

Male–male relationships

We know that adult male capuchins are not as closely bonded to one another as are the adult females, but the question remains as to whether the relationships among adult males are basically hierarchical and competitive, or egalitarian and affiliative. Here there seem to be strong species, ecological, and site differences, so let us take up the question for each species in turn.

C. apella at Manu, Peru are reported to have rather despotic alpha males who monopolize mating and exhibit high rates of aggression toward subordinate males that remain at the periphery of the group (Janson 1986a, b). In turn, these alpha males do not receive aid from subordinate males during defense of the group against predators and against males from other groups. However, Izawa (1980) reported low rates of intra-male aggression in *C. apella* at his site (La Macarena, Colombia), and J. Lynch (pers. comm.) reported low rates of agonism between *C. apella* males at Caratinga, Brazil. Furthermore, van Schaik and van Noordwijk (1989) reported that male *C. apella* do cooperate in defense against predators even at Manu. In his more recent work with *C. apella* at Iguazú Falls, Argentina, Janson (1998b) found that estrous female capuchins were less active in their pursuit of the alpha male than at Manu, and in socially unstable groups, Iguazú females mated promiscuously with every adult male in the group. Janson attributes these variable degrees of alpha male monopolization of mating to differences in the ability of alpha males to monopolize and provide food sources for the females and their young. In captivity, male *C. apella* live together very affably as cage-mates, and present a much more tolerant, relaxed picture of male–male social relationships than the original descriptions from Manu. Several unrelated males can live together amiably for years without a single fight (although rank reversals are tense times with possible wounding). And alpha males can retain their rank even when they are older and smaller than the other adult males. Apart from times of group instability (when new individuals are introduced to groups for example; Cooper *et al.* 2001) captive adult male *C. apella* play, groom and sit in contact, generally exhibiting a strong social interest and dependence on one another.

C. olivaceus in nature are fairly similar to the original descriptions of *C. apella* at Manu. Each group of *C. olivaceus* is reported to have a highly central alpha male who receives almost all the mating invitations from females and who keeps other males on the periphery (Robinson 1988b). Although there may be more than one male present in each group, these groups are functionally uni-male social systems (X. Valderrama pers. comm.). During the mating season, females actively solicit and mate only with the group's alpha male. In sharp contrast to their affiliative interactions with the alpha male, females avoid interactions with the subordinate males, which are typically agonistic. Thus, female preference for the alpha male may be even more skewed in this species than in *C. apella*, and males other than the alpha males are not believed to sire any offspring (O'Brien 1991). Subordinate *C. olivaceus* males will cooperate with the alpha male to defend the group, although what benefits they obtain from this cooperation remain to be determined.

C. albifrons present some contrast to the two species described above in that adult males are cohesive and aggregate in the center of the group, where they take turns feeding, and exhibit low rates of agonism and high rates of cooperation, especially in defense of the group (Defler 1982, Janson 1986a). According to Janson (1986a) mating opportunities and access to food is more equitably distributed among *C. albifrons* than *C. apella* at Manu. A somewhat egalitarian mating system among males is also the case among *C. capucinus*. In every *C. capucinus* group there is a clear alpha male, but the degree to which he asserts his dominance is highly variable. Although it is likely that the alpha male fathers more infants than do the other resident males (at least at Santa Rosa; Jack and Fedigan, unpublished data), subordinates are all seen to mate, sometimes in clear view of the alpha male, with no retaliation. Perry (1998b) characterized relationships among male *C. capucinus* as tense yet cooperative and Rose (1998) found that although male white-faced capuchins seldom groom or associate affiliatively in the course of normal day-to-day intra-group life, they band together reliably in the face of extra-group danger. However, in some years and some groups, the males of a group do form close affiliative relations (Jack 2001, Fedigan, unpublished data, Perry, unpublished data). Rose argued that male *C. capucinus* are capable of both strong alliance and lethal violence toward other males. The issue of lethal violence among males will be further described below under "Social structure" and male take-overs of groups.

Male–female relationships

As Rose (1998) noted, male and female capuchins have markedly asymmetrical relationships – females frequently groom males with little or no reciprocation, and females direct most of their avoidance and submissive behaviors toward males. On the other hand, adult male capuchins only rarely direct physical aggression toward females and most are quite tolerant of females feeding and resting in close proximity. Miller (1998b) witnessed a fatal attack by an adult male *C. olivaceus* on an adult female that may have taken place during a male takeover of a social group (see "Social structure" below on male takeovers) but this is, as of yet, a unique observation. O'Brien (1991) observed two groups of *C. olivaceus* over a 15-month period at Masaguaral and although he found that adult males often supplant females during foraging, he reported no cases of serious aggression directed by males toward females. Janson (1986b) found that female *C. apella* receive relatively little aggression from any adult males. However, he reported that male *C. albifrons* sometimes force an unwilling female to mate (1986b).

Although individual adult males usually dominate individual adult females in all species, such male dominance over females can be counteracted by female–female coalitions. For example, in *C. capucinus*, two females can evict any adult male from a feeding tree. Furthermore, females are not acquiescent in their relationships to males. In *C. capucinus*, females direct more aggression toward males than they receive from them, although this agonism is always in the form of threats rather than physical aggression (Perry 1997, Rose 1998). In *C. capucinus*, *C. olivaceus* and *C. apella* groups, the alpha female may rank immediately below the alpha male and above the other adult males of the group (Robinson 1988a, Fedigan 1993, Rose 1998) or she may rank below all the adult males (S. Perry, pers. comm.).

Perry (1997) has pointed out that even though the different reproductive strategies of male and female primates set the stage for conflicts between the sexes, each sex can also provide important benefits to the other, especially through the reciprocal expression of affiliative behaviors. Several researchers have looked directly and indirectly at the benefits and costs that male and female capuchins bestow upon one another. The costs mainly take the form of aggression and foraging disruptions directed by males toward females. As noted above, male capuchins seldom inflict wounds on females. However, they do often supplant adult females and younger members of the group from choice feeding sites, and most agonism between the sexes takes place in a foraging context. Adult males may also pose a threat to the females' infants – the potential and evidence for infanticide in capuchins is described later in this chapter.

There are several benefits that males and females may confer on each other. It was noted earlier that females groom males more than vice versa (except for *C. apella* at La Macarena, Izawa 1980) but do not always seem to receive agonistic support from males in return. Rose (1998b) found that at Santa Rosa the males that are groomed most often by the females are the same ones that perform the most vigilance. This is possibly a case where there is some "exchange" of female grooming for male vigilance. In addition, it has been documented in all four capuchin species that males are more vigilant than females (de Ruiter 1986, van Schaik and van Noordwjik 1989, Fragaszy 1990a, Rose and Fedigan 1995). Males are also more active in mobbing and other forms of defense against predators and incursions by other groups and non-group males (Robinson 1988b, O'Brien 1991, Perry 1996b). These conspicuous behaviors of males directed to factors outside the group, fit the model of males as "hired guns" in the group (Cowlishaw 1998, Rose 1998) and suggest that these are significant defense compensations to females for the foraging costs that males impose on them. Furthermore, Rose (1994b, 1998) found that adult male *C. capucinus* perform important infant care behaviors – not only rescuing them from danger, but also carrying them more often than do females who are not mothers of young infants. O'Brien

(1991) found that *C. olivaceus* males also provide some infant care. And Janson (1984) argued that in *C. apella*, alpha males allow immatures and their mothers access to an important food source (clusters of *Scheelia* nuts) that they would not otherwise be able to obtain because the initial nuts removed from the cluster require strength to extract. It would be easy for the alpha male to monopolize access to this resource, but instead he tolerates the co-feeding of females and their young. Finally, Robinson (1988a) noted that male *C. olivaceus* are the primary participants in intergroup encounters and the groups that win such encounters gain feeding advantages that apply to the young and females as well as the males of the group. So on the one hand, males may sometimes disrupt the feeding of females and their young in the short term, but they also are powerful and positive influences on the long-term availability and protection of resources to the other members of their groups.

Adult-immature relationships

As in most primate species, the keystone relationship in capuchins is the one between a mother and her infant, and mothers are the primary caretakers of young infants (see Chapter 6). Female capuchins, like female catarrhines, exhibit some degree of variation in the amount of care and energy they expend on their infants, but for the first 3 months of life or so, maternal care is dominant (Escobar-Páramo 1989, Mitchell 1989, Fragaszy *et al.* 1991, O'Brien and Robinson 1991). However, in both captive and natural settings (Calle 1990a, Valenzuela 1993), other group members approach to nuzzle, sniff and visually inspect new infants from the day of birth, often emitting a distinctive low-volume guttural vocalization while they do so. Manson *et al.* (1999) referred to such inspections as a form of "infant handling" and he concluded that in *C. capucinus*, such infant handling occurs between those female dyads that frequently groom and form coalitions with one another.

In nature, infant *C. capucinus* and *C. olivaceus* spend 80–100% of their time in contact with their mother's body during the first 2–3 months of life. Although other group members may express interest in inspecting the infant, forms of alloparenting such as transfer onto their bodies does not occur until the third month of life, and then becomes relatively common from 3–6 months of age (Mitchell 1989, O'Brien and Robinson 1991). Infant *C. apella* older than 3 months are often found near the alpha male, both during resting and foraging periods (Calle 1990b, Escobar-Parama 1989).

In the wild, capuchins are noteworthy among primates (excepting monogamous species) for the high level of alloparental care they exhibit, especially for the amount of time group members other than the mother carry the infants. The differences are less striking in captivity; in one study, for example, squirrel monkey infants spent as much time on someone other than the mother as did capuchin infants over the first 6 months of life (Fragaszy *et al.* 1991). The high degree of allocarrying by wild capuchins may reflect the particular mode of life of these monkeys. Capuchins forage in a very dispersed manner and infants may become widely separated from their mothers and find themselves unable to cross gaps alone. When this occurs, the infant will hitch a ride on the nearest monkey who almost always readily carries the infant along the group's trajectory to its next stopping point. Alpha males, peripheral males, juvenile males, immatures only just larger than the rider – all will carry an infant in distress without protest. Only mothers with their own small infants tend to refuse to give a ride to a hitchhiker. Although almost everyone in the group will give an infant an occasional ride, female relatives, especially older siblings who are nulliparous, are the most common alloparents (Escobar-Páramo 1989, O'Brien and Robinson 1991).

If capuchins are unusual for their general levels of alloparenting, they are perhaps unique among primates in the extent to which females will nurse each other's infants. Allonursing has been reported in three of the four species (all except *C. albifrons* for which we do not have any relevant data) and this seems to be a widespread pattern both in captivity and in the wild. The willingness of a lactating female to allow another infant to suckle from her is sometimes (and sometimes not) related to rank or close kinship of the infant's mother (O'Brien 1988, O'Brien and Robinson 1991, Perry 1995; Weaver 1999). Rose (1998) reported that an orphaned infant in Santa Rosa was able to survive for several months after the death of its mother by suckling from several other lactating females in the group and was eventually adopted by an adult female. Perry (1996a) has argued that allonursing may be an evolutionary stable strategy in that the cost of giving a small amount of milk to another infant is far less than the potential benefit if one's own infant is saved from starvation by another lactating female during an accidental separation or after maternal death. She also pointed out that females that do not nurse others' infants ("cheaters") are easily detected because infants stage tantrums when denied access to any female's nipple (and perhaps even selectively to non-mothers; Weaver 1999).

Why then is allonursing not more common in other primate species? Perhaps capuchins are unusual in the extent to which two behavioral patterns co-occur: (1) they disperse widely during foraging and (2) they commonly carry each other's infants during the first year of life. Thus, it frequently happens that an infant is far from its mother when it *wants* to nurse and it occasionally happens that a female will be separated from her group for more than a day. Under these conditions, allonursing would be a singularly adaptive strategy. Furthermore, capuchins are a relatively slow-reproducing and large-brained species for their body size (see Chapters 4 and 5) and this leads to high parental investment in offspring. As pointed out by Strier (1999), high post-natal investment in young and high risk of infant mortality may be related to the high levels of allomothering reported for capuchins. At the proximate level, the generally high level of tolerance by adults to virtually all forms of contact by infants promotes acceptance of suckling by any infant on the part of an adult female. In this sense, allonursing is one more element in a general pattern of high social tolerance, particularly towards infants.

There is only a very small literature on juvenile capuchins in the wild and their relationships with the rest of the group. O'Brien and Robinson (1993) reported that juvenile *C. olivaceus* are usually found together in an age-class cluster during daily activities and that is also our experience with juveniles in *C. capucinus* groups (MacKinnon 1995). O'Brien and Robinson also reported that rank relations among juveniles predicts their adult dominance ranks. Juveniles are mainly ignored and/or tolerated by adult males but may be supported in their agonistic interactions by adult females. O'Brien (1993b) described juveniles as seeking to develop social relations with adults by grooming them more than they are groomed in return, and by providing coalitionary aid during agonistic encounters and by providing allomothering services. Welker *et al.* (1992a) argued that in captive *C. apella* the mother is the most attractive social partner for a juvenile, but in free-ranging *C. olivaceus* and *C. capucinus*, the most attractive social partners during the daylight hours are other juveniles.

The most extensive study thus far of relationships between immature and adult capuchins in the wild was carried out by MacKinnon (2002). She found that in *C. capucinus*, infant and juvenile interactions with adult males are mostly affiliative in nature and that immatures strongly favor interacting with the alpha male of their group or their mother's favorite male social partner (which is sometimes the same male). Although adult males do not initiate that many interactions with immatures, the latter do show a lot of interest in the former. Juveniles intently watch and follow adult males, especially when the latter are vigilant or eating foods that require extraction and processing. Juveniles also initiate play with adult males, who sometimes respond positively, and they seek to groom adult males.

SOCIAL STRUCTURE

Do capuchins live in multi-male, multi-female, age-graded or uni-male social systems?

As noted earlier, each *C. apella* and *C. olivaceus* group contains an alpha male who is preferentially solicited by females for sexual interaction and who (at some sites) readily supplants other resident males from space and food. There are other males in these groups, but they are subordinate and often peripheral and younger or smaller than the alpha male. This would seem to be an age-graded male system, or functionally speaking a uni-male system since it is thought that only the alpha males breed and father young (Escobar-Páramo 2000, X. Valderrama, pers. comm.). However, we await results of further DNA analyses to confirm these conclusions based on observational and preliminary genetic data (see Chapter 12). Furthermore, there is variability among sites, between wild and captive groups, and even among individuals in the extent to which *C. apella* alpha males seem to "throw their weight around" in the wider social and foraging contexts and in whether or not the males cooperate to drive off predators and competitors. In captivity, *C. apella* present a more tolerant, relaxed picture of alpha male behavior than they do from many field studies.

In contrast to the two species considered above, both *C. albifrons* and *C. capucinus* have social systems in which more than one adult male is socially and reproductively active in each group. These two species also have easily recognized alpha males in each group, but they are not the object of exclusive female choice and they do form cooperative alliances with other males to defend their groups. Thus, it is fair to say that *C. albifrons* and *C. capucinus* live in multi-male groups in which all adult males have access to estrous females (whether this is equal access is not yet clear), and in which they cooperate actively in group defense, they look for and retrieve lost males, and they sometimes transfer groups together.

These different descriptions of modal male behavior are no doubt variants of the same fundamental capuchin pattern that lies along a continuum from what we might call alpha male "despotism" (or intolerance) to more relaxed tolerance of other males. When the alpha male is more intolerant and monopolizes mating and feeding to the detriment of subordinate males, thereby minimizing the possibilities of male alliances (i.e., *C. apella* at Manu), we characterize the resulting social system as uni-male (although there are peripheral males hanging out on the edges of groups). When the balance shifts such that the alpha male is more tolerant of the feeding and mating behaviors of the other males in his group – possibly in return for their assistance in group defense – (e.g., *C. albifrons* and *C. capucinus*), we characterize the social system as multi-male. Why would the balance shift in different directions? At least in *C. capucinus*, it seems possible that alpha males are forced to tolerate the feeding and breeding of other males in their groups by the constant and very real threat of invasion from non-resident males (see below). At Santa Rosa, *C. capucinus* alpha males turn over as often as every 3 years, whereas in *C. olivaceus* at Masaguaral, one alpha male held his position for more than 12 years (O'Brien 1991). Only more long-term field data from the other species of capuchins will reveal the underlying longitudinal dynamics of male–male relationships that affect the proximate patterns of everyday dominance we observe in our groups.

Are capuchins female-bonded?

Wrangham (1980) first defined female-bonded primates as those species exhibiting female philopatry and male dispersal. In such societies, he described females as developing strong and discriminatory bonds with kin-related females, as revealed by frequent association, grooming and agonistic alliances. They band together against members of other groups in order to defend the resources (food, water, space) they need to survive and rear their young. Often they develop dominance hierarchies based largely on matrilineal alliances. Females typically take an active role in intergroup agonism and decide the direction and timing of group movement. Males (except for brothers) seldom associate or groom each other and they resist the immigration of new males. Adult males are usually individually dominant over females. Thus, Wrangham hypothesized, intergroup competition over food is the main driving force behind female bonding within groups.

We can see from the information presented in this chapter that many of the characteristics of Wrangham's female-bonded groups are accurate descriptors of capuchin social relationships. Female *C. capucinus*, *C. olivaceus*, *C. albifrons* and possibly *C. apella* do form strong bonds with other females in their groups, develop dominance hierarchies, and powerfully influence the direction and timing of group movement. Males are usually dominant over females and they are generally less affiliative with one another. But resident males (at most sites) will band together readily to resist the immigration of new males. However, some aspects of Wrangham's model have not yet been documented for capuchins (i.e., whether affiliative relations and dominance hierarchies are based on matrilineal alliances). And at least one premise of Wrangham's model is not universal and possibly not common in capuchins – female capuchins seldom participate in encounters between groups, contrary to what we would expect if intergroup encounters were mainly directed toward resource defense. In *C. capucinus*, all adult females and immatures tend to avoid intergroup encounters (Perry 1996b, Fedigan, unpublished data), whereas in *C. apella* the alpha female sometimes participates, but the other females do not (M. Di Bitetti, pers. comm.).

One reason for low female participation in intergroup encounters is suggested by van Schaik's (1989) model of female relationships. He proposed that for females, competition *within* groups is more important in determining the nature of female–female relationships than is competition *between* groups. In particular, he proposed that nepotistic bonding results from high levels of *within* group feeding competition over small food patches, and serves to regulate agonistic encounters among the females of a group. As Perry (1996b) noted, females in this type of social system are not generally expected to participate extensively in intergroup encounters because only high-ranking females would experience much benefit from participation.

Another reason that female capuchins seldom participate in intergroup encounters is that the conflict between groups may not be so much over resources as it is over access to females. In this case, the agonism is not so much *between groups* as *between clumps of males* resident in the ranges of different female kin-groups. As noted above, studies of all four capuchin species in the wild have shown that adult males are consistently more vigilant than are adult females. Some of this vigilance appears to be directed toward predators, but much of it is also directed at possible encroachments by non-resident males (Rose and Fedigan 1995). Recently, theorists (Janson 2000, Treves 1998, van Schaik and Kappeler 1997) suggested that minimizing the possibility of infanticide by non-resident males may be the most important selective force for year-round primate social bonding – both female bonding and close associations between females and the resident male(s) who father their infants.

Long-term studies of *C. capucinus* at Santa Rosa and Lomas Barbudal are revealing a pattern in which periodic violent incursions of adult males in twos and threes result in extensive fighting and wounding. During these episodic events, the resident males are usually evicted, adult females may be wounded and small dependent infants are sometimes bitten and killed (Fedigan 1993, unpublished data, Perry 1998a, Rose 1998). As Perry (1998a) has pointed out, the threat of invasion by foreign males plays a large role in shaping the relationships among resident males. In particular, it may explain why alpha males tolerate and even encourage other males to remain in their groups. One of the paradoxes of male behavior in *C. capucinus* is that the alpha male may repeatedly supplant a subordinate male from food, and even direct physical aggression toward him, but if the subordinate male then leaves the group, the alpha male will spend hours lost-calling and looking for his wandering group member, and when he finds him, will perform a reunion display, and attempt to lead him back to the group (see descriptions in Perry 1998b and Rose 1998). It seems that the alpha male must balance the tendency to monopolize food and estrous females against his need for male allies in times of trouble. At Santa Rosa, groups with more males generally win intergroup encounters and defend themselves better against male invasions (Rose and Fedigan 1995), but Perry did not find this to be the case at the nearby site of Lomas Barbudal (Perry 1996b).

Male capuchins, especially adolescent ones, sometimes transfer between groups singly and without much aggression. But at least in *C. capucinus*, an alternate male strategy for gaining access to female mates is apparently to stage aggressive invasions and take-overs of groups. At Santa Rosa, we have experienced such massive violent influxes of males into our study groups approximately every 3–4 years since we began observations in 1983. What are the repercussions to female white-faced capuchins of this male reproductive strategy? They lose the resident males with whom they have formed affiliative bonds, they suffer wounding and they may lose their young infants. Their groups and typical foraging patterns are thrown into disarray as males rampage through the forest, looking for one another. No wonder that females and juveniles stay away from inter-group encounters!

On the other hand, both Perry and Rose describe females as playing important roles when males are fighting by throwing their allegiance behind either the resident male, or one of the invading males – grooming and affiliating with the male of their choice. During times of male upheaval, it is also common for high-ranking females to lead the group far away from male fighting, and even to evade the encroaching males who seem to be searching for them. Once the resident males have been soundly defeated or the invaders routed, the females return to their normal ranging patterns and begin to initiate contact with winners of the male–male competition. Even though male take-overs have only been extensively described for *C. capucinus*, females have been seen to initiate sexual behavior with recently inaugurated alpha males in *C. capucinus*, *C. olivaceus* (O'Brien 1991) and *C. apella* in the wild (Janson 1984) as well as *C. apella* in captivity (Cooper *et al.* 2001, Fragaszy *et al.* 1994b, Visalberghi, unpublished data). It is both during the invasion and during the immediate aftermath that small infants, fathered by former resident males, are particularly vulnerable to harm. Females attempt to defend themselves and their infants and will band together to drive off males, but occasionally their infants are bitten by adult males and die from their wounds. The coincidence of infant wounding, death and disappearance with male rank reversals has been reported in *C. apella* (Izawa 1994) and *C. olivaceus* (O'Brien 1991). At Santa Rosa, we have directly observed two infants to be killed by males, and inferred seven other cases from strong circumstantial evidence. At Masaguaral, researchers observed three cases of infanticide in *C. olivaceus* (Valderrama *et al.* 1990). At both of these sites, infants are usually thought to be killed by unrelated males, but may be occasionally killed by their putative fathers. Again, we await the DNA evidence to know for certain. Whether these infants are deliberately targeted by the males, or are the victims of redirected aggression, is unknown and probably irrelevant to a female with a dead infant. As Janson (2000) has pointed out, selection may have operated on females to produce two "counterstrategies" to minimize the risk of infanticide: they can form strong alliances with other females and with resident males to protect their infants, and they can mate promiscuously to confuse the issue of paternity. Female capuchins employ both of these counterstrategies, sometimes even at the same time (Manson *et al.* 1997).

Why are capuchins female-bonded?

Female bonding in capuchins is quite an anomalous social pattern among platyrrhines, most of which are female-dispersed rather than female philopatric. One very good reason for female capuchins to form strong alliances with other females is to reduce the negative consequences to themselves and their offspring of male–male sexual competition. They can also form strong bonds with the top-ranking males of their groups and live in year-round groups with these males to diminish the possibility of infanticide by encroaching males. Finally, female bonding in capuchins may serve some of the same functions it is thought to serve in catarrhines, namely to help individual females defend vital resources against neighboring groups of capuchins. These hypothesized functions are not mutually exclusive.

Strier (1999) has also considered the "*Cebus* anomaly" of female bonding in a comparison of Neotropical monkeys and argued that there may be ecological reasons for the variation in the exact pattern of female-bonding found in different species of capuchins. She suggested that female *C. capucinus* may not need to engage in cooperative resource defense because of their opportunistic, extractive, generalized foraging strategy and their ability to shift their diets onto "fall back" foods when preferred items are low in abundance. On the other hand, in those cases where food competition is strong, females may participate in intergroup encounters and preferentially mate and affiliate with the alpha male in order to increase their access to food, as in *C. apella* at Manu (Janson 1985) and *C. olivaceus* at Masaguaral (O'Brien 1991). Strier also argued that the slow-paced life history pattern of capuchins compared with other Neotropical monkeys (i.e., long inter-birth intervals, slow development; see Chapters 4 and 6) may encourage the system of cooperative infant rearing by group females that is exemplified by their extensive carrying, rescuing and nursing of infants other than their own.

Izar (in press), analyzing patterns of female dispersal in *Cebus apella* in Carlos Botelho State Park, Sào Paulo, Brazil, notes that competition among females in that site is strong and related to the abundance and distribution of food. Both males and females disperse at this site, and females in particular disperse during periods of significantly lower per capita energy intake (i.e., when fruit is scarce). Her findings support the views of van Schaik (1989) and Sterck *et al.* (1997) that contest and scramble competition within and between groups can vary independently, and Isbell and van Vuren's (1996) idea that subordinate females are more likely to disperse and enter small groups. Izar concludes that patterns of female social relationships that she observed are not sufficiently explained by ecological models alone, and that social benefits provided by the dominant male may temper the ecological costs to females of remaining in a large group.

SUMMARY

We know that capuchins have a rich repertoire of visual, vocal, olfactory and tactile signals and we are slowly developing a body of data to help us understand their communication patterns. Although this genus of New World monkey does not exhibit the frequent agonistic behavior and rigid dominance hierarchies characteristic of the more familiar macaques and baboons of the Old World, capuchins do engage in noisy altercations during which participants seek to raise support from bystanders. Coalitionary support during agonism is common in capuchins, but reconciliation after agonism is not so easily recognized. Most affiliative behavior is expressed through proximity and grooming interactions, and grooming is not just hygienic but serves a number of social functions. We have tantalizing clues (particularly from captive studies) that kinship and friendship structure affiliative patterns, but need further long-term field research to document that these are the lynchpins of social structure in this genus. Affiliative female–female and female–male relationships are more prominent than relationships among males. Male capuchins are capable of both strong alliances and lethal violence in their relationships with other males, and with females and immatures. Infant capuchins are frequently cared for by alloparents and even suckled by allomothers. Juveniles tend to stay close to other juveniles in age-class clusters, but they are also highly attracted to adult males. In terms of their social system, different capuchin species fall along a continuum from uni-male to multi-male. They show many of the characteristics of female-bonded primate species. Maturing males transfer groups (usually non-aggressively when they are younger but aggressively when they are in their prime), whereas females almost always stay in their natal group. During male invasions of reproductive groups, the resident male(s) may be driven out, small infants may be killed and resident females may be wounded. Female capuchins form strong bonds with resident males and with other females in their groups, probably to counteract such violent male competitive strategies and to safeguard their access to resources defended by the group.

12 • Erotic artists. Sexual behavior, forms of courtship and mating

I'd always considered myself free from prejudices. But when, many years ago, I saw a capuchin female courting two males in turn – both rather reluctant – I couldn't believe my eyes. Though I was convinced that females are not sexually passive I would have never imagined that a capuchin could be so assertive in her solicitations, struggling for hour upon hour, day after day to get the male's attention. She flirted, she insisted, and after much frustration, she harassed one and then the other. The amazing variety of facial expressions, vocalizations, and charming gestures that she directed at him had little apparent effect; he just ignored her or turned away. That the courted male turned her down for days on end was also beyond my imagination. When close to ovulation, females of many animal species become more attracted to males and are said to be in "estrus." The word derives from the Latin term *oestrus* defining a genus of flies whose larvae move under the skin of their host animals, producing an insatiable desire to scratch to reduce the itch. Female capuchins seem to experience an irresistible desire to mate truly worthy of the label "oestrus".

Sometimes males can be as enthusiastic about mating as females. J. Anderson related a story to me about an old capuchin male he observed mating in a tree. Enraptured, the male did not realize that, while holding the female in the ventro-ventral position, he was slowly slipping from his perch, and no longer had a solid hold. Suddenly he fell to the ground, four meters below. Anderson thought he might have died from the fall. But after 10 long seconds, the old male just got back onto his feet and with his usual tottering gait clambered back up the tree to his waiting partner and resumed mating.

The vignette reflects what those who study *Cebus apella* often witness: an astonishing richness of sexual behavior, especially in females (Figure 12.1). Therefore, it may seem odd that details of capuchins' sexual behavior are known from just a few studies. We are just beginning to understand how capuchins' sexual behavior is related to fertility and conception, and we have data only on two species. As shown in Table 12.1, reports from the wild mainly refer to *C. capucinus* and *C. apella*; laboratory studies are limited to the latter species. Only very recently has the sexual behavior of tufted capuchins of both sexes been described in relation to the female hormonal cycle (Carosi and Visalberghi 2002, Carosi *et al.* 1999) and levels of testosterone and cortisol of males (Lynch *et al.* 2002). Most of the information we present in this chapter, unless otherwise specified, derives from the ongoing studies of the *C. apella* groups hosted in Visalberghi's laboratory in Rome. The results of these studies accord with what others (e.g., M. Carosi, M. Di Bitetti, D. Fragaszy, C. Janson, J. Lynch and C. Welker; see Table 12.2b) have observed in other groups of tufted capuchins, but we do not know how *C. apella* differs from other species in the genus in these matters.

FEMALE SEXUALITY

Beach (1976) distinguished three components in the sexuality of female mammals: attractivity, proceptivity and receptivity. Attractivity refers to the non-behavioral features of females (changes in color/morphology, odor, etc.) that arouse the males' sexual interest, and therefore increase the likelihood they will mate. Proceptivity refers to the behavioral patterns displayed by females in order to initiate and maintain sexual interaction with males. Receptivity refers to the willingness of females to permit intromission, copulation and ejaculation. We describe the sexual behavior of female capuchins using Beach's framework.

Figure 12.1. Tufted capuchin female showing chest rubbing and grin as it appears in the *Histoire naturelles des Mammifères* of Etienne Geoffroy Saint-Hilaire (1772–1844) and Georges Frederic Cuvier (1773–1838). This was the first large treatise on mammals published in Europe. Notice the stone held in the left foot. This drawing was probably made from a living specimen because it illustrates species-typical behavioral traits, such as courtship behavior (possibly towards a person) and propensity to manipulate objects.

Attractivity

The reproductive status of female capuchins, as in most New World primate species, is not evident from changes in body color or morphology (for an exception in this regard, see *Alouatta palliata*; Glander 1980). Moreover, capuchins do not seem to behave in ways aimed at depositing scent (as for example squirrel monkeys, marmosets and tamarins do by rubbing specific parts of the body on a substrate or by urinating) and only in some capuchin species do males sniff the female's urine. In short, non-behavioral features (changes in color/morphology, odor, etc.) that arouse the male capuchins' sexual interest appear limited or absent. However, the reproductive status of females is not at all concealed; it is very obviously signaled by means of proceptive behavior.

Proceptivity

During the periovulatory period, females exhibit remarkable proceptive behaviors, some of which correspond closely to the cyclicity of urinary progesterone (Carosi *et al.* 1999). Proceptivity in most females is impossible to overlook, though sometimes the proceptive behavior of the lowest ranking female of a group may pass unnoticed, because of her tendency to withdraw from the rest of the group. Though psychosocial stress may constrain proceptivity, the finding that fertility is relatively uniform among females in a group (Fragaszy and Adams-Curtis 1998) suggests that such stress does not suppress ovarian cycles, as is the case for callitrichids under certain conditions (Abbott *et al.* 1993)

Capuchins display proceptivity using a rich and varied behavioral repertoire consisting of facial expressions, vocalizations, gestures and body postures (Figure 12.1). The details of the repertoire vary somewhat across species. Table 12.2a presents the sexual ethogram of *Cebus apella* and Table 12.2b lists the occurrence of these behaviors in the other species of *Cebus*. Proceptive behaviors are exhibited with higher frequencies throughout the periovulatory phase than in the non-periovulatory phase. Variability in proceptive behavior is the norm: the behavioral elements of the sexual repertoire are combined in ways that may differ among females, across cycles of the same female and in relation to the partner (Carosi and Visalberghi 2002).

Female *C. apella* display proceptive behaviors for several consecutive days. Proceptivity, especially in its first days, is not a continuous state; therefore, depending on the number of hours the female was under continuous observation, proceptivity has been variously estimated to last from 1–4 days (Carosi *et al.* 1999), 3–5 days (Visalberghi and Dal Secco, unpublished. data), 5–6 days (Janson 1984), to 1–7 days with strong seasonal variation (longer proceptivity during conceptive season) and longer proceptivity in adults than in subadults (Lynch 2001). When she is proceptive, the female actively follows a target male (in most cases the dominant male in her group) toward which she directs most, if not all of

Table 12.1. *Recent studies on the sexual behavior of* Cebus *ordered by year. For each study we provide reference; species; type of study (field or laboratory study); whether behavioral data were collected; length of the ovarian cycle and basis for its assessment (detection of menstruation and/or hormonal profile) in females; whether hormonal data were collected and sample type for males.*

Reference	Species	Type of study	Behavior	Length of ovarian cycle	Assessment source	Male hormones (source)
Wright and Bush 1977	*C. apella*	lab	no	21.14	Menstruation/vaginal smears	no
Nagle *et al.*(1979)	*C. apella*	lab	no	21.0 ± 1.1	Hormones from plasma and menstruation/vaginal smears	no
Nagle and Denari (1982)	*C. apella*	lab	no	20.8 ± 1.2	Hormones from plasma and menstruation/vaginal smears	Testosterone (plasma)
Nagle and Denari (1983 review)	*C. apella*	lab	no	20.8 ± 1.2	Menstruation and hormones from plasma and urine (one cycle)	Testosterone (plasma)
Janson (1984)	*C. apella*	field	yes	no		no
Janson (1986a)	*C. apella* *C. albifrons*	field	yes	no		no
Visalberghi and Welker (1986)	*C. apella*	lab	yes	no		no
Welker *et al.* 1990	*C. apella*	lab	yes	no		no
Di Giano *et al.* 1992	*C. apella*	lab	no	range 19–22	Hormones from plasma and saliva	no
Philips *et al.* 1994	*C. apella*	lab	yes	no		no
Linn *et al.* 1995	*C. apella*	lab	yes	20.8 ± 1.2	Menstruation/vaginal swabs	no
Manson *et al.* 1997	*C. capucinus*	field	yes	no		no
Carosi *et al.* 1999; Carosi and Visalberghi 2002	*C. apella*	lab	yes	20.6 ± 1.6	Hormones from feces and urine	no
Lynch *et al.* 2002; Lynch 2001	*C. apella*	field	yes	no		Testosterone (feces) and cortisol
Recabarren *et al.* 2000	*C. apella*	lab	yes (only mating)	no	Hormones from plasma	no

Table 12.2a. *Behaviors observed during sexual interactions in captivity*[a,d] *with comments and indications if they occur outside of a sexual context*

Eyebrow raising

Definition. Eyebrows are raised up and backwards (see Figure 12.2, top left and middle) and the fur over the crown is flattened. Performed by both sexes.

Comments. As a result of flattening of the fur on the forehead this area appears paler (see also "forehead raise", Weigel 1978). Most of the time A[a] displays this facial expression while watching R.

Occurrence in other contexts in C. apella. It is common between all sex age classes in social affiliative interactions and play. Observed after an aggression either between former opponents (see also Verbeek and de Waal 1997), or directed by a third party toward the loser of an aggressive interaction. The eyebrow raise in sexual contexts is distinguishable from the aggressive eyebrow "flash", because in sexual contexts the eyebrow raise is sustained for long periods of time and may be more exaggerated (J. Lynch pers. comm.)

Eyebrow raising with grin and vocalization

Definition. Eyebrows are raised (as above). The monkey has a grin on its face: the corners of the mouth are rhythmically drawn backwards with closed jaws, usually baring some teeth – often the lateral ones (Figure 12.2 top left). Female's vocalization is a more or less continuous soft whistle that easily turns into a hoarse whine. Di Bitetti (2001b) describes it as a high frequency low amplitude descending call which is repeated insistently. Performed by both sexes (male's vocalization may be different from female's).

Comments. Most of the time, A performs these behaviors while watching R. The movement of the corners of the mouth is synchronous with the emission of vocalizations. The vocalization seems associated with this facial expression, i.e., if the grin is present vocalizations are present. However, it may sometimes be very soft, and therefore in the wild not audible to the observer even at close range (Janson 1984). Many authors (Janson 1984, Linn *et al.* 1995, Oppenheimer 1977, Phillips *et al.* 1994, Weigel 1978), have used the terms "grimace" and "grin" as synonyms. In tufted capuchins "grimace" or "silent bared teeth" display (see Figure 11.1 top left) is a submissive facial expression (similar to Old World monkeys grimace) in which the lips are more retracted then in "grin", teeth and gums may sometimes be exposed, usually the eyebrows are not raised, and the mouth corners are drawn backwards, without rhythmic movements. Therefore we distinguish grin and grimace.

Mutual gaze

Definition. A and R maintain mutual eye contact for at least 2–3 seconds (Figure 12.2 bottom left and right). It involves eyebrow raising. Performed by both sexes.

Comments. The monkeys may move (still raising their eyebrows) while mutual gazing. However, regardless of the increased or decreased distance between them, they try to regain mutual gaze. The mutual gaze can last for several minutes, being occasionally interrupted for a few seconds (to change spatial position or to glance at an external event). It is usually accompanied by one or all the following behavioral patterns: eyebrow raising with grin and vocalizations, head tilting and chest rubbing. Interestingly in primates, capuchins included, threats may involve gaze and eye contact as well (aggressive stares); however, in the agonistic context the gaze is usually accompanied by other specific agonistic behavioral components.

Occurrence in other contexts in C. apella. Play, affiliative solicitations between adults and infants.

Head tilting

Definition. The head is tilted to one side (approx. 45 degrees; also called head cocking). The head may gently change side every few seconds (Figure 12.2 top left). Performed by both sexes.

Comments. Most commonly occurs in males during post-copulatory display, in association with the male sitting upright, swaying from side to side, and holding one hand tight to the chest (Lynch 2001).

Occurrence in other contexts in C. apella. The prominent use is undoubtedly the sexual context. However, Weigel (1979) describes its usage in affiliative, appeasement, reassurance contexts and during maternal solicitations and play. Observed after an aggression either between former opponents (see also Verbeek and de Waal 1997), or directed by a third party toward the loser of an aggressive interaction.

Table 12.2a. *(cont.)*

Chest rubbing
Definition. Hand(s) are slowly rubbed back and forth on the fur of one's own chest. Palms can be flattened and fingers extended or the hands can be closed in a fist (Figure 12.2 top right). The movements are usually performed upward and/or downward and repeated several times in a row. In other cases the hand can gently open and close on the fur. During chest rubbing the movement of the hand can stop while the hand stays still on the chest. Performed by both sexes.
Comments. Occasionally, chest rubbing is performed on the partner's chest (Visalberghi pers. observ., *C. apella* colony in Kassel, Germany). Janson (1984) proposed a relation between chest rubbing and a scent-producing gland. Epple and Lorenz (1967) described a throat and an epigastric gland in a male specimen.
Occurrence in other contexts in C. apella. Frequent in affiliative contexts in all age-classes. Observed after an aggression either between former opponents, or directed by a third party toward the loser of an aggressive interaction. Throat rubbing on substrates has occasionally been observed mainly performed by alpha males (M. Carosi pers. obs.; J. Lynch pers. comm.). At Caratinga, *C. apella* males perform vigorous displays of chest rubbing on tree trunks; while holding the trunk in their arms, they rub chest and neck up and down in rapid motion on the trunk, sometimes for at least 30 seconds or so.

Touching-and-running
Definition. A approaches R, quickly pushes R's body with its hand (often the arm is outstretched) and runs away. Performed by the female.
Comments. Sometimes the female touches the male and turns her head, as if starting to run away, without in fact running away. In these cases, if running away occurs shortly after, it appears to be prompted by a sudden body movement of the male. Rarely, in this approach-quick-withdrawal pattern, nuzzling or a light bite on male's body have been observed instead of touching. The female may also quickly pull the male's tail, or fur on back or flanks (a female was observed dragging the male on his back around the cage floor by holding and pulling his tail, Visalberghi pers. obs.). One subordinate female tended to push objects that were in the cage close to the male, sometimes "touching" the male by means of the object launched toward him. At Caratinga, females do not touch-and-run towards males; however, females approach, run towards, jump over, run past, follow, re-approach and leave target males (J. Lynch pers. comm.).
Occurrence in other contexts in C. apella. The touching-and-running behavior performed by the female in a sexual context resembles the behavior performed by both sexes during playful interactions as an invitation to play (Guidi 1992). Weaver (1999) considers passing while tail tapping as a playful approach.

Nuzzling
Definition. A gently and very quickly contacts the body of R with its face.
Comments. This behavior is performed almost exclusively by the female and almost always toward the male. It requires the female staying in close proximity with the male slightly longer than in touching-and-running. Very occasionally it is directed toward the male's ano-genital area; the usual target areas (recipient's head, shoulder and back) are not related to where scent glands have been described (see above, Epple and Lorenz 1967).

Tense Arm(s)
Definition. A slowly moves/stretches one or both arms toward R, without contacting R. Also called extended arms. Individuals are in proximity usually seated, facing and looking at each other, raising the eyebrows. Performed by both sexes. Frequently males and sometimes females perform a "tense arm" display after copulation, with one arm bent so the hand was held rigidly to the chest while staring at and vocalizing toward the recipient, but this is not accompanied by the other arm outstretched towards the recipient (J. Lynch pers. comm.).
Comments. Apparently this behavior is not aimed at touching the recipient's body, rather it seems to signal readiness to contact.
Occurrence in other contexts in C. apella. Observed after an aggression directed by a third party toward the loser of an aggressive interaction.

Body touching
Definition. A and R are close. A watches R, and A's hand slowly stretches out and contacts R for at least a few seconds (Figure 12.2 bottom left). This is performed by both sexes.

Table 12.2a. *(cont.)*

Comments. Reported in an immature *Cebus albifrons* approaching an adult male (Oppenheimer 1977).
Occurrence in other contexts in C. apella. As an invitation to play. In affiliative contexts involving infants.

Frontal posture

Definition. A is usually seated facing R and keeps one hand on its chest while the other is outstretched toward R. The outstretched arm may be also raised to various extents. Performed by the female.
Comments. The orientation of the body (towards the recipient of this signal) and the raising of the arm (when it occurs) may indicate readiness for contact. This posture, when performed by the male, is often associated with genital display (see below). An upright posture is common in males after copulation, in association with head-tilting and tense-arm on chest behaviors (J. Lynch pers. obs.).
Occurrence in other contexts in C. apella. Females often direct this body posture toward their infant, who generally reacts by going toward its mother and suckling. Observed after an aggression either between former opponents, or directed by a third party toward the loser of an aggressive interaction. In infants, an upright sitting posture ("bunny sit posture") when accompanied by eye contact with a conspecific, functions as a friendly invitation to interact. (E. Visalberghi pers. obs.; Weaver 1999).

Back posture

Definition. A is crouched with its back toward R. A looks back at R while keeping one hand on its chest and the other on the ground for support. Performed by both sexes.

Backing into lap

Definition. A backs into R's lap. Mounting may or may not occur. Performed by both sexes, however, the actor is usually the female.

Genital display

Definition. The male's penis is erect and highly visible. Depending on the male's posture, four types of displays are observed. (1) Classic genital display: the male is in front of the female, his legs spread and often performs "Chest rubbing". (2) Genital display with frontal posture: only one hand performs chest rubbing while the other arm is outstretched as in the "Frontal posture". (3) Supine genital display: the male lies on its back in front of or to one side of the female and often performs "Chest rubbing". (4) Bipedal genital display: the male stands bipedal.
Comments. Typically performed by males who are not the target of female's courtship (usually the subordinates, see Discussion). Tufted capuchins have a long nail-shaped penis (Visalberghi pers. comm.) and *Cebus* and *Saimiri* have a longer baculum than that of other New World monkeys (Dixson 1987). Subordinate males use genital display to invite the female to mate. (Apparently, the dominant male does not need to invite the female to mount, since she is proceptive and receptive to him). At Caratinga, adult males performed genital displays most frequently after copulation, either holding their penis in their hand and staring at the target female, or sitting with legs splayed in the direction of the female, whereas juvenile males used "splayed-legs" reclining genital displays in tense play or dominance interactions in the context of intergroup encounters (J. Lynch pers. obs.). The bipedal genital display is observed more often in infants and juveniles. Although a full erection of the clitoris has been observed in infant and juvenile females, in our study we never observed genital display in adult females (but see Freese and Oppenheimer 1981). Youngsters use a "spread leg display, body position exposing the genitals because legs are extended and stretched to either side" (Weaver 1999, p. 190) as a submissive signal.

Mounting attempt

Definition. A tries to mount R, but R moves away. Performed by both sexes, however, the actor is usually the male.

Mounting

Definition. A mounts R in a position which allows for intromission. Thrusting usually occurs. A mounting bout starts when A gains a mounting position and ends when it dismounts. Bouts can be isolated or form a mounting sequence. Performed by both sexes.
Comments. Thrusts occur during most of the time the male spends in mounting position, intromission, however, can only be inferred. Capuchins may ejaculate after one mount as well as after several mounts. In addition to the dorso-ventral mount, ventro-ventral (Figure 12.2 bottom middle), and latero-ventral mounts have been observed and

Table 12.2a. *(cont.)*

filmed (Visalberghi and Welker 1986, Visalberghi and Carosi 1997, C. Tomaz pers. comm.). It is possible that the occurrence of ventro-ventral mounting (observed in captivity but not yet in the wild) is related to the presence in the laboratory of large and flat surfaces on which to mate. During mounting the partners vocalize and mutual gaze is very common.

Females also perform mounting on adult males (see also Janson's "reverse mounting", 1984). J. Lynch (pers. comm.) has observed that during multiple mount matings, the male and the female usually took turns mounting one another, up to seven times back and forth. Females usually stay on the male's back in a position resembling that of an infant on its mother. However, the female can also take up a more caudal mounting position, perform pelvic thrusts and may rub her genitals on the male's fur, as if masturbating. Female mounting in alternation with male mounting has been reported for *Cebus capucinus* (Oppenheimer 1968, 1973 in Oppenheimer 1977). Linn *et al.* (1995) report female-female mounts (when at least one female was proceptive) were sometimes interrupted by the dominant male.
Occurrence in other contexts in C. apella. Male-male mounts may occur during reunion displays following separation or situations of excitement or tension (see also Matheson *et al.* 1996). In Visalberghi's groups at reunions the subordinate male(s) always mounted the older dominant one (total number of observations N = 13, Carosi and Visalberghi unpublished data; also observed by G. Linn pers. comm.). Adults of both sexes may mount juveniles of both sexes, and vice versa.

Ejaculation
Definition. The discharge of semen.
Comments. When ejaculation occurs inside the female's vagina it cannot be directly observed. Nevertheless, it is possible to infer its occurrence on the basis of other observable events. Videotape analysis suggested that ejaculation occurs during a reduction or pause in the rhythm of deep thrusts and few seconds of body rigidity (but a reduction or pause in the rhythm of thrusts does not always correspond to ejaculation). Muscular contractions of the male hips often occur, and the presence of fresh semen may be observed on male and female genital areas. After ejaculation, individuals of both sexes groom their genitals and may eat solidified clumps of the semen. Sometime the female reaches deeply with her fingers inside the vagina, extracts the semen and eats it. In the wild, ejaculate is rarely seen after copulation, but both males and females may perform genital inspections after copulation (J. Lynch pers. obs.).

Turning head and reaching back
Definition. A, while mounted by R, twists the torso and turns its head back to look at R and reaching out with its arm and hand around the shoulder (or head, or neck) of R, gently touches (never grasps) R's fur (Fig. 12.2 bottom right). Performed by the female and rarely by the male when mounted by the female.
Comments. Turning the head may occur without reaching back, but usually not vice versa. In most cases, mutual gaze also occurs.

Mounting vocalization
Definition. A harsh sound together with a continuous squeal may occur during mounting. Performed by the female. Di Bitetti (2001) describes the former as a rapid series of short notes containing a lot of noise; this noisier vocalization differs from the other more typical vocalization made during courtship and mating.
Comments. These vocalizations are emitted by the female when she is mounted by the male and they may last for the whole duration of the mount. The male also vocalizes during copulation and post-copulatory display, but sometimes his soft vocalizations are overshadowed by the females' louder ones. Mounts can also be silent.

Inspection of the partner's genitals
Definition. A inspects (by sniffing, touching) the genitals of R. Although both sexes commonly inspect their own genitals especially after ejaculation, inspection of the partner's genitals is rare.
Comments. Also reported by Phillips *et al.* (1994) and J. Lynch (pers. comm.) for the female.
Occurrence in other contexts in C. apella. Adults of both sexes often inspect infants' genitals.

Masturbation
Definition. Rubbing one's own genitals with hands (usually males), or against a substrate (usually females). Performed by both sexes, but rare.

Table 12.2b. *Occurrence of sexual behaviors in wild* C. apella *and in other species of* Cebus[b].

Behaviors Listed in Table 12.2a	Sex	Captive *C. apella*	Wild *C. apella* Iquazú	Wild *C. apella* caratinga	Wild *C. capucinus*	Wild *C. olivaceus*	Captive *C. olivaceus*	Wild *C. albifrons*
Eyebrow raising	M	Yes[c]	Yes	Yes	No	Yes	No	No
	F	Yes[c]	Yes	Yes	No	Yes	No	No
Eyebrow raising, grin and vocalization	M	Yes	Yes	Yes	No	Yes	only grin	No
	F	Yes	Yes	Yes	No	Yes		No
Mutual gaze	M	Yes[c]	Yes	Yes	Yes	No?	Yes	Yes? Brief
	F	Yes[c]	Yes	Yes	Yes	No?	Yes	Yes? Brief
Head tilting	M	Yes[c]	Yes	Yes	Yes	Yes	Yes	No?
	F	Yes[c]	Yes	Yes	Yes	Yes	Yes	No?
Chest rubbing	M	Yes[c]	Yes	Yes[c]	No	No?	No	No
	F	Yes[c]	Yes	No	No	No?	No	No
Touching-and-running	M	No[c]	No	No	No	Yes	Yes	No
	F	Yes[c]	Yes	No	No	Yes	Yes	No
Nuzzling	M	No/very rare	No	No	No	No?	No	No
	F	Yes	?	No	No	No?	No	No
Tense Arm(s)	M	Yes[c]	?	No	No	No?	Yes	No
	F	Yes[c]	?	No	Rare	No?	Yes	No
Body touching	M	Yes	Yes	No	No	Yes	Yes	Yes
	F	Yes	Yes	No	No	Yes	Yes	Yes
Frontal posture	M	Yes[c]	Yes	No	No	?	No	Yes
	F	Yes[c]	Yes	No	No	?	No	Yes
Back posture	M	Yes	?	No	No	?	Yes	?
	F	Yes	Yes	No	Very rare	?	Yes	?
Backing into lap	M	Rare	?	No	No	No?	Yes	?
	F	Yes	Yes	Yes	No	Yes	Yes	?
Genital displaying	M	Yes[c]	?	Yes[c]	No	Yes	No	No
	F	No[c]	?	No	No	No	No	No

Mounting attempt	M	Yes[c]	Yes	Rare	Yes	Yes	Yes	Yes
	F	Rare[c]	Rare	No	No	No?	No	No
Mounting	M	Yes	Yes	Yes	Yes	Yes	Yes	Yes
	F	Yes	Yes	Yes	No	Yes	No	No
Turning head and	M	Yes	Yes	Yes (turn head)	Yes	Yes	Yes (only	No
reaching back	F	Yes	Yes	Yes	Yes	Yes	male-male)	No
Mounting vocalization	M	No	?	Yes	No	Yes	No	?
	F	Yes	Yes	Yes	No	Yes	No	?
Genital inspection	M	Very rare	Rare	Yes	Rare	No	—	?
	F	Rare	Rare	Yes	Rare	No	—	Yes?
Masturbation	M	Rare	N?	No	No	No	—	Yes
	F	Very rare	?	No	No	No	—	No?
Other behaviors (those not listed in Table 12.2a)								
Duck face	M	No	No	No	Yes	—	—	No
	F	No	No	No	Yes	—	—	No
Sniffing urine	M	No	One case	Yes	Yes	Yes	—	Yes
	F	No	No		Yes	Yes	—	No
Pirouette	M	Yes[c]	No	No	Yes	—	—	No?
	F	Yes[c]	No	No	Yes	—	—	No?
Looking between legs	M	Yes[c]	?	No	Yes	—	—	?
	F	Yes??[c]	?	No	Yes	—	—	?

[a] A stands for actor, R for recipient.

[b] Data for captive *C. apella* (Carosi and Visalberghi 2002 and unpublished data, D. Fragaszy pers. obs., G. Linn pers. comm.); for wild *C. apella* (Janson 1984, 1986a, b, pers. comm.; M. Di Bitetti, pers. comm., Lynch 2001, pers. comm.); for wild *C. capucinus* (Manson *et al.* 1997, L. Fedigan, pers. obs.); for wild *C. albifrons* (Janson 1984, 1986a, b, pers. comm.); for wild *C. olivaceus* (Robinson 1979, pers. comm.); for captive *C. olivaceus* (M. Dubois, pers. comm.). When sources disagree, we chose to list the behavior as occurring.

[c] Indicates the behavior also occurs outside the sexual context.

[d] Definitions of the behaviors updated from Carosi and Visalberghi (2002) with permission from John Wiley & Sons Inc.

Figure 12.2. Sketches of a few body postures and facial expressions of *C. apella* during sexual interactions. Top left, a female performing *eye-brow raising with grin and vocalization*, and *head tilting*. Top middle, a male responding to the female with the same facial expression and (top right) performing *chest rubbing*. Bottom left, a female and a male performing *mutual gaze* and *touching each other's body*. Bottom middle, the male and the female are engaged in a *ventro-ventral mount*. Bottom right, the female *turns and reaches back* while the male mounts her; she is also grinning and vocalizing. (Drawings by Andy de Paoli from videotapes).

her solicitations. Initially (for hours and/or days), the solicited male does not reciprocate and tends to avoid the female by leaving as soon as she approaches him. Later, the target male starts to respond to the female's solicitations with behaviors similar to hers (see Table 12.2a). At this point, mutual sexual interest becomes evident and mating occurs.

Janson (1984) reports that at Manu the female actively maintains proximity with the dominant male and he is her closest neighbor for 64–95% of the time in all the days in which the female is proceptive, except the last day. The dominant male actively follows the female only on the day before the last one. On the last day of proceptivity the female does not associate with the dominant male anymore and she often solicits other males; the latter respond to her more readily than the dominant male does.

Phases of courtship

The sexual interactions of tufted capuchins go through four phases that are characterized by different assortments of sexual behaviors and spacing between male and female. In Phase 1 the target male ignores or avoids the female; in Phase 2 he reciprocates her solicitations. In Phase 3, the pair mate and possibly the male ejaculates. Phase 4 follows ejaculation; the partners are still involved in sustained reciprocal exchange of sexual behaviors. Phases have variable duration (from minutes to hours) and Phase 1 and 2 are usually repeated multiple times. Phase 1 is prominent in the initial part of the female's proceptive period (which may last from 1 up to several days, see below and Figure 12.5); the other phases occur later in the female's proceptive period. Next, we describe in detail the sexual interactions of *C. apella* and a typical courtship and mating sequence in *C. apella* (see Figure 12.3) as Visalberghi and colleagues repeatedly observed them.

Phase 1: female's solicitation

Phase 1 is unilateral in the sense that the female shows sexual interest toward the male but he does not reciprocate behaviorally. This phase is characterized by the female visually monitoring the male, vocalizing insistently (Di Bitetti 2001a), seeking his proximity and displaying persistent eyebrow raising in his direction. The male often ignores her solicitations for hours and occasionally shows avoidance (by breaking the proximity she has achieved) and may also threaten her. Consequently, the female is often nearly simultaneously soliciting and avoiding the male. She actively moves in space so as to intercept the male's gaze while raising her eyebrows and sometimes she gets close to him and, achieving contact, she touches him quickly and runs away, as if fearful of his reactions. Sometimes she manages to contact his body briefly with her muzzle. Meanwhile she repeatedly rubs her chest while sitting erect, vocalizes and sometimes tilts her head on one side while gazing at him. Eventually, she triggers the male's attention, and he starts to respond positively to her overtures.

Phase 2: mutual courtship

Phase 2 is bilateral in the sense that the male reciprocates the female's behavior. This change in the male's behavior occurs for no apparent reason that we have yet discerned. Mutual gaze marks the "switch" from Phase 1 to Phase 2. In Phase 2, the female stops touching the male and then running away. In addition to mutual gaze, the behaviors typical of this phase are eyebrow raising with grin and vocalization, chest rubbing, head tilting, tense arms, frontal posture and touching the other's body. Proximity is achieved and broken by both sexes equally often; the male and the female take turns in following and moving away from the partner, apparently each willing to move away if followed. If this does not occur they are as likely to change roles and go back to the partner or follow him/her. This looks like a "dance" between "magnets" whose polarity changes after having approached or having moved away a few meters (probably more than a few meters in larger spaces). As soon as one approaches, the other moves away, and as soon as he/she has moved off, he/she pirouettes 180°, facing the partner again. He/she then regains mutual gaze as soon as possible and displays chest rubbing, head tilting and the other behaviors mentioned above. In the meanwhile the follower has stopped and reciprocates. At this point the "dance" may start again with the same or opposite roles, or otherwise the partners may approach one another and display behaviors that imply body contact. In fact, it appears obvious that moving away is done to prompt the partner to follow and not to escape from the partner. The female in both phases and the male in Phase 2 pay little attention to other group members; they seem totally absorbed by what they are doing.

Phase 3: mating

In Phase 3, the partners mate while continuing to perform the same set of behaviors described for Phase 2. The male achieves one or more mounting bouts that are accompanied by distinctive vocalizations from both partners. These vocalizations are louder than those previously emitted in conjunction with grin and eyebrow raising. If ejaculation occurs then Phase 4 follows. If ejaculation does not occur there is usually a period in which the partners are less sexually attracted to each other, and then the typical patterns of the previous phases (especially Phase 2) may resume.

Mounting is often preceded by the frontal posture, the back posture and the "backing into lap" behaviors (see Table 12.2, for descriptions). Genital inspection before mating does not occur. Both partners try to position themselves correctly in relation to the other for intromission. However, since they are both active in pursuing a copulatory position a mismatch may occur for a few seconds. At least in captivity, where stable surfaces are common, capuchins mate in different positions (see below). Pelvic thrusts are always present. While the male mounts the female they both display grins, eyebrow raising and vocalizations; often the female turns her head back and looks at the male; mutual gaze occurs when the male intercepts her gaze. Mounting bouts can be isolated or form mounting sequences. Ejaculation usually occurs after a sequence of mounts. Females sometimes mount the male and perform pelvic thrusts.

Phase 4: post-copulatory display

Phase 4 starts after ejaculation. After ejaculation has occurred, the partners may continue to interact as in Phase 2, i.e., they both display to each other the same

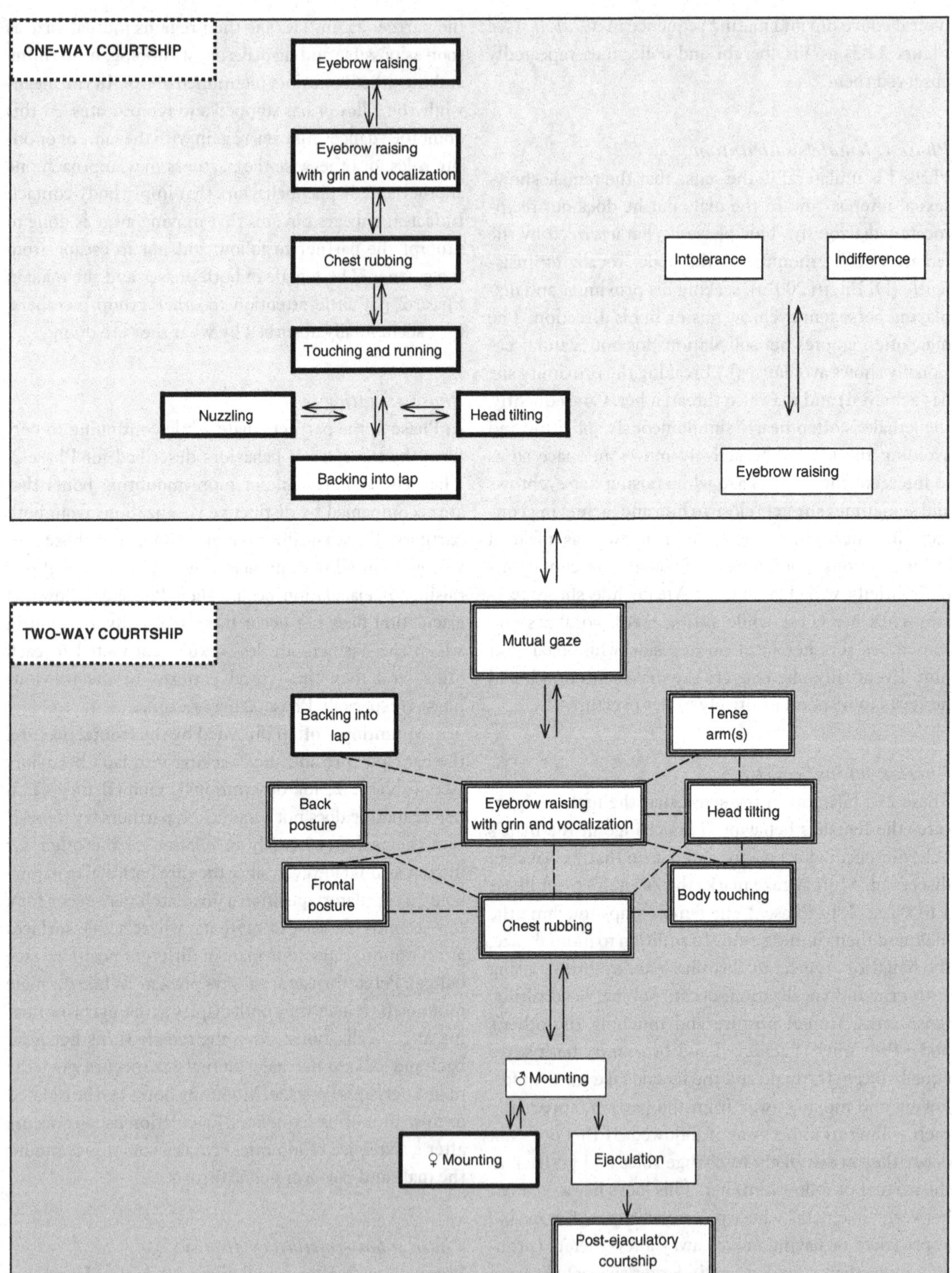
ONE-WAY COURTSHIP
Eyebrow raising
Eyebrow raising with grin and vocalization
Chest rubbing
Touching and running
Nuzzling
Head tilting
Backing into lap
Intolerance
Indifference
Eyebrow raising
TWO-WAY COURTSHIP
Mutual gaze
Backing into lap
Tense arm(s)
Back posture
Eyebrow raising with grin and vocalization
Head tilting
Frontal posture
Body touching
Chest rubbing
♂ Mounting
♀ Mounting
Ejaculation
Post-ejaculatory courtship

patterns of sexual behavior usually used to solicit mating. The sexual solicitations performed by the male can be even more intense than before ejaculation. This phase on average lasts 20 minutes but may last more than 40 minutes (Dal Secco 2000). After these post-ejaculatory sexual interactions, the interest of both partners fades. After the last mounting bouts of an episode in which ejaculation has not occurred, sexual solicitations between partners on average last just a few minutes.

At Caratinga, the suite of behaviors that most commonly occurred after the last mount and during post-copulatory display were described by Lynch (pers. comm.) as characterized by male 'chutter' vocalizations and female cries. The male and female stared into one another's eyes with eyebrows raised, and the female curled up and remained stationary, while the male moved around her, approaching and leaving, sitting up on his haunches with one hand placed on his chest, and rotating his head from side to side. At this time, males might also curl up briefly; males could sway from side to side; give genital displays towards females; massage or inspect their penis; turn around in circles; run quickly past, away from, and back towards the female in a running display; jump over the female; or quickly turn away from her and turn back, repeatedly. These displays could last for over 20 minutes (Lynch 2001).

Overall, *C. apella* exhibits the richest set of proceptive behaviors ever described among nonhuman primates (for a comparison with those of other primate species, see Dixson 1998a). Several other features also contribute to make sexual behavior in *C. apella* very unusual among primates: the initial reluctance of the male to mate, the variety of ways in which the two sexes organize their sexual behaviors in the flow of interactions that leads to mating, the extended lengths of time during which they are exhibited and, last but not least, the intense "courtship" after ejaculation.

Species differences

The extent to which the above descriptions hold true for the other *Cebus* species still needs to be determined. Only *C. apella* has been extensively studied, and primarily in captivity (see Table 12.2b). Field observations afford some comparisons, however. In *C. capucinus*, the eyebrow raising and grin are replaced by the "duck face" in which lips are protruded. Manson *et al.* (1997) describe a "courtship dance" for *C. capucinus* that includes "duck-face" pirouettes (with piloerection in the male), reciprocal slow motion chases and many of the behaviors listed for *C. apella*. The estrous female *C. albifrons* solicits males intermittently; these are suspected to be briefer in duration and behaviorally less rich than those of *C. apella* (Janson 1986a, C. Janson, pers. comm.). We have no specific reports about mating behavior in *C. olivaceus*.

Species of capuchins apparently differ in the extent to which the male sniffs the female's urine or other vaginal or perineal secretions. Male *C. capucinus* observed by Manson *et al.* (1997) were "extremely interested in female urine and frequently stopped to sniff locations in which females had been sitting" (p. 777). Male *C. olivaceus* investigate the sites where females have performed urine washing (Robinson 1979) and male *C. albifrons* often sniff the female's urine (Janson 1986a). This behavior is rare in *C. apella* although males sometimes monitor either the branch that a female was sitting on, or on which she had urine washed (C. Janson pers. comm., Lynch 2001). It is possible that in *C. apella* the female's proceptive behavior has such an overwhelming role in

Figure 12.3. A typical courtship and mating sequence in *C. apella*. The complexity of their interactions prevented us from providing in this chart all the variability observed. The behaviors of the one-way courtship of Phase 1 are in the box above, whereas those of the two-way courtship and mating (Phases 2–4) are in the box below. Female behaviors are framed by a thick line, male behaviors by a thin line and behaviors performed by both sexes by a multiple line. The order in which the behaviors are listed top to bottom indicates the order in which the behaviors are most commonly observed. Connecting arrows in both directions between behaviors indicate that the temporal sequence is variable. One-way courtship (box above) precedes two-way courtship (box below). The courtship behaviors reported in the box below lack a discernible temporal sequence. To indicate the simultaneous occurrence of some of them we used connecting dashed lines. Mutual gaze precedes, and perhaps triggers, the other courtship behaviors in the lower box. Mounting, ejaculation and post-ejaculatory courtship are usually preceded by the courtship behaviors listed above. Female eyebrow raising usually occurs throughout the whole courtship sequence and mutual gaze throughout the two-way courtship (see text for details). (from Carosi and Visalberghi 2002, *American Journal of Physical Anthropology*, Copyright © 2002 John Wiley & Sons Inc).

communicating her physiological state that it replaces the role played by olfactory cues. However, more experimental work is necessary before we can be sure what role olfaction plays in sexual interactions in any species of *Cebus*.

Receptivity and mating

The proceptive female *C. apella* takes the sexual initiative. Usually the female solicits one male with whom she will eventually mate; however, solicitations and mating (in full view of other group members) with more than one male within the same periovulatory period and/or in different ones have also been reported (Janson 1984, Lynch 1998, Visalberghi and Welker 1986). Janson (1984) observed a female mating with four different males in 10 minutes! Males (usually subordinate ones) that witness long bouts of solicitations not directed to them require fewer solicitations to mate. Apparently, the proceptive female is receptive to the male she is soliciting. However, she is not universally receptive to all males. In captivity, females housed with one or more males may even solicit males in another cage, and if separated only by wire mesh, mate with males in an adjacent cage (Cooper *et al.* 2001, Welker 1992). When this occurs, male members of her group do not actively seek to mate with her, unless she directly solicits them too.

Forced copulations have not been observed in wild (C. Janson, pers. comm.) or captive *C. apella* (Visalberghi, unpublished. data). During 18 months (800 hours of observations) G. Linn (pers. comm.) observed three instances in which a sexually aroused subordinate male (who had been solicited by the dominant female and did not reciprocate, possibly because he was inhibited by the presence of the dominant male) directed aggression at a subordinate female until she remained stationary, at which point he mounted her. Janson (1986a, p. 171) reports that the male *C. albifrons* "tries to mount the female even if she is not proceptive" or when she is "unwilling" (Janson 1986b, p. 51) and that sometimes females also refuse to mate with the dominant male. The male may then proceed and, regardless of her screams, literally raise up the female's hindquarters to try to achieve intromission (C. Janson, pers. comm.). Forced copulations have not been observed in *C. capucinus* at Santa Rosa (L. Fedigan, pers. obs.), but Baker (1998) reports that after a male and a pregnant female jointly threatened a human observer, the male grabbed the female and began to mount her twice while the female struggled to get away.

ENDOCRINE CORRELATES OF SEXUAL BEHAVIOR

Small-bodied New World primates (e.g., squirrel monkeys, marmosets) have higher levels of circulating testosterone and ovarian steroids than most Old World primates (Coe *et al.* 1992, Dixson 1998a); this is partially true for progesterone and testosterone levels in capuchins, but not for estrogen (see Coe *et al.* 1992 for the values of many species, *C. apella* included). Early studies on reproductive physiology of capuchin monkeys were carried out by Hamlett (1939) and more recently and systematically by Nagle and co-workers. Though they studied females, Nagle and Denari (1982, 1983) also assessed males' plasma testosterone levels and found that testosterone shows striking variations within and between males and no consistent pattern throughout the year. Fecal testosterone concentrations in wild male *Cebus apella* at Caratinga showed a striking seasonal pattern, with a high peak in testosterone in all adult males during the early dry season when all females conceived, and relatively low levels of testosterone for all males throughout the remainder of the year (Lynch *et al.* 2002).

The ovarian cycle of the female *C. apella* has been characterized in several ways: (a) by measurements of luteinizing hormone (LH) from plasma (Nagle and Denari 1983a); (b) by measurements of 17-ß estradiol and progesterone from plasma (Nagle *et al.* 1979, Nagle and Denari 1982, Recabarren *et al.* 2000) and from urine (only one cycle, Nagle and Denari 1983a), and (c) by measurements of progestins in urine and feces (20 cycles; Carosi *et al.* 1999). In addition, vaginal swabs can detect the very minimal amount of blood loss due to menstruation (menstruation lasts 2.8 ± 0.4 days, range 1–5 days) and provide a fairly accurate estimate of menstrual cycle length in capuchins. Recabarren *et al.* (2000) provide extensive information on plasma progesterone, estradiol and prolactin profiles in nursing and non-nursing females following delivery of an infant. Nagle and Denari (1982, 1983) report that the length of the menstrual cycle ranges between 18 and 23 days, with a mean value of 20.8 ± 1.2 days, including a follicular phase of 8.3 ± 1.2 days and a luteal phase of 11.7 ± 0.7 days (Nagle and Denari 1982, 1983). Carosi *et al.* (1999) provide

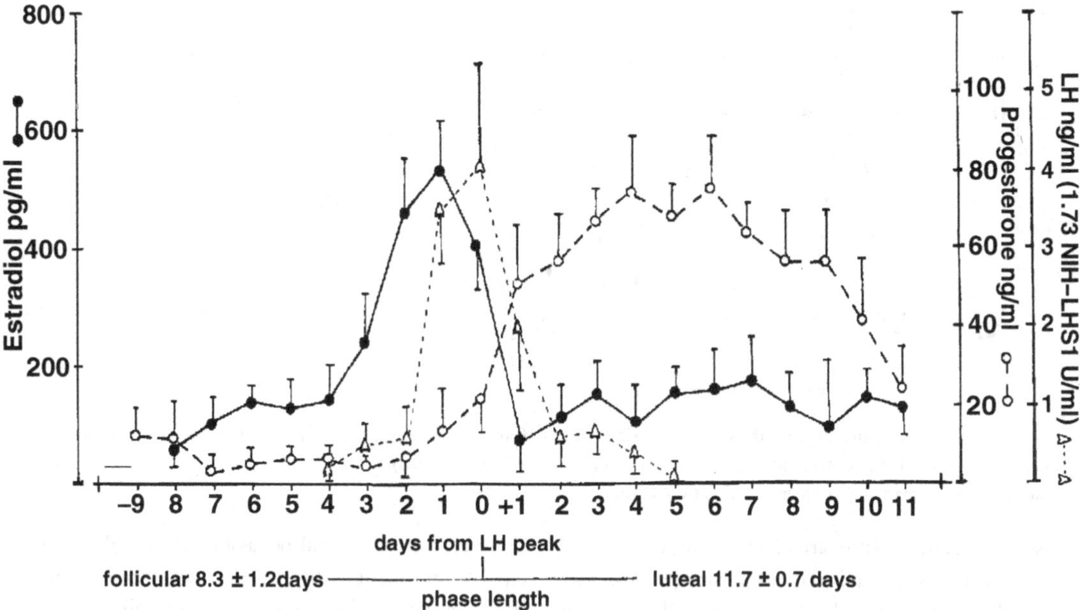

Figure 12.4. Mean (±) s.e.m. concentrations of estradiol (solid circle), progesterone (open circle) and immunoreactive LH (open triangle) in plasma samples taken daily from six *Cebus apella* during an entire menstrual cycle. Day 0 represents the day with the highest LH concentration in each cycle and data are centered around it. (Figure courtesy of Carlos Nagle).

similar values: the mean cycle length was of 20.6 ± 1.6 days (range 17–24 days), comprising a follicular phase of 6.1 ± 1.0 days (range 4–8 days) and a luteal phase of 14.5 ± 1.0 days.

Nagle and Denari (1983, Nagle *et al.* 1979, see Figure 12.4) report that plasma concentrations of 17-ß estradiol increase gradually during the first days of the cycle (from 50 to about 150 pg/ml) and have a rapid increase to a peak of 540 pg/ml on days 7–10, falling again with the start of the luteal phase to the same levels observed during the early follicular phase (about 150 pg/ml). During the follicular phase the plasma concentration of progesterone is very low (5 ng/ml); its concentration rises significantly on the day after the LH peak (60–100 ng/ml.) Two to three days before the onset of menstruation, progesterone concentrations decline to 10–20 ng/ml. At mid-cycle the plasma LH levels are about 8–10 times higher than during the rest of the cycle. The day of the LH peak usually occurs one day after the estradiol surge. Whereas progesterone levels in the urine reflect the pattern of progesterone in the plasma, the levels of pregnanediol are practically undetectable in urine.

Nagle *et al.* (1980) investigated the relationship between plasma concentrations of estradiol and progesterone and follicular development and ovulation, which they observed directly by means of serial laparoscopic examinations of the ovaries. They demonstrated that ovulation is preceded, within 10–24 hours, by an estradiol peak (3–5 times higher than baseline levels) and by a rise in progesterone. In addition, ovulation is concomitant with a decrease in estradiol and a continuing rise in progesterone.

Progesterone profiles have also been successfully obtained non-invasively from saliva (Di Giano *et al.* 1992) and from urine and feces (Carosi *et al.* 1999). Di Giano *et al.* (1992) measured progesterone in saliva and in the blood plasma in normally cycling and in ovariectomized females and found that the levels of progesterone in the saliva follow a pattern similar to that in the plasma, reflecting the free steroid portion. In normally cycling females, the concentrations of progesterone in the saliva were highly correlated with those found in the plasma, and the concentration of progesterone in the saliva accounted for 6.5% and 3.2% of that measured in plasma for the follicular and luteal

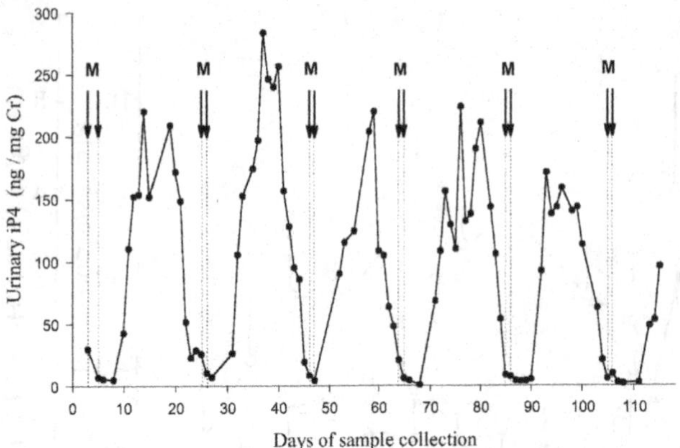

Figure 12.5. Immunoreactive urinary progesterone during five consecutive menstrual cycles of an individual tufted capuchin female. The arrows mark days of menstruation (M). (From Carosi *et al.* (1999). *Hormones and Behavior*, Copyright © 1999 Academic Press).

phases, respectively. In ovariectomized females, which have consistently low salivary and plasma progesterone levels, the values showed a lower, but still significant correlation.

Carosi *et al.* (1999) determined that the estrogen measurements in urine and fecal samples showed high intra- and inter-individual day-to-day variability whereas those of progestins were less variable. Therefore, they used the significant rise in urinary progesterone (P4), coupled with data on menstruation, as an indicator of the periovulatory phase (see Figure 12.5). They found a strong correlation between urinary and fecal progestins (progesterone and pregnanediol), a finding that will enable researchers to use fecal samples (that are easier to collect than urine, blood or saliva) in future studies of reproductive cycles.

Proceptivity and ovulatory cycle

Behavioral and physiological data collected in parallel on the same individual capuchins are available from two studies. In the first, Linn *et al.* (1995) investigated sexual solicitation (female's raising of the eyebrows and "grimace"), copulation and other social behaviors in relation to menstrual cycle (detected by vaginal swabbing) in females living in social groups (total number of cycles = 182). They found that female solicitation and mating were significantly more frequent at the time when ovulation was estimated to occur. In the second study, Carosi *et al.* (1999) investigated the cyclicity of females' hormonal changes and the cyclicity of three clusters of sexual behaviors extracted by a Principal Component Analysis. The cyclic rise in urinary iP4 levels (indicating the periovulatory phase) matched the cyclicity of one of the behavior clusters. This cluster includes sexual behaviors such as "eyebrow raising with vocalization," "touching and running," "nuzzling" and, to a lesser extent "head cocking." Using the same data set, Carosi and Visalberghi (2002) showed that "eyebrow raising," "eyebrow raising, grin and vocalization," "mutual gaze," "head cocking," "chest rubbing," "touching and running," "nuzzling" and "tense arm" have significantly higher frequencies in the periovulatory period than in the non-periovulatory period (see Table 12.2a for a description of these behaviours.

Regardless of the duration of the proceptive periods, the day in which urinary progesterone rises is the last day of proceptivity (± 1 day) (Carosi *et al.* 1999). Thus, rising progesterone signals a change in ovarian function. This finding matches correlated changes in progestins and non-behavioral sexual cues, such as sexual swelling in other species (Graham *et al.* 1973, Heistermann *et al.* 1996, Wildt *et al.* 1977). Because the rise in P4 occurs on the same day of LH peak (Nagle and Denari 1983), which in primates precedes ovulation by 12–24 hours, it is very likely that the day of ovulation is also the day of the urinary progesterone rise (since the urinary rise lags behind that in plasma by 12–24 hours). These findings suggest that ovulation in *C. apella* may occur at the onset of the proceptive period (for shorter periods) or at the end of the proceptive period (for longer periods): the apparent pattern is that, the longer the proceptive period, the more the P4 rise is shifted towards the end of it.

Lactation and fertility

Recabarren *et al.* (2000) studied the impact of lactation upon fertility in female *Cebus apella*. Females in their study, after delivery of an infant, were kept separated from males until vaginal smears indicated that their ovarian cycle had resumed. They found that nursing capuchin females show a period of lactational infertility of 159 ± 9 days in which cyclic ovarian function ceased and an additional period of residual infertility of 302 ± 23 days when cycles were present. In this latter period, despite the presence of males and the occurrence of mating, females did not conceive. The period of lactational amenorrhea for non-nursing females (females that lost or rejected their infants at birth) lasted 43 ± 6 days and the residual infertility lasted 153 ± 8 days.

On the one hand, the periods of lactational amenorrhea and residual infertility documented by Recabarren *et al.* would combine to produce an average interbirth interval of approximately 2 years, as is noted in many colonies and in natural populations. On the other hand, we have several documented instances of females in our laboratories rearing an infant and giving birth to another infant within 1 year, and at least one case of a female conceiving about 2 months after the loss of her infant in the perinatal period. It seems likely that the period of lactational amenorrhea documented by Recabarren *et al.* (2000) is more consistent across individuals and settings than the period of residual infertility; the latter may well be influenced by the presence of adult males in the group. In Recabarren *et al.*'s study, males were not housed together with females until researchers determined that females had resumed their ovarian cycling.

MALE SEXUALITY

Mating

Males' mating consists of mounting bouts with pelvic thrusts interspersed among a continuous flow of reciprocal solicitations. Mounting bouts can be isolated or can occur in sequence. The duration of a mounting bout, which starts when the male gains a position that enables intromission and ends when he dismounts, lasts from a few seconds up to four minutes (Janson 1984, Visalberghi and Moltedo, unpublished). Duration varies within and across individuals, is greater for dominant than subordinate males, and is greater in adult males when the female is in the periovulatory phase (averaging 93 seconds) than in the non-periovulatory phase (averaging 55 seconds). Mounting bouts of sexually immature individuals, on the other hand, are much shorter (less than 20 seconds) and do not vary systematically in relation to the female's ovarian phase (Carosi and Visalberghi 2002 and unpublished). Two subordinate adult males, mating with receptive females in the periovulatory phase while the dominant male was absent from the group, still averaged much shorter mounting bouts (24 seconds) than two dominant males (92 seconds) (Visalberghi and Moltedo, unpublished data). Janson (1984) reports similar results for wild tufted capuchins (dominant males' mounting bouts averaged 96 seconds; subordinate males' averaged 22 seconds); moreover, subordinate males in Janson's study spent much less time interacting with the female per copulation than did the dominant male (average of 2.8 *vs.* 12.7 minutes).

The duration of capuchins' mounts is shorter than that reported for woolly monkeys (3–14 minutes; Nishimura 1988) or muriqui (6–18 minutes; Strier 1992), both species in which ejaculation is usually achieved in one mounting bout. On the other hand, in species which perform, or can perform, more than one mounting bout before ejaculation, mount duration is usually shorter than what has been observed in capuchins (Hrdy and Whitten 1987). For example, copulations in chimpanzees, bonobos and patas monkeys (*Erythrocebus patas*) last an average of 5–10 seconds (Kano 1992, Loy 1975 cited in Loy 1981, Tutin and McGinnis 1981).

Ejaculation

Ejaculation is the most important event in male sexuality for reproductive success. But ejaculation is not easy to assess, especially in the field; even the most accurate studies may not distinguish between intromission with and without ejaculation (e.g., Dixson 1995, 1997). Primatologists have most often scored mating, copulation and mounting without providing operational definitions for events and without clarifying whether they encompass ejaculation or not (e.g., de Waal 1988, Linn *et al.* 1995, Loy 1981, Nadler and Collins 1991, Strier and Ziegler 1997, but see Kano 1992). Therefore it is not surprising that what little we know about ejaculation in *Cebus* comes from laboratory studies where close observation of the partners allows us to infer ejaculation (see ejaculation in Table 12.2, for details).

Dewsbury and Pierce (1989) reviewed the copulatory patterns of primates and divided species according to several criteria including (1) whether males are able to ejaculate only after more than one mount and (2) whether males may resume copulatory activity within 1 hour after ejaculation. In *C. apella* ejaculation may occur both after a single mounting bout and after a sequence of bouts, though the latter pattern is more common (Carosi and Visalberghi 2002) and males do not resume mating within 1 hour. Let us discuss these points in further detail.

The post-ejaculatory interval is defined as the period of sexual inactivity, during which sexual arousal is reduced following ejaculation (Dixson 1998a). After ejaculation the *C. apella* male does not mount the female for at least several hours or days, as in species with a single ejaculatory pattern (see Dixson 1998a, p. 117). However, shortly after ejaculation, male courtship towards the female is indistinguishable from that before ejaculation (see Phase 4 above). Dal Secco and Visalberghi (2001) noticed that the longer the male solicits the female after copulation, the later she inspects her genitals and extracts from her vagina bits of solidified sperm. They hypothesized that post-ejaculatory courtship is a male strategy to increase the chances of fertilization. Dixson and Anderson (2002) found that seminal coagulation and the formation of copulatory plugs is best developed in genera where sperm competition is most prevalent (that is, those in which females mate with multiple partners). Capuchins were not among the genera they considered; however, the post-ejaculatory courtship, the quick coagulation of the semen after ejaculation (Nagle and Denari 1982), as well as other features of mating behavior discussed later in this chapter (such as the size of the seminal vesicles and the evidence that females court, mate and are fertilized by more than one male) all suggest that capuchins fit the pattern proposed by Dixson and Anderson for species with substantial sperm competition.

When proceptive females were observed for 7–8 hours a day during the entire proceptive period, the solicited (dominant) male was observed to ejaculate on average 0.7 times per female's proceptive period; the subordinate male was never observed to ejaculate, though sometimes he mated (Dal Secco 2000). Although ejaculation cannot be consistently determined in the wild, Lynch (submitted) found evidence for limited ejaculatory frequency in male *C. apella* and argues that a similar trend might be present in *C. olivaceus* and *C. capucinus*. She notes that squirrel monkeys, which are closely related to capuchins (Chapter 1), also have a relatively low rate of ejaculation according to Clewe and DuVall (1966). In this respect, *Cebus* and *Saimiri* are very different from other primate species with a multi-male, multi-female mating system (where mean hourly frequency of ejaculation is 0.8–0.9), or with monogamous–polygynous species (where mean hourly frequency of ejaculation is 0.1) (Dixson 1998a, Figure 8.18, p. 241).

The low rate of ejaculation of *C. apella* is surprising for a species that has characteristics commonly found in species in which male–male competition is high, as we will show next. The structure of the genitals and gonads reflects their function and is related to the species' mating system. In most species that live in multi-male groups, the males' reproductive organs are compatible with frequent copulations and sperm competition. In fully mature male *Cebus*, the seminal vesicles are of medium size as demanded by frequent copulations and for the production of a coagulum (Dixson 1998b). The seminal fluid of *C. apella* has a concentration of spermatozoa varying between 38×10^6 and 30×10^7/ml with a high percentage of motility and an acceptable live/dead ratio in all samples, and conception occurs following an average of 1.7 matings (Nagle and Denari 1983). As an indirect measure of sperm production, the testes weight/body weight ratio, expressed as a percentage, can be used (Møller 1988); this value for *C. apella* is 0.35%, a percentage similar to those reported for species such as macaques, baboons and chimpanzees that live in multi-male groups (Harcourt *et al.* 1981).

It is thus possible that sperm is a costly and limited resource for the *C. apella* male, and that he tries to allocate it in the best way to promote fertilization. In fact, we noticed that although males frequently mount females in the non-periovulatory period, ejaculation has been scored only in the periovulatory period, mostly in the final days (Carosi *et al.* 1999, Visalberghi, unpublished data). As shown in Figure 12.6, the eight ejaculations observed by Carosi *et al.* (1999) occurred 2 days before the urinary P4 rise (N = 2), one day before the P4 rise (N = 3), on the day of the P4 rise (N = 2) and one day

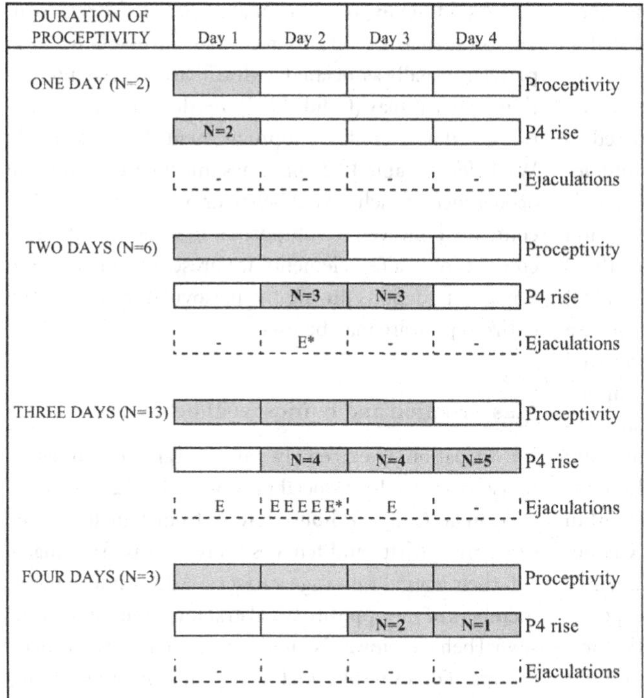

Figure 12.6. Temporal relations among proceptive behavior, day in which the significant rise in P4 occurred, and ejaculation. The rise in P4 tended to occur in the last 1 or 2 days of the proceptive period, or the day after it ended. The duration of proceptivity ranged from 1–4 days, as indicated by the gray shading in the proceptivity line (N indicates the number of cases in which proceptivity lasted the number of days indicated). The number of times (N) in which the significant rise of P4 occurred during each day of proceptivity is indicated in the gray shaded boxes in the P4 rise line. Each ejaculation observed is indicated by the letter E; E* indicates that the ejaculation occurred on the day of the P4 rise. (From Carosi *et al.* (1999). *Hormones and Behavior*, Copyright © 1999 Academic Press.)

after the P4 rise (N = 1). (Note that as mentioned earlier, the day of ovulation is probably the day of urinary P4 rise.)

Timing of mating has significant consequences for reproductive success. Janson (1984) argued that the dominant males have higher reproductive success because they copulate mostly in the middle of the female's proceptive period, which is usually slightly before ovulation (Carosi *et al.* 1999). In humans, according to Barrett and Marshall (1969), the best timing to mate and conceive precedes ovulation; the peak of fertility occurs 2 days before ovulation. If we assume this to be the case for capuchins as well, then mating before the occurrence of ovulation would increase likelihood of conception compared with mating at or following ovulation. Unfortunately, the data available for capuchins describing the temporal pattern of sperm migration through the female genital tract do not consider the time of ovulation (Ortiz *et al.* 1995). In any case, we hypothesize that the male's reluctance to accept the female's courtship and to mate promotes ejaculation near the time of ovulation, and that the male's post-ejaculation courtship functions to enhance the likelihood of fertilization.

Occurrence of sexual and mating behavior outside the periovulatory period

In capuchins, as in many primate species, copulations occur outside the females' periovulatory phase, including during pregnancy and postpartum amenorrhea (Dixson 1998a). For example, female *C. capucinus* studied by Manson *et al.* (1997) exhibited courtship behavior in all reproductive states, although males were more likely to respond to females' solicitations when the females were potentially ovulating. Despite males' selectivity, most of the mounts documented by Manson *et al.* (1997) involved females who could not conceive, and about two-thirds of observed mounts occurred in socially tense situations and during play. Sexual interactions occur in *C. apella* also during non-conceptive periods (Visalberghi and Welker 1986) and post-conception proceptivity has been reported (Phillips *et al.* 1994) Welker *et al.* 1983). Fragaszy and Adams Curtis (1998) report that postpartum proceptivity is common within a week following birth but that no infants were conceived as a result of mating during that period in their sample.

Capuchins commonly copulate during group formation in captive situations. When new individuals are introduced to an established group, fertile as well as non-conceptive females may begin to solicit new males within 20 minutes from their introduction (*C. apella*: Cooper *et al.* 2001, Fragaszy *et al.* 1994a, Welker 1992, Visalberghi, pers. obs.). In several New World primate species (e.g., *Saguinus oedipus*: Brand and Martin 1983, for a review see Dixson 1998a), copulations occur throughout the ovarian cycle as well as during pregnancy when males and females (not belonging to the same group) are paired. When paired for a 10-minute test with a subordinate male from their own group, female *C. apella* behaved variably: five cycling females mated only during the proceptive period (Moltedo 1998) whereas one female who was pregnant, and later lactating, mated almost every day (Visalberghi and Carosi, unpubl. data). A similar phenomenon may occur in nature: Manson *et al.* (1997) report that four out of six females (three of which were already pregnant) solicited and copulated with an immigrant male *C. capucinus* shortly after he joined the study group.

Most of the published studies concerning male mounting do not take into account the extent to which courtship included the entire set of behaviors present in the periovulatory phase, or whether intromission and ejaculation were performed during mounting. In short, the studies often neglected to assess the extent to which the sexual behavior performed by the two sexes was similar to the sexual behavior they exhibit when the female is in the periovulatory phase. It is possible that courtship between males and non-conceptive females does not include the entire repertoire and that ejaculation by the male is less common than when the female is conceptive, but at present these ideas are merely educated guesses.

As in many other primate species (see Dixson 1998a), sexual interactions in *C. apella* involve behavioral patterns usually associated with affiliative social interactions, such as play (Guidi 1992), mother–infant interactions, and post-conflict rapprochement (Verbeek and de Waal 1997). Table 12.2a provides information about the occurrence of each sexual behavior of *C. apella* in other contexts. However, it is important to stress that in these contexts only a few elements are present whereas in the sexual interactions during the periovulatory period the entire repertoire may be used.

Masturbation and homosexual behaviors

Masturbation is extremely rare in captive capuchins housed in mixed-sex social groups (Table 12.2a). However, male *C. apella* housed away from females sometimes masturbate and females housed away from males rub their genitalia on cage surfaces. Adults housed with members of the opposite sex also show some elements of sexual behavior towards same-sex partners in their social groups. For example, males may mount other males, and females other females. After short separations from each other, male *C. apella* embrace each other (Matheson *et al.* 1996, Visalberghi and Moltedo, unpublished data; see Chapter 11). Sometimes after this affiliative behavior, one male invites the other to mount him (Visalberghi and Moltedo unpublished data). Male *C. capucinus* perform a "wheeze" dance, similar to a courtship dance except for the "wheeze" vocalization, that terminates with the males mounting and thrusting on one another. During mounts both males usually stop wheezing and make the "duck face" (in which lips are protruded), the typical facial expression of courtship in this species (Manson *et al.* 1997). Cooper *et al.* (2001) observed a courtship-like sequence between two males when one was introduced to the other's (mixed-sex) group. Behavior during these socially tense conditions incorporates many of the affiliative and dyadically coordinated elements of sexual behavior, although it is clearly not "sexual" in a procreative sense.

MATE CHOICE AND ORGANIZATION OF MATING

Species of capuchins differ in the extent to which females prefer the dominant male for mating. Female *C. capucinus* females may solicit and copulate with lower ranking males (Fedigan 1993, Oppenheimer 1968, Perry 1997). Similarly, in *C. albifrons* the dominant male does not achieve the largest share of matings and mating is distributed more equitably among males than in *C. apella* (Janson 1986a, b). Female *C. olivaceus* actively solicit and mate only with the dominant male (X. Valderrama, pers. comm.) and the dominant male seems to be responsible for most matings both in large and small groups (Robinson 1988b). Thus it seems that, in *C. apella* and in *C. olivaceus*, female mate choice is oriented towards the dominant male more than in the other species.

Tufted capuchin females take the sexual initiative and direct most of their solicitations to a target male. Both in the wild and in captivity, females direct their courtship to the dominant male in the vast majority of cases. When females court some other male, solicitations and mating may occur in full view of the dominant male without his intervention (Phillips *et al.* 1994; Visalberghi and Welker 1986, Visalberghi pers. observation). When a male other than the dominant is the target, he acts much like the dominant by initially showing little interest and avoidance towards the soliciting female (Janson 1984, Visalberghi and Welker 1986). Our impression is that when chosen by the female, a male can afford to be reluctant since she will keep soliciting him until he reciprocates and mates. It is possible that by doing so the male is "selecting" the best time for fertilization. Males who are not chosen by the female are not reluctant, and instead exploit any opportunity to mate without going through the usual long courtship. Males therefore adopt different mating strategies according to whether the female chooses them as partners.

In the Manu National Park, dominant male *C. apella* have a higher copulatory rate (1.5 copulations per estrus; note that here copulation does not imply ejaculation) than do adult subordinate males (0.7 copulations per estrus). Moreover, the subordinate males copulate during the last days of proceptivity, and occasionally on the first days of proceptivity (Janson 1984), days when the female is less likely to conceive. Although each estrous female copulates with the dominant male, not all subordinate males mate with each female. Welker *et al.* (1990) report that female *C. apella* living in a large captive group (21–34 individuals) directed 81% of their solicitations (N = 738) to the dominant male, 14% to the four other adult males and 5% to the two subadult males. The dominant male of this group accounted for 52% of the copulations observed in the group (N = 159), the other adult males for 36% and the subadult males for 12%. The males other than the dominant "copulated proportionally twice as often as they were observed to receive courtship" (p. 167). Similarly, Linn *et al.* (1995) report that the dominant males in four different groups received 84% of the solicitations and accounted for 58% of 251 copulations. The dominant males copulated with high-ranking females, but not subordinate females, significantly more frequently than subordinate males did, even though high- and low-ranking females directed the same proportion of solicitations at dominant males (Linn *et al.* 1995). In contrast, the 12 copulations observed by Phillips *et al.* (1994) were similarly distributed among dominant and subordinate males. Finally, Janson (1998, p. 160) observed "extreme female mating promiscuity" in socially unstable groups of *C. apella* in Iguazú whereas in socially stable groups in the same area the females behaved like those living in the Manu National Park. Overall, it seems clear that under most conditions, in groups of *C. apella*, the dominant male has a primary role in mating. However, paternity data are needed to determine whether, and to what extent, males have differential reproductive success.

Social influences on mating in *C. apella*

Sexual interactions between proceptive females and subordinate males are inhibited by the presence of the dominant male more than they are actively stopped by his intervention. In tufted capuchins studied in Manu, the dominant male infrequently aggressed against subordinate males while they were interacting sexually with females, and such aggression was more likely at the end of the proceptive period (Janson 1984). Subordinate males were very tolerant of one another and never directed aggression toward the dominant male. Similarly, Phillips *et al.* (1994) did not observe male tufted capuchins interfering with each others' sexual interactions in captive breeding groups. However, Linn *et al.* (1995), studying captive groups with vasectomized males (and therefore constantly cycling females) describes both male–male and female–female interference with sexual activities.

Thus it seems that, under captive conditions at least, social status can influence the mating behavior of both male and female. Moltedo (1998, Visalberghi and Moltedo 2001) studied this phenomenon experimentally by observing sexual interactions between proceptive females and subordinate males in the group, and while housed together for short periods of time apart from their social group. Although the female did not solicit the subordinate male before separation (and vice versa), during separation, they readily interacted sexually. At reunion, the dominant male was not aggressive towards the female or the subordinate male; instead, his sexual interest towards the female tended to increase. These findings suggest that the presence of the dominant male can inhibit sexual interactions between females and subordinate males. As suggested by Linn *et al.* (1995), social influences affect females of high and low rank differently. Subordinate females appear to take more advantage than dominant females of the absence of particular group members to mate with the dominant male; conversely, when the dominant male is absent, the dominant females are sexually more active than the subordinate ones (Visalberghi and Moltedo, unpublished data).

The qualities of males that affect their attractiveness to females include age, relatedness to the female, rank, ability to form alliances, display frequency, services and resources provided to females, and resources controlled by males (Robinson 1982b). The ability of the male to control access to food resources has received particular attention. Janson (1984) noted a positive correlation between the male's ability to control access to food resources and a female's willingness to mate with him among tufted capuchins in Manu. In Iguazú, Argentina, where food sources cannot be controlled by any individual, the preference of females for the dominant male is less apparent, and the dominant male more actively seeks proximity with proceptive females (Janson 1998a). Another idea, suggested by O'Brien (1991), is that the female may reduce the risk of the dominant male attacking her infant by mating with him. As in many other species, infanticide has been documented in capuchins (*C. olivaceus*: Valderrama *et al.* 1990, *C. capucinus*: Rose 1994b, *C. apella*: Fragaszy and Adams-Curtis 1998, Welker 1992; see Chapter 10 and Janson and van Schaik 2000). Unfortunately, we do not have the evidence needed to evaluate the generality of any of these explanations for females' mating preferences. Moreover, we should keep in mind that (a) a female's behavior may change throughout the ovarian cycle, (b) she may respond opportunistically to the absence of the dominant male, (c) within-sex competition for access to a male can affect females' behavior, and finally (d) familiarity from birth of the female with a particular male seems to decrease her attraction towards him (see below).

REPRODUCTIVE SUCCESS AND MATING

"Paternity assignment on the basis of [observed] mating is bound to be inaccurate" (Dixson 1998a, p. 76). Paternity studies using DNA extracted non-invasively from hair and fecal samples have been conducted for groups of *C. apella,* both in the wild (Iguazú, Argentina and La Macarena, Colombia) and in captivity (Poolesville, National Institutes of Health, USA) by Escobar-Páramo (1999), for *C. capucinus* in Santa Rosa, Costa Rica (Jack and Fedigan in press), and in Hato Masaguaral, Venezuela, for *C. olivaceus* (Robinson *et al.* 2000). Although the data are not yet available for *Cebus apella*, for both *C. capucinus* and *C. olivaceus* there is a measureable reproductive advantage to the dominant male. In *C. capucinus*, for example, the dominant males in two groups sired 63% of the offspring, other males sired 16%, and paternity could not be determined for 21% of the 19 offspring (Jack and Fedigan in press).

There may be no advantage of dominance for siring offspring in captivity. The paternity of six infants born in two groups of tufted capuchins in which there were two and three adult males respectively was assessed by Ferrucci, Romano, Derme and Nicolai (unpublished data). These researchers compared the C-banding of the chromosomes of the offspring and of the adult males of the groups. Though behavioral observations showed that all the females consistently chose the dominant male as target of their solicitations, the banding analysis showed that only two out of six infants were sired by the dominant males (see also Welker 1992).

Inbreeding avoidance

Emigration from the natal group, short tenure for resident males, and lower sexual attraction between familiar individuals all can reduce inbreeding (Dixson 1998a). In *Cebus*, males usually emigrate and immigrant males can evict resident males (see Chapter 4). Robinson (1988b) estimated that in *C. olivaceus* the breeding residency of a male in a large group is at least 9 years, while the tenure of a male in a small group is about half as long. Thus females living in large groups may mate with their father if adult males remain in the group longer than 4–5 years. One possible mechanism to avoid inbreeding in these cases is reduced attraction between resident males and females that have grown up with these males, or between females and their sons. Evidence of this proximal mechanism comes from both the laboratory and the field. In a captive group, a female *C. apella* had four male offspring sired by a male who was then removed. During the following years her proceptive behaviors towards her (adult) sons were almost non-existent, although when her sons were 6–10 years old she gave birth to another infant (Visalberghi, unpublished). Welker *et al.* (1990) report that a male who was otherwise a common sexual partner of females in his natal group, never mated with his mother. In a large group of *C. apella* at Iguazú over a 4-year period, the dominant male was chosen by, and only copulated with, the older (>14 years old) females of the group but was never chosen by, or copulated with, the younger ones (<10 years old), which this male had probably sired (Di Bitetti and Janson 2001). Genetic analyses confirmed that the dominant male had not sired the offspring of the younger females (Escobar-Páramo 1999).

SUMMARY

Our study of sexual behavior in capuchins is just beginning; we have some detailed information at present about just one species (*C. apella*). We can, however, describe in broad terms the key features of courtship and mating behaviors in these monkeys. One striking feature is that ovulation is signaled most overtly by means of strikingly variable, prolonged and insistent proceptive behavior by females. Females may also emit olfactory cues about their reproductive status, but these have not been investigated yet. A second striking feature is that, although the female capuchin typically chooses to court one male for most of an estrus period, she seems to struggle for his attention; surprisingly, he responds very reluctantly to her advances. A third striking feature is that males, after ejaculation, continue to court the female vigorously for some time. A fourth noteworthy feature is that females will court males and copulate with them even when they are not able to conceive. Finally, although females predominantly choose to court the dominant male of their social group, they will mate with other males, and the dominant male may not have a strong reproductive advantage in the group (although studies on this topic are just beginning). This suite of characters, taken collectively, is unique among New World monkeys, so far as is now known.

Capuchins exhibit multiple mounts before a single ejaculation and a refractory period of several hours following ejaculation, and thus a low rate of ejaculation. This character, together with the size of the testes (moderately large for their body size), the character of courtship (elaborate and prolonged, even after ejaculation), and the common observation that females remove solidified ejaculate from the vagina with their fingers some time after copulation, suggest that male capuchins compete with each other reproductively through the mechanisms of female choice and sperm competition. In light of this possibility, the male's reluctance to accept the female's courtship and to mate may result in ejaculation occurring close to the time of ovulation, and post-ejaculation courtship may promote fertilization. These ideas require empirical examination; at this point we merely suggest them as plausible hypotheses.

There are many important features of sexual behavior in capuchins that we need to explore. For example, we do not know if ejaculation is limited to the peri-ovulatory period, even though mounting is not. Nor do we know how frequencies of copulation, ejaculation and time of ejaculation relate to paternity (that is, to a male's reproductive success). We do not know precisely what features of a male make him attractive to a female, nor why females' courtship behavior varies in the ways we have observed. These and other basic features of reproductive behavior remain to be ascertained.

13 • Learning together. Socially biased learning

Cammello, an adult male capuchin, has discovered how to use a stick to push a peanut out of a tube! He does it enthusiastically whenever he sees a peanut inside the tube. Sometimes, just for fun, he inserts a peanut shell in the tube himself and then uses his tool to get it. Brahms, an adult female capuchin, carefully watches her companion using the stick proficiently over and over again. Her face is just above the tube and from her vantage point she can attentively monitor the peanut moving inside the tube as it is pushed by Cammello's magic wand. Although there are plenty of sticks, Brahms does not take a turn herself at using one. When Brahms occasionally succeeds in stealing the peanut, Cammello is very tolerant; he just waits for the next one to appear in the tube (placed there by the experimenter). Is Brahms just exploiting Cammello's gentle nature, or is she unable to use the stick? We decide to present the tube and stick to her without Cammello. It is immediately obvious that by herself Brahms is completely lost. She wants to get the peanut, but she does not even try to use the stick for this purpose. We wonder what she was looking at when Cammello was solving the task. The videotapes show us that Brahms was looking at . . . the peanut! Her attention was not selectively focused on the events related to moving the peanut, such as the insertion of the stick in the tube. She paid as much attention to relevant events, such as the insertion of the stick, as to irrelevant events, such as holding the stick. Even after witnessing Cammello use the stick to solve the problem 50–70 times, she did not use the stick herself. Brahms was not an exception in this respect; over the years we have become convinced that adult capuchins do not learn how to use a tool by watching others, whether they watch another do so ten, a hundred, or even a thousand times.

In Chapters 9 and 10 we discussed how capuchins perceive and learn by looking, exploring and manipulating their surroundings, and by detecting the consequences of their own actions. Now we turn our attention to social learning, and our first task is to dispel two common misconceptions about social learning in monkeys. A common notion about social learning is that it involves learning how to do something by watching others, that is, by imitative learning (see Table 13.1 for definitions of these and other terms related to social learning). This view is deeply rooted in what we think we (adult humans) do when, in order to learn something, we watch more knowledgeable individuals and try to learn from them. However, as suggested in the vignette above, capuchins do not learn how to do something by watching others. In fact, imitation is rare in all nonhuman primates, even the apes (Tomasello and Call 1997). But even if the common saying "monkey see, monkey do" is a cliché with no basis in fact, social learning can still contribute importantly in various ways to many behaviors in group-living animals; as we explain below, it is likely to influence feeding, predator and prey identification, and general exploratory activity in capuchin monkeys, and can result in traditions.

A second common misconception is that "social learning" involves special learning processes that are not shared with "individual learning." This is not so, with the possible exception of imitation, which, as we pointed out above, is not seen in capuchin monkeys anyway. It is clearer and more accurate to describe learning that occurs in a social context as "socially-biased learning." The actions of the others can alter the individual's interest in certain events or places, and its evaluation of these places and events as attractive or fear-evoking, for example. These influences impact how the individual goes about exploring or accommodating to the situation

Table 13.1. *Definitions of social learning processes mentioned in the text.*

Process	Definition	Source
Social facilitation	An increase in the frequency of a behavior pattern (already in the observer's repertoire) in the presence of others displaying the same behavior pattern at the same time	Clayton 1978
Stimulus enhancement	As a result of some act being done in a particular place, or to a particular object, there is an "enhancement of the particular limited aspect of the total stimulus situation to which the response is to be made"	Spence 1937, p. 821
Local enhancement	"Apparent imitation resulting from directing the animal's attention to a particular object or part of the environment"	Thorpe 1963, p. 134
Imitative learning	It occurs when (1) something C (the copy of the behavior) is produced by an organism; (2) where C is similar to something else M (the Model's behavior); (3) observation of M is necessary for the production of C (above baseline levels of C occurring spontaneously); (4) C is designed to be similar to M; (5) the behavior C must be a novel behavior, not already organized in that precise way in the organism's repertoire	Visalberghi and Fragaszy 1990a
Emulation	Emulation learning in which an observer sees a demonstrator manipulating an object (or solving a task) and so learns new things about the dynamic affordance of that object (or that task) which they might have discovered on their own	Tomasello 1999
Object movement re-enactment	"An observer learns how an object, or the various parts of an object move". Hence, if the movement is associated with a reward the observer will be motivated to try to re-create that movement	Custance *et al.* 1999, p. 13
Tradition (including social convention)	Enduring behavior patterns shared among members of a group that depend to a measurable degree on social contribution for their acquisition	Fragaszy and Perry 2003b

Figure 13.1. Adult capuchin monkeys tolerate youngsters inspecting at close range items that they hold. Here three monkeys closely inspect a potential food item. (Photograph courtesy of Amy Galloway).

itself. Thus socially biased learning is distinguished by the context in which learning occurs and not by distinctive cognitive processes.

THE FOUNDATIONS OF SOCIALLY BIASED LEARNING: SOCIAL TOLERANCE AND INTEREST IN OTHERS' ACTIVITIES

According to Coussi-Korbel and Fragaszy (1995), a naïve individual may learn something general about an activity, such as the timing (through behavioral coordination in time) or the location (behavioral coordination in space), or sometimes both (behavioral coordination in time and space) by observing and acting with others. The third situation (coordination in time and space) provides the best opportunity to pick up a wide range of information, because the learner is in the right place at the right time to experience the widest range of elements with another at close range. But to coordinate action with another in both time and space requires that the other individual tolerates the learner's close presence. It follows that social dynamics within a group (the social relations of individuals with each other) constrain who learns with whom, when and in what circumstances (Coussi-Korbel and Fragaszy 1995). In particular, the degree of tolerance among group members towards each other influences the extent to which an individual is able to approach group members and/or to match (in space and/or time) its own activities with those of group members, such as feeding together (van Schaik 2003). A second factor that constrains socially biased learning is the degree of interest that one individual has in other individuals (we shall call this social orientation).

Capuchins are both very socially oriented and quite tolerant of one another (Figure 13.1). Solitary life is seldom a realistic option for capuchins and social companionship is actively pursued throughout life. As an indication that social partners are important to capuchins, consider that when forced to choose, pair-housed capuchins prefer their familiar companion over food, even following lengthy periods (up to 22 hours) of food deprivation (Dettmer and Fragaszy 2000). Still, social relations within groups are highly organized (see Chapter 11). Although youngsters are able to move the most freely in the group, others interact most often with specific partners (e.g., same-sex peers). Captive capuchins most often express affiliation (proximity, contact, grooming, feeding, food transfers and alliance formation) with individuals to whom they are related through their mothers (Martin 1994, Welker *et al.* 1990, 1992a, b). Females maintain stronger affiliative relations with their mother than do males (Welker *et al.* 1992b) and in general female matrilineal relations remain important across the lifespan (see Chapter 11). The importance of these matters here is that social affiliations are likely to affect the degree of social influence individuals have on one another; individuals are more likely to learn in the presence of those with whom they are closely affiliated.

Beyond tolerance and affiliation, the third characteristic necessary for social learning is an individual's interest in the locations and objects acted upon by companions, and companions' expression of emotion (e.g., contentment or fear). Such interest leads to phenomena usually called local and stimulus enhancement (see Table 13.1). Enhancement can be conceived as "value added" to an object or place as the outcome of a companions' movement to (or in) that place or manipulation of that object. Notice that enhancement does not necessarily lead to the observer performing the same behaviors (Zuberbühler *et al.* 1996).

Altogether, these three processes make it possible for individuals to acquire shared preferences, perceptions and skills. If individual A spends time near individual B, attends to the activities and emotional expressions of B, and is tolerated by B, then A has increased chances of experiencing some of the same things B is attending to, perceiving or doing. Just by being with B, A may experience, more or less, what B is experiencing. If A joins B, acts in the same area on the same materials, and experiences the same conditions and opportunities, it is likely that A's behavior will eventually become similar to B's (see Huffman and Hirata 2003 for a similar view; Figure 13.2). Learning in this way should reduce the time required by a naïve individual to become as efficient or proficient as its more knowledgeable companions compared with learning completely on its own.

In this chapter we describe social influences on a variety of behavioral domains: feeding, food choice, actions with objects (including using objects as tools), reactions to other species as predators or prey, and social conventions. We begin with two caveats: first, we often have better information on the time frame, direction and scale of social influence on learning than we do on exactly what processes are mediating that influence. Second, it is devilishly difficult to test hypotheses about social learning with data obtained from observing monkeys in natural settings. To do so requires data on how much time individuals spend with others in particular circumstances where there is something specific to learn (e.g., to eat a specific item), and on the consequences for learning by being with specific others (Fragaszy and Perry 2003b). It is not easy to collect these detailed data in the field (MacKinnon 2002) but it can be done (Russon 2003). Identifying differences across groups in certain behaviors is suggestive evidence of social learning (van Schaik *et al.* 2003, Whiten *et al.* 2001), but it is far more compelling to document social influence during learning, as there are always alternative explanations for differences across groups. We present some data bearing on social learning in wild capuchin monkeys, but we have more opportunities to collect such data with captive monkeys, where we know which problems are new for them, we can observe every time that individuals encounter events and objects in various social circumstances, and we can stage controlled comparisons. Thus, most of what we report about social learning in capuchin monkeys has been learned from studies with captive monkeys, and most of these studies have been conducted with tufted capuchins.

Figure 13.2. A mother licking a stick with apple sauce on it tolerates her son as he attempts to get some of her food for himself. Adults routinely permit youngsters to take food from them, however, they do not actively give them food. (Drawing by Stephen Nash from a photograph by Elisabetta Visalberghi.)

FOOD CHOICE

In principle, learning what to eat from others could be safer and occur more quickly than learning on ones' own, especially if there is a chance that one could ingest something poisonous or suffer harm from chemical or mechanical defenses when trying to eat unfamiliar items. Thus researchers have been interested in the possibility that young capuchins learn what to eat from others. The monkeys' tolerance towards each other lends plausibility to the idea: although wild capuchins compete for food resources, individuals sometimes tolerate others in close proximity and even allow others (especially immature individuals) to inspect their food and collect pieces that fall nearby (Izawa 1980, Janson 1996, Perry and Rose 1994). For example, at Raleighvallen, Boinski *et al.* (2001, p. 748) observed that, while a juvenile female was striking a hard fruit on a branch, "every blow was calmly and deliberately delivered, as were the slow, smooth, and careful motions with which the pyxidium [a hard part of the fruit] was rotated during the intervals of inspection. During the course of processing the fruit, at least four other immatures scrutinized every one of this juvenile female's movements. Their faces and gaze were so closely and intently fixated on the juvenile female's actions that the pyxidium seemed imminently to crash down on one of her audience's heads." This particular example concerns juveniles attracted to another juvenile; the same kind of event occurs with an adult actor and a juvenile audience.

Tolerance towards others during feeding seems greater in captivity (de Waal *et al.* 1993, Fragaszy *et al.* 1997a) and in semi-captivity, where monkeys are provisioned (Izar and Sato 1997), than in natural settings. In nature, capuchin monkeys past infancy forage on their own, several meters from each other, far more often than in close proximity to one another. In some contrast, capuchins in captivity seem to be "cohesive feeders that tolerate other group members to a fairly great extent regardless of kinship status, age or gender" (Furbush 1994, p. 44; see also Fragaszy *et al.* 1994a, Rogers 1996). It is not uncommon to see an animal collect a piece of food and then approach another and sit near it while eating.

Tolerance and interest towards others' food results in coordination among individuals in time and space during feeding and this coordination permits individuals to bias each other's feeding activities. For example, if an individual is eating a food novel to another, tolerance, interest and coordination in space and time will result in exposure of the naïve individual to the novel food. Naïve individuals could learn about new foods in the presence of more knowledgeable individuals if (a) others permit them to collect or otherwise sample some of the food, (b) the others' behavior alters the naïve individual's perception of the value of the item, and/or (c) the others' eating motivates the naïve individual to eat. As we shall see, there is experimental evidence that all three mechanisms are at work in capuchin monkeys.

Transfer of food

Captive monkeys take or collect food from a maternal relative more commonly than from unrelated individuals (Martin 1994). Thierry *et al.* (1989) report that the transfer of food (and objects) is especially frequent among young individuals and that the item is more likely to pass from one to another if it is more difficult for the possessor to escape with it (i.e., it is large rather than small) and valuable (i.e., a piece of food is more valuable than an inedible object). Although transfers can be achieved by aggression, the majority are peaceful.

In large captive spaces, capuchin monkeys rarely transfer whole food items; collection of fallen/discarded pieces from others is far more common. Transfer of food also readily occurs when two capuchins are kept side by side in small cages separated by a mesh partition and one individual is provided with food, well out of reach of the other monkey (de Waal *et al.* 1993). The possessor of the food usually approaches the wire mesh with the food and commences to eat in the typically messy style of capuchins. The companion then reaches through the mesh to take bits and pieces of fallen or discarded food, much as happens when the monkeys are freer in a larger space. In these circumstances, adult females are likely to sit near others (and thus enable others to take food) reciprocally (that is, when female A has food she sits as close to female B as often as B sits to A when B has food). Males are not as reciprocal towards others in their behavior; they sit near the mesh no matter how the other has behaved toward them recently (de Waal 1997). Although de Waal *et al.* (1993) labeled these behaviors as "voluntary food sharing," de Waal now more parsimoniously labels them "facilitated taking" because of "the active seeking of proximity in which food transfers can take place and the possessor's otherwise general passivity in the sharing process" (de Waal 2000, p. 14).

Several primate species transfer food, and a very few actually share food (for a review see Feistner and McGrew 1989). Two hypotheses that have been proposed to explain the occurrence of food transfer are relevant for social learning. The food scarcity/difficulty hypothesis proposes that adults should permit infants to take food from them if the foods are difficult or

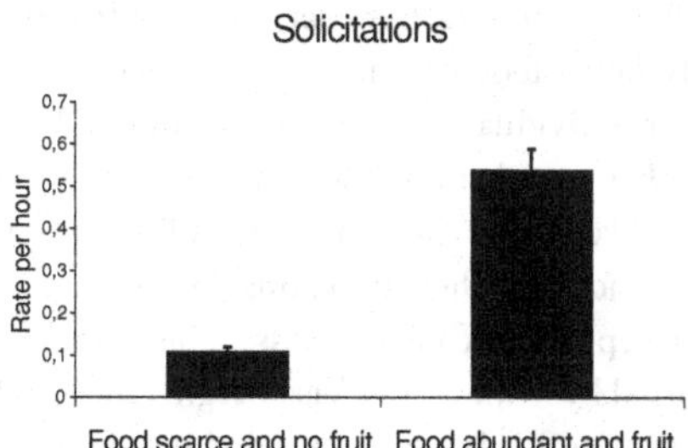

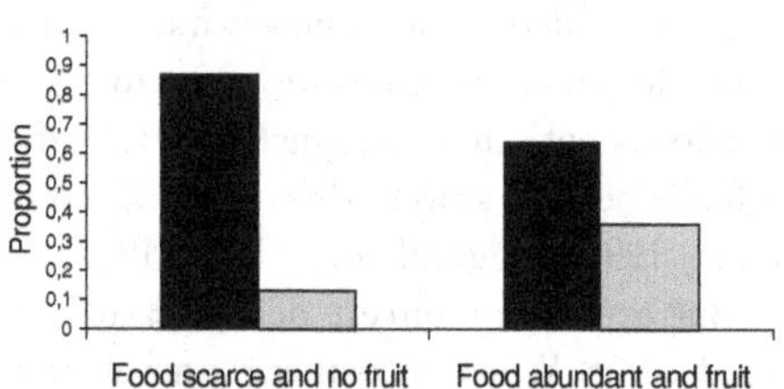

Figure 13.3. Captive capuchins solicit food from others (express close interest in another's food) significantly more often when the food is high quality and abundant than when it is low quality and scarce. Their expressions of interest are tolerated by the owner of the food in the majority of instances, but especially when the food is of lower quality. Left, average hourly rate per focal animal of food solicitations in the two experimental conditions used by Furbush (1994). Right, proportions of solicitations tolerated (black) and resisted (stippled) by focal animals in these two conditions. The overall proportion of solicitations tolerated was significantly greater than the proportion of solicitations resisted. A greater proportion of solicitations were tolerated when food was low quality and scarce than when high quality and abundant. (Redrawn from Furbush 1994.)

impossible for the latter to acquire otherwise (Silk 1978). The food learning hypothesis postulates that transfers of food between more knowledgeable individuals (adults) and less knowledgeable ones (infant and juveniles) result in the latter learning to ingest suitable food items (e.g., King 1994a, b, 1999). Three studies have tested these hypotheses in capuchins.

The first study tested the food difficulty hypothesis by observing groups of capuchins provisioned with abundant quantities of whole, unshelled pecan nuts (*Carya illinoensis*) or commercial pellets (Fragaszy *et al.* 1997a). Youngsters tried to bite open the nuts and banged them on hard surfaces, exactly as adults do, but mostly without success. Youngsters also frequently approached their more proficient companions and tried to take unshelled nuts or parts of nuts from them. Youngsters, regardless of whether or not they were able to shell nuts, were nearly universally tolerated by older (successful) individuals whom they could closely monitor and from whom they could take or collect bits of this highly desirable food. More dramatically, infants sometimes ate directly from their mother's hand. Youngsters attempted to inspect and to take others' pellets as well as nuts. However, they inspected and took nuts significantly more often than pellets. The adults permitted infants and juveniles to take food from them whether or not the younger individuals could open the nuts themselves; adults were not monitoring the young animals' proficiency. Thus the strong version of the food difficulty hypothesis (which places the responsibility for food transfer on the adult) is not supported, but the infants' efforts produce a similar result: infants attempt to take foods they cannot acquire otherwise.

The other two studies addressed the food scarcity hypothesis. Furbush (1994) investigated whether the quality and abundance of food affected attempts by others to take food and the outcome of such attempts. Two large captive groups of capuchins were tested when only leftovers of pellets from a previous feeding were available (Condition 1), and when both pellets and fruit were freshly provided and abundant (Condition 2). In Condition 2 fruit was present in such abundance that it was not finished during the observation period. In this latter condition, capuchins more frequently attempted to take food, and a greater proportion of their attempts were rebuffed, than in Condition 1 (Figure 13.3). Later work showed that capuchins attempt to take fruit from others more often than chow regardless of its abundance (Rogers 1996). The food owner generally tolerates others attempting to take its food, but permits others to take food more often when the food is less preferred than when it is more preferred. Thus the food scarcity hypothesis was not supported by either study.

Overall, it seems that neither the food scarcity/difficulty nor the food learning hypotheses account for the pattern of food transfer in captive capuchins. Instead it seems that a special value is added to a food held by another individual, regardless of whether or not the same food is freely available elsewhere. This would suggest that stimulus enhancement is taking place, i.e., a food is more attractive if it is possessed by another group member.

Learning about new foods

Throughout its lifetime each monkey must learn to recognize edible items. Capuchins are known to exploit new foods or foods not eaten by capuchins elsewhere when staple foods become scarce (Brown and Zunino 1990, Fernandes 1991, Langguth and Alonso 1997, Siemers 2000). But capuchin monkeys do not casually accept new food items. When encountering novel items that might be edible, wild capuchins may ignore them, or they may sample them to a limited extent (Visalberghi *et al.* 2003). They can be described as moderately neophobic toward potential new foods. This cautious attitude toward foods may be an adaptive strategy for a generalist taxon such as *Cebus* because it allows the gradual introduction of new foods into the diet and reduces the risk of poisoning by ingestion of large amounts of toxic compounds (Glander 1982, Milton 1993, Visalberghi 1994).

One reason that we and others have thought that social context could aid naïve monkeys to learn about a novel food is that capuchin monkeys show such tolerance towards others during feeding, and individuals frequently closely attend to the foods of others. Attending to others while they eat a food novel to the observer could promote acceptance of novel foods in capuchins in at least three ways. First, the naïve individual might seek information from others, watching if they eat the food. If so, then one should expect a "learning interaction" with a more knowledgeable individual to occur before ingestion of a novel food by the viewer and the pattern of learning about food and food acceptance to differ between individuals encountering foods alone or with companions. The second mechanism is simpler; it could be that seeing others eating a new item promotes interest or acceptance of that particular item by the viewer. A third mechanism is even simpler: seeing others eat promotes eating in the viewer, without regard to the specific items eaten (social facilitation; see Table 13.1).

We have the most positive evidence for the third mechanism. For example, satiated capuchins recommence eating if familiar individuals begin eating (Galloway 1998). Monkeys more readily eat novel foods when other monkeys nearby are eating (Galloway 1998, Visalberghi and Fragaszy 1995) than without group members nearby (Visalberghi and Addessi 2000a). After just a few encounters, however, consumption of the previously novel food reaches similar levels regardless of whether individuals previously encountered it alone or with other members of their group (Visalberghi *et al.* 1998). It seems that the presence of group members, and the fact that they are eating, promote an individual's acceptance of novel foods. This idea is strongly supported by Addessi and Visalberghi (2001), who evaluated the feeding behavior of individual capuchin monkeys encountering a novel food alone, with a partner when only it could eat, or when a fellow member of its group was also eating. In the presence of a group member, individuals ate 22% more (by weight) of a novel food than when they were alone. When the group member was present and eating, the individuals ate 64% more than when they encountered the food alone.

There is also some support for the second mechanism, that seeing others eat promotes interest in the food that the other is eating, especially if the food is novel. Capuchins show interest (inspection at distances of less than 5 cm of another's food) in other individuals' novel food far more often than familiar food (Fragaszy *et al.* 1997b). This holds true when a food with a familiar aspect has a novel odor (Drapier *et al.* 2003). Interest in others' novel food is especially evident among immature animals (Fragaszy *et al.* 1997a), even though juvenile and infant capuchins in captivity do not display neophobia; they are as likely to eat novel as familiar foods (Fragaszy *et al.* 1997b).

Although the close attention that monkeys show towards others' food suggests that they might learn about that specific food, this does not seem to influence what they eat next. Capuchin monkeys consume more of a novel food when others are also eating (measured as the estimated duration of feeding and as weight of food consumed; Visalberghi and Adddessi 2000a), but the amount (g) of food ingested is not affected by whether the observer eats food of the same color or a strikingly different color from what the others are eating (Visalberghi and Addessi 2001). A similar result is obtained when the demonstrator can choose between two foods (of the same two colors as the novel foods presented to the observer) and the demonstrator eats *only one* of them. Even when the observer witnesses the demonstrator choosing just the food of one color, it does not preferentially sample the food with the same color as that chosen by the demonstrator (Visalberghi and Addessi, in press). The results are clear: seeing its companions eating motivates a capuchin to eat foods available to it, including novel ones, but the monkey does not selectively eat foods that match in color the food that other monkeys are eating. Visalberghi *et al.* (2003, in press) showed that capuchins' preferences for seven initially novel foods encountered in all binary combinations correlated positively with glucose and fructose content and negatively with fiber content. After further experience consuming the seven foods, however, preferences shifted to correlate with energy content. Individuals eating the foods alone shifted preferences in the same direction and magnitude as individuals that encountered the foods together with group members eating the same foods. Thus social partners did not influence the development of preferences towards new foods.

Nor do capuchins rely on others' behavior to assess current palatability of familiar food (Visalberghi and Addessi 2000b). Capuchins, both individually and within their group, encountered familiar palatable food for five sessions, then the same food rendered unpalatable by the addition of white pepper for five sessions, and then the familiar palatable form again in the final five sessions. Capuchins, both as solitary individuals and in a group, quickly responded to the change in palatability; there was no difference in behavior between those that encountered the altered foods by themselves or in the group. As an aside, during this study, as in previous studies concerning acceptance of novel foods by captive capuchin monkeys (Fragaszy *et al.* 1997b, Visalberghi and Addessi 2000b, Visalberghi and Fragaszy 1995, Visalberghi *et al.* 1998, unpublished data), we have never observed a mother (or a more knowledgeable individual) preventing or attempting to prevent its offspring (or a naïve individual) from eating unpalatable or novel foods, nor encouraging the offspring to eat a novel food (see Rapaport and Ruiz-Miranda 2002 for rare cases of tutoring in wild golden lion tamarins; see Boesch and Boesch-Achermann 2000, Matsuzawa *et al.* 2001 for a discussion of these issues in chimpanzees).

In sum, experimental results do not support the hypothesis that naïve individuals learn to recognize new foods by taking food from others, nor by monitoring others' behavior towards potential food items. Nevertheless, the naïve individual's eating is facilitated by the feeding activities of its companions in such a way that individuals can learn to eat novel foods more readily with others than if they encounter these same items alone.

FEEDING TRADITIONS IN WILD CAPUCHINS

Field researchers who have studied *C. capucinus* in Costa Rica at three ecologically similar sites (Santa Rosa, Lomas Barbudal and Palo Verde) over many years looked for interpopulation variability in foraging behavior as suggestive evidence of traditions in these monkeys (Panger *et al.* 2002). Out of 61 foods that were common to all three populations, 20 (16 plants and four animals) were processed differently in at least one of the sites by at least one individual. For example, in some (but not all) of the sites capuchins have been seen to wrap *Automeris* caterpillars or *Sloanea terniflora* fruits in leaves (which can reduce contact with chemical and mechanical defenses) before rubbing these foods against a substrate. Monkeys at all the sites eat these foods. For some less common feeding techniques (such as wrapping leaves around caterpillars), monkeys who shared these techniques spent more time in proximity to one another than monkeys who did not use these techniques. This correlational evidence suggests that close associates influence each other's behavior. Similarly, Boinski *et al.* (2003) suggest that social context likely aids tufted capuchins in Raleighvallen, Suriname to develop skills at pounding open hard fruits, an activity to which members of this population devote much time. Forthcoming data concerning the links among social affiliations, timing and context of skill development by young monkeys in this population will be very welcome.

Capuchins at the three Costa Rican field sites mentioned above also differ in the range of prey taken and the extent to which they hunt vertebrate species (Rose *et al.* 2003). Looking only at the prey species that are common at all three sites, the number of species that are hunted at Palo Verde is much smaller than the number taken at Santa Rosa, with Lomas Barbudal being intermediate. Capuchins at Lomas typically kill squirrels by biting them on the neck prior to consuming them, whereas Santa Rosa capuchins begin eating small squirrels while they are still alive and attempt to kill larger squirrels through a variety of less effective techniques. According to Perry *et al.* (2003b), neck biting by the capuchins at Lomas Barbudal is a good candidate for a behavior maintained at least in part through social contributions to learning (in other words, it is likely to be a tradition). Whereas the techniques used at Santa Rosa correspond to typical behaviors with live prey seen in capuchins elsewhere, neck biting seems to be a more precise and more effective technique. Outside of squirrel hunting, other prey capture techniques do not seem to differ among the populations at the different field sites.

ACTIONS ON OBJECTS, INCLUDING USING OBJECTS AS TOOLS

General social influence

Captive capuchins are extremely interested in manipulating objects, whether familiar or novel. Wild capuchins are also interested in objects, though cautious towards novel ones (Visalberghi *et al.* 2003). Of course social dynamics within a group affect which individuals are likely to manipulate objects provided to captive groups. When only a single object is provided to a group of capuchins, the adult males are likely to monopolize it at first (Byrne and Suomi 1996). This results in a group of immatures (and some adults) attentively monitoring the "owner" and moving in synchrony with the possessor. If the object is abandoned, even briefly, it is instantly taken by one of these followers, that becomes the new owner. Adult capuchins are especially tolerant of young individuals attempting to take objects from them (Thierry *et al.* 1989) and even uninteresting objects (e.g., wooden blocks), particularly when scarce, are highly desirable to others by virtue of the owner's interest in them (Visalberghi 1988). Thus the enhancement of interest in something held by another, evident with food, is also evident with inedible objects.

Social influences can interfere with exploration and learning also. One common form of interference is that lower-ranking individuals' access to objects can be limited by others. For example, when a group of capuchins was presented with a pan of water and sand-coated fruit, all capuchins readily washed the fruit in the water, with the exception of a single female who continued to eat sandy, unwashed fruit. When she was given the pan of water and sand-coated fruit without her group members, she immediately washed the fruit in the water (Visalberghi and Fragaszy 1990a). At the other end of the social spectrum, sometimes the power to monopolize or expropriate resources from others interferes with learning. For example, when an individual can take some or all of a food item opened up by another, they are less likely to learn how to open the item themselves (but see Caldwell and Whiten 2003). This is commonly called "scrounging" (Giraldeau and Lefebvre 1986).

THE BENEFIT OF A DEMONSTRATOR

Early investigators studied the imitative skills of capuchins (Haggerty 1909, Thorndike 1901) and the interest in this topic has not faded since then. Researchers interested in whether capuchins are able to imitate novel behaviors (see Table 13.1 for a definition), present the monkeys with "demonstrators" (capuchins or humans) performing tasks requiring improbable actions. If the naïve individuals adopt the same (or a very similar) behavior as that exhibited by the demonstrators, the claim for imitation is strong (so long as appropriate experimental controls are also conducted). Otherwise, behavioral coordination in space and/or time and less direct social learning processes, such as those we have discussed as occurring during feeding, are more parsimonious explanations of a benefit for learning from watching another (see Table 13.1).

Visalberghi and Fragaszy (1990b, 2002) and Fragaszy and Visalberghi (1989; see Figure 13.4) reviewed studies which used *post hoc* inspection to evaluate social influences on mastering a new skill (Antinucci and Visalberghi 1986, Visalberghi 1987, Visalberghi and Fragaszy 1990a, Westergaard and Fragaszy 1987a), as well as studies that considered social influences more directly (Fragaszy and Visalberghi 1989, Visalberghi 1990a), and studies where the behavior of observers in relation to the behavior of a model was analyzed in detail (Adams-Curtis and Fragaszy unpublished data, Visalberghi and Trinca 1989). They found no evidence of imitation. However, other positive biases, such as stimulus enhancement and behavioral coordination in space and time, do seem to contribute to the acquisition of skills by naïve individuals, and negative biases (such as inhibition and exploitation) limit an individual's learning. We illustrate these findings below with three studies.

Figure 13.4. Capuchins do not learn how to use an object as a tool by watching another do so. Here, an adult male capuchin uses a wooden block to crack open a nut. Other group members closely watch his activities. Despite attentive monitoring of the skillful actor, and abundant opportunities to handle nuts and potential tools, the observers did not learn how to crack open a nut with the tool over the course of the experiment (Visalberghi 1987). (Drawing by Stephen Nash from a photograph by Elisabetta Visalberghi.)

Learning sequences of actions: the latch puzzle

As described in Chapter 9, Adams-Curtis (1988, Adams-Curtis and Fragaszy 1995) presented a three-step mechanical puzzle to a captive group of capuchins. The puzzle could be solved to reveal a raisin by performing three actions in a fixed sequence: pull out a pin, lift a hasp and rotate a hinge. All these actions are individually easy for capuchins. The puzzle was presented in a test area that only two or three individuals could occupy concurrently. One individual (a juvenile female; hereafter the model) out of 12 animals in the group was proficient at solving the task, solving it more than 300 times over several test sessions. The researchers monitored the activities of the skilled individual and her companions in the test area. In particular, they focused on how and in what order the monkeys contacted the different parts of the puzzle.

Juveniles' interest in the apparatus was greater in the presence of the model than when in the test area alone, but this effect was not evident in adults. According to Adams-Curtis and Fragaszy (1995), three nonexclusive hypotheses could account for this result: (1) juveniles may be generally more attentive to activities of other individuals than are adults, (2) juveniles are more tolerated by others and therefore can approach them more than adults, and (3) juveniles are more cohesive among themselves (and, in this case, with the juvenile model). All three hypotheses are supported by evidence derived from related studies. However, although the other juveniles watched the model solve the puzzle many times, there was no evidence that they matched either the order in which the model handled the different parts or the behavior the model directed to each part.

Learning to insert a stick in a tube

"It seems obvious that discovering the affordance of a tool is most easily learned from a demonstration by another person . . . What is learned from observation is not how to manipulate the tool, but what the tool might

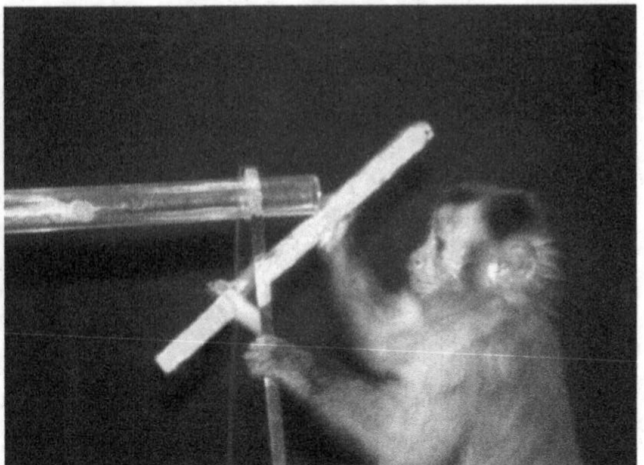

Figure 13.5. Tube task. Left, a monkey using the stick to get a peanut out of the tube is closely observed by a member of its group. Right, afterwards, the same individual fails to use the stick correctly despite having watched the skillful monkey solving this problem many times. The unsuccessful capuchin continued to contact the tube with the stick in the same manner as it did before observing many solutions performed by the other. (Photograph by Elisabetta Visalberghi.)

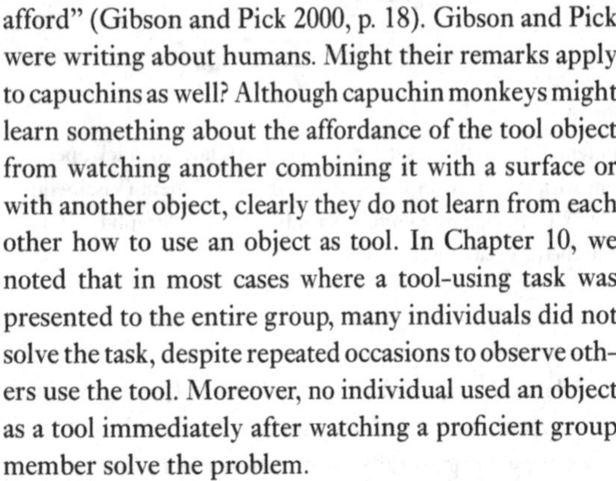

afford" (Gibson and Pick 2000, p. 18). Gibson and Pick were writing about humans. Might their remarks apply to capuchins as well? Although capuchin monkeys might learn something about the affordance of the tool object from watching another combining it with a surface or with another object, clearly they do not learn from each other how to use an object as tool. In Chapter 10, we noted that in most cases where a tool-using task was presented to the entire group, many individuals did not solve the task, despite repeated occasions to observe others use the tool. Moreover, no individual used an object as a tool immediately after watching a proficient group member solve the problem.

This last point is illustrated in the experiment presenting capuchin monkeys with sticks and a baited tube (the tube task; Visalberghi 1993b; see Chapter 10), and in a less formal tone in the vignette that opened this chapter. In this experiment, three out of the six capuchins tested spontaneously solved the task within 2 hours of presentation. Despite their interest in the food inside the tube, the other three capuchins (one adult and two juveniles) did not use the stick to push out the food when tested alone even after many trials. To assess how a model's actions could influence theirs, these three monkeys were presented with the tube task again in what we called "lessons." During the lessons, the three monkeys had the opportunity to watch models insert the stick into the tube and push out the reward (Figure 13.5). Researchers videotaped the proceedings. The three pupils witnessed 57, 75 and 75 solutions respectively. To make a long story short, although the pupils had ample opportunity to watch the models using the stick to push food out of the tube, none of them solved the problem, although the monkeys did contact the tube with the stick significantly more often during and after, as compared with before, the lessons. During the lessons they did not look selectively at the events relevant for solution (i.e., insertion of the stick in the tube and pushing the reward *vs.* holding the stick and eating the reward). Nor did they improve the orientation of the stick in relation to the opening of the tube after watching the models (Figure 13.5). Although increased activity combining the stick with the tube did not lead the monkeys to success in this experiment, in principle, success could be fostered by increases in such exploratory activity. In this sense, observing others using an object to solve a problem might bias an individual's exploration and favor the discovery of the actions that result in solution, but this idea should be examined in new studies.

Learning to operate a dispenser

If the activity of others supports learning by naïve individuals, then youngsters should acquire skills more readily in supportive social circumstances, when adults are acting proficiently, than they would if encountering the same challenges on their own. Fragaszy *et al.* (2002 and unpubl.) examined the contribution of social context to acquisition of specific skills in 20 infant and young juvenile tufted capuchin monkeys (*Cebus apella*) living in

Table 13.2. *Number of youngsters obtaining juice in one of two ways when adults in their groups modeled one method of solution*

	Wheel	Lever
Group 1		
Adults	Not Available	5 out of 6
Infants	2 out of 6	4 out of 6
Group 2		
Adults	5 out of 5	Not Available
Infants	9 out of 14[a]	2 out of 14[a]

[a] Significant Within Group Difference ($P < .05$), Fisher's Exact test.

two groups. In the first phase of the experiment, the youngsters encountered an apparatus containing juice in an area that the adults could not enter (a crèche). The youngsters could obtain juice in two ways: by putting a finger into a small opening and turning a finger wheel, or by pushing a lever down, then releasing it. None of them solved the task in the crèche using either method in 20 half-hour sessions. In the second phase, Fragaszy *et al.* (2002) presented the same apparatus to the same infants in two sites per group for 22 half-hour sessions; one apparatus was in the crèche with both solutions enabled, and another was in the group enclosure with either the lever enabled (Group 1), or the finger wheel enabled (Group 2). Most adults learned to obtain juice within these sessions using the single technique enabled by their apparatus. Four of the six youngsters that could be around adults demonstrating the lever solution (youngsters in Group 1) solved the lever task, but only two used the finger wheel in the crèche. Youngsters in Group 2 observed the finger wheel instead of the lever solution. Only two of the 14 youngsters in this group solved the lever in the crèche, but nine solved the task by using the finger wheel, the solution demonstrated by the adults. Thus immature animals did not manage to solve these tasks on their own, but did learn to do so when they had adult models. Moreover, when they had adult models, infants tended to adopt the solution style of adults in their groups (see Table 13.2.)

These results, although suggestive, are not compelling evidence that the adults' activity helped the youngsters to learn how to operate the dispenser. We still need to confirm that young naïve individuals that have not previously encountered the apparatus will benefit from the adult's activity while they learn to solve the problem. That is, we must rule out that experience with the apparatus alone can account for the rapid acquisition of the solution once the adults begin to demonstrate solutions. At this writing, Fragaszy *et al.* are in the process of conducting a replication with a new cohort of infants that from the beginning encounter the apparatus together with proficient adults. If these infants master the problem more quickly than the first cohort of infants, it will be stronger evidence that social context contributed to learning in this situation.

Temporal and spatial influences on juveniles' exploratory activities

Coordination with others in space and time is an obvious requirement for animals that live in a group (Boinski 1993, Boinski and Garber 2000; see Chapter 2). Earlier in this chapter, we mentioned the framework proposed by Coussi-Korbel and Fragaszy (1995) that considers social influences on learning in terms of the coordination of behaviors between two individuals in space and time. Fragaszy *et al.* (1994b) examined behavioral coordination in time and/or space among naïve young capuchins and older ones able to solve two tool-using tasks, and the relation between coordination and mastering the task. The dipping task consisted of a box from which yogurt could be obtained by inserting a straw or stick into small holes. The pounding task involved walnuts partly imbedded in a wooden base. The walnuts could be opened by striking them with a metal object (see Figure 10.5). In each session, one of the two tasks was available to all group members and a duplicate was placed in a separate compartment (crèche) that only juveniles could enter (adults were too large to fit through the door). The apparatus were presented first in the crèche and then in the home cage (sequential condition) or in the crèche and in the home cage at the same time (simultaneous condition). This design afforded an opportunity to examine coordination of behavior by youngsters with respect to adults.

The results revealed considerable variation among juveniles, the age group presumed to be most susceptible to positive social influences. Most juveniles explored the apparatus more frequently when adults had a similar apparatus at the same time. Older individuals preferred the tool site in the crèche whereas younger ones were usually at the same tool site as the adults. Despite their proximity to adults at the apparatus, juveniles were still less likely to solve the task than adults. While in the crèche, juveniles, regardless of their proficiency, did not visually monitor the activity of the adult models, nor did they match their behavior with those of the adults. In fact, the best predictors of success were age and individual practice. These results suggest that adults' behavior facilitated the activity of youngsters at the same time (resulting in coordination in time) even if youngsters did not necessarily coordinate their activity in space with adults.

Resende *et al.* (2003) describe another way that juveniles may loosely coordinate their activity with adults in time and space in a way that supports acquiring problem-solving skills. In their study of semi-free tufted capuchins in a large park outside São Paulo, youngsters combined social play with exploratory activities at sites where adults were cracking palm nuts using stones and stone anvils. As the youngsters get older, they spend proportionally less time playing socially and more time attempting to crack nuts, eventually (at a few years of age) succeeding at doing so. At first, the social element of play appears to motivate their presence and activity at the nut-cracking sites more than their interest in the food. The parallels with the socially tolerant manner in which young chimpanzees gradually acquire skill at cracking nuts using stones and anvils are striking (Inoue-Nakamura and Matsuzawa 1997).

Opening a latched box

Although in a variety of studies conducted in the 1980s and early 1990s capuchins did not learn to produce specific sequences of actions or to produce specific orientations of objects by observing other monkeys, researchers remained hopeful that the monkeys might be able to reproduce actions modeled explicitly (and therefore very clearly) by a human. To this end, Custance *et al.* (1999) modeled actions on an object to 11 capuchins reared by humans (and thus most likely to attend to a human demonstrator), and carefully monitored the monkeys' subsequent actions with the box. The transparent box, closed by a "barrel latch" or by a "bolt latch," contained a visible food reward. Each latch on the box could be opened with two distinct methods, and each method required two related actions. The researchers demonstrated a different method to each of two groups of monkeys. In the "barrel latch" task, half of the monkeys witnessed a pin at the front of the box being *turned* several times and then a handle *turned*. The remaining half saw the pin being *spun* and then the handle *pulled*. In the "bolt latch" task, half of the monkeys observed two rods at the top of the box being *poked*, and *pushed*, while the other half observed the rods being *twisted* and *pulled*.

The key question was whether the monkeys would perform the actions they had seen performed by a human demonstrator more frequently than the actions that they had not witnessed. In a similar task, children and chimpanzees did preferentially perform the same actions as the human demonstrator (Whiten *et al.* 1996). Two scorers, working from videotapes and naïve as to which actions each monkey had observed, were asked (1) to decide which of the two demonstrations each monkey had witnessed, and their confidence in the decision; and (2) to count the actions on the pins and handles (e.g., poking *vs.* twisting, pushing *vs.* pulling etc.) that each monkey performed.

The monkeys did not differentially reproduce the demonstrated action to open the "barrel" latch, and the scorers were unable to infer reliably which action the monkey had observed. Similarly, when the demonstrator had opened the "bolt" latch, the monkeys did not reproduce the demonstrated technique. However, for this latch, the scorers were able to infer reliably which demonstration the monkey had witnessed. Therefore Custance *et al.* recorded the location on the box (front *vs.* back of the box) where the monkeys concentrated their efforts on the latch, the directions the rod moved, and from where the rod was removed (back *vs.* front). They found that the groups differed in how often they (a) pulled the rod from the front or back of the box; (b) pushed the rod from the front side of the box (but not from the back side), and (c) removed the rod from the front or back side of the box. It should be noted that pulling or pushing, followed by removing, constitute behavioral sequences in which actions are dependent on one another. For example, if a monkey pushes the rod from the front of the box, it is likely to pull it and remove it from the back of the box. The spatial features of the monkeys' actions (i.e., whether they occurred in the front or back of the box) appear to have been the cues that allowed the scorers to discriminate between the experimental groups (e.g., the scorers could have discriminated between groups on the basis of which side of the box the rods were removed).

Custance *et al.* (1999) label the phenomenon of reproducing the direction of the rods' movement as "object movement re-enactment." They suggest that this, or "perhaps simple imitation" (p. 21) of the model, may account for their results, but they could not distinguish between the two explanations in their study. The notion that the monkeys acted to move an object at the front or back of the box (the side where they saw it move) is plausible. The foraging actions of capuchins incorporate many forms of poking, pushing or pulling, and attending to the spatial relations of an object in relation to the substrate, such as front *vs.* back, in such situations seems potentially very useful. This form of social influence is like that seen by Hemery *et al.* (1998, see also Visalberghi and Fragaszy 2002) when capuchins acted to bring together objects that the experimenter had handled and brought together, and (at least a few times) moved an object to reproduce its movement to a new position. Capuchins that are attentive to human demonstrators seem able to capture some of the spatial relations between objects produced by a human's actions with those objects. However, because the frequencies of the modeled techniques were not affected by the demonstrations for either latch, we conclude that the monkeys in Custance *et al.*'s (1999) study did not imitate the human model.

FEARFUL REACTIONS

In Chapter 9, we mentioned that one group of capuchins reacted fearfully to one object but not another object, and that the group next door responded in opposite fashion. Visalberghi (unpublished data) observed a similar phenomenon: one group of capuchins, upon encountering a novel object for them (a very large zucchini, weighing about 5–6 kg) responded first by exploring and contacting the zucchini, followed by harsh manipulation aimed at dismantling the zucchini to eat it. A second group responded in a completely different manner. An adult male cautiously approached the zucchini, then swiftly jumped back, gave alarm calls and showed facial expressions of fear. Immediately afterward, other members of the group began to behave as he did, and none of them contacted or ate the zucchini. After several minutes, to prevent further distress, the experimenter took the zucchini out of the cage. The following day, when the experimenter walked by each group's cage holding the zucchini, the first group responded with interest and food vocalizations, whereas the second one responded fearfully and threatened the vegetable.

It is evident that most individuals in the second group quickly learned to fear the zucchini by observing one individual's fearful reaction towards it. This phenomenon is actually widespread in the animal kingdom, and has been the subject of some interesting research. For example, Mineka *et al.* (1984) briefly exposed young macaques that did not fear snakes to adult conspecifics that displayed fear toward a snake in their presence. The young macaques developed an intense and long-lasting fear of snakes from this experience. Mineka *et al.* (1984) describe this process as observational conditioning. Observational conditioning occurred with other stimuli as well, but the procedure worked better to produce "vicarious fear" when the stimuli had certain perceptual features related to natural predators. For example, toy snakes were effective stimuli whereas plastic flowers were not (Cook and Mineka 1990).

Suboski (1990) suggested that vertebrates (and invertebrates) can acquire emotional responses to a stimulus from observing a conspecific displaying emotion to the same stimulus. In this kind of learning, which Suboski called "releaser-induced recognition learning," when the experienced individual reacts to the stimulus, its emotional response elicits a similar response in naïve observers. The observers then associate their own response to the stimulus and learn to react in the same manner. Mobbing predators, food recognition and many other species-typical behaviors may be passed from individual to individual in this way. There is currently a debate as to the details of this conditioning process (cf. Mineka and Cook 1993 and Suboski 1990), but the general idea of the process is not in dispute.

SOCIAL CONVENTIONS IN WILD CAPUCHINS

One of the most exciting new findings with capuchins in the domain of social learning is that wild capuchins can develop social conventions. A social convention is

Figure 13.6. Sucking another's tail is a peculiar social behavior occurring among some *C. capucinus*. Santa Rosa National Park, Costa Rica. (Photograph courtesy of Katherine Jack.)

a well-defined arbitrary way of interacting with at least one other individual (like giving "high fives" to your friends). Pairs of white-faced capuchins sometimes perform unusual affiliative behaviors (Perry *et al.* 2003a, b), including hand sniffing, sucking body parts, and games that provide some kind of risk, such as inserting one's fingers into another individual's mouth or allowing another individual to repeatedly "bite" one's hand. During hand sniffing, one monkey inserts his/her fingers in the other's nose, or cups a hand/foot over the nose and mouth of another capuchin (see Chapter 11). During body sucking, one monkey sucks another individual's fingers, toes, ears or tail for prolonged periods of time (see Figure 13.6). These behaviors, often performed reciprocally, occur in a relaxed social context, such as during grooming sessions or play. These behaviors are not typical in all groups all the time. Rather, they emerge within a group, spread to a number of individuals, and then disappear. Extinction of these behaviors usually occurs when more active practitioners emigrate or die. Because these behaviors emerge independently in more than one group, minor differences in their execution may be present across groups.

Perry *et al.* (2003a, b) noted that those capuchins that spend time together during relaxed activities are more likely to exhibit these conventional behaviors. It is believed that capuchins derive pleasure from these activities (capuchins inhale deeply and have trance-like facial expressions during hand-sniffing). However, these activities also involve a certain amount of discomfort. The biological function of such social conventions among group members is difficult to establish but two hypotheses have been proposed. Perry *et al.* (2003a, b) propose that they serve (a) to test social bonds, because an individual is likely to find the behavior pleasurable only if it has a positive bond with its partner (see Zahavi 1977); or (b) as interaction rituals (see Collins 1981, 1993), in which individuals agree on a specific activity to perform and on playing their respective roles.

These behaviors remind us of circular reactions (*sensu* Piaget, Piaget 1952) because the behavioral schemata produced in order to receive pleasurable and interesting proprioceptive feedback are exhibited repeatedly. We know little of how these conventions originate and spread among group members. Perhaps, by chance an individual may have discovered the pleasure of sucking its tail and repeated it over time. But how did the behavior become shared with another? Perhaps when an individual is the recipient of this kind of behavior, the sensations are also pleasurable, and it is subsequently motivated to reproduce the sensations on its own body. To do so means taking on the active role; then one is prepared to take on the active role and direct these behaviors at another's body. For example, an individual has similar sensations in its mouth whether sucking its own tail or another's (although the feedback obtained from its tail differs in the two cases). Eventually body parts involved as effectors and receptors of the action belong to the individual or to its partner, in all possible combinations. This speculative scenario is drawn from observations of individual captive chimpanzees acquiring new social conventions (Tomasello and Call 1997). According to Tomasello *et al.* (1989), individuals acquire conventional gestures more easily when they themselves have participated in producing them (e.g., babies learn to raise their arms to be picked up). In contrast, when chimpanzees only see gestures exchanged between third parties there is no evidence that observers learn to make or respond to the gestures.

SUMMARY

In this chapter we examined the various ways in which social context biases capuchins' learning. Capuchins exhibit three characteristics that make them good candidates for social learning: they are relatively tolerant of each other, they are interested in each other and they are interested in food and other objects held by another. These characteristics are especially obvious in captivity. For example, while feeding, capuchin monkeys permit companions to remain close and to collect food or even to take food from them directly. Infants in particular are tolerated by others and allowed to take food. Infants are especially interested in others' novel foods, although they will eat novel foods before an adult has done so. Observing others eat motivates the observer to eat, but the observer monkey does not selectively or preferentially eat what the others are eating. This loosely coordinated pattern is sufficient for an individual to learn what is palatable through its own explorations in the right neighborhood (that is, where and when others are eating), but it does not suggest that capuchins directly recognize palatable foods from others' eating. They do, however, learn what to fear from watching others react fearfully to specific objects or events.

The picture is similar in the domain of problem-solving, where we have quite a bit of experimental data now. Watching another monkey solve a problem (usually related to obtaining food) promotes the observing monkey's interest in the places and objects where the other worked or is working. However, watching others is not sufficient for a capuchin to learn a sequence of actions or to produce specific relations between objects. Thus, as for feeding, the benefits to the observer in learning to solve the problem itself are indirect. Capuchins, as far as is now documented, do not learn by imitating other capuchins, nor do they replicate actions demonstrated to them by humans. It seems that perception of their own action during solution of a problem is fundamental for learning how to act with objects, and that social context can best support learning in these situations by promoting effective action. Overall, it is more accurate and more biologically meaningful to think of capuchin monkeys as learning *with* another rather than *from* another (Fragaszy and Perry 2003b).

The emerging evidence of social conventions and perhaps traditions in foraging actions will no doubt provoke continuing interest in this facet of capuchins' lives. Examination of these issues in wild monkeys takes dedication. We need longitudinal data on identified individuals documenting behavioral changes over time, the social context in which the pertinent behaviors are practiced, and the long-term social relationships that form the background for all of capuchins' activities. We are just beginning this task.

Epilogue
The (In)Complete capuchin

We have worked from many directions at putting together the puzzle that capuchins present to the life scientist. It is clear that we are farther along at understanding some dimensions of capuchins than others. Taxonomy is still a work in progress, and we have hardly touched genetic correlates of taxonomic entities. Evolutionary patterns and phylogenetic relations with other primates remain more or less a mystery. Functional aspects of neural organization, sensation, perception and skill development have been studied but there is still much left to learn about these topics. The unusual aspects of capuchins' social behavior, including the variations in sexual behavior in all the species and the social dynamics of groups, particularly those surrounding male take-overs in white-faced capuchins, have hardly been touched. The relations between capuchins and other species (plant and animal) have not been much studied, but clearly capuchins impact the communities in which they reside. And so on – there is much left wanting about the available information for nearly every topic that we have reviewed. Moreover, we have not covered some topics at all – for example, the burgeoning literature in biomedical and neuroscience fields in which capuchin monkeys represent either a convenient model for humans, or provide a useful comparative reference point.

One could say that we are asking for too much, wanting to know more about all these facets of one genus. On the other hand, how else are we to understand, truly and deeply understand, a genus? If we do not aim for this goal, we would be poor biologists. Besides, wanting to know more is an intrinsic part of being a scientist. Therefore we make no apologies. Instead, we hope that we have instilled in others a respect for these fascinating monkeys, and a desire to learn more about them.

To our colleagues who work with capuchins, we add an additional note. With a few exceptions, we included literature that appeared up through May, 2002. If we had continued to incorporate the latest findings past that time, we would never have finished the book! We want to express our sincerest thanks to our many generous colleagues who offered materials, and our apologies if we were not able to incorporate them from limitations of time or space.

Appendix I. Plant foods eaten by capuchins (*Cebus apella* [1], *C. albifrons* [2], *C. capucinus* [3], and/or *C. olivaceus* [4])

Family[a]	Genus	Species	Flowers	Fruits/Seeds	Leaves	Shoots	Other	Unknown/Unspecified
Acanthaceae	*Mendoncia*	*littoralis*		3				3
		sp.	1	1, 3			1	
Amaryllidaceae	*Hymenocallis*	*venezuelensis*				4	4	4
Anacardiaceae	*Anacardium*	*excelsum*		3			3	
	Astronium	*graveolens*		3	3			
		*occidental**		3				
	Mangifera	*indica*		3				
	Spondias	*mombin*	3, 4	1, 2, 3, 4	3		3	3, 4
		purpurea	3	3	3		3	
		radlkoferi		3	3			3
	Tapirira	*guianensis*		1				
Annonaceae	*Annona*	*acuminata*						3
		densicoma		1				
		glabra		3				
		hayesii						3
		jahnii		4				4
		muricata	3	3			3	
		purpurea	3	3			3	
		reticulata	3	3			3	
		sericea		1				
		spraguei		3				3
		sp.		1, 3				4
	Desmopsis	*panamensis*		3				
	Duguettia	*calycina*		1				
		insculpta		1				
		riparia		1				
		quitarensis		1, 2				

Family[a]	Genus	Species	Flowers	Fruits/Seeds	Leaves	Shoots	Other	Unknown/Unspecified
		surinamensis		1				
		sp.		1				
	Guatteria	*sp.*		1, 2				
		graciliipes		1				
		schomburgkiana		1				
	Malmea	*aff. lucida*		2				
		diclina		1				
	Oxandra	*acuminata*		1				
		asbeckii		1				
		espintana		1				
	Rollinia	*exsucca*						2
	Rollinea	*sericea*		1				
	Sapranthus	*palanga*		3				
	Unonopsis	*matthewsii*		2				
	Unknown	*sp.*		1				
	Xylopia	*aromatica*		1				
		ligustrifolia		1				
Apocynaceae	*Ambelania*	*acida*		1				
	Aspidospermum	*megalocarpon*		3				
	Bonafousia	*sananho*		1				
	Couma	*guianensis*		1				
	Forsteronia	*spicata*						3
	Geissospermum	*laeris**		1				
		sericeum		1				
	Lacmellea	*cf. aculeata*		1				
		panamensis		3				
	Odontadenia	*punticulosa*		3				
		sp.		1				
	Pacouria	*guianensis*		1				
	Parahancornia	*fasciculata*		1				
	Rauvolfia	*praecox*		1, 2				
		tetraphylla		3				
	Rhabdadenia	*biflora*	3					
	Stemmadenia	*donnell-smithii*		3				
		obovata	3	3				
	Tabernaemontana	*arborea*		3				
		chrysocarpa		3			3	
		undulata	1	1				

Family[a]	Genus	Species	Flowers	Fruits/Seeds	Leaves	Shoots	Other	Unknown/Unspecified
	Thevetia	*ahouai*		3				
		nitida		3				
Aquifoliaceae	*Ilex*	*sp.*		3				
Araceae	*Anthurium*	*crassinervium*		1				
		harrisii		1				
		kunthii		2				
	Aroid	*sp.*		1				
	Heteropsis	*flexuosa*		1				
		jenmani		1				
		oblongifolia		1			1	
	Monstera	*adansonii*		3				
		pertusa		3			1	
		sp.	1	1, 3				
	Philodendron	*acutatum*		1				
		appendiculattum		1				
		bipinnatifidum		1				
		corcovadense		1				
		crassinervium		1				
		eximium		1				
		insigne			1			
		linnaei			1			
		obliquifolium		1				
		*omequilaterum**		3				
		cf. peatum			1			
		sp.	1	3				3
	Syngonium	*angustatum*		3	3			
		podophyllum		1, 2				
		sp.	1					
	Unknown	*sp.*		1, 2	1, 2			
Araliaceae	*Dendropanax*	*arboreus*		3				3
	Sciadodendron	*excelsum*		3				
Arecaceae	*Acrocomia*	*vinifera*		3				
	Astrocaryum	*aculeatum*		1				
		chambira	1				1	
		farinosum	1	1				
		gynacanthum		1				
		munbaca		1				
		paramaca	1	1				
		sciophilum	1					

Family[a]	Genus	Species	Flowers	Fruits/Seeds	Leaves	Shoots	Other	Unknown/Unspecified
		sp.		1		1	1	
		standleyanum		3				3
	Attalea	*maripa*		1				2
	Bactris	*acanthocarpoides*		1				
		*balanoidea**				3		
		elegans		1				
		guineensis					3	
		major		1				
		minor		3				1
		sp		3		3		
	Cocos	*nucifera*		3		3		
	Copernicia	*tectorum*	4	4	4			4
	Desmoncus	*isthmius*		3		3		
	Elaeis	*guianensis*		3				
		oleifera		3				
	Euterpe	*edulis*		1				
		oleracea	1	1				
		precatoria		1		1	1	
		sp.	1	1, 2				
	Hyospathe	*sp.*						
	Iriartea	*exorrhiza*		1				
		sp.		1	1		1	
		ventricossa		1				
	Jessenia	*bataua*		1				
		polycarpa	1	1			1	
	Mauritia	*flexuosa*		2		2		2
	Maximiliana	*maripa*		1				1
	Oenocarpus	*bacaba*		1				
		mapora					1	
		panamanus		3				3
	Orbignya	*cohune*		3				
	Palm	*sp.*		1				
	Paurotis	*wrightii*	3	3		3		
	Roystonea	*oleracea*			1			2
	Scheelea	*attaleoides*						1
		magdalenica		3				
		rostrata		3				
		sp.		1, 2		1	1	
		zonensis		3				3

Family[a]	Genus	Species	Flowers	Fruits/Seeds	Leaves	Shoots	Other	Unknown/Unspecified
	Socratea	*elegans*					1	
		exorrhiza		1				
	Syagrus	*inajai*		1				
	Unknown	*sp.*		1	1		1	
Asclepiadaceae	*Asclepias*	*curassavica*		3				
	Marsdenia	*undulata*		4				4
	Matelea	*maritima*						4
Bignonaceae	*Adenocalymna*	*inundatum*		1				
	Amphilophium	*paniculatum*		1				
	Bignonia	*sp.*	1	3				
	Crescentia	*cujete*	3					
	Cydista	*aequinoctialis*	3					
	Jacaranda	*copaia*		1, 3				
	Macfadyena	*uncata*	4					4
	Pachyptera	*hymenaea*			3			
	Pithecoctenium	*crucigerum*		1, 3				
	Pleonotoma	*clematis*		1				
	Potamoganos	*microcalyx*		1				
	Pyrostegia	*sp.*				1		
	Schlegelia	*ramizii*		1				
	Tabebuia	*guayacan*		3				
		neochrysantha		3				
		ochracea	3	3	3			
		serratifolia		1				
		sp.	3					
	Unknown	*sp.*	3	1				
Bixaceae	*Cochlospermum*	*vitifolium*	3	3				4
	Xylosma	*flexuosum*		3				
Bombacaceae	*Bombacopsis*	*quinata*		3			3	
	Catostemma	*sp.*		1				
	Hampea	*appendiculata*		3				
	Ochroma	*limonensis*		3				
		pyramidale	3					
	Pachira	*cf. dolichocalyx*		1				
		insignis	1	1				
	Quararibea	*asterolepis*		3				3
		cordata		1, 2			1, 2	
		rhombifolia		1, 2				

Family[a]	Genus	Species	Flowers	Fruits/Seeds	Leaves	Shoots	Other	Unknown/Unspecified
		wittii		1, 2				
	Scleronema	*sp.*			1			
Boraginaceae	*Cordia*	*bicolor*		3				
		collococca		4				4
		fulva		1				
		lasiocalyx		3				
		lomatoloba		1				
		macrostachya		1				
		nitida		3				
		nodosa		1				
		panamensis		3				
		polycephala		4				4
		sp.		1				
	Tournefortia	*cuspidata*		1				
		hirsutissima		3				
Bromeliaceae	*Aechmea*	*distichantha*			1		1	
		melinonii			1			
		setigera	3					
	Ananas	*cf. nanus*		1				
	Bromelia	*chrysantha*	4	4	4			4
		karatas		3				
		pinguin		3	3	3		4
		plumieri		4				4
		sp.			3	3		
	Guzmania	*lingulata*			1			
	Streptocalyx	*longifolius*			1			
	Tillandesia	*caput-medusae*			3			
		circinnata					3	
		*distichanta**			1	1		
		ixioides			1	1		
		maxima			1	1		
		monadelpha			1			
		pulchella			1	1		
		sp.	1, 3					
	Vriesea	*pleiosticha*			1			
		tucumanensis			1	1		
Burseraceae	*Bursera*	*simaruba*	3	3			3	
		tomentosa		3				

Family[a]	Genus	Species	Flowers	Fruits/Seeds	Leaves	Shoots	Other	Unknown/Unspecified
	Crepidospermum	*rhoifolium*		1				
	Protium	*hostmanii*		1				
		neglectum		1				
		panamense		3				3
		sp.		1, 3				
	Tetragastris	*altissima*		1				
		panamensis		1, 3			1	
		sp.		1				
Cactaceae	*Acanthocereus*	*pentagonus*	3					
	Epiphyllum	*phyllanthus*		2				3
		sp.		1, 3				
	Hylocereus	*costaricensis*		3				
		polyrhizus	4	4	4			4
Capparaceae	*Capparis*	*baducca*		3				
		cocolobifolia	4	4				4
		frondosa		3				3
		odoratissima		4				4
	Crateva	*tapia*		3				
Caricaceae	*Carica*	*sp.*		1				
	Jacaratia	*digitata*		2				
		spinosa		1				
Caryocaraceae	*Caryocar*	*glabrum*		1				
Cecropiaceae	*Cecropia*	*glaziovi*		1				
		insignis		3				
		aff. Leucophae		1, 2				
		longipes		3				
		obtusaa		1				
		peltata	3	3				4
		sp.	1, 3	1, 3, 4				1
	Coussapoua	*angustifolia*		1				
		latifolia		1				
		sp.		1				
	Pourouma	*cecropiifolia*		2				
		guianensis		1, 3				
		melinonii		1				
		cf. minor		1				
		mollis		1				
Celastraceae	*Goupia*	*glabra*		1				

Family[a]	Genus	Species	Flowers	Fruits/Seeds	Leaves	Shoots	Other	Unknown/Unspecified
	Maytenus	*magnifolia*		2				
		cf. myrsinoides		1				
Chrysobalanaceae	*Chrysobalanus*	*icaco*		3				
	Couepia	*caryophylloides*		1				
	Hirtella	*racemosa*		3				
		sp.		1				
		triandra		3				3
	Licania	*apetala*		4				4
		arborea		3				
		discolor		1				
		heteromorpha		1				
		jimenezii		1				
		platypus		3				3
	Unknown	*sp.*					1	
Clusiaceae	*Calophyllum*	*brasiliense*		2, 3				
		longifolium		3				3
	Clusia	*grandiflora*		1				
		parviflora		1				
		scrobiculata		1				
		suborbicularis		3				
	Garcinia	*benthamiana*		1				
	Guttiferae	*sp.*		1				
	Moronobea	*coccinea*	1					
	Rheedia	*acuminata*		1, 2, 3				
		benthamiana		1				
		edulis		3				
		madruno		3				
	Symphonia	*globulifera*	1, 3					
	Tovomita	*stylosa*		3				
Combretaceae	*Combretum*	*assimile*					1, 2	
		farinosum	3					
		fruticosum	3, 4					4
		sp.		1, 2				
Commelinaceae	*Dichorisandra*	*hexandra*		3				
Compositae	*Mikania*	*tonduzii*	3					
Connaraceae	*Connarus*	*panamensis*						3
		perrottetii		1				
		sp.		1				

Family[a]	Genus	Species	Flowers	Fruits/Seeds	Leaves	Shoots	Other	Unknown/Unspecified
		turczinanowii		3				
		venezuelanus		4				4
	Rourea	*glabra*		3			3	
Convolvulaceae	*Dicranostyles*	*ampla*		1				
		cf. villosus		1				
	Maripa	*glabra*		1				
		panamensis		3				
		cf. peruviana		1				
Costaceae	*Costus*	*scaber*	1					
Cucurbitaceae	*Cayaponia*	*ophthalmica*		1				
		sp.		1, 2				
	Gurania	*suberosa*		3				
	Luffa	*operculata*		4				4
	Melothria	*pendula*		3				
		trilobata		4				4
	Psiguria	*triphylla*		1				
	Unknown	*sp.*		1				1
Cyclanthaceae	*Asplundia*	*sp.*	1	1				
	Stelestylis	*surinamensis*	1					
	Unknown	*sp.*		1			2	
Cyperaceae	*Scleria*	*secans*	3	3				
		setuloso-ciliata		4				4
Dilleniaceae	*Curatella*	*americana*		3				
	Davilla	*kunthii*					3	
	Doliocarpus	*dentatus*		3				
		major		3				3
		multiflorus		3				
		olivaceus						3
		sp.		3				
	Pinzona	*coriacea*		1				
	Tetracera	*volubilis*	4	3, 4				4
Ebenaceae	*Diospyros*	*ierensis*		4				4
		nicaraguensis		3				
		subrotata		1				
		sp.		1				
Elaeocarpaceae	*Muntingia*	*calabura*		3				
	Sloanea	*guianensis*		1, 2				
		cf. obtusifolia		1, 2				

Family[a]	Genus	Species	Flowers	Fruits/Seeds	Leaves	Shoots	Other	Unknown/Unspecified
		tenuiflora		3				
		terniflora		3				
		sp.			1			
Ericaceae	*Satyria*	*cerander*	1					
Erythroxylaceae	*Erythroxylum*	*havanense*		3				
		*piletarianum**		1				
Euphorbiaceae	*Amanoa*	*guianensis*		1				
	Croton	*tragioides*	3					
		sp.		3				
	Dalechampia	*scandens*		4				4
	Euphorbia	*schlechtendalii*	3					
	Glycydendron	*amazonicum*			1			
	Gymnanthes	*lucida*		3				
	Hura	*crepitans*		3				1, 3
	Hyeronima	*alchorneoides*		1				
		laxiflora		3				3
	Mabea	*fistulifera*		1			1	
		occidentalis		3				
	Margaritaria	*nobilis*		3, 4				3, 4
	Plukenetia	*polyadenia*					1	
	Pogonophora	*schoumburgkiana*		1				
	Richeria	*cf. racemosa*		1				
	Sapium	*aereum*		1				
		caudatum		3				3
		thelocarpum		3				3
	Sebastiana	*confusa*		3				
	Unknown	*sp.*		1, 3				4
Fagaceae	*Quercus*	*costaricensis*		3				
		oleoides		3				
Flacourtiaceae	*Banara*	*kuhlmanii*				1		
	Casearia	*aculeata*		3				
		arborea		3				3
		corymbosa		3				
		decandra						
		fasciculata						
		guianensis		3				
		javitensis				1	1	
		mariquiensis		1				

Family[a]	Genus	Species	Flowers	Fruits/Seeds	Leaves	Shoots	Other	Unknown/Unspecified
		ulmifolia		1				
		sp.	3	1, 3				
	Hecatostemon	*completus*		4				4
		guazumaifolius						4
	Hasseltia	*floribunda*		3				
	Laetia	*procera*		1				
		thamnia		3				
	Lindackeria	*laurina**		3				
	Mayna	*parviflora**		1				
	Prockia	*crucis*		3				
	Zuelania	*guidonia*		3				
Gesneriaceae	*Drymonia*	*serrulata*						3
Gramineae	*Bambusa*	*guadua*			1			
	Elytrostachys	*typica*			1			
	Gynerium	*sagittatum*					1	
	Lasiacis	*anomala*		4	4			4
		divaricata		3				
		sorghoidea						3
		sp.		3				
	Olyra	*latifolia*		3				
		sp.						4
	Panicum	*sp.*		1				
	Pharus	*glaber*			1			
	Phragmites	*australis*	3					
	Unknown	*sp.*		3	1	1		3
	Zea	*mays*		1, 2, 3				
Heliconiaceae	*Helioconia*	*imbricata*	3	3		3		
		sp.						1, 2, 3
Hippocrateaceae	*Salacia*	*cf. cordata*		1				
		impressifolia		3				
	Tontelea	*cylindrocarpa*		1				
		scandens		1				
Humiriaceae	*Humiria*	*balsamifera*		1				
Icacinaceae	*Calatola*	*venezuelana*		2				
	Dendrobangia	*boliviana*		1				
Lacistemataceae	*Lacistema*	*aggregattum*		3			3	
Laranthaceae	*Phoradendron*	*sp.*		4				
Lauraceae	*Aniba*	*riparia*		1			1	

Family[a]	Genus	Species	Flowers	Fruits/Seeds	Leaves	Shoots	Other	Unknown/Unspecified
	Beilschmiedia	*pendula*		3				
	Cryptocaria	*moschata*		1				
	Endlicheria	*paniculata*		1				
	Laurus	*latifolia*		3				
	Nectandra	*sp.*		1				
	Ocotea	*cernua*		3				
		veraguensis		3				
	Persea	*americana*		1				
	Phyllostemonodaphne	*geminiflora*		1				
	Unknown	*sp.*		1, 2				
Lecythidaceae	*Couratari*	*stellata*						1
	Eschweilera	*coriacea*		1				
		corrugata		1				
		simiorum		1				
		subglandulosa		1				1
	Gustavia	*augusta*	1	1				
		superba		3		3	3	3
	Lecythis	*corrugata*	1	1				
		pedicellata		1				
		pisonis		1				
		zabucajo		1				
Leguminosae	*Abarema*	*jupunba*		1				
	Acacia	*collinsii*	3	3			3	
		cornigera					3	
		sp.	3	3			3	
	Albiaia	*polycephala*		1				
	Albizia	*adinocephala*		3				
		caribaea			3			
		guachapele			4			4
	Anadenanthera	*colubrina*		1				
		peregrina		1				
	Andira	*coriacea*		1				
	Apuleia	*leiocarpa*	1			1		
		glabra		3				
	Bauhinia	*ungulata*	3					
	Cassia	*grandis*	3	3				4
		undulata		3				
	Centrosema	*pubescens*	4	4				4
	Cesalpinia	*bonduc*		1				

Family[a]	Genus	Species	Flowers	Fruits/Seeds	Leaves	Shoots	Other	Unknown/Unspecified
	Clitoria	*arborescens*	3	3				
		javitensis	1, 3					
	Copaifera	*epunctata*		4			4	
		guianensis		1				
		langsdorfii		1				
		officinalis		4				4
	Dioclea	*reflexa*		3				
		wilsonii						3
	Dipteryx	*panamensis*		3				
	Dussia	*discolor*		1				
	Entada	*polystachya*			4			4
	Enterolobium	*cyclocarpum*		3				
		mongollo				1		
	Eperua	*falcataa*	1	1				
	Erythrina	*costaricensis*						3
		panamensis		3				
	Gliricidia	*sepium*				3	3	
	Hymenaea	*altissima*		1				
		courbaril	3	1, 3, 4	3		3	4
	Inga	*acreana*		1				
		acrocephala		1				
		alba		1				
		auristellae		1				
		*bourgonii**		1			1	
		cf. capitata		1			1	
		cayennensis		1				
		cf. chartaceae		1				
		cinnamomea		1				
		cocleensis		3			3	
		edulis		1, 3			1	3
		fagifolia		3				3
		fanchoniana		1				
		fastuosa		1				
		goldmanii		3				
		gracilifolia		1				
		huberi		1				
		jenmani		1				
		lateriflora		1				
		laurina						2

Family[a]	Genus	Species	Flowers	Fruits/Seeds	Leaves	Shoots	Other	Unknown/Unspecified
		leiocalycina		1			1	
		marginata		1, 2, 3				
		mathewsiana		1, 2				
		melinonis		1				
		minutula		3				
		nobilis		1			1	
		paraensis		1				
		pavoniana		1				
		pezizifera		1, 3			1	
		punctata		1, 3				
		retinocarpa		1				
		rubiginosa		1			1	
		samialata		1				
		sapindoides		3				
		sertulifera						1
		splendens		1				
		sp.		1, 2, 3			1	2
		stipularis		1			1	
		*suwicu**						1
		thibaudiana		1			1	
		tonduzii	1	1			1	
		umbellifera		3				
		unknown		1				
		venosa						2
		vera		3			3	
		*veragauensis**		3				
	Lecointea	*peruviana*		2				
	Leucaena	*leucocephala*		3			3	
	Lonchocarpus	*costaricensis*	3					
		sp.		3				
	Machaerium	*moritzianum*			4			4
		setulosum					3	
	Macrolobium	*augustifolium*		1				
	Macuna	*sp.*	1	1				
	Mucuna	*puriens**		3				
	Pentaclethra	*macroloba*						2
	Phaseolis	*sp.*	3	3				
	Pithecellobium	*daulense*			4		4	4
		guaricense		4				4

Family[a]	Genus	Species	Flowers	Fruits/Seeds	Leaves	Shoots	Other	Unknown/Unspecified
		jupunba						2
		saman		3, 4				4
		tortum						4
	Prioria	*copaifera*		3				
	Senna	*alata*		1				
		multijuga		1				
	Stryphnodendron	*sp.*						1
	Swartzia	*benthamiana*		1				
		cubensis		3				
		flaemingii *var. psilonema*		1				
		myrtifolia		1				
	Tamarindus	*indica*		1				
	Zygia	*latifolia*		1				
		racemosa		1				
Loganiaceae	*Strychnos*	*asperula*		1				
		erichsonii		1				
		mitscherlichii		1				
		sp.		1				
		tomentosa		1				
Loranthaceae	*Phoradendron*	*quadrangulare*	3					
		sp.		3, 4				
	Psittacanthus	*sp.*		1				4
Malpighiaceae	*Banisteria*	*sp.*		1				
	Bunchosia	*biosellata*		3				
		cornifolia		3				
		sp.		3				
	Byrsonima	*coriacea*						2
		crassifolia		3			3	
	Malpighia	*emarginata*		4				4
Malvaceae	*Malvaviscus*	*arboreus*	3	3				
		sp.	3	3				
	Wissadula	*periplocifolia*		4				4
Marantaceae	*Calathea*	*sp.*	1	1			1	
	Ischnosiphon	*pruinosus*					3	
		puberulus					1	
	Maranta	*sp.*					1	
	Thalia	*geniculata*	4	4				

Family[a]	Genus	Species	Flowers	Fruits/Seeds	Leaves	Shoots	Other	Unknown/Unspecified
Marcgraviaceae	*Marcgravia*	*macrocarpa*		2				
		nepenthoides		3				
		polyantha		1				
	Norantea	*guianensis*	1					
		sp.		1				
	Souroubea	*guianensis*		1				
Melastomataceae	*Bellucia*	*axinanthera*		2				
	Clidemia	*septuplinervia*		3				
	Conostegia	*bracteata*		3				
		cinnamomea		3				
		xalapensis		3				
	Miconia	*affinis*		3				
		argentea		3				
		hondurensis						
		impetiolaris		2				
		parvifolia		3				
		prasina		3				
		pusilliflora		1				
		splendens		1				
		sp.		2				
	Mouriri	*huberi*		1				
		myrtilloides		3				
		sp.		1				
	Ossaea	*quinquenervia*		3				
	Topobaea	*praecox*						3
	Unknown	*sp.*		1, 2			1	
Meliaceae	*Carapa*	*guianensis*		1				
		procera		1				
	Cedrela	*odorata*		3			3	3
	Guarea	*gomma*		1				
		grandifolia		3				
		guara						2
		guidonia		1			1	
		kunthiana		1				
		sp.						1
	Trichilia	*cipo*		3				
		micrantha		1				
		pleeana		1				
		poeppigii		1				

Family[a]	Genus	Species	Flowers	Fruits/Seeds	Leaves	Shoots	Other	Unknown/Unspecified
		quadrijuga		1				
		sp.		2				
		trifolia		3, 4				4
		tuberculata						3
Menispermaceae	*Abuta*	*cf. bullata*		1				
		cf. obovata		1				
		cf. rufescens		1				
		grandifolia		1				
		selloana		1				
	Anomospermum	*cf. chloranthum*						
		grandifolium		1				
		reticulatum		1				
	Cissampelos	*tropaeolifolia*		3				
	Hyperbaena	*sp.*		1				
Monimiaceae	*Siparuna*	*ariane*		1				
		pauciflora		3				
Moraceae	*Artocarpus*	*altilis*	3	3				
	Brosimum	*alicastrum*		1, 2, 3				3
		bernadetteae		3				
		guianense		1				
		lactescens		1, 2				
		parinarioides		1				1
		sp.		1, 2				
	Castilla	*elastica*	3	3				
	Chlorophora	*tinctoria*		3, 4				4
	Clarisia	*racemosa*		1, 2				
	Coussapoa	*sp.*		2				
	Ficus	*cf. amazonica*		1, 2				
		broadwayi		1				
		cf. casapeinsis		1, 2				
		colubrinae		3				
		costaricana		3				
		dugandii		3				
		erythrosticta		1, 2				
		cf. expansa		1, 2				
		gommellera		1				
		guianensis						2
		insipida		1, 2, 3				3
		killipii		1, 2				

Family[a]	Genus	Species	Flowers	Fruits/Seeds	Leaves	Shoots	Other	Unknown/Unspecified
		*matthewsii**		1				
		maxima		1				
		obtusifola		3				3
		perforata		1, 2, 3				
		pertusa		2, 4				4
		popenoei		3				
		cf. regularis		1				
		sanguinosa		1				
		velutina					1	
		yoponensis		3				
	Helicostylis	*tomentosa*		1				
	Maclura	*tinctoria*		3				
	Maquira	*costaricana*		3				
		guianensis		1				
	Naucleopsis	*guianensis*		1				
	Olmedia	*asperuia*		3				
	Perebea	*guianensis*		1				
		mollis		1				
	Poulsenia	*armata*		3				
	Pseudolmedia	*laevis*		1				
	Sorocea	*cf. briquetii*						1, 2
		guimelliniana				1	1	
		sp.		1, 2, 3, 4	3			1, 4
		trigona		1, 2				
		trigonata		3, 4				4
	Trophis	*americana*						4
		racemosa		3				
Musaceae	*Musa*	*sapientum*		3				
		sp.	3	3		3	3	
		velutina		1				
Myristaceae	*Iryanthera*	*juruensis*		1			1	
		sp.						3
	Otoba	*parvifolia*		1				
		*bicuiba**		1				
	Virola	*gardneri*				1		
		melinonii		1			1	
		*michelii**		1				
		nobilis		3				
		panamensis		3				

Family[a]	Genus	Species	Flowers	Fruits/Seeds	Leaves	Shoots	Other	Unknown/Unspecified
		sebifera						3
		surinamensis		1			1	1, 3
Myrsinaceae	*Ardisia*	*revoluta*		3				
Myrtaceae	*Calycorectes*	*bergi*		1				
	Calyptranthes	*sp.*		1				
	Campomanesia	*neriiflora**		1				
		sp.		1				
	Eugenia	*aeruginea*		3				
		cf. acrensis		1				
		cf. punicifolia		1				
		choapamensis		3				
		cowannii		1				
		feijoi		1				
		nesiotica		3				
		oerstediana		3				
		patrisii		1				
		salamensis		3				
		sp.		1				
	Gomidesia	*crocea*		1				
	Marlieria	*sp.*		1				
	Myrcia	*amazonica*		1				
		ancepsis		1				
		pyrifolia		1				
	Psidium	*guajava*		3				
		sartorianum		3				
		sp.		3				
	Syzygium	*aqueum*	3	3				
Nyctaginaceae	*Andradeae*	*floribunda*					1	
	Guapira	*standleyana*		3				
	Neea	*amplifolia*		3				
		*chlorantha**		1				
		floribunda		1				
		psychotrioides		3				
	Pisonia	*macranthocarpa*						3
Ochnaceae	*Ouratea*	*guildingi*		4				4
		nitida		3				
Olacaceae	*Cathedra*	*acuminata*		1				
	Dulacia	*guianensis*		1				
	Heisteria	*silvianii*		1				

Family[a]	Genus	Species	Flowers	Fruits/Seeds	Leaves	Shoots	Other	Unknown/Unspecified
Orchidaceae	*Cattleya*	*skinneri*						3
	Oncidium	*carthaginense*			4			4
		cebolleta	4		4			4
Passifloraceae	*Dilkea*	*sp.*		1				
	Passiflora	*alata*		1				
		ambigua		3				
		cirrhiflora		1				
		coccinea	1	1				
		crenata		1				
		fuchsiiflora		1				
		garckei		1				
		laurifolia	1	1			1	
		*passiflora**		1				
		platyloba		3				
		*serratiflora**		3				
		serrulata		4				4
		trisetosa		3				
		vitifolia		3				
		sp.		1, 3				
	Unknown	*sp.*		2				
Phytolaccaceae	*Trichostigma*	*octandrum*		1, 2				
Piperaceae	*Piper*	*hostmannium*					1	
		marginatum					3	
		sp.		3	3			
Polygalaceae	*Bredemayera*	*sp.*		1			1	
	Moutabea	*guianensis*		1				
Polygonaceae	*Coccoloba*	*caracasana*		1, 4				4
		manzinellensis		3				
		*panimensis**		3				3
		uvifera		3				
		venosa		3				
		sp.		1				
Quiinaceae	*Lacunaria*	*crenata*		1				
	Quiina	*peruviana*		2				
Rhamnaceae	*Hovenia*	*dulcis*		3				
	Karwinskia	*calderonei*		3				
	Krugiodendron	*ferreum*		3				
	Rhamnidium	*sp.*		1				
	Zizyphus	*cinnamomum*		1				
		saeri						4

Family[a]	Genus	Species	Flowers	Fruits/Seeds	Leaves	Shoots	Other	Unknown/Unspecified
Rhizophoraceae	*Cassipourea*	*guianensis*		3				
	Rhizophora	*mangle*	3	3	3		3	
Rubiaceae	*Alibertia*	*edulis*		3				
	Amaioua	*guianensis*		1				
	Calycophyllum	*candidissimum*		3				
	Chiococca	*alba*		3				
	Chomelia	*spinosa*		3, 4				4
	Coussarea	*curvigemmia*		3				3
		impetiolaris		3				
		*voceurosa**		1				
	Duroia	*aquatica*		1				
		eriopila		1				
		longiflora		1				
		velutina		1			1	
	Faramea	*multiflora*		1				
		occidentalis		1, 3				
	Genipa	*americana*		2, 3, 4				4
		caruto		3				
	Guettarda	*deamii*		3				
		divaricata		4				4
		foliacea						3
		macrosperma		3				
	Hamelia	*patens*		3				
	Pentagonia	*macrophylla*	3					
		pubescens		3				
	Posoqueria	*latifolia*		3				
	Psychotria	*anceps*		1, 4				4
		grandis						3
		horizontalis		3				
		mapourioides		1				
		suterella		1				
		sp.		3				
		sp.2		1				
	Randia	*armata*		1, 3				3
		echinocarpa		3				
		hebecarpa		4				4
		subcordata		3				
		thurberi		3				
		venezuelensis		4				4

Family[a]	Genus	Species	Flowers	Fruits/Seeds	Leaves	Shoots	Other	Unknown/Unspecified
	Tocoyena	*guianensis*		1				
		pittieri		3				
	Unknown	*sp.*		1				
Rutaceae	*Citrus*	*sp.*		3			1	
	Zanthoxylum	*culantrillo*		4				4
		setulosum		3				
Sapindaceae	*Allophylus*	*cobbe*		4				4
		occidentalis		3				
		psilospermus		3				
		scrobiculatus		1				
	Cupania	*fulvida*		3				
		guatemalensis		3				
		latifolia		3				
		sylvatica		3				
	Dipterodendron	*costaricene*		3				
	Melicocca	*bijuga*						4
	Melicoccus	*bijugatus*		3				
	Paullinia	*acuminata*		1				
		cf. anodonta		1				
		capreolata		1				
		cururu		3, 4				4
		fibrigera		3				
		fuscescens		3				
		glomerulosa						3
		hystrix		1				
		pinnata		3				
		sp.		1, 2, 3			1	
	Talisia	*micrantha*		1				
		micrantha		1				
		microphylla		1				
		nervosa		3				
		sp.		1				
Sapotaceae	*Chrysophyllum*	*auratum*		1				
		cainito		3				
		lucentifolium		1				
		mexicanum		3				
		panamene		3				
		aff. pomiferum		1				

Family[a]	Genus	Species	Flowers	Fruits/Seeds	Leaves	Shoots	Other	Unknown/Unspecified
		sp.	1					
		sp.2		1, 3				
		viride		1				
	Manilkara	*bidentata*		1				1
		chicle	3	3				
		huberi		1				
		zapota	3	3				
		sp.		1				
	Mastichodendron	*capiri*		3				
	Micropholis	*cayennensis*	1	1				
		guynensis		1				
		melinoniana		1				
		obscura		1				
		venulosa		1				
		sp.		1				
	Pouteria	*cf. anibaifolia*		1				
		cf. boliviana		1				
		coriacea		1				
		cf. ephedrantha		2				
		guianensis		1				
		hispida		1				
		cf. ulei		2				
		cf. venulosa		1				
		melanopoda		1				
		*poulatta**		1				
		sagotiana		1				
		sapotea						3
		singularis		1				
		sp.		1, 3				
		torta		1				
		unilocularis		3				
	Prieurella	*cuneifolia**		1				
	Sapotaceae	*sp.*		1				
	Unknown			1				
Scitamineae	*Costus*	*laevis*	3					
		spicatus	3					
	Renealmia	*cernua*	3					
	Unknown	*sp.*		1				

Family[a]	Genus	Species	Flowers	Fruits/Seeds	Leaves	Shoots	Other	Unknown/Unspecified
Simaroubaceae	*Quassia*	*amara*	3	3				1
		cedron		1				
		glauca		3				
Solanaceae	*Cestrum*	*sp.*		3				
	Markea	*sp.*	1					
	Solanum	*antillarum*		3				
		hazenii		3				
		quitoense		3				
		sp.		1				
	Unknown	*sp.*	1					
Sterculiaceae	*Guazuma*	*tomentosa*		3, 4				4
		ulmifolia		3				
	Herrania	*kanukansis*		1				
	Sterculia	*apetala*		3, 4			3	4
		caribaea						2
		excelsa		1				
		frondosa		1				
		sp.		1				
	Theobroma	*cacao*		1, 2				
		subincana		1				
		velutinum		1				
Strelitzaceae	*Phenakospermum*	*guyanense*	1	1			1	
		sp.						1
Theophrastaceae	*Clavija*	*tarapotana*		2				
	Jacquinia	*nervosa*		3				
		pungens		3				
Tiliaceae	*Apeiba*	*aspera*		3				
		echinata		1			1	
		glabra		1				
		membranacea		1, 3				
		tibourbou		1, 3				
	Hasseltia	*floribunda*		3				
	Luehea	*candida*	3	3				
		seemannii						3
		speciosa	3	3			3	
		sp.			3			
	Muntingia	*calabura*		3				
	Ochroma	*limonensis*		3				
		pyramidale	3					

Family[a]	Genus	Species	Flowers	Fruits/Seeds	Leaves	Shoots	Other	Unknown/Unspecified
Ulmaceae	*Ampelocera*	*sp.*		2				
	Celtis	*iguanea*		1, 2				
	Trema	*micrantha*		3				
Ulticaceae	*Bagassa*	*guianensis*		1				
Urticaceae	*Urera*	*caracasana*		1, 2				
		eggersii		1				
Verbenaceae	*Aegiphila*	*elata*		3				
	Avicennia	*germinans*		3				
	Lantana	*camara*		3				
	Petrea	*aspera*		1				
	Vitex	*capitata*		4				4
		compressa		1, 4				4
		orinocensis		4				4
		triflora		1				
Violaceae	*Hybanthus*	*anomalus*		3				
		prunifolius		3				
	Leonia	*glycycarpa*		1, 2				
	Rinorea	*riana*		1				
		squamata					3	
Vitaceae	*Cissus*	*alata*		4				4
		rhombifolia						3
		sicyoides		2, 4				3, 4
		ulmifolia		1, 2				
		sp.		3				

Sources:

Cebus apella [1]: S. Boinski (pers. comm.), Brown & Zunino (1990), Freese & Oppenheimer (1981), Izar (1999), Izawa (1979), Rimoli (2001), Terborgh (1983), Zhang (pers. comm.)

C. albifrons [2]: Freese (1977), Freese & Oppenheimer (1981), K. Phillips (pers. comm.), Terborgh (1983)

C. capucinus [3]: Baker (1998), Buckley (1983), Chapman & Fedigan (1990), Fedigan (unpublished), Freese & Oppenheimer (1981), Gilbert, Stouffer and Stiles (unpublished), Mitchell (1979), Phillips (1994),

C. olivaceus [4]: Freese & Oppenheimer (1981), Robinson (1986), Miller (1997)

[a]Genera noted in original sources as belonging to more than one family (e.g., Mimosaceae and Leguminosae) have been pooled into one family in this table.

Note: Taxonomy is according to The Plant Names Project (2003). International Plant Names Index. Published on the Internet; http://www.ipni.org [accessed March 12, 2004]. Species that could not be identified in this index are marked with an*.

Appendix II. Sites of long-term field observations of capuchin monkeys listed in the bibliography

Country:	*Location*:	*Sources*:
Argentina	Iguazú National Park	Brown, Di Bitetti, Janson
Brazil	Caratinga Biological Station (now Reserva Particular do Patrimônio Natural (RPPN) Feliciano Miguel Abdala)	Lynch, Rimoli
	Carlos Botelho State Park, São Paulo	Izar
Colombia	Macarena Biological Station, Tinigua National Park	Izawa
Costa Rica	Corcovado National Park	Boinski
	Curu Wildlife Reserve	Baker
	La Selva Biological Station	Campbell
	La Suerte Biological Field Station	Garber
	Lomas Barbudal Biological Reserve	Gros-Louis, Perry, Manson
	Santa Rosa National Park	Chapman, Fedigan, Jack, MacKinnon, Rose
French Guyana	Nouragues Field Station for the Study of Tropical Forest	Zhang
Honduras	Trujillo – now Laguna Guaimoreto Wildlife Reserve	Buckley
Nicaragua	Ometepe Biological Station	Garber
Panama	Smithsonian Tropical Research Institute, Barro Colorado Island	Freese, Mitchell, Oppenheimer, Phillips
Peru	Cocha Cashu Biological Station, Manu National Park	Janson, Terborgh
Suriname	Raleighvallen Nature Reserve	Boinski, Fleagle, Mittermeier, van Roosmalen
Trinidad	Bush Bush Wildlife Reserve	Phillips
Venezuela	Hato Masaguaral	de Ruiter, Fragaszy, O'Brien, Robinson, Valderrama
	Hato Piñero	Miller

Appendix III. Blood biochemistry, hematological and serum protein parameters for males and females (a) and juveniles and adults (b)

(a)

	Females			Males			
Blood parameter	N[a]	Mean ± S.D.	Range	N	Mean ± S.D.	Range	U-test
Calcium	20	9.18±0.61	7.95–10.7	16	9.02±0.37	8.55–10	N.S.[b]
AST	20	48.18±10.77	23–66.5	16	42.22±8.78	29.5–58	p<0.05
ALT	20	36.73±18.78	17–108	16	34.38±7.97	24–48	N.S.
GGT	20	95.53±26.05	63–160	16	71.06±14.32	34–94.5	p<0.01
Alkaline phosphatase	20	135.65±108.46	43.5–410	16	145.22±61.12	39–244	N.S.
Phosphorus	20	3.43±0.77	2.3–5.0	16	2.95±0.83	1.9–4.9	p=0.05
Glucose	20	98.5±41.97	35–192	16	76.22±29.09	39–164	p=0.05
Total protein	20	7.30±0.55	6.3–8.35	16	7.25±0.5	6.6–8.6	N.S.
Urea nitrogen	20	16.38±5.23	8.5–31	16	13.44±4.82	7.0–26	p<0.05
Creatinine	20	0.61±0.12	0.45–0.94	16	0.71±0.13	0.41–0.91	p<0.01
CPK	20	377.5±224.2	131–964	16	364.88±276.69	115–1009	N.S.
Total bilirubin	20	052±0.05	0.5–0.7	16	0.51±0.05	0.5–0.7	N.S.
Cholesterol	20	162.13±53.55	82–272	16	136±26.44	98–184	N.S.
Triglycerides	20	92.7±41.18	48–221	16	89.53±26.41	63.5–176	N.S.
Magnesium	20	2.2±0.27	1.63–2.63	16	2.15±0.34	1.33–2.49	N.S.
LDH	20	284.85±126.17	91–677	16	263.19±63.94	183.5–414	N.S.
Uric acid	20	3.39±1.24	0.84–5.88	16	3.46±0.88	1.79–4.73	N.S.
Amylase*	17	132.09±48.18	53–207	13	138.92±56.77	88–254	N.S.
Albumin*	17	4.54±0.23	4.2–5	13	4.57±0.44	4.1–5.57	N.S.
Lipase*	17	81.88±30.79	51–166	13	81.58±16.05	52–110	N.S.
Sodium*	17	146.79±2.17	142–151	13	147.15±3.57	139–152.5	N.S.
Potassium*	17	4.13±0.5	3.45–5.1	13	4.43±0.57	3.6–5.5	N.S.

(cont.)

Blood parameter	Females			Males			
	N[a]	Mean ± S.D.	Range	N	Mean ± S.D.	Range	U-test
Hematological parameter							
Erythrocytes	20	5.43±0.62	3.76–6.31	16	6.27±0.44	5.63–7.31	p<0.01
Leukocytes	20	7.21±2.09	3.3–12	16	7.72±1.84	4.3–12.15	N.S.
Hemoglobin	20	13.39±1.28	9.6–15.25	16	15.01±0.78	13.8–16.2	p<0.01
Hematocrit	20	40.94±5.72	27.6–51.75	16	45.27±3.60	37.65–52.7	p<0.01
Neutrophils	20	59.1±9.65	41.5–79	16	63.16±5.98	51–72	N.S.
Basophils	20	0	0	16	0	0	N.S.
Eosinophils	20	0.03±0.11	0–0.5	16	0	0	N.S.
Lymphocytes	20	41.63±10.64	21–63	16	36.78±5.96	28.49	N.S.
Monocytes	20	0	0	16	0	0	N.S.
*MCV**	*17*	70.41±2.3	67–76	13	69.53±1.85	66–72.5	N.S.
*MCH**	*17*	24.63±1.23	22–28	13	24.54±0.80	23–26	N.S.
*MCHC**	*17*	34.8±2.15	28–36.67	13	35.03±1.51	31.67–38	N.S.
*Platelet**	*17*	388.44±65.29	264.33–491.5	13	388.27±68.75	299–522	N.S.
Serum protein parameter							
Albumin	20	63.92±3.49	58.1–71.3	16	63.85±4.91	54.1–69.7	N.S.
α 1	20	2.15±0.45	1.5–2.9	16	2.0±0.53	1.35–3.3	N.S.
α 2	20	5.7±2.65	2.6–10.7	16	5.65±2.05	2.9–9.7	N.S.
β	20	14.58±3.45	8.2–22.1	16	15.27±3.16	9.45–22	N.S.
γ	20	13.66±3.47	7.6–20.4	16	13.23±2.83	9.95–18.6	N.S.
A/G Ratio*	17	1.57±0.15	1.28–1.83	13	1.55±0.25	1.09–1.99	N.S.

(b)

	Juveniles			Adults			
Blood parameter	N	Mean ± S.D.	Range	N	Mean ± S.D.	Range	U-test
Calcium	9	9.52±0.6	8.9–10.7	27	8.65±0.54	7.95–10.1	p<0.05
AST	9	51.83±5.01	47–63	27	43.43±10.74	23–66.5	p<0.05
ALT	9	34.89±7.44	21.5–48	27	35.94±16.7	17–108	N.S.
GGT	9	77.56±9.74	63–91	27	87.02±27.68	34–160	N.S.
Alkaline phosphatase	9	242.61±99.93	79–410	27	105.67±53.31	39–244	p<0.01
Phosphorus	9	4.11±0.85	2.95–5.0	27	2.92±0.56	1.9–4.3	p<0.01
Glucose	9	125±43.52	73–192	27	76.46±27.40	35–174	p<0.01
Total protein	9	7.16±0.47	6.3–7.65	27	7.31±0.55	6.45–8.6	N.S.
Urea nitrogen	9	15.94±6.71	8.0–31	27	14.78±4.71	7.0–26	N.S.
Creatinine	9	0.64±0.13	0.49–0.94	27	0.66±0.14	0.41–0.91	N.S.
CPK	9	445.72±231.49	185–898	27	347.28±248.9	115–1009	N.S.
Total bilirubin	9	0.51±0.04	0.5–0.62	27	0.51±0.05	0.5–0.7	N.S.
Cholesterol	9	166.89±50.56	107.5–253.5	27	145.06±42.68	82.5–272	N.S.
Triglycerides	9	98.06±31.65	67–172.5	27	89.04±36.27	48–221	N.S.
Magnesium	9	2.16±0.37	1.63–2.53	27	2.18±0.29	1.33–2.62	N.S.
LDH	9	294.22±57.68	211.5–379	27	268.89±113.85	91–677	N.S.
Uric acid	9	4.05±1.09	2.89–5.88	27	3.21±1.01	0.84–4.7	N.S.
Amylase*	9	137.22±44.45	79.5–203.5	21	134.12±54.90	53–254	N.S.
Albumin*	9	4.52±0.15	4.3–4.75	21	4.57±0.39	4.1–5.57	N.S.
Lipase*	9	76.28±15.13	52–96	21	84.10±28.33	51–166	N.S.
Sodium*	9	146.33±2.78	142–149.5	21	147.21±2.84	139–152.5	N.S.
Potassium*	9	4.24±0.74	3.45–5.5	21	4.27±0.46	3.55–5.1	N.S.
Erythrocytes	9	5.96±0.38	5.41–6.57	27	5.76±0.76	3.76–7.31	N.S.
Leukocytes	9	6.58±1.51	4.3–8.75	27	7.72±2.05	3.3–12.15	N.S.
Hemoglobin	9	14.31±0.69	13.3–15.2	27	14.04±1.51	9.6–16.2	N.S.

(*cont.*)

	Juveniles			Adults			
Blood parameter	N	Mean ± S.D.	Range	N	Mean ± S.D.	Range	U-test
Hematocrit	9	44.37±4.98	37.65–51.75	27	41.02±9.06	4.55–52.7	N.S.
Neutrophils	9	56.39±5.46	46.5–63.5	27	62.41±8.7	41.5–79	p<0.05
Basophils	9	0	0	27	0	0	N.S.
Eosinophils	9	0	0	27	0	0	N.S.
Lymphocytes	9	44.1±5.71	36.5–53	27	37.93±9.55	21–63	p<0.05
Monocytes	9	0	0	27	0	0	N.S.
*MCV**	9	68.87±1.65	66–71.5	21	70.52±2.15	67–76	N.S.
*MCH**	9	24.89±0.58	24–26	21	24.62±1.21	22–28	N.S.
*MCHC**	9	36.26±0.70	35.67–38	21	34.75±1.37	31.5–37	N.S.
*Platelet**	9	415±61.02	324.33–522	21	376.95±65.58	264.33–492.33	N.S.
Serum protein parameter							
Albumin	9	66.91±3.56	59.45–71.3	27	62.88±3.84	54.1–69.25	p<0.01
α 1	9	2.32±0.39	1.8–2.8	27	2.0±0.5	1.35–3.3	p<0.05
α 2	9	4.26±2.04	2.9–9.6	27	6.15±2.31	2.6–10.7	p<0.05
β	9	15.95±1.92	12.2–19.05	27	14.53±3.6	8.2–22.1	N.S.
γ	9	10.56±2.41	7.6–15.65	27	14.44±2.79	9.95–20.4	p<0.01
A/G Ratio*	9	1.69±0.15	1.5–1.99	21	1.51±0.2	1.09–1.83	N.S.

Values were registered during two annual check-ups. From Riviello and Wirz 2001, except parameters marked with asterisks refer to unpublished values.

[a] N, number of individuals.

[b] N.S., not significant.

Appendix IV. Literature resources related to management of captive capuchins[1]

Animal Welfare Information Center, United States Department of Agriculture (January, 22, 1991). Appendix 1 in *Nonhuman Primate Management Plan*. Retrieved June 01, 2002 from http://oacu.od.nih.gov/regs/primate/primex.htm **H**

(January, 22, 1991). Chapter 3 in *Nonhuman primate management plan*. Retrieved June 01, 2002 from http://oacu.od.nih.gov/regs/primate/primex.htm **J**

Baer, J. F. 1998. A veterinary perspective of potential risk factors in environmental enrichment. In *Second Nature: Environmental Enrichment for Captive Animals*, ed. D. J. Shepherdson, J. D. Mellen and M. Hutchins, pp. 277–302. Washington, DC: Smithsonian Institution Press. **A, D, J**

Bayne, K. and Novak, M. 1995. Behavioral disorders. In *Nonhuman Primates in Biomedical Research: Biology and Management*, ed. B. T. Bennett, C. R. Abee and R. Henrickson, pp. 485–501. New York: Academic Press. **B, F, G**

Cicmanec, J. L., Hernandez, D. M., Jenkins, S. R., Campbell, A. K. and Smith, J. A. 1979. Hand-rearing infant callitrichidae (*Saguinus* spp. and *Callithrix jacchus*), owl monkeys (*Aotus trivirgatus*), and capuchins (*Cebus albifrons*). In *Nursery Care of Nonhuman Primates*, ed. G. C. Ruppenthal, pp. 307–313. New York: Plenum Press. **F**

Coates, M. 1999. Nutrition and feeding. In *The UFAW Handbook on the Care and Management of Laboratory Animals, 7th Edition. Volume 1, Terrestrial vertebrates*, ed. T. Poole, pp. 45–61. London: Blackwell Science. **D**

Coe, C. L. 1991. Is social housing of primates always the optimal choice? In *Through the Looking Glass*, ed. M. A. Novak and A. J. Petto, pp. 78–93. Washington, DC: American Psychological Association. **B**

Cooper, M., Bernstein, I., Fragaszy, D. and de Waal, F. 2001. Integration of new males into four social groups of tufted capuchins (*Cebus apella*). *International Journal of Primatology* **22**, 663–683. **C**

Crockett, C. M. 1998. Psychological well-being of captive nonhuman primates: Lessons from laboratory studies. In *Second Nature: Environmental Enrichment for Captive Animals*, ed. D. J. Shepherdson, J. D. Mellen and M. Hutchins, pp. 129–153. Washington, DC: Smithsonian Institution Press. **B, G**

Dettmer, E. and Fragaszy, D. 2000. Determining the value of social companionship to captive tufted capuchin monkeys (*Cebus apella*). *Journal of Applied Welfare Science* **3**, 293–304. **B**

Fortman, J. D., Hewett, T. A. and Bennett, B. T. 2002. *The Laboratory Nonhuman Primate*. Boca Raton, FL: CRC Press. **A, D, E, G, L, M, N**

Fragaszy, D. M., Baer, J. and Adams-Curtis, L. 1994. Introduction and integration of stangers into social groups of tufted capuchins. *International Journal of Primatology* **15**, 399–420. **C**

Laule, G. M. 1999. Training laboratory animals. In *The UFAW Handbook on the Care and Management of Laboratory Animals, 7th Edition. Volume 1, Terrestrial Vertebrates*, ed. T. Poole, pp. 21–28. London: Blackwell Science. **K**

Laule, G. and Desmond, T. 1998. Postive reinforcement training as an enrichment strategy. In *Second Nature: Environmental Enrichment for Captive Animals*, ed. D. J. Shepherdson, J. D. Mellen and M. Hutchins, pp. 302–314. Washington, DC: Smithsonian Institution Press. **K**

Mason, W. A. 1991. Effects of social interaction on well-being: Developmental aspects. *Lab Animal Science* **41**, 323–328. **B**

McKenzie, S. M., Chamove, A. S., and Feistner, A. T. C. 1986. Floor-coverings and hanging screens alter arboreal monkey behavior. *Zoo Biology* **5**, 339–348. **A, G**

Mineka, S. and Kihlstrom, J. 1978. Unpredictable and uncontrollable events. *Journal of Abnormal Psychology* **87**, 256–271. **J**

National Research Council 1998. New world monkeys: Cebids. In *The Psychological Well-being of Nonhuman Primates*, pp. 80–90. Washington, DC: National Academy Press. **A, D, E, G**

Reinhardt, V. 1997. Training nonhuman primates to cooperate during blood collections: A critical review. *Animal Welfare* **4**, 221–238. **K**

1997. Training nonhuman primates to cooperate during handling procedures: A review. *Animal Technology* **48**, 55–73. **K**

University of Wisconsin, Madison, Primate Info Net, Wisconsin Primate Research Center (February 15, 2002). Capuchin monkey basic information and diet. Retrieved June 01, 2002 from the University of Wisconsin Website: http://www.primate.wisc.edu/pin/capuchin.html **D**

Visalberghi, E. and Anderson, J. R. 1999. Capuchin monkeys. In *The UFAW Handbook on the Care and Management of Laboratory Animals, 7th Edition. Volume1: Terrestrial Vertebrates*, ed. T. Poole, pp. 601–611. London: Blackwell Science. **A, C, E, F, G, L, M**

Widowski, T. 1990. The evaluation and promotion of well-being in farm animals and laboratory primates: Common problems in contemporary animal care. In *Well-Being of Nonhuman Primates in Research*, ed. J. A. Mench and D. P. L. Krulisch, pp. 19–26. Bethesda, MD: Scientists Center for Animal Welfare. **G**

Wolfe, T. L. 1985. Laboratory animal technicians: Their role in stress reduction and human-animal bonding. *Vet Clinics North American Small Animal Practice* **15**, 449–454. **I, J**

1991. Psychological well-being: The billion-dollar solution. In *Through the Looking Glass*, ed. M. A. Novak and A. J. Petto, pp. 125–128. Washington, DC: American Psychological Association. **G**

ENDNOTE

1 **Subject Codes**: **A** – Physical Environment, **B** – Social Environment, **C** – Introduction/Integration, **D** – Nutrition, **E** – Reproduction, **F** – Infant Care, **G** – Well-being, **H** – Abnormal Behavior, **I** – Personnel, **J** – Stress, **K** – Training and Handling, **L** – Health and Disease, **M** – Record Keeping, **N** – Risks and Hazards.

References

Abbott, D. H., Barrett, J. and George, L. 1993. Comparative aspects of the social suppression of reproduction in female marmosets and tamarins. In *Marmosets and Tamarins: Systematics, Behaviour and Ecology*, ed. A. B. Rylands, pp. 151–163. Oxford: Oxford University Press.

Aboitiz, F. 1996. Does bigger mean better? Evolutionary determinants of brain size and structure. *Brain, Behaviour, and Evolution* **47**, 225–245.

Adams-Curtis, L. 1988. Behavior by capuchin monkeys (*Cebus apella*) toward a sequential puzzle. Master's thesis, Washington State University.

1990. Conceptual learning in capuchin monkeys. *Folia Primatologica* **54**, 129–137.

Adams-Curtis, L. E. and Fragaszy, D. M. 1994. Development of manipulation in capuchin monkeys during the first six months. *Developmental Psychobiology* **27**, 123–136.

1995. Influence of a skilled model on the behavior of conspecific observers in tufted capuchin monkeys (*Cebus apella*). *American Journal of Primatology* **37**, 65–71.

Adams-Curtis, L. E., Fragaszy, D. and England, N. 2001. Prehension in infant capuchins (*Cebus apella*) from six weeks to twenty-four weeks: video analysis of form and symmetry. *American Journal of Primatology* **52**, 55–60.

Addessi, E. 2002. Il ruolo delle influenze sociali sul comportamento alimentare nel cebo dai cornetti (*Cebus apella*). Dissertation, Università degli Studi di Roma "La Sapienza".

Addessi, E. and Visalberghi, E. 2001. Social facilitation of eating novel food in tufted capuchin monkeys (*Cebus apella*): input provided by group members and responses affected in the observer. *Animal Cognition* **4**, 297–303.

Adolph, K. E., Eppler, M. A. and Gibson, E. J. 1993. Development of perception of affordances. In *Advances in Infancy Research (Vol. 8)*, ed. C. Rovee-Collier and L. P. Lipsitt, pp. 51–98. Norwood, NJ: ABLEX Publishing Corporation.

Agoramoorthy, G. and Hsu, M. J. 1995. Population status and conservation of red howling monkeys and white-fronted capuchin monkeys in Trinidad. *Folia Primatologica* **64**, 158–162.

Aiello, L. C. and Wheeler, P. 1995. The expensive-tissue hypothesis: the brain and the digestive system in human and primate evolution. *Current Anthropology* **36**, 199–221.

Amaral, J. M. J. and De Yong, D. 1999. O comportamento de macaco prego (*Cebus apella*) em uma mata urbana de Riberão Preto, SP. *Livro de Resumos of the IX Congresso Brasileiro de Primatologia. Sociedade Brasileira de Primatologia 25–30 July 1999, Santa Teresa (ES) Brasil*, pp. 30–31.

Anapol, F. and Lee, S. 1994. Morphological adaptation to diet in playrrhine primates. *American Journal of Physical Anthropology* **94**, 239–261.

Anderson, J. R. 1990. Use of objects as hammers to open nuts by capuchin monkeys (*Cebus apella*). *Folia Primatologica* **54**, 138–145.

Anderson, J. 1994. The monkey in the mirror: a strange conspecifics. In *Self-Awareness in Animals and Humans: Developmental Perspectives*, ed. S. T. Parker, R. W. Mitchell and M. L. Boccia, pp. 315–329. Cambridge: Cambridge University Press.

1996. Chimpanzees and capuchin monkeys: comparative cognition. In *Reaching into Thought. The Minds of the Great Apes*, ed. A. Russon, K. Bard and S. Parker, pp. 23–55. Cambridge: Cambridge University Press.

1998. Sleep, sleeping sites, and sleep-related activities: awakening to their significance. *American Journal of Primatology* **46**, 63–75.

2001. Self and others in nonhuman primates: a question of perspective. *Psychologia* **44**, 3–16.

2002. Tool-use, manipulation and cognition in capuchin monkeys (*Cebus*). *New Perspectives in Primate Evolution and Behaviour*, ed. C. Harcourt, pp. 127–146. Otley, UK: Westbury Publishing.

Anderson, J. R. and Henneman, M. C. 1994. Solutions to a tool-use problem in a pair of *Cebus apella*. *Mammalia* **58**, 351–361.

Anderson J. R. and Visalberghi, E. 1991. Capacités cognitives des primates non humain: implications pour l'élevage en captivité. *Sciences et Techniques de l'Animal de Laboratoire* **1**, 163–171.

Anderson, J. R., Degiorgio, C., Lamarque, C. and Fagot, J. 1996. A multi-task assessment of hand lateralization in capuchin monkey (*Cebus apella*). *Primates* **37**, 97–103.

Anderson, S. 1997. Mammals of Bolivia – taxonomy and distribution. *Bulletin of the American Museum of Natural History* **231**, 1–652.

Andrade de Costa, B. L. S. and Hokoc, J. N. 1995. Photoreceptor mosaic and estimate of visual acuity in the cebus monkey. *Society Neuroscience Abstracts* **21**, pt. 2, 1175.

Ankel, F. 1965. Der Canalis sacralis als Indikator für die länge der caudalregion der Primaten. *Folia Primatologica* **3**, 263–276.

1972. Vertebral morphology of fossil and extant primates. In *The Functional and Evolutionary Biology of Primates*, ed. R. Tuttle, pp. 223–240. Chicago: Aldine Atherton.

Antinucci, F. 1989. *Cognitive Structure and Development in Nonhuman Primates*. Hillsdale, NJ: Lawrence Erlbaum Associates.

Antinucci, F. and Visalberghi, E. 1986. Tool use in *Cebus apella*: a case study. *International Journal of Primatology* **7**, 351–363.

Aquino, R. and Encarnación, F. 1994. Primates of Peru/Los Primates del Perú. *Primate Report* **40**, 1–127.

Araujo Santos, F. G., Bicca-Marques, J. C., Calegaro-Marques, C., de Farias, E. M. P. and Azevedo, M. A. O. 1995. On the occurrence of parasites in free-ranging callitrichids. *Neotropical Primates* **3**, 46–47.

Armstrong, E. and Falk, D. 1982. *Primate Brain Evolution: Methods and Concepts*, pp. xiii, 332. New York: Plenum Press.

Armstrong, E. and Shea, M. A. 1997. Brains of New World and Old World monkeys. *New World Primates: Ecology, Evolution, and Behavior*, ed. W. G. Kinzey, pp. 25–44. New York: Aldine de Gruyter.

Aureli, F. and de Waal, F. B. M. 2000. *Natural Conflict Resolution*. Berkeley: University of California Press.

Ausman, L. M., Powell, E. M. Mercado, D. L., Samonds, K. W., el Lozy, M. and Gallina, D. L. 1982. Growth and developmental body composition of the cebus monkey (*Cebus albifrons*). *American Journal of Primatology* **3**, 211–227.

Austad, S. N. and Fisher, K. E. 1992. Primate longevity: its place in the mammalian scheme. *American Journal of Primatology* **28**, 251–261.

Bachevalier, J. 1990. Ontogenetic development of habit and memory formation in primates. In *The Development and Neural Basis of Higher Cognitive Functions*, ed. A. Diamond, pp. 1–19. New York: New York Academy of Sciences.

Baddely, A. 1995. Working memory. In *The Cognitive Neurosciences*, ed. M. Gazzaniga, pp. 755–764. Cambridge, MA: MIT Press.

Baker, M. 1992. Capuchin monkeys (*Cebus capucinus*) and the ancient Maya. *Ancient Mesoamerica* **3**, 219–228.

1996. Fur rubbing: use of medicinal plants by capuchin monkeys (*Cebus capucinus*). *American Journal of Primatology* **38**, 263–270.

1998. Fur rubbing as evidence for medicinal plant use by capuchin monkeys (*Cebus capucinus*): ecological, social and cognitive aspects of the behavior. Doctoral Dissertation, University of California, Riverside.

Baldwin, J. D. and Baldwin, J. I. 1972. The ecology and behavior of squirrel monkeys (*Saimiri oerstedii*) in a natural forest in western Panama. *Folia Primatologica* **18**, 161–184.

Baldwin, J. D. and Baldwin, J. I. 1976. Primate populations in Chiriqui, Panama. In *Neotropical Primates: Field Studies and Conservation*, ed. R. W. Thorington and P. G. Heltne, pp. 20–31. Washington, DC: National Academy of Sciences.

Baldwin, J. D. and Baldwin, J. I. 1981. The squirrel monkeys, genus *Saimiri*. In *Ecology and Behavior of Neotropical Primates (Vol. 1)*, ed. A. F. Coimbra-Filho and R. A. Mittermeier, pp. 277–330. Rio de Janeiro: Academia Brasileira de Ciercias.

Balestra, R. and Bastos, R. P. 1999. Interaçõoes sociais entre macacos pregos (*Cebus apella*) em área sob influencia antrópica. *Livros de Resumos, IX Congresso Brasileiro de Primatologia*, p. 44.

Ballard, D. H., Hahoe, M. M., Pook, P. K. and Rao, R. P. N. 1997. Deictic codes for the embodiment of cognition. *Behavioral and Brain Sciences* **20**, 723–767.

Barnard, P. 1960. *Monkey in the House*. London: Cassell & Company.

Barrett, J. C. and Marshall, J. 1969. The risk of conception on different days of the menstrual cycle. *Population Studies* **23**, 455–461.

Barrows, E. 2001. *Animal Behavior Desk Reference, 2nd Edition*. Boca Raton, FL: CRC Press.

Barros, R. S., Galvao, O. F. and McIlvane, W. J. 2002. Generalized identity matching-to-sample in *Cebus apella*. *The Psychological Record* **52**, 441–460.

Barton, R. A. 1999. The evolutionary ecology of the primate brain. In *Comparative Primate Socioecology*, ed. P. C. Lee, pp. 167–203. Cambridge: Cambridge University Press.

Barton, R. A. and Harvey, P. H. 2000. Mosaic evolution of brain structure in mammals. *Nature* **405**, 1055–1058.

Barton, R. A., Purvis, A. and Harvey, P. H. 1995. Evolutionary radiation of visual and olfactory brain systems in primates, bats and insectivores. *Philosophical Transactions of the Royal Society of London* **B348**, 381–392.

Bartus, R. T. and Dean, R. L. 1988. Effects of cholinergic and adrenergic enhancing drugs on memory in aged monkeys. In *Current Research in Alzheimer Therapy*, ed. E. Giacobini, pp. 179–190. New York: Taylor and Francis.

Bartus, R. T., Dean, R. L. and Beer, B. 1980. Memory deficits in aged cebus monkeys and facilitation with central cholinomimetics. *Neurobiology of Aging* **1**, 145–152.

1983. An evaluation of drugs for improving memory in aged monkeys: implications for clinical trials in humans. *Psychopharmacology Bulletin* **19**, 168–184.

Bartus, R. T., Dean III, R. L., Beer, B. and Lippa, A. S. 1982. The cholinergic hypothesis of geriatric memory dysfunction. *Science* **217**, 408–417.

Bauchot, R. 1979a. Encephalic indexes and interspecific distances by the insectivora and the primates. I. Encephalon and telencephalon. *Mammalia* **43**, 173–189.

1979b. Encephalic indexes and interspecific distances by the insectivora and the primates. II. Diencephale and thalamus. *Mammalia* **43**, 407–426.

Bauchot, R. and Stephan, H. 1966. Données nouvelles sur l'encéphalisation des insectivores et des prosimiens. *Mammalia* **30**, 160–196.

1969. Encéphalisation et niveau évolutif chez les simiens. *Mammalia* **33**, 225–275.

Beach, F. A. 1976. Sexual activity, proceptivity and receptivity in female mammals. *Hormones and Behavior* **7**, 481–494.

Beck, B. B. 1980. *Animal Tool Behavior*. New York: Garland Press.

Beck, B., Kleiman, D., Dietz, J., Castro, I., Carvalho, C., Martins, A. and Rettberg-Beck, B. 1991. Losses and reproduction in reintroduced golden lion tamarins *Leontopithecus rosalia*. *Dodo* **27**, 50–61.

Belt, T. 1874/1985. *The Naturalist in Nicaragua*. Chicago: University of Chicago Press.

Bergeron, R. J., Lin, R. and McManis, J. S. 1992. Structural alterations in desferrioxamine compatible with iron clearance in animals. *Journal of Medical Chemistry* **35**, 4739–4744.

Bergeson, D. 1996. The positional behavior and prehensile tail use of *Alouatta palliata*, *Ateles geoffroyi*, and *Cebus capucinus*. Doctoral Dissertation, Washington University, St. Louis.

Bernstein, I. S. 1965. Activity patterns in a *Cebus* monkey group. *Folia Primatologica* **3**, 211–224.

1966. Analysis of a key role in a capuchin (*C. albifrons*) group. *Tulane Studies in Zoology* **13**, 49–54.

Bernstein, N. 1996. *Dexterity and its Development*. Ed. and translated by M. L. Latash and M. T. Turvey. Hillsdale, NJ: Lawrence Erlbaum Associates.

Bidell, T. R. and Fischer, K. W. 1994. Developmental transitions in children's early on-line planning. In *The Development of Future-oriented Processes*, ed. M. M. Haith, J. B. Henson, R. J. Roberts Jr. and B. F. Pennington, pp. 141–176. Chicago: University of Chicago Press.

Biegert, J. 1961. Volarhart der Haende und Fuesse. In *Primatologia. Handbuch der Primatenkunde* (*Vol. 2 Teil 2 Leif 3*), ed. H. Hofer, A. H. Schultz and D. Starck, pp. 1–326. Basel: S. Karger.

1963. The evaluation of characteristics of the skull, hands, and feet for primate taxonomy. *Viking Fund Publications in Anthropology* **37**, 116–145.

Bierens de Haan, J. A. 1931. Werkzeuggebrauch und Werkzugherstellung bei einem nierderen Affen (*Cebus hypoleucus* Humb). *Zeitschrift Vergleichende Physiologie* **13**, 639–695.

Bierregaard Jr., R. O., Gascon, C., Lovejoy, T. E. and Mesquita, R. 2001. *Lessons from Amazonia: The Ecology and Conservation of a Fragmented Forest*. New Haven: Yale University Press.

Bodini, R. and Pérez-Hernández, R. 1987. Distribution of the species and subspecies of cebids in Venezuela. *Fieldiana, Zoology, New Series* **39**, 231–244.

Boesch, C. and Boesch H. 1989. Hunting behavior of wild chimpanzees in the Taï National Park. *American Journal of Physical Anthropology* **78**, 547–573.

Boesch, C. and Boesch-Achermann, H. 2000. *The Chimpanzees of the Taï Forest*. Oxford: Oxford University Press.

Boesch, C., Marchesi, N., Fruth, B. and Joulian, F. 1994. Is nut cracking in wild chimpanzees a cultural behaviour? *Journal of Human Evolution* **26**, 325–338.

Boher-Benetti, S. and Cordero-Rodríguez, G. A. 2000. Distribution of brown capuchin monkeys (*Cebus apella*) in Venezuela: a piece of the puzzle. *Neotropical Primates* **8**, 152–153.

Boinski, S. 1988. Use of a club by a wild white-faced capuchin (*Cebus capucinus*) to attack a venomous snake (*Bothrops asper*). *American Journal of Primatology* **14**, 177–180.

1989. Why don't *Saimiri oerstedii* and *Cebus capucinus* form mixed-species groups? *International Journal of Primatology* **10**, 103–114.

1993. Vocal coordination of troop movement among white-faced capuchin monkeys, *Cebus capucinus*. *American Journal of Primatology* **30**, 85–100.

1996. Vocal coordination of troop movement in squirrel monkeys (*Saimiri oerstedi* and *S. sciureus*) and white-faced capuchins (*Cebus capucinus*). In *Adaptive Radiation of Neotropical Primates*, ed. M. A. Norconk, A. L. Rosenberger and P. A. Garber, pp. 251–269. New York: Plenum Press.

Boinski, S. and Campbell, A. F. 1995. Use of trill vocalizations to coordinate troop movement among white-faced capuchins: a second field test. *Behaviour* **132**, 875–901.

1996. The huh vocalization of white-faced capuchins: a spacing call disguised as a food call? *Ethology* **102**, 826–840.

Boinski, S. and Fragaszy, D. M. 1989. The ontogeny of foraging in squirrel monkeys, *Saimiri oerstedi*. *Animal Behavior* **37**, 415–428.

Boinski, S. and Garber, P. 2000. *On the Move: How and why Animals Travel in Groups*. Chicago: Chicago University Press.

Boinski, S. and Scott, P. E. 1988. Association of birds with monkeys in Costa Rica. *Biotropica* **20**, 136–143.

Boinski, S., Gross, T. S. and Davis, J. K. 1999b. Terrestrial predator alarm vocalizations are a valid monitor of stress in captive brown capuchins (*Cebus apella*). *Zoo Biology* **18**, 295–312.

Boinski, S., Quatrone, R. P. and Swartz, H. 2000. Substrate and tool use by brown capuchins in Suriname: ecological contexts and cognitive bases. *American Anthropologist* **102**, 741–761.

Boinski, S., Quatrone, R. P., Sughrue, K., Selvaggi, L., Henry, M., Stickler, C. and Rose, L. 2003. Do brown capuchins socially learn foraging skills? In *The Biology of Traditions*, ed. D. M. Fragaszy and S. Perry, pp. 365–390. Cambridge: Cambridge University Press.

Boinski, S., Swing, S. P., Gross, T. S. and Davis, J. K. 1999a. Environmental enrichment of brown capuchins (*Cebus apella*): behavioral and plasma and fecal cortisol measures of effectiveness. *American Journal of Primatology* **48**, 49–68.

Boinski, S., Treves, A. and Chapman, C. 2000. A critical evaluation of the effects of predators on primates: Effects on group travel. In *On the Move: How and why Animals Travel in Groups*, ed. S. Boinski and P. Garber, pp. 43–72. Chicago: Chicago University Press.

Bolanowski, S. J. Jr., Gescheider, G. A., Verrillo, R. T. and Checkosky, C. M. 1988. Four channels mediate the mechanical aspects of touch. *Journal of the Acoustical Society of America* **84**, 1680–1694.

Bolen, R. H. and Green, S. M. 1997. Use of olfactory cues in foraging by owl monkeys (*Aotus nancymai*) and capuchin monkeys (*Cebus apella*). *Journal of Comparative Psychology* **111**, 152–158.

Bortoff, G. A. and Strick, P. L. 1993. Corticospinal terminations in two New-World primates: further evidence that corticomotoneural connections provide part of the neural substrate for manual dexterity. *Journal of Neuroscience* **13**, 5105–5118.

Bouvier, M. 1986. Biomechanical scaling of mandibular dimensions in New World monkeys. *International Journal of Primatology* **7**, 551–567.

Bradshaw, J. and Rogers, L. 1993. *The Evolution of Lateral Asymmetries, Language, Tool use, and Intellect*. San Diego, CA: Academic Press.

Brand, H. M. and Martin, R. D. 1983. The relationship between urinary estrogen excretion and mating behavior in cotton-topped tamarins, *Saguinus oedipus oedipus*. *International Journal of Primatology* **4**, 275–290.

Bronowski, J. 1973. *The Ascent of Man*. Boston: Little and Brown.

Brook, B. W., O'Grady, J. J., Chapman, A. P., Burgman, M. A., Akçakaya, H. R. and Frankham, R. 2000. Predictive accuracy of population viability analysis in conservation biology. *Nature* **404**, 385–387.

Brown, A. D. 1986. Biogeografia historica y a diversificación de los primates. Historia biogeogrífica del noroeste Argentino. *Boletin Primatológico Argentino* **4**(1), 53–85.

1989. Distribución y conservación de *Cebus apella* (Cebidae: Primates) en el noroeste Argentino. In *La*

Primatología en Latinoamérica, ed. C. J. Saavedra, R. A. Mittermeier and I. B. Santos, pp. 159–164. Washington, DC: World Wildlife Fund – US.

1990. Prioridades de conservación del subtrópico húmedo de la Argentina. *Boletin Primatológico Latinoaméricano* **2**(1), 48–61.

Brown, A. D. and Colillas, O. J. 1984. Ecología de *Cebus apella*. In *A Primatologia no Brasil*, ed. M. T. de Mello, pp. 301–312. Brasília: Sociedade Brasileira de Primatologia.

Brown, A. D. and Zunino, G. E. 1990. Dietary variability in *Cebus apella* in extreme habitats: evidence for adaptability. *Folia Primatologica* **54**, 187–195.

Brown, A. D., Chalukian, S. C., Malmierca, L. M. and Calillas, O. J. 1986. Habitat structure and feeding behavior of *Cebus apella* (Cebidae) in El Rey National Park, Argentina. In *Current Perspectives in Primate Social Dynamics*, ed. D. M. Taub and F. A. King, pp. 137–151. New York: Van Nostrand Reinhold Co.

Buckley, J. S. 1983. The feeding behavior, social behavior, and ecology of the white-faced monkey, *Cebus capucinus*, at Trujillo, Northern Honduras. Doctoral Dissertation, University of Texas at Austin.

Buffon, G. L. 1770. *Histoire naturelle générale et particulier*. Paris: de l'Imprimerie Royale.

Burdyn, L. E. Jr. and Thomas, R. K. 1984. Conditional discrimination with conceptual simultaneous and successive cues in the squirrel monkey (*Saimiri sciureus*). *Journal of Comparative Psychology* **98**, 405–413.

Burton, G. 1993. Non-neural extensions of haptic sensitivity. *Ecological Psychology* **5**, 105–124.

Butchart, S. H. M., Barnes, R., Davies, C. W. N., Fernandez, M. and Seddon, N. 1995. Observations of two threatened primates in the Peruvian Andes. *Primate Conservation* **16**, 15–19.

Byrne, G. 1993. Differences between firstborn and later-born *Cebus apella* infants' interactions with mothers. *American Association of Zoological Parks and Acquaria Regional Conference Proceedings*, 761–766.

Byrne, G. and Suomi, S. J. 1995. Development of activity patterns, social interactions and exploratory behavior in infant tufted capuchins. *American Journal of Primatology*, **35**, 255–270.

1996. Individual differences in object manipulation in a colony of tufted capuchins. *Journal of Human Evolution* **31**, 259–267.

1998. Relationship of early infant state measures in predicting behavior over the first year of life in infant tufted capuchins (*Cebus apella*). *American Journal of Primatology* **44**, 43–56.

1999. Social separation in infant *Cebus apella*: patterns of behavioral and cortisol response. *International Journal of Developmental Neuroscience* **17**, 265–274.

Byrne, G., Abbott, K. M. and Suomi, S. J. 1996. Reorganization of dominance rank among adult females in a captive group of tufted capuchins (*Cebus apella*). *Laboratory Primate Newsletter* **35**, 1–4.

Byrne, R. and Whiten, A. 1988. *Machiavellian Intelligence. Social Expertise and the Evolution of Intellect in Monkeys, Apes, and Humans*. Oxford: Clarendon Press.

Caldwell, C. A. and Whiten, A. 2003. Scrounging facilitates social learning in common marmosets, *Callithrix jacchus*. *Animal Behaviour* **65**, 1085–1092.

Calle, Z. 1990a. A field observation of infant development and social interactions of a wild black-capped capuchin (*Cebus apella*) female and infant at La Macarena (Colombia). *Field Studies of New World Monkeys*, La Macarena, Colombia **4**, 1–8.

1990b. A field study of the social interactions of one year old black-capped capuchins (*Cebus apella*) at La Macarena (Colombia). *Field Studies of New World Monkeys*, La Macarena, Colombia **4**, 9–26.

Carlson, M. 1984a. Development of tactile discrimination capacity in Macaca mulatta. II. Effects of partial removal of primary somatic sensory cortex (SmI) in infants and juveniles. *Developmental Brain Research* **16**, 83–101.

1984b. Development of tactile discrimination capacity in *Macaca mulatta*. III. Effects of total removal of primary somatic sensory cortex (SmI) in infants and juveniles. *Developmental Brain Research* **16**, 103–117.

Carlson, M. and Nystrom, P. 1994. Tactile discrimination capacity in relation to size and organization of somatic sensory cortex in primates: I. Old-World prosimian, *Galago*; II. New-World anthropoids, *Saimiri* and *Cebus*. *Journal of Neuroscience* **14**, 1516–1541.

Carlson, M., Huerta, M. F., Cusick, C. G. and Kaas, J. H. 1986. Studies on the evolution of multiple somatosensory representations in primates: the organization of anterior parietal cortex in the New World callitrichid, *Saguinus*. *Journal of Comparative Neurology* **246**, 409–426.

Carosi, M. and Haines, M. 1999. The occurrence of a bone-like structure in the clitoris of female tufted capuchins (*Cebus apella*). *American Journal of Primatology* **49**, 41.

Carosi, M. and Rosofsky, A. 1999. Urine washing in tufted capuchins (*Cebus apella*): relationship with air temperature and relative humidity in indoor and outdoor conditions. *American Journal of Primatology* **49**, 41.

Carosi, M. and Visalberghi, E. 2002. An analysis of tufted capuchin (*Cebus apella*) courtship and sexual behavior repertoire: changes throughout the female cycle and female inter-individual differences. *American Journal of Physical Anthropology* **118**, 11–24.

Carosi, M., Gerald, M., Ulland, A. and Suomi, S. (2000) Virilized external genitalia in female tufted capuchins (*Cebus apella*). *American Journal of Primatology* **51**, 50.

Carosi, M., Heistermann, M. and Visalberghi, E. 1999. Display of proceptive behaviors in relation to urinary and fecal progestin levels over the ovarian cycle in female tufted capuchin monkeys. *Hormones and Behavior* **36**, 252–265.

Carvalho, Jr. O. de, Pinto, A. C. B. and Galetti, M. 1999. New observations on *Cebus kaapori* (Queiroz 1992) in eastern Brazilian Amazonia. *Neotropical Primates* **7**, 41–42.

Case, R. 1992. *The Mind's Staircase*. Hillsdale, NJ: Lawrence Erlbaum Associates.

Case, R. and Okamoto, Y. 1996. The role of central conceptual structures in the development of children's thought. *Monographs of the Society for Research in Child Development* **61(1–2)**, 1–265.

Casey, D. E. 1984. Tardive dyskinesia – animal models. *Psychopharmacology Bulletin* **20**, 376–379.

Cassidy, J. and Shaver, P. 1999. *Handbook of Attachment: Theory, Research, and Clinical Applications*. New York: Guilford Press.

Chalmeau, R., Visalberghi, E. and Gallo, A. 1997. Capuchin monkeys, *Cebus apella*, fail to understand a cooperative task. *Animal Behaviour* **54**, 1215–1225.

Chang, H. T. and Ruch, T. C. 1947. Morphology of the spinal cord, spinal nerves, caudal plexus, tail segmentation, and caudal musculature of the spider monkey. *Yale Journal Biology Medicine* **19**, 345–377.

Chapman, C. A. 1986. Boa constrictor predation and group response in white-faced *Cebus* monkeys. *Biotropica* **18**, 171–172.

1987a. Foraging strategies, patch use, and constraints on group size in three species of Costa Rican primates. Doctoral Dissertation, University of Alberta.

1987b. Flexibility in diets of three species of Costa Rican primates. *Folia Primatologica* **29**, 90–105.

1988. Patterns of foraging and range use by three species of neotropical primates. *Primates* **29**, 177–194.

1989. Primate seed dispersal: the fate of dispersed seeds. *Biotropica* **21**, 148–154.

1990. Ecological constraints on group size in three species of neotropical primates. *Folia Primatologica* **55**, 1–9.

1995. Primate seed dispersal: coevolution and conservation implications. *Evolutionary Anthropology* **4**, 74–82.

Chapman, C. A. and Chapman, L. 2000. Determinants of group size in primates: the importance of travel costs. In *On the Move. How and Why Animals Travel in Groups*, ed. S. Boinski and P. A. Garber, pp. 24–42. Chicago: University of Chicago Press.

Chapman, C. A. and Fedigan, L. M. 1990. Dietary differences between neighboring *Cebus capucinus* groups: local traditions, food availability or responses to food profitability? *Folia Primatologica* **54**, 177–186.

Chapman, C. A. and Peres, C. A. 2001. Primate conservation in the new millennium. *Evolutionary Anthropology* **10**, 16–33.

Chevalier-Skolnikoff, S. 1989. Spontaneous tool use and sensorimotor intelligence in *Cebus* compared with other monkeys and apes. *Behavioral and Brain Sciences* **12**, 561–627.

1990. Tool use by wild *Cebus* monkeys at Santa Rosa National Park, Costa Rica. *Primates* **31**, 375–383.

Chivers, D. and Hladik, M. 1980. Morphology of the gatrointestinal tract in primates: comparisons with other mammals in relation to diet. *Journal Morphology* **166**, 337–386.

Christel, M. 1993. Grasping techniques and hand preferences in hominoidea. *Hands of Primates*, ed. H. Preuschoft and D. J. Chivers, pp. 91–108. New York: Springer-Verlag.

Christel, M. and Fragaszy, D. 2000. Manual function in *Cebus apella*. Digital mobility, preshaping, and endurance in repetitive grasping. *International Journal of Primatology* 21, 697–719.

Clayton, D. A. 1978. Socially facilitated behavior. *Quarterly Review of Biology* **53**, 373–392.

Clewe, T. H. and DuVall, W. M. 1966. Observations on frequency of ejaculation of squirrel monkeys, *Saimiri sciureus*. *American Zoologist* **6**, 602 (Abstract 411).

Cochrane, M. A., Alencar, A., Schulze, M. D., Souza Jr., C. M., Nepstad, D. C., Lefebvre, P. and Davidson, E. A. 1999. Positive feedbacks in the fire dynamic of closed canopy tropical forests. *Science* **284**, 1832–1835.

Coe, C. L., Savage, A. and Bromley, L. J. 1992. Phylogenetic influences on hormone levels across the Primate order. *American Journal of Primatology* **28**, 81–100.

Coe, M. D. 1978. Supernatural patrons of Maya scribes and artists. In *Social Process in Maya History*, ed. N. Hammond, pp. 327–346. New York: Academic Press.

Coimbra-Filho, A. F. and Câmara, I. de G. 1996. *Os Limites Originais do Bioma Mata Atlântica na Região Nordeste do Brasil.* Rio de Janeiro: Fundação Brasileira para a Conservação da Natureza (FBCN).

Coimbra-Filho, A. F., Rocha e Silva, R. and Pissinatti, A. 1991. Acerca da distribuicao geografica original de *Cebus apella xanthosternos* Wied 1820 (Cebidae, Primates). *A Primatologia no Brasil* **3**, 215–224.

Coimbra-Filho, A. F., Rylands, A. B., Pissinatti, A. and Santos, I. B. 1991/1992. The distribution and conservation of the buff-headed capuchin monkey, *Cebus xanthosternos*, in the Atlantic forest region of eastern Brazil. *Primate Conservation* **12–13**, 24–30.

Cole, T. M. III. 1992. Postnatal heterochrony of the masticatory apparatus in *Cebus apella* and *Cebus albifrons*. *Journal of Human Evolution* **23**, 253–282.

Collins, R. 1981. On the microfoundations of macrosociology. *American Journal of Sociology* **86**, 984–1014.

1993. Emotional energy as the common denominator of rational action. *Rationality and Society* **5**, 203–230.

Colombo, M. and D'Amato, M. R. 1986. A comparison of visual and auditory short-term memory in monkeys (*Cebus apella*). *Quarterly Journal of Experimental Psychology* **4**, 425–448.

Colombo, M. and Frost, N. 2001. Representation of serial order in humans: a comparison to the findings with monkeys (*Cebus apella*). *Psychonomic Bulletin and Review* **8**, 262–269.

Committee on Well-Being of Nonhuman Primates, Institute for Laboratory Animal Research, National Research Council, 1998. *The Psychological Well-Being of Nonhuman Primates*. Washington, DC: National Academy Press.

Cook, M. and Mineka, S. 1990. Selective associations in the observational conditioning of fear in rhesus monkeys. *Journal of Experimental Psychology: Animal Behaviour Processes* **16**, 372–389.

Cooper, L. R. and Harlow, H. F. 1961. Note on a *Cebus* monkey's use of a stick as a weapon. *Psychological Reports* **8**, 418.

Cooper, M. A., Bernstein, I. S., Fragaszy, D. M. and de Waal, F. B. M. 2001. The integration of new males into four social groups of tufted capuchin monkeys (*Cebus apella*). *International Journal of Primatology* **22**, 663–683.

Corballis, M. 1991. *The Lopsided Ape: Evolution of the Generative Mind*. Oxford: Oxford University Press.

Cormier, L. A. 2000. 2002. Monkey as food, monkey as child: Guajá symbolic cannibalism. In *Implications of Human and Nonhuman Primate Connections*, ed. A. Fuentes and L. Wolfe, pp. 63–84. Cambridge: Cambridge University Press.

2003. *Kinship with Monkeys. The Guaja Foragers of Eastern Amazonia*. New York: Columbia University Press.

Corradini, P., Recabarren, M., Serron-Ferrer, M. and Parraguez, V. H. 1998. Study of prenatal growth in the capuchin monkey (*Cebus apella*) by ultrasound. *Journal of Medical Primatology* **27**, 287–292.

Costello, M. B., and Fragaszy, D. M. 1988. Prehension in *Cebus* and *Saimiri*: 1. Grip type and hand preference. *American Journal of Primatology* **15**, 235–245.

Coussi-Korbel, S. and Fragaszy, D. 1995. On the relation between social dynamics and social learning. *Animal Behaviour* **50**, 1441–1453.

Cowlishaw, G. 1998. The role of vigilance in the survival and reproductive strategies of desert baboons. *Behaviour* **135**, 431–42.

Cowlishaw, G. and Dunbar, R. 2000. *Primate Conservation Biology*. Chicago: University of Chicago Press.

Crawford, M. P. 1937. The cooperative solving of problems by young chimpanzees. *Comparative Psychology Monographs* **14**, 1–88.

Croat, T. B. 1975. Phenological behavior of habitat and habitat classes on Barro Colorado Island. *Biotropica* **7**, 270–277.

Cropp, S., Boinski, S. and Li, W.-S. 2002. Allelic variation in the squirrel monkey X-linked color vision gene: Biogeographical and behavioral correlates. *Journal of Molecular Evolution* **54**, 734–745.

Cruz Lima, E. 1945. *Mammals of Amazonia. I. General Introduction and Primates*. Rio de Janeiro: Oficina Gráfica Macia Ltda.

Cummins, S. E. 1999. Detection of environmental constraints in a tool-use task by tufted capuchin monkeys (*Cebus apella*). Master's Thesis, University of Georgia.

Cummins-Sebree, S. E. and Fragaszy, D. M. 2001. The right stuff: Capuchin monkeys perceive affordances of tools. In *Studies in Perception and Action VI*, ed. G. Burton and R. Schmidt, pp. 89–92. Hillsdale, NJ: Lawrence Erlbaum Associates.

Cummins-Sebree, S., Fuller, A. and Fragaszy, D. 2000. It's all about getting your goodies: Capuchins' success with their selected tools in a food-retrieving task. Abstracts, Animal Behavior Society, Morehouse College, p. 19.

Custance, D., Whiten, A. and Fredman, A. 1999. Social learning of an artificial fruit task in capuchin monkeys (*apella*). *Journal of Comparative Psychology* **113**, 13–23.

Cuvier, F. 1824. *Histoire Naturelle des Mammiferes*. Paris: A. Belen.

D'Amato, M. R. 1988. A search for tonal pattern perception in Cebus monkeys: Why monkeys can't hum a tune. *Music Perception* **5**, 453–480.

1991. Comparative cognition: Processing of serial order and serial pattern. In *Current Topics in Animal Learning: Brain, Emotion, and Cognition*, ed. L. Dachowski and C. Flaherty, pp. 165–185. Hillsdale, NJ: Lawrence Erlbaum Associates.

D'Amato, M. R. and Buckiewicz, J. 1980. Long-delay, one-trial conditioned preference and retention in monkeys (*Cebus apella*). *Animal Learning and Behavior* **8**, 359–362.

D'Amato, M. R. and Colombo, M. 1985. Auditory matching-to-sample in monkeys (*Cebus apella*). *Animal Learning and Behavior* **13**, 375–382.

1988a. On tonal pattern perception in monkeys (*Cebus apella*). *Animal Learning and Behavior* **16**, 417–424.

1988b. Representation of serial order in monkeys (*Cebus apella*). *Journal of Experimental Psychology: Animal Behavior Processes* **14**, 131–139.

1989. Serial learning with wild card items by monkeys (*Cebus apella*): Implications for knowledge of ordinal position. *Journal of Comparative Psychology* **103**, 252–261.

1990. The symbolic distance effect in monkeys (*Cebus apella*). *Animal Learning and Behavior* **18**, 133–140.

D'Amato, M. R. and Salmon, D. P. 1984. Cognitive processes in cebus monkeys. In *Animal Cognition*, ed. H. L. Roitblat, T. G. Bever and H. S. Terrace, pp. 149–168. Hillsdale, NJ: Lawrence Erlbaum Associates.

D'Amato, M. R., Buckiewicz, J. and Puopolo, M. 1981a. Long-delay spatial discrimination learning in monkeys (*Cebus apella*). *Bulletin of the Psychonomic Society* **18**, 85–88.

D'Amato, M. R., Salmon, D., Loukas, E. and Tomie, A. 1985. Symmetry and transitivity of conditional relations in monkeys (*Cebus apella*) and pigeons (*Columba livia*). *Journal of Experimental Analysis of Behavior* **44**, 35–47.

1986. Processing of identity and conditional relations in monkeys (*Cebus apella*) and pigeons (*Columba livia*). *Animal Learning and Behavior* **14**, 365–373.

D'Amato, M. R., Salmon, D. and Puopolo, M. 1981b. Long-delay visual discrimination learning in monkeys (*Cebus apella*). *Bulletin of the Psychonomic Society* **18**, 89–91.

Daegling, D. J. 1992. Mandibular morphology and diet in the genus *Cebus*. *International Journal of Primatology* **13**, 545–570.

Dahl, J. F. 1986. The status of howler, spider and capuchin monkey populations in Belize, Central America. *Primate Report* **14**, 161.

Dal Secco, V. 2000. Il comportamento sessuale del cebo dai cornetti, *Cebus apella* (L., 1758): studio delle dinamiche tra procettività della femmina e risposta del maschio. Tesi di Laurea, Università degli Studi di Roma "La Sapienza". Facoltà di Scienze Matematiche, Fisiche e Naturali.

Dal Secco, V. and Visalberghi, E. 2001. Male response to female proceptivity in tufted capuchin monkeys (*Cebus apella*). *Folia Primatologica* **72**, 130–131.

Dampier, W. 1697. *A New Voyage Round the World*. London: J. Knapton, at the Crown in St. Paul's Church-yard.

Darwin, E. 1794. *Zoonomia or Laws of Organic Life*. London: J. Johnson.

Darwin, F. (ed.) 1888. *Life and Letters of Charles Darwin*, Vol. III. New York: Appleton and Company.

De Lillo, C. and Visalberghi, E. 1994. Transfer index and mediational learning in tufted capuchins (*Cebus apella*). *International Journal of Primatology* **15**, 275–286.

De Lillo, C., Aversano, M., Tuci, E. and Visalberghi, E. 1998. Spatial constraints and regulatory functions in monkeys' (*Cebus apella*) search. *Journal of Comparative Psychology* **111**, 82–90.

De Lillo, C., Visalberghi, E. and Aversano, M. 1997. The organization of exhaustive searches in a "patchy" space by capuchin monkeys (*Cebus apella*). *Journal of Comparative Psychology* **111**, 82–90.

de Oviedo, F. G. 1526/1996. *Sumario de la Natural Historia de las Indias*. Biblioteca Americana – Fondo de Cultura Económica. First edition, Toledo, Spain, 1526.

de Palermo, K. E., Carbonetto, C. H., Malchiodi E. L., Margni, R. A. and Falalasca, C. A. 1988. Humoral and cellular parameters of the immune system of *Cebus apella* monkeys: cross reactivity between monkey and human immunoglobulins. *Veterinary Immunology and Immunopathology* **19**, 341–349.

de Ruiter, J. 1986. The influence of group size on predator scanning and foraging behavior of wedge-capped capuchin monkeys (*Cebus olivaceus*). *Behaviour* **98**, 240–258.

De Valois, R. L. 1971. Vision. In *Behavior of Nonhuman Primates: Modern Research Trends*, Vol. 3, ed. A. M. Schrier and F. Stollnitz, pp. 107–157. New York: Academic Press.

de Waal, F. B. M. 1988. The communicative repertoire of captive bonobos (*Pan paniscus*) compared to that of chimpanzees. *Behaviour* **106**, 183–251.

1989. *Peace-making among primates*. Cambridge, MA: Harvard University Press.

1997. Food transfers through mesh in brown capuchins. *Journal of Comparative Psychology* **111**, 370–378.

1999. Cultural primatology comes of age. *Nature* **399**, 635–636.

2000. Attitudinal reciprocity in food sharing among brown capuchins. *Animal Behaviour* **60**, 253–261.

de Waal, F. B. M. and Berger, M. L. 2000. Payment for labour in monkeys. *Nature* **404**, 563.

de Waal, F. B. M. and Davis, J. M. 2003. Capuchin cognitive ecology: Cooperation based on projected returns. *Neuropsychologia* **41**, 221–228.

de Waal, F. B. M. and Luttrell, L. M. 1986. The similarity principle underlying social bonding among female rhesus monkeys. *Folia Primatologica* **46**, 215–234.

de Waal, F. B. M., Luttrell, L. M. and Canfield, M. E. 1993. Preliminary data on voluntary food sharing in brown capuchin monkeys. *American Journal of Primatology* **29**, 73–78.

Deacon, T. 1990a. Fallacies of progression in theories of brain size evolution. *International Journal of Primatology* **11**, 193–236.

1990b. Problems of ontogeny and phylogeny in brain-size evolution. *International Journal Primatology* **11**, 237–282.

Dean, W. 1995. *With Broadaxe and Firebrand: The Destruction of the Brazilian Atlantic Forest*. Berkeley, CA: University of California Press.

Debyser, I. W. J. 1995. Platyrrhine juvenile mortality in captivity and in the wild. *International Journal of Primatology* **16**, 909–933.

Defler, T. R. 1979a. On the ecology and behavior of *Cebus albifrons* in Eastern Colombia: I. Ecology. *Primates* **20**, 475–490.

1979b. On the ecology and behavior of *Cebus albifrons* in Eastern Colombia: II. Behavior. *Primates* **20**, 491–502.

1982. A comparison of intergroup behavior in *Cebus albifrons* and *C. apella*. *Primates* **23**, 385–392.

1985. Contiguous distribution of two species of Cebus monkeys in El Tuparro National Park, Colombia. *American Journal of Primatology* **8**, 101–112.

Defler, T. R. and Hernández-Camacho, J. I. 2002. The true identity and characteristics of *Simia albifrons* Humboldt, 1812: description of neotype. *Neotropical Primates* **10**, 49–64.

Deputte, B. L. and Busnel, M. 1997. An example of a monkey assistance program: P. A. S. T.- the French project of simian help to quadriplegics. A response to Iannuzzi and Rowan's (1991) paper on ethical issues in animal-assisted programs. *Anthrozoös* **10**, 76–81.

Deputte, B. L., Vrot, M., Pierre, G., Bellec, S. and Jouanjean, A. 1995. Development of manipulation in a brown capuchin raised in a human family: changes in social facilitation with age. *Folia Primatologica* **64**, 77.

Deshaies, N., Deputte, B. L., Cohalion-Buisson, Baran, M. and Baudoin, C. 1996. Interpretation of brown capuchin vocalization by human subjects. *Folia Primatologica* **67**, 87–88.

Dettmer, E. and Fragaszy, D. 2000. Determining the value of social companionship to captive capuchin monkeys (*Cebus apella*). *Journal of Applied Animal Welfare Science* **3**, 293–304.

Dewsbury, D. A. and Pierce, J. D. 1989. Copulatory patterns of primates as viewed in broad mammalian perspective. *American Journal of Primatology* **17**, 51–72.

Di Bitetti, M. S. 1997. Evidence for an important social role of allogrooming in a platyrrhine primate. *Animal Behaviour* **54**, 199–211.

2001a. Home range use by the tufted capuchin monkey, *Cebus apella nigritus*, in a subtropical rainforest of Argentina. *Journal of Zoology (London)* **253**, 33–45.

2001b. Food-associated calls in the tufted capuchin monkey (*Cebus apella*). Doctoral Dissertation, State University of New York at Stony Brook.

In press. Food-associated calls of tufted capuchin monkeys (*Cebus apella nigritus*) are functionally referential signals. *Behaviour*.

Di Bitetti, M. S. and Janson, C. H. 2000. When will the stork arrive? Patterns of birth seasonality in neotropical primates. *American Journal of Primatology* **50**, 109–130.

2001. Reproductive socioecology of tufted capuchins (*Cebus apella nigritus*) in northeastern Argentina. *International Journal of Primatology* **22**, 127–142.

Di Bitetti, M. S., Luengos Vidal, E. M., Baldovino, M. C. and Benesovsky, V. 2000. Sleeping site preferences in tufted capuchin monkeys, *Cebus apella nigritus*. *American Journal of Primatology* **50**, 257–274.

Di Giano, L., Nagle, C. A., Quiroga, S., Paul, N., Farinati, Z., Torres, M. and Mendizabal, A. F. (1992) Salivary progesterone for the assessment of the ovarian function in the capuchin monkey (*Cebus apella*). *International Journal of Primatology*, **13**, 113–123.

Dixson, A. F. 1987. Baculum length and copulatory behavior in primates. *American Journal of Primatology* **13**, 51–60.

1995. Sexual selection and ejaculatory frequencies in primates. *Folia Primatologica* **64**, 146–152.

1997. Evolutionary perspectives on primate mating systems and behavior. *Annals of the New York Academy of Sciences* **807**, 42–61.

1998a. *Primate Sexuality*. New York: Oxford University Press

1998b. Sexual selection and evolution of the seminal vesicles in primates. *Folia Primatologica* **69**, 300–306.

Dixson, A. L. and Anderson, M. J. 2002. Sexual selection, seminal coagulation and copulatory plug formation in primates. *Folia Primatologica* **73**, 63–69.

Dobroruka, L. J. 1972. Social communication in the brown capuchin (*Cebus apella*). *International Zoological Yearbook* **12**, 43–49.

Dominy, N. J. and Lucas, P. W. 2001. Ecological importance of trichomatic vision to primates. *Nature* **410**, 363–365.

Drapier, M., Addessi, E. and Visalberghi, E. 2003. The response of tufted capuchin monkeys (*Cebus apella*) to foods flavored with familiar and novel odor. *International Journal of Primatology* **24**, 295–315.

Dubois, M., Gerard, J. F., Sampaio, E., Galvão, O. & Guilhem, C. 2001. Spatial facilitation in a probing task in *Cebus olivaceus*. *International Journal of Primatology* **22**, 991–1006.

Dubois, M., Sampaio, E., Gerard, J. F., Quenette, P. Y. and Muniz, J. 2000. Location-specific responsiveness to environmental perturbations in wedge-capped capuchins (*Cebus olivaceus*). *International Journal of Primatology* **21**, 85–102.

Dunbar, R. I. M. 1991. Functional significance of social grooming in primates. *Folia Primatologica* **57**, 121–131.

2001. Brains on two legs; group size and the evolution of intelligence. In *Tree of Origin,* ed. F. de Waal, pp. 175–191. Cambridge, MA: Harvard University Press.

Dunn, F. L. 1968. The parasites of *Saimiri* in the context of platyrrhine parasitism. In *The Squirrel Monkey*, ed. L. A. Rosenblum and R. W. Cooper, pp. 31–68. New York: Academic Press.

Dyke, B., Gage, T. B., Alford, P. L., Swenson, B. and Williams-Blangero, S. 1995. Model life table for captive chimps. *American Journal of Primatology* **37**, 25–37.

1989. Order primates. *Mammals of the Neotropics. Vol. 1. The Northern Neotropics: Panama, Colombia, Venezuela, Guyana, Suriname*, pp. 233–261. Chicago: University of Chicago Press.

Eisenberg, J. F. and Redford, K. H. 1999. *Mammals of the Neotropics. Vol. 3. The Central Neotropics: Ecuador, Bolivia, Peru, Brazil.* Chicago: University of Chicago Press.

Elias, M. F. 1977. Relative maturity of cebus and squirrel monkeys at birth and during infancy. *Developmental Psychobiology* **10**, 519–528.

Elias, M. F. and Samonds, K. W. 1973. Exploratory behavior of cebus monkeys after having been reared in partial isolation. *Child Development* **44**, 218–220.

1974. Exploratory behavior and activity of infant monkeys during nutritional and rearing restriction. *American Journal of Clinical Nutrition* **27**, 458–463.

Elliot, D. G. 1913. *A Review of the Primates. Monograph Series vol II*. New York: American Museum of Natural History.

Elliott, J. and Connolly, K. 1984. A classification of manipulative hand movements. *Developmental Medicine and Child Neurology* **26**, 283–296.

Elman, J. L., Bates, E. A., Johnson, M. H., Karmiloff-Smith, A., Parisi, D. and Plunkett, K. 1996. *Rethinking Innateness*. Cambridge, MA: MIT Press.

Emmons, L. H. and Feer, F. 1997. *Neotropical Rainforest Mammals. A Field Guide*. Chicago: University of Chicago Press.

Encarnación, F. and Cook, A. G. 1998. Primates of the tropical forest of the Pacific coast of Peru: The Tumbes Reserved Zone. *Primate Conservation* **18**, 15–20.

Epple, G. 1986. Communication by chemical signals. In *Comparative Primate Biology, Vol. 2, Part A. Behavior, Conservation, and Ecology*, ed. G. Mitchell and J. Erwin, pp. 531–580. New York: Alan Liss.

Epple, G. and Lorenz, R. 1967. Vorkommen, Morphologie und Funktion der Sternaldrüse bei den Platyrrhini. *Folia Primatologica* **7**, 98–126.

Erickson, O. E. 1948. The morphology of the forelimb of the capuchin monkey, *Cebus capucinus* (L.). Doctoral Dissertation, Harvard University.

Escobar-Páramo, P. 1989. The development of the wild black-capped capuchin (*Cebus apella*) in La Macarena, Colombia. *Field Studies of New World Monkeys, La Macarena, Colombia* **2**, 4–56.

1999. Inbreeding avoidance and the evolution of male mating strategies. Doctoral Dissertation, State University of New York at Stony Brook.

2000. Microsatellite primers for the wild brown capuchin monkey *Cebus apella*. *Molecular Ecology* **9**, 107–118.

Eudey, A. A. (compiler). 1987. *Action Plan for Asian Primate Conservation: 1987–91*. Gland, Switzerland: The World Conservation Union (IUCN).

Falk, D. 1980. Comparative study of the endocranial casts of New and Old World monkeys. *Evolutionary Biology of the New World Monkeys and Continental Drift*, ed. R. Ciochon and A. Chiarelli, pp. 275–292. New York: Plenum.

Falk, D. and Gibson, K. R. 2001. *Evolutionary Anatomy of the Primate Cerebral Cortex*. Cambridge: Cambridge University Press.

Fedigan, L. M. 1990. Vertebrate predation in *Cebus capucinus*: meat eating in a neotropical monkey. *Folia Primatologica* **54**, 196–205.

1993. Sex differences and intersexual relations in adult white-faced capuchins (*Cebus capucinus*). *International Journal of Primatology* **14**, 853–877.

Fedigan, L. M. and Jack, K. 2001. Neotropical primates in a regenerating Costa Rican dry forest: a comparison of howler and capuchin population patterns. *International Journal of Primatology* **22**, 689–713.

Fedigan, L. M. and Rose, L. M. 1995. Interbirth interval variation in three sympatric species of neotropical monkey. *American Journal of Primatology* **37**, 9–24.

Fedigan, L. M. and Zohar, S. 1996. Sex differences in mortality of Japanese macaques: 21 years of data from the Arashiyama West population. *American Journal of Physical Anthropology* **102**, 161–175.

Fedigan, L. M., Rose, L. M. and Morera, R. A. 1996. See how they grow. Tracking capuchin monkey (*Cebus capucinus*) populations in a regenerating Costa Rican dry forest. In *Adaptive Radiations of Neotropical Primates*, ed. M. A. Norconk, A. L. Rosenberger and P. A. Garber, pp. 289–307. New York: Plenum Press.

Feistner, A. T. C. and McGrew, W. C. 1989. Food-sharing in primates: a critical review. In *Perspectives in Primate Biology Vol. 3*, ed. P. K. Seth and S. Seth, pp. 21–36. New Delhi: Today and Tomorrow's Printers.

Felleman, D. J., Nelson, R. J., Sur, M. and Kaas, J. H. 1983. Representations of the body surface in areas 3b and 1 of postcentral parietal cortex of cebus monkeys. *Brain Research* **268**, 15–26.

Fernandes, E. B. M. 1991. Tool use and predation of oysters (*Crassostrea rhizophorae*) by the tufted capuchin, *Cebus apella apella*, in brackish water mangrove swamp. *Primates* **32**, 529–531.

Ferrari, S. F. and Diego, V. H. 1995. Habitat fragmentation and primate conservation in the Atlantic Forest of eastern Minas Gerais, Brazil. *Oryx* **29**, 192–196.

Ferrari, S. F. and Lopes, M. A. 1996. Primate populations in eastern Amazonia. In *Adaptive Radiations of Neotropical Primates*, ed. M. A. Norconk, A. L. Rosenberger and P. A. Garber, pp. 53–67. New York: Plenum Press.

Ferrari, S. F. and Queiroz, H. L. 1994. Two new Brazilian primates discovered, endangered. *Oryx* **28**, 31–36.

Finlay, B. L. and Darlington, R. B. 1995. Linked regularities in the development and evolution of mammalian brains. *Science* **268**, 1578–1584.

Finlay, B. L., Darlington, R. B. and Nicastro, N. 2001. Developmental structure in brain evolution. *Behavioral and Brain Science* **24**, 263–308.

Finlay, B. L., Hersmann, and Darlington, R. B. 1998. Patterns of vertebrate neurogenesis and the paths of vertebrate evolution. *Brain, Behavior and Evolution* **52**, 232–242.

Fischer, K. W. and Bidell, T. R. 1998. Dynamic development of psychological structures in action and thought. In *Handbook of Child Psychology: Vol. 1. Theoretical Models of Human Development (Fifth edition)*, ed. W. Damon (series editor) and R. M. Lerner (volume editor), pp. 467–561. New York: John Wiley & Sons.

Fleagle, J. G. 1999. *Primate Adaptation and Evolution (Second edition)*. New York: Academic Press.

Fleagle, J. G. and Kay, R. F. 1997. Platyrrhines, catarrhines, and the fossil record. In *New World Primates. Ecology, Evolution and Behavior*, ed. W. G. Kinzey, pp. 3–23. New York: Aldine.

Fleagle, J. G. and Mittermeier, R. A. 1980. Locomotor behavior, body size, and comparative ecology of seven Surinam monkeys. *American Journal of Physical Anthropology* **52**, 301–314.

Fleagle, J. G. and Samonds, K. 1975. Physical growth of cebus monkeys (*Cebus albifrons*) during the first year of life. *Growth* **39**, 35–52.

Fleagle, J. G. and Schaffler, M. B. 1982. Development and eruption of the mandibular cheek teeth in *Cebus albifrons*. *Folia Primatologica* **38**, 158–169.

Fleagle, J. G., Mittermeier, R. A. and Skopec, A. L. 1981. Differential habitat use by *Cebus apella* and *Saimiri sciureus* in Central Suriname. *Primates* **22**, 361–367.

Fobes, J. and King, J. 1982. Auditory and chemoreceptive sensitivity in primates. *Primate Behavior*, ed. J. Fobes and J. King, pp. 245–270. New York: Academic Press.

Fontaine, R. P. 1980. Observations on the foraging association of double-toothed kites and white-faced capuchin monkeys. *Auk* **97**, 94–98.

1994. Play as physical flexibility training in five ceboid primates. *Journal of Comparative Psychology* **108**, 203–212.

Ford, S. M. 1986. Systematics of New World monkeys. In *Comparative Primate Biology, Volume 1: Systematics, Evolution and Anatomy*, ed. D. Swindler, pp. 73–135. New York: Alan R. Liss.

1990. Platyrrhine evolution in the West Indies. *Journal of Human Evolution* **19**, 237–254.

1994. Evolution of sexual dimorphism in body weight in platyrrhines. *American Journal Primatology* **34**, 221–244.

Ford, S. M. and Corruccini, R. 1985. Intraspecific, interspecific, metabolic and phylogenetic scaling in platyrrhine primates. In *Size and Scaling in Primate Biology*, ed. W. Jungers, pp. 401–433. New York: Plenum.

Ford, S. M. and Davis, L. 1992. Systematics and body size: implications for feeding adaptations in New World monkeys. *American Journal of Physical Anthropology* **88**, 415–468.

Ford, S. M. and Hobbs, D. 1996. Species definition and differentiation as seen in the postcranial skeleton of *Cebus*. In *Adaptive Radiations of Neotropical Primates*, ed. M. Norconk, A. Rosenberger, and Garber, pp. 229–249. New York: Plenum.

Fragaszy, D. M. 1986. Time budgets and foraging behavior in wedge-capped capuchins (*Cebus olivaceus*): age and sex differences. In *Current Perspectives in Primate Social Dynamics*, ed. D. Taub and F. King, pp. 159–174. New York: Van Nostrand.

1989. Activity states and motor activity in an infant capuchin monkey (*Cebus apella*) from birth through 11 weeks. *Developmental Psychobiology* **22**, 141–157.

1990a. Sex and age differences in the organization of behavior in wedge-capped capuchins, *Cebus olivaceus*. *Behavioral Ecology* **1**, 81–94.

1990b. Early behavioral development in capuchins (*Cebus*). *Folia Primatologica* **54**, 119–128.

1990c. Sensorimotor development in hand-reared and mother-reared tufted capuchins: A systems perspective on the contrasts. In *"Language" and Intelligence in Monkeys and Apes: Comparative Developmental Perspectives*, ed. S. T. Parker and K. R. Gibson, pp. 172–204. Cambridge: Cambridge University Press.

1995. State organization and activity in infant cebid monkeys (*Cebus* and *Saimiri*) in two rearing conditions. *International Journal of Comparative Psychology* **8**, 150–167.

Fragaszy, D. M. and Adams-Curtis, L. E. 1991. Generative aspects of manipulation in tufted capuchin monkeys (*Cebus apella*). *Journal of Comparative Psychology* **105**, 387–397.

1997. Developmental changes in manipulation in tufted capuchins from birth through two years and their relation to foraging and weaning. *Journal of Comparative Psychology* **111**, 201–211.

1998. Growth and reproduction in captive tufted capuchins (*Cebus apella*). *American Journal of Primatology* **44**, 197–203.

Fragaszy, D. M. and Bard, K. 1997. Comparison of development and life history in *Pan* and *Cebus*. *International Journal of Primatology* **18**, 683–701.

Fragaszy, D. M. and Boinski, S. 1995. Patterns of individual diet choice and efficiency of foraging in wedge-capped capuchin monkeys (*Cebus olivaceus*). *Journal of Comparative Psychology* **109**, 339–348.

Fragaszy, D. M. and Cummins, S. E. 1999. Prediction of movement by capuchins. *American Journal of Primatology* **49**, 53.

Fragaszy, D. M. and Mason, W. A. 1983. Comparisons of feeding behavior in captive squirrel and titi monkeys (*Saimiri sciureus* and *Callicebus moloch*). *Journal of Comparative Psychology* **97**, 310–326.

Fragaszy, D. M. and Mitchell, S. 1990. Hand preference and performance on unimanual and bimanual tasks in capuchin monkeys (*Cebus apella*). *Journal of Comparative Psychology* **104**, 275–282.

Fragaszy, D. M. and Perry, S. 2003a. *The Biology of Traditions. Models and Evidence*. Cambridge: Cambridge University Press.

2003b. Towards a biology of traditions. In *The Biology of Traditions. Models and Evidence*, ed. D. Fragaszy and S. Perry, pp. 1–32. Cambridge: Cambridge University Press.

Fragaszy, D. M. and Visalberghi, E. 1989. Social influences on the acquisition and use of tools in tufted capuchin monkeys (*Cebus apella*). *Journal of Comparative Psychology* **103**, 159–170.

1990. Social processes affecting the appearance of innovative behaviours in capuchin monkeys. *Folia Primatologica* **54**, 155–165.

Fragaszy, D. M., Adams-Curtis, L. E., Baer, J. F. and Carlson-Lammers, R. 1989. Forelimb dimensions and goniometry of the wrist and fingers in tufted capuchin monkeys (*Cebus apella*): developmental and comparative aspects. *American Journal of Primatology* **17**, 133–146.

Fragaszy, D. M., Baer, J. and Adams-Curtis, L. 1991. Behavioral development and maternal care in tufted capuchins (*Cebus apella*) and squirrel monkeys (*Saimiri sciureus*) from birth through seven months. *Developmental Psychobiology* **24**, 375–393.

1994a. Introduction and integration of strangers into captive groups of tufted capuchins (*Cebus apella*). *International Journal of Primatology* **15**, 399–420.

Fragaszy, D. M., Feuerstein, J. M. and Mitra, D. 1997a. Transfers of food from adults to infants in tufted capuchins (*Cebus apella*). *Journal of Comparative Psychology* **111**, 194–200.

Fragaszy, D. M., Galloway, A., Johnson-Pynn, J. and Brakke, K. 2002. The sources of skill in seriating cups in children, monkeys and apes. *Developmental Science* **5**, 118–131.

Fragaszy, D. M., Johnson-Pynn, J., Hirsh, E. and Brakke, K. 2003. Strategic navigation of two-dimensional alley mazes: comparing capuchin monkeys and chimpanzees. *Animal Cognition* **6**, 149–160.

Fragaszy, D., Landau, K. and Leighty, K. 2002. Inducting traditions in captive capuchins: Part 1. [Abstract] Caring for Primates. Abstracts of the XIXth Congress. The International Primatological Society, pp. 317–318. Beijing: Mammalogical Society of China.

Fragaszy, D. M., Visalberghi, E. and Galloway, A. T. 1997b. Infant tufted capuchin monkeys' behaviour with novel foods: opportunism, not selectivity. *Animal Behaviour* **53**, 1337–1343.

Fragaszy, D. M., Vitale, A. F. and Ritchie, B. 1994b. Variation among juvenile capuchins in social influences on exploration. *American Journal of Primatology* **32**, 249–260.

Frankie, G. W., Baker, H. G. and Opler, P. A. 1974. Comparative phenological studies of trees in tropical wet and dry forests in the lowlands of Costa Rica. *Journal of Ecology* **62**, 881–919.

Freese, C. H. 1976. Censusing *Alouatta palliata*, *Ateles geoffroyi*, and *Cebus capucinus* in the Costa Rican dry forest. In *Neotropical Primates: Field Studies and Conservation*, ed. R. W. Thorington and P. G. Heltne, pp. 4–9. Washington, DC: National Academy of Science.

1977. Food habits of white-faced capuchins *Cebus capucinus* L. (Primates: Cebidae) in Santa Rosa National Park, Costa Rica. *Brenesia* **10**, 43–56.

Freese, C. H. and Oppenheimer, J. R. 1981. The capuchin monkey, genus *Cebus*. In *Ecology and Behavior of Neotropical Primates, Vol. 1*, ed. A. F. Coimbra-Filho and R. A. Mittermeier, pp. 331–390. Rio de Janeiro: Academia Brasileira de Ciencias.

Fujita, K. 2001. How do capuchin monkeys (*Cebus apella*) complete occluded figures? *Advances in Ethology* **36**, 158.

Fujita, K., Kuroshima, H. and Asai, S. 2003. How do tufted capuchin monkeys (*Cebus apella*) understand causality involved in tool use? *Journal of Experimental Psychology* **29**, 233–242.

Fujita, K., Kuroshima, H. and Masuda, T. 2002. Do tufted capuchin monkeys (*Cebus apella*) spontaneously deceive opponents? A preliminary analysis of an experimental food-competition contest between monkeys. *Animal Cognition* **5**, 19–25.

Fulton, J. and Dousser de Barenne, J. 1933. The representation of the tail in the motor cortex of primates, with special reference to spider monkeys. *Journal of Cellular Comprehensive Physiology* **2**, 399–426.

Furbush, S. A. R. 1994. Feeding behavior in captive *Cebus apella*: the effects of age, sex and kinship. Masters Thesis, University of Georgia.

Galetti, M. 1990. Predation on the squirrel, *Sciurus aestuans* by capuchin monkeys, *Cebus apella*. *Mammalia* **54**, 152–154.

Galetti, M. and Pedroni, F. 1994. Seasonal diet of capuchin monkeys (*Cebus apella*) in a semideciduous forest in south-east Brazil. *Journal of Tropical Ecology* **10**, 27–39.

Galliari, C. A. 1985. Dental eruption in captive-born *Cebus apella*: from birth to 30 months old. *Primates* **26**, 506–510.

Galloway, A. T. 1998. Social inducement of feeding and food preferences in pair-housed capuchin monkeys, *Cebus apella*. Doctoral Dissertation, University of Georgia.

Garber, P. and Dolins, F. 1997. Testing learning paradigms in the field: Evidence for use of spatial and perceptual information and rule-based foraging in wild moustached tamarins. In *Adaptive Radiations of Neotropical World Primates*, ed. M. Norconk, A. Rosenberger and P. Garber, pp. 201–216. New York: Plenum.

Garber, P. A. and Brown, E. 2002. Experimental field study of tool use in wild capuchins (*Cebus capucinus*): learning by association or insight? *American Journal of Physical Anthropology* Supplement **34**, 74–75.

Garber, P. A. and Lambert, J. E. 1998. Primates as seed dispersers: ecological processes and directions for future research. *American Journal of Primatology* **45**, 3–8.

Garber, P. A. and Paciulli, L. 1997. Experimental field study of spatial memory and learning in wild capuchin monkeys (*Cebus capucinus*). *Folia Primatologica* **68**, 236–253.

Garber, P. A. and Rehg, J. A. 1999. The ecological role of the prehensile tail in white-faced capuchins (*Cebus capucinus*). *American Journal of Physical Anthropology* **110**, 325–339.

Garber, P. A. and Rehg, J. A. 2000. The ecology of group movement: evidence for the use of spatial, temporal, and social information in some primate foragers. In *On the Move: How and Why Animals Travel in Groups*, ed. S. Boinski and P. A. Garber, pp. 261–298. Chicago: University of Chicago Press.

Garcez, L. M., Silveira, F. T., El Harith, A., Lainson, R. and Shaw, J. J. 1997. Humoral responses of *Cebus apella* (Primates: Cebidae) to infections of *Leishmania* (*Leishmania*) *amazonensis*, *L.* (*Viana*) *lainsoni* and *L.* (*V.*) *brasiliensis* using the direct agglutination test. *Acta Tropica* **68**, 65–76.

Gärdenfors, U. 2000. Population viability analysis in the classification of threatened species: problems and potentials. *Ecological Bulletins* **48**, 181–190.

2001. Classifying threatened species at national versus global levels. *Trends in Ecology and Evolution* **16**, 511–516.

Garner, R. 1892. *The Speech of Monkeys*. New York: Charles L. Webster.

1900. *Apes and Monkeys: Their Life and Language*. Boston: Ginna Co.

Gascon, C., Bierregaard, Jr., R. O., Laurance, W. F. and Rankin-de-Merona, J. 2001. Deforestation and forest fragmentation in the Amazon. In *Lessons from Amazonia: The Ecology and Conservation of a Fragmented* Forest, ed. R. O. Bierregaard Jr., C. Gascon, T. E. Lovejoy and R. Mesquita, pp. 21–30. New Haven: Yale University Press.

Gautier-Hion, A., Duplantier, J.-M., Quris, R., Feer, F., Sourd, C., Decoux, J.-P., Dubost, G., Emmons, L., Erard, C., Hecketsweiler, P., Roussilhon, C. and Thiollay, J.-M. 1985. Fruit characters as a basis of fruit choice and seed dispersal in a tropical forest vertebrate community. *Oecologia* **65**, 324–337.

Gebo, D. L. 1992. Locomotor and postural behavior in *Alouatta palliata* and *Cebus capucinus*. *American Journal of Primatology* **26**, 277–290.

Geoffroy Saint-Hilaire, E. and Cuvier, F. 1824. *Histoire Naturelles des Mammifères*. Paris: Publisher unknown.

German, R. 1982. The functional morphology of caudal vertebrae in New World monkeys. *American Journal Physical Anthropology* **58**, 453–459.

Gibson, B. M. 2001. Cognitive maps not used by humans (*Homo sapiens*) during a dynamic navigational task. *Journal of Comparative Psychology* **115**, 397–402.

Gibson, E. J. and Pick, A. D. 2000. *An Ecological Approach to Perceptual Learning and Development*. New York: Oxford University Press.

Gibson, J. J. 1966. *The Senses Considered as Perceptual Systems*. Boston: Houghton-Mifflin.

1979. *The Ecological Approach to Visual Perception*. Boston: Houghton Mifflin.

Gibson. K. R. 1986. Cognition, brain size and the extraction of embedded food resources. In *Primate Ontogeny, Cognition and Social Behaviour*, ed. J. G. Else and P. C. Lee. Cambridge: Cambridge University Press.

1990. Tool use, imitation, and deception in a captive cebus monkey. In *"Language" and Intelligence in Monkeys and Apes*, ed. S. Parker and K. Gibson, pp. 205–218. Cambridge: Cambridge University Press.

Giraldeau, L. A. and Lefebvre, L. 1986. Exchangeable producer and scrounger roles in a captive flock of feral pigeons: a case for the skill pool effect. *Animal Behaviour* **34**, 797–783.

Glander, K. E. 1980. Reproduction and population growth in free-ranging mantled howling monkeys. *American Journal of Physical Anthropology* **53**, 25–36.

1982. The impact of plant secondary compound on primate feeding behavior. *Yearbook of Physical Anthropology* **25**, 1–18.

Gomes, U. R., Pessoa, D. M. A., Tomaz, C. and Pessoa, V. F. 2002. Color vision perception in the capuchin monkey (*Cebus apella*): A re-evaluation of procedures using

Munsell papers. *Behavioural Brain Research* **129**, 153–157.

Gonzalez-Kirchner, J. P. and Sainze de la Maza, M. 1998. Primate hunting by Guayami Amerindians in Costa Rica. *Human Evolution* **13**, 15–19.

Goodman, M., Porter, C. A., Czelusniak, J., Page, S. L., Schneider, H. Shoshani, J., Gunnell, G. and Groves, C. P. 1998. Toward a phylogenetic classification of primates based on DNA evidence complemented by fossil evidence. *Molecular Phylogenetic Evolution* **9**, 585–598.

Gould, S. J. 1977. *Ontogeny and Phylogeny*. Cambridge: Harvard University Press.

Graham, C. E., Keeling, M., Chapman, C., Cummins, L. B. and Haynie, J. 1973. Method of endoscopy in the chimpanzee: relations of ovarian anatomy, endometrial histology and sexual swelling. *American Journal of Physical Anthropology* **38**, 211–216.

Greenfield, P., Nelson, K. and Saltzman, E. 1972. The development of rulebound strategies for manipulating seriating cups: a parallel between action and grammar. *Cognitive Psychology* **3**, 291–310.

Greenlaw, J. S. 1967. Foraging behavior of the double-toothed kite in association with white-faced monkeys. *Auk* **64**, 596–597.

Gros-Louis, J. 2002. Contexts and behavioral correlates of trill vocalizations in wild white-faced capuchin monkeys. *American Journal of Primatology* **57**, 189–202.

Groves, C. P. 2001. *Primate Taxonomy*. Washington, DC: Smithsonian Institution Press.

Guidi, C. 1992. Il comportamento di gioco nel cebo dai cornetti (*Cebus apella*). Un esperimento sull'acquisizione dell'uso di strumenti. Università di Roma "La Sapienza", Italia, Tesi di laurea.

Haggerty, M. E. 1909. Imitation in monkeys. *Journal of Comparative Neurology and Psychology* **19**, 337–445.

Hakeem, A. Sandoval, R. G., Jones, M. and Allman, J. 1996. Brain and life span in primates. In *Handbook of the Psychology of Aging* (Fourth Edition) ed. J. E. Birren and K. W. Schaie, pp. 78–104. San Diego: Academic Press.

Hall, K. and Fedigan, L. M. 1997. Spatial benefits afforded by high rank in white-faced capuchins. *Animal Behaviour* **53**, 1069–1082.

Hamlett, G. W. D. 1939. Reproduction in American monkeys. I. Estrous cycles, ovulation and menstruation in *Cebus*. *Anatomical Record* **73**, 171–187.

Harada, M. L. and Ferrari, S. F. 1996. Reclassification of *Cebus kaapori* Queiroz, 1992, based on new specimens from eastern Pará, Brazil. Abstracts, XVIth Congress of the International Primatological Society, Madison, WI, USA, pp. 729.

Harada, M. L., Schneider, H., Schneider, M. P. C., Sampaio, I., Czelusniak, J. and Goodman, M. 1995. DNA evidence on the phylogenetic systematics of New World monkeys: support for the sister grouping of *Cebus* and *Saimiri* from two unlinked nuclear genes. *Molecular Phylogenetic Evolution* **4**, 331–349.

Harcourt A. H., Harvey, P. H., Larson, S. G. and Short, R. V. 1981. Testes weight, body weight, and breeding system in primates. *Nature* **293**, 55–57.

Hare, B., Addessi, E., Call, J., Tomasello, M. and Visalberghi, E. 2003. Do monkeys (*Cebus apella*) know what conspecifics do and do not see? *Animal Behaviour* **65**, 131–142.

Harlow H. F. 1951. Primate learning. In *Comparative Psychology*, ed. C. P. Stone, pp. 183–238. New York: Prentice-Hall.

Harlow, H. and Settlage, P. 1934. Comparative behavior of primates: VII. Capacity of monkeys to solve patterned string tests. *Journal of Comparative Psychology* **18**, 423–435.

Hartwig, W. C. 1994. Patterns, puzzles and perspectives on platyrrhine origins. In *Integrative Paths to the Past: Paleoanthropological Advances in Honor of F. Clark Howell*, ed. R. S. Corruccini and R. L. Ciocchon, pp. 69–93. New Jersey: Prentice Hall.

Hartwig, W. C. 1996. Perinatal life history traits in New World monkeys. *American Journal of Primatology* **40**, 99–130.

Harvey, P. H. and Clutton-Brock, T. H. 1985. Life history variation in primates. *Evolution* **39**, 559–581.

Harvey, P. H. and Krebs, J. 1990. Comparing brains. *Science* **249**, 140–146.

Hauser, M. 1997. Artifactual kinds and functional design features: what a primate understands without language. *Cognition* **64**, 285–308.

Heinrich, B. 2000. Testing insight in ravens. In *The Evolution of Cognition*, ed. C. Heyes and L. Huber, pp. 289–305. Cambridge, MA: MIT Press.

Heistermann, M., Möhle, U., Vervaecke, H., van Elsacker, L. and Hodges, J. K. 1996. Application of urinary and fecal steroid measurements for monitoring ovarian function and pregnancy in the bonobo (*Pan paniscus*) and evaluation of perineal swelling patterns in relation to endocrine events. *Biological Reproduction* **55**, 844–853.

Held, J. R. and Wolfe, T. L. 1994. Imports: current trends and usage. *International Journal of Primatology* **34**, 85–96.

Hemery, C., Deputte, B. and Fragaszy D. M. 1998. Human-socialized capuchins match objects but not actions. Abstracts of the XVII Congress of the International Primatological Society, p. 45. University of Antananarivo, Madagascar.

Henzi, S. P. and Barrett, L. 1999. The value of grooming to female primates. *Primates* **40**, 47–60.

Hernandez-Camacho, J. and Cooper, R. W. 1976. The nonhuman primates of Colombia. In *Neotropical Primates. Field Studies and Conservation*, ed. R. W. Thorington, Jr. and P. G. Heltne, pp. 35–69. Washington, DC: National Academy of Sciences.

Hershkovitz, P. 1949. Mammals of northern Colombia. Preliminary report No. 4: Monkeys (Primates) with taxonomic revisions of some forms. *Proceedings of the United States National Museum* **98**, 323–427.

Hervé, N. and Deputte, B. L. 1993. Social influence in manipulations of a capuchin monkey raised in a human environment: A preliminary case study. *Primates* **34**, 227–232.

Hien, E. and Deputte, B. L. 1997. Influence of a capuchin monkey companion on the social life of a person with quadriplegia: an experimental study. *Anthrozoös* **10**, 101–107.

Hill, K. and Padwe, J. 2000. Sustainability of Aché hunting in the Mbaracayu Reserve, Paraguay. In *Hunting for Sustainability in Tropical Forests*, ed. J. G. Robinson and E. L. Bennett, pp. 79–105. New York: Columbia University Press.

Hill, W. C. O. 1960. *Primates. Comparative Anatomy and Taxonomy. IV. Cebidae. Part A*. Edinburgh: University of Edinburgh Press.

Hilton-Taylor, C. 2002. *2002 IUCN Red List of Threatened Species*. Gland: The World Conservation Union (IUCN). Website: <www.redlist.org>.

Hinde, R. A. 1983. *Primate Social Relationships. An Integrated Approach*. Sunderland, MA: Sinauer Associates, Inc.

Hirsch, J. and Coxe, W. 1958. Representation of cutaneous tactile sensibility in cerebral cortex of *Cebus*. *Journal of Neurophysiology* **21**, 481–498.

Hladik, C. M. 1981. Diet and the evolution of feeding strategies among forest primates. In *Omnivorous Primates: Gathering and Hunting in Human Evolution*, ed. R. S. O. Harding and G. Teleki, pp. 215–254. New York: Columbia University Press.

Hladik, C. M. and Hladik, A. 1969. Raports trophiques entre vegetation et primates dans la forest de Barro Colorado (Panama). *Terre et Vie* **116**, 25–117.

Hladik, C. M., Hladik, A., Bousset, J., Valdebouze, P., Viroben, G. and Delort-Laval, J. 1971. Le regime alimentaire des Primates de L'ile de Barro-Colorado (Panama). *Folia Primatologica* **16**, 85–122.

Hladik, M. and Simmen, B. 1996. Taste perception and feeding behavior in nonhuman primates and human populations. *Evolutionary Anthropology* **5**, 58–71.

Hobhouse, L. 1915. *Mind In Evolution*. London: Macmillan.

Hofstadter, D. 1982. Variations on a theme as the essence of imagination. *Scientific American* **247**, 20–29.

Hollister, N. 1914. Four new mammals from tropical America. *Proceedings of the Biological Society of Washington* **27**, 103–106.

Holloway, R. L. 1979. Brain size, allometry and reorganization towards a synthesis. In *Development and Evolution of Brain Size*, ed. M. E. Hahn, pp. 61–88. New York: Academic Press.

Hook-Costigan, M. and Rogers, L. 1996. Hand preferences in New World primates. *International Journal of Comparative Psychology* **4**, 173–207.

Hopkins, W. D. 1996. Chimpanzee handedness revisited. 55 years since Finch (1941). *Psychonomic Bulletin and Review* **3**, 449–457.

Hopkins, W. D. and Pearson, K. 2000. Chimpanzee (*Pan troglodytes*) handedness: Variability across multiple measures of hand use. *Journal of Comparative Psychology* **114**, 126–135.

Howe, H. F. 1980. Monkey dispersal and waste of a neotropical fruit. *Ecology* **61**, 944–959.

1984. Implications of seed dispersal by animals for tropical reserve management. *Biological Conservation* **30**, 264–281.

Hrdy, S. B. and Whitten, P. L. 1987. Patterning of sexual activity. In *Primate Societies*, ed. B. B. Smuts, D. L. Cheney, R. M. Seyfarth, R. W. Wrangham and T. T. Struhsaker, pp. 370–384. Chicago: University of Chicago Press.

Hubrecht, R. C. 1986. Operation Raleigh primate census in the Maya Mountains, Belize. *Primate Conservation* **7**, 15–17.

Huffman, M. A. and Hirata, S. 2003. Biological and ecological foundation of primate behavioral traditions. In *The Biology of Traditions. Models and evidence*, ed. D. Fragaszy and S. Perry, pp. 267–296. Cambridge: Cambridge University Press.

Humle, T. and Matsuzawa, T. 2002. Ant-dipping among the chimpanzees of Bossou, Guinea, and some comparisons with other sites. *American Journal of Primatology* **58**, 133–148.

Hylander, W. L. 1979. Manibular function in *Galago crassicaudatus* and *Macaca fascicularis*: an *in vivo* approach to stress analysis of the mandible. *Journal of Morphology* **159**, 253–296.

1984. Stress and strain in the mandibular symphysis of primates: a test of competing hypotheses. *American Journal of Physical Anthropology* **64**, 1–46.

1985. Mandibular function and biomechanical stress and scaling. *American Zoologist* **25**, 315–330.

1988. Implications of *in vivo* experiments for interpreting the functional significance of "robust" australopithecine jaws. In *Evolutionary History of the Robust Australopithecines*, ed. F. E. Grine, pp. 55–83. New York: Aldine de Gruyter.

Inoue-Nakamura, N. and Matsuzawa, T. 1997. Development of stone tool use by wild chimpanzees (*Pan troglodytes*). *Journal of Comparative Psychology* **111**, 159–173.

Isbell, L. and van Vuren, D. 1996. Differential costs of locational and social dispersal and their consequences for female group-living primates. *Behaviour* **133**, 1–36.

Itakura, S. and Anderson, J. R. 1996. Learning to use experimenter-given cues during an object-choice task by a capuchin monkey. *Current Psychology of Cognition*, **15**, 103–112.

IUCN 2001. *IUCN Red List Categories and Criteria. Version 3.1*. Gland, Switzerland: The World Conservation Union (IUCN), Species Survival Commission (SSC).

Iwaniuk, A. N., Pellis, S. M. and Whishaw, I. Q. 1999. Is digital dexterity really related to corticospinal projection?: A re-analysis of the Heffner and Masterton data set using modern comparative statistics. *Behavioral Brain Research* **101**, 173–187.

Izar, P. 1999. Aspectos de ecologia e comportamento de um grupo de macacos-pregos (*Cebus apella*) em área de Mata Atlantica, São Paulo. Unpublished doctoral dissertation, Universidade de São Paulo.

In press. Female social relationships of *Cebus apella nigritus* in southeastern Atlantic forest: An analysis through ecological models of primate social evolution. *Behaviour*.

Izar, P. and Sato, T. 1997. Influência de abundância alimentar sobre a estrutura de espaçamento interindividual e relações de dominância em um grupo de macacos-prego (*Cebus apella*). In *A Primatologia no Brasil, Vol. 5*, ed. S. F. Ferrari and H. Schneider, pp. 249–267. Editora Universitária – UFPA. Belém.

Izawa, K. 1978. Frog-eating behavior of wild black-capped capuchin (*Cebus apella*). *Primates* **19**, 633–642.

1979. Foods and feeding behavior of wild black-capped capuchins (*Cebus apella*). *Primates* **20**, 57–76.

1980. Social behavior of the wild black-capped capuchin (*Cebus apella*). *Primates* **21**, 443–467.

1988. Preliminary report on social changes of black-capped capuchins (*Cebus apella*). *Field Studies of New World Monkeys, La Macarena, Colombia* **1**, 13–18.

1990a. Social changes within a group of wild black-capped capuchins (*Cebus apella*) in Colombia (II). *Field Studies of New World Monkeys, La Macarena, Colombia* **3**, 1–5.

1990b. Rat predation by wild capuchins (*Cebus apella*). *Field Studies of New World Monkeys, La Macarena, Colombia* **3**, 19–24.

1992. Social changes within a group of wild black-capped capuchins (*Cebus apella*). III. *Field Studies of New World Monkeys, La Macarena, Colombia* **7**, 9–14.

1994. Social changes within a group of wild black-capped capuchins IV. *Field Studies of New World Monkeys, La Macarena, Columbia*, **9**, 15–21.

1997. Social changes within a group of wild black-capped capuchins, V. *Field Studies of New World Monkeys, La Macarena, Colombia* **11**, 1–18.

1999. Social changes within a group of wild black-capped capuchins, VI. *Field Studies of New World Monkeys, La Macarena, Colombia* **13**, 1–13.

Izawa, K. and Mizuno, A. 1977. Palm-fruit cracking behavior of wild black-capped capuchins (*Cebus apella*). *Primates* **18**, 773–792.

Jack, K. M. 2001. Life history patterns of male white-faced capuchins (*Cebus capucinus*): male-bonding and the evolution of multimale groups. Doctoral Dissertation, University of Alberta.

Jack, K. M. and Fedigan, L. M. In press (a). Male dispersal patterns in white-faced capuchins (*Cebus capucinus*). Part 1: patterns and causes of natal emigration. *Animal Behaviour*.

Jack, K. M. and Fedigan, L. M. In press (b). Male dispersal patterns in white-faced capuchins (*Cebus capucinus*). Part 2: patterns and cases of secondary dispersal. *Animal Behaviour*.

Jacobs, G. H. 1994. Variations in primate color vision: mechanisms and utility. *Evolutionary Anthropology* **3**, 196–205.

1996. Primate photopigments and primate color vision. *Proceedings of the National Academy of Sciences of the USA* **93**, 577–581.

Jacobs. G. H. 1997. Color vision polymorphisms in New World monkeys: Implications for the evolution of primate trichromacy. In *New World Primates: Ecology, Evolution, and Behavior*, ed. W. G. Kinzey, pp. 45–74. New York: Aldine de Gruyter.

1998. A perspective on color vision in platyrrhine monkeys. *Vision Research* **38**, 3307–3313.

1999. Prospects for trichromatic color vision in male *Cebus* monkeys. *Behavioural Brain Research* **101**, 109–112.

Jalles-Filho, E. 1995. Manipulative propensity and tool use in capuchin monkeys. *Current Anthropology* **36**, 664–667.

Jalles-Filho, E., da Cunha R G.T. and Salm, R. A. 2001. Transport of tools and mental representation: is capuchin monkey tool behavior a useful model for Plio-Pleistocene hominid technology? *Journal of Human Evolution* **40**, 365–377.

Janson, C. H. 1983. Adaptation of fruit morphology to dispersal agents in a neotropical rainforest. *Science* **219**, 187–189.

1984. Female choice and mating system of the brown capuchin monkey *Cebus apella* (Primates: Cebidae). *Zeitschrift für Tierpsychologie* **65**, 177–200.

1985. Aggressive competition and individual food consumption in wild brown capuchin monkeys (*Cebus apella*). *Behavioral Ecology and Sociobiology* **18**, 125–138.

1986a. The mating system as a determinant of social evolution in capuchin monkeys (*Cebus*) In *Primate Ecology and Conservation*, ed. J. G. Else and P. C. Lee, pp. 169–179. Cambridge: Cambridge University Press.

1986b. Capuchin counterpoint: divergent mating and feeding habits distinguish two closely related monkey species of the Peruvian forest. *Natural History* **95**, 45–52.

1988. Food competition in brown capuchin monkeys (*Cebus apella*): quantitative effects of group size and tree productivity. *Behaviour* **105**, 53–76.

1990a. Ecological consequences of individual spatial choice in foraging groups of brown capuchin monkeys, *Cebus apella*. *Animal Behaviour* **40**, 922–934.

1990b. Social correlates of individual spatial choice in foraging groups of brown capuchin monkeys, *Cebus apella*. *Animal Behaviour* **40**, 910–921.

1996. Toward an experimental socioecology of primates: examples from Argentine brown capuchin monkeys (*Cebus apella nigrivittatus*). In *Adaptive Radiations of Neotropical Primates*, ed. M. Norconk, P. Garber and A. Rosenberger, pp. 309–325. New York: Plenum Press.

1998a. Capuchin counterpoint. Divergent mating and feeding habits distinguish two closely related monkey species of the Peruvian forest. In *The Primate Anthology. Essays on Primate Behavior, Ecology and Conservation from Natural History*, ed. R. L. Ciochon and R. A. Nisbett, pp. 153–160. Upper Saddle River, NJ: Prentice Hall.

1998b. Experimental evidence for spatial memory in foraging in wild capuchin monkeys (*Cebus apella*). *Animal Behaviour* **55**, 1229–1243.

2000. Primate socioecology: the end of a golden age. *Evolutionary Anthropology* **9**, 73–86.

Janson, C. H. and Boinski, S. 1992. Morphological and behavioral adaptations for foraging in generalist primates: the case of the Cebines. *American Journal of Physical Anthropology* **88**, 483–498.

Janson, C. H. and Di Bitetti, M. S. 1997. Experimental analysis of food detection in capuchin monkeys: effects of distance, travel speed, and resource size. *Behavioral Ecology and Sociobiology* **41**, 17–24.

Janson, C. H., Terborgh, J. and Emmons, L. H. 1981. Non-flying mammals as pollinating agents in the Amazonian forest. *Biotropica* **13**, 1–6. Supplement: Reproductive Botany.

Janson, C. H. and van Schaik, C. P. 1993. Ecological risk aversion in juvenile primates: Slow and steady wins the race. In *Juvenile Primates: Life History, Development, and Behavior*, ed. M. E. Pereira and L. A. Fairbanks, pp. 57–74. New York: Oxford University Press.

Janzen, D. H. 1973. Sweep samples of tropical foliage insects: effects of seasons, vegetation types, time of day and insularity. *Ecology* **54**, 687–708.

1983. Dispersal of seeds by vertebrate guts. In *Coevolution*, ed. J. Futuyma and M. Slatkin, pp. 232–262. Sunderland: Sinauer Associates.

Jenkins, F. A. and Krause, D. W. 1983. Adaptations for climbing in North American multituberculates (*Mammalia*). *Science* **220**, 712–715.

Jerison, H. J. 1973. *Evolution of the Brain and Intelligence*. New York: Academic Press.

Johnson, M. 1989. *The Body in the Mind*. Chicago: University of Chicago Press.

Johnson-Pynn, J. and Fragaszy, D. 2001. Do apes and monkeys rely upon conceptual reversibility? A review of studies using seriated nesting cups in children and nonhuman primates. *Animal Cognition* **4**, 315–324.

Johnson-Pynn, J., Fragaszy, D. M., Hirsh, E., Brakke, K. and Greenfield, P. 1999. Strategies used to combine seriated cups by chimpanzees (*Pan troglodytes*), bonobos (*Pan paniscus*), and capuchins (*Cebus apella*). *Journal of Comparative Psychology* **113**, 137–148.

Jones, C. G., Lawton, J. H. and Shachak, M. 1994. Organisms as ecosystem engineers. *Oikos* **69**, 373–386.

Jones, C. G., Lawton, J. H. and Shachak, M. 1997. Positive and negative effects of organisms as physical ecosystem engineers. *Ecology* **78**, 1946–1957.

Jorgensen, M. 1994. Investigating the antecedents of self-recognition using a video-task paradigm in capuchins (*Cebus apella*) and chimpanzees (*Pan troglodytes*). Doctoral Dissertation, University of California, Riverside.

Jorgensen, M., Suomi, S. and Hopkins, W. 1995. Using a computerized testing system to investigate the preconceptual self in nonhuman primates and humans. In *The Self in Infancy: Theory and Research*, ed. P. Rochat, pp. 243–256. New York: Elsevier.

Julliot, C. and Simmen, B. 1998. Food partitioning among a community of neotropical primates. *Folia Primatologica* **69**, 43–44.

Jungers, W. L. and Fleagle, J. G. 1980. Postnatal growth allometry of the extremities in *Cebus albifrons* and *Cebus apella*: a longitudinal and comparative study. *American Journal of Physical Anthropology* **53**, 471–478.

Kano, T. 1992. *The Last Ape. Pygmy Chimpanzee Behavior and Ecology*. Stanford: Stanford University Press.

Kay, R. F. 1981. The nut-crackers – a new theory of the adaptations of the Ramapithecinae. *American Journal of Physical Anthropology* **5**, 141–151.

1990. The phyletic relationships of extant and fossil Pithecinae (Platyrrhini, Anthropoidea). *Journal of Human Evolution* **19**, 175–208.

Kay, R. F., Madden, R. H., van Schaik, C. and Higdon, D. 1997. Primate species richness is determined by plant productivity: implications for conservation. *Proceedings of the National Academy of Science USA* **94**, 13023–13027.

Kellman, P. and Atterberry, M. 1998. *The Cradle of Knowledge. Development of Perception in Infancy*. Cambridge, MA: MIT Press.

King, B. J. 1986. Extracting, foraging, and the evolution of primate intelligence. *Human Evolution* **4**, 361–372.

1994a. *The Information Continuum: Social Information Transfer in Monkeys, Apes, and Humans*. Santa Fe, NM: SAR Press.

1994b. Primate infants as skilled information gatherers. *Pre- and Peri-natal Psychology Journal* **8**, 287–307.

1999. New directions in the study of primate learning. In *Mammalian Social Learning. Comparative and Ecological Perspectives*, ed. H. O. Box, and K. R. Gibson, pp. 17–32. Cambridge: Cambridge University Press.

King, J. E. 1982. Complex learning by primates. In *Primate Behavior*, ed. J. L. Fobes and J. E. King, pp. 327–360. New York: Academic Press.

Kinzey, W. 1974. Ceboid models for the evolution of hominoid dentition. *Journal of Human Evolution*, **3**, 193–203.

Kinzey, W. G. 1982. Distribution of primates and forest refuges. In *Biological Diversification in the Tropics*, ed. G. T. Prance, pp. 455–482. New York: Columbia University Press.

1997. *New World Primates. Ecology, Evolution and Behavior*. New York: Aldine Press.

Kirkwood, J. and Stathatos, K. 1992. *Biology, Rearing, and Care of Young Primates*. Oxford: Oxford University Press.

Klahr, D. 1994. Discovering the present by predicting the future. In *The Development of Future-oriented Processes*, ed. M. Haith, J. Benson, R. Roberts Jr. and B. Penningon, pp. 177–218. Chicago: University of Chicago Press.

Klein, L. L. and Klein, D. J. 1976. Neotropical primates: aspects of habitat usage, population density and regional distribution in La Macarena, Colombia. In *Neotropical Primates: Field Studies and Conservation*, ed. R. W. Thorington and P. G. Heltne, pp. 70–78. Washington, DC: National Academy of Sciences.

Klüver, H. 1933. *Behavior Mechanisms in Monkeys*. Chicago: Chicago University Press.

1937. Re-examination of implement-using behavior in a Cebus monkey after an interval of three years. *Acta Psychologica* **2**, 347–397.

Köhler, W. 1925/1976. *The Mentality of Apes*. New York: Liveright.

Konstant, W. R. 1996/1997. Funding for primate conservation: where has it originated? *Primate Conservation* **17**, 30–36.

Kummer, H. 1995. *In Quest of the Sacred Baboon*. Princeton: Princeton University Press.

Kuroshima, H., Fujita, K., Fuyuki, A. and Masuda, T. 2002. Understanding of the relationship between seeing and knowing by tufted capuchin monkeys (*Cebus apella*). *Animal Cognition* **5**, 41–48.

Kuypers, H. G. J. M. 1981. Anatomy of the descending pathways. In *Handbook of Physiology, Section I, The Nervous System, Motor Control, Part I*, ed. V. B. Brooks, pp. 597–666. Bethesda: American Physiological Society.

Lacreuse, A. and Fragaszy, D. M. 1997. Manual exploratory procedures and asymmetries for a haptic search task: A comparison between capuchins (*Cebus apella*) and humans. *Laterality* **2**, 247–266.

1999. Left hand preferences in capuchins (*Cebus apella*): role of spatial demands in manual activity. *Laterality* **4**, 65–78.

2003. Tactile exploration in nonhuman primates. In *Touching for Knowing*, ed. Y. Hatwell, A. Streri and E. Gentaz, pp. 221–234. Amsterdam: John Benjamins. (First published 2000 by Press Universitaires de France, Paris, as *Toucher pour Connaître*).

Langer, J. 1980. *The Origins of Logic: 6 to 12 Months*. New York: Academic Press.

1986. *The Origins of Logic: One to Two Years*. New York: Academic Press.

2000. The heterochronic evolution of primate cognitive development. In *Biology, Brains, and Behavior*, ed. S. Parker, J. Langer and M. McKinney, pp. 215–235. Santa Fe: School of American Research Press.

Langguth, A. and Alonso, C. 1997. Capuchin monkeys in the Caatinga: tool use and food habits during drought. *Neotropical Primates* **5**, 77–78.

Laska, M. 1996. Manual laterality in spider monkeys (*Ateles geoffroyi*) solving visually and tactually guided food reaching tasks. *Cortex* **32**, 717–726.

Lauer, U., Anke, T. and Hansske, F. 1991. Antibiotics from basidiomycetes.XXXVIII.2-methoxy-5-methyul-1,4-benzoquinone, a thromboxane A_2 receptor antagonist from *Lentinus adhaerens*. *Journal of Antibiotics* **44**, 59–65.

Lavallee, A. C. 1999. Capuchin *(Cebus apella)* tool use in a captive naturalistic environment. *International Journal of Primatology* **20**, 399–414.

Leca, J. B., Fornasieri, I. and Petit, O. 2002. Aggression and reconciliation in *Cebus capucinus*. *International Journal of Primatology* **23**, 979–998.

Lederman, S. and Klatzky, R. 1987. Hand movements: a window into haptic object recognition. *Cognitive Psychology* **19**, 342–368.

Lee, P. C. 1996. The meanings of weaning: growth, lactation, and life history. *Evolutionary Anthropology* **5**, 87–96.

Leeuwenberg, F. J. and Robinson, J. G. 2000. Traditional management of hunting by a Xavante community in Central Brazil: the search for sustainability. In *Hunting for Sustainibility in Tropical Forests*, ed. J. G. Robinson and E. L. Bennett, pp. 375–394. New York: Columbia University Press.

Lehman, S. M. 2000. Primate community structure in Guyana: a biogeographic analysis. *International Journal of Primatology* **21**, 333–351.

Leichnetz, G. R. and Gonzalo Ruiz, A. 1996. Prearcuate cortex in *Cebus* monkey has cortical and subcortical connections like the macaques' frontal eye field and projects to fastigial-recipient oculomotor-related brain-stem nuclei. *Brain Research Bulletin* **41**, 1–29.

Leighty, K. and Fragaszy, D. 2003. Joystick acquisition in tufted capuchins (*Cebus apella*). *Animal Cognition* **6**, 141–148.

Leighty, K., Fragaszy, D. M., Byrne, G. and Lussier, I. In press. The incidence of twinning in captive Cebus monkeys (*Cebus apella*). *Folia Primatologica*.

Lejeune, C., Installé, S., Houbeau, G. and Mercier, M. 1998. Etudes des interactions homme-animal en situation d'apprentissage instrumental dans le cadre du projet d'aide simienne. *XXXème Colloque de la Société Française pour l'Étude du Comportement Animal, Louvain-La-Neuve, 18–20 mai*.

Lemelin, P. 1995. Comparative and functional morphology of the prehensile tail in New World monkeys. *Journal of Morphology* **224**, 1–18.

Leutenegger, W. 1970. Beziehungen zwischen Neugeborenengrösse und dem Sexualdimorphismus am Becken bei simischen Primaten. *Folia Primatologica* **12**, 224–235.

Levy, L. E. and Bodini, R. 1986. Study of infant behavior in capuchin monkeys (*Cebus nigrivittatus*) in captivity. *Primatologia no Brasil* **2**, 151–161.

Lifshitz, K., O'Keeffe, R. T., Lee, K. L., Linn, G. S., Mase, D., Avery, J., Ee-Sing Lo and Cooper, T. B. 1991. Effects of extended depot fluphenazine treatment and withdrawal on social and other behaviors of *Cebus apella* monkeys. *Psychopharmacology* **105**, 492–500.

Lifshitz, K., O'Keeffe, R. T., Linn, G. S., Lee, K. L. Camp-Bruno, J. A. and Suckow, R. F. 1997. Effects of dopamine antagonists on *Cebus apella* monkeys with previous long-term exposure to fluphenazine. *Biological Psychiatry* **41**, 657–667.

Limongelli, L., Boysen, S. and Visalberghi, E. 1995. Comprehension of cause and effect relationships in a tool-using task by common chimpanzees (*Pan troglodytes*). *Journal of Comparative Psychology* **109**, 18–26.

Limongelli, L., Sonetti, M. G. and Visalberghi, E. 1994. Hand preference of tufted capuchins (*Cebus apella*) in tool-using tasks. In *Current Primatology, Vol. III: Behavioural Neuroscience, Physiology and Reproduction*, ed. J. R. Anderson, J. J. Roeder, B. Thierry and N. Herrenschmidt, pp. 9–15. Strasbourg: Université Louis Pasteur.

Linares, O. J. 1998. *Mamíferos de Venezuela*. Caracas: Sociedad Conservacionista Audubon de Venezuela.

Linn, G. and Javitt, D. C. 2001. Phencyclidine (PCP)-induced deficits of prepulse inhibition in monkeys. *Neurophysiology, Basic and Clinical* **12**, 117–120.

Linn, G., Lifshitz, K., O'Keeffe, R. T., Lee, K. and Camp-Lifshitz, J. 2001. Increased incidence of dyskinesias and other behavioral effects of re-exposure to neuroleptic treatment in social colonies of *Cebus apella* monkeys. *Psychopharmacology* **153**, 285–294.

Linn, G. S., Mase, D., Lafrancois, D., O'Keeffe, R. T. and Lifshitz, K. 1995. Social and menstrual cycle phase influences on the behavior of group-housed *Cebus apella*. *American Journal of Primatology* **35**, 41–57.

Linn, G. S., O'Keeffe, R. T., Schroeder, C., Lifshitz, K. and Javitt, D. C. 1999. Behavioral effects of chronic phencyclidine in monkeys. *NeuroReport* **10**, 2789–2793.

Lockman, J. 2000. A perception-action perspective on tool use development. *Child Development* **71**, 137–144.

Longino, J. T. 1984. True anting by the capuchin, *Cebus capucinus*. *Primates* **25**, 243–245.

Lönnberg, E. 1939. Remarks on some members of the genus *Cebus*. *Arkiv för Zoologi, Stockholm 31A* **23**, 1–24.

Lorenz, K. 1949. *Er redete mit dem Vieh, den Vögeln und den Fischen*. Wien: Borotha-Schoeler (Eng. trans., *King Solomon's Ring*, 1952).

Loy, J. 1975. The copulatory behavior of adult male patas monkeys, *Erythrocebus patas*. *Journal of Reproductive Fertility* **45**, 193–195.

1981. The reproductive and heterosexual behaviours of adult patas monkeys in captivity. *Animal Behaviour* **29**, 714–726.

Ludes, E. and Anderson, J. R. 1995. "Peat-bathing" by captive white-faced capuchin monkeys (*Cebus capucinus*). *Folia Primatologica* **65**, 38–42.

Ludes, E. and Anderson, J. R. 1996. Comparison of the behaviour of captive white-faced capuchin monkeys (*Cebus capucinus*) in the presence of four kinds of deep litter. *Applied Animal Behaviour Science* **49**, 293–303.

Ludes-Fraulob, E. and Anderson, J. 1999. Behaviour and preferences among deep litters in captive capuchin monkeys (*Cebus capucinus*). *Animal Welfare* **8**, 127–134.

Lumer, H. and Schultz, A. 1947. Relative growth of the limb segments and tail in *Ateles geoffroyi* and *Cebus capucinus*. *Human Biology* **19**, 53–67.

Lynch, J. W. 1998. Mating behavior in wild tufted capuchins (*Cebus apella)* in Brazil's Atlantic forest. *American Journal of Physical Anthropology* **26**, 153.

2001. Male behavior and endocrinology in wild tufted capuchin monkeys, *Cebus apella nigritus*. Doctoral dissertation, University of Wisconsin.

Lynch, J. W. and Rimoli, J. 2000. Demography of one group of tufted capuchin monkeys (*Cebus apella nigritus*) at the Estacao Biologica de Caratinga, Minas Gerais, Brazil. *Neotropical Primates* **8**, 44–49.

Lynch, J. W., Ziegler, T. E. and Strier, K. B. 2002. Individual and seasonal variation in fecal testosterone and cortisol levels of wild male tufted capuchin monkeys, *Cebus apella nigritus*. *Hormones and Behavior* **41**, 275–287.

Mace, G. M. and Lande, R. 1991. Assessing extinction threats: toward a re-evaluation of IUCN threatened species categories. *Conservation Biology* **5**, 148–157.

MacKinnon, K. C. 1995. Age differences in foraging patterns and spatial associations of the white-faced capuchin (*Cebus capucinus*) in Costa Rica. Master's thesis, University of Alberta.

2002. Social development of wild white-faced capuchin monkeys (*Cebus capucinus*) in Costa Rica: an examination of social interactions between immatures and adult males. Doctoral Dissertation, University of California, Berkeley.

MacPhee, R. D. E., Horovitz, I, Arredondo, O. and Jimenez Vasquez, O. 1995. A new genus for the extinct Hispaniolan monkey *Saimiri bernensis* Rimoli 1977, with notes on its systematic position. *American Museum Novitates* **3134**, 1–21.

MacPhee, R. D. E. and Rivero de la Calle, M. 1996. Accelerator mass spectrometry C14 age determination for the alleged "Cuban spider monkey", *Ateles (=Montaneia) anthropomorphus*. *Journal of Human Evolution* **30**, 89–94.

MacPhee, R. D. E. and Woods, C. A. 1982. A new fossil cebine from Hispaniola. *American Journal of Physical Anthropology* **58**, 419–436.

Mannu, M. 2002. O uso espontâneo de ferramentas por macacos-prego (*Cebus apella*) em condições de semi-liberdade: Desçricão e demografia. Unpublished doctoral dissertation, University of São Paulo.

Manson, J. H. 1999. Infant handling in wild *Cebus capucinus*: testing bonds between females? *Animal Behaviour* **57**, 911–921.

Manson, J. H., Perry, S. and Parish, A. R. 1997. Nonconceptive sexual behavior in bonobos and capuchins. *International Journal of Primatology* **18**, 767–786.

Manson, J. H., Rose, L. M., Perry, S. and Gros-Louis, J. 1999. Dynamics of female–female relationships in wild *Cebus capucinus*: data from two Costa Rican sites. *International Journal of Primatology* **20**, 679–706.

Mantecon, M. A. F. de, Mudry de Pargament, M. D. and Brown, A. D. 1984. *Cebus apella* de Argentina, distribución geográfica, fenotipo y cariotipo. *Revista do Museu Argentino de Ciéncias Naturales "Bernadino Rivadavia", Zoologia* **13**, 399–408.

Marchal, P. and Anderson, J. 1993. Mirror-image responses in capuchin monkeys (*Cebus capucinus*): Social responses and use of reflected environmental information. *Folia Primatologica* **61**, 165–173.

Marineros, L. and Gallegos, F. M. 1998. *Guía de Campo de los Mamíferos de Honduras*. Tegucigalpa, Honduras: Instituto Nacional de Ambiente y Desarrollo (INADES).

Martin, D. A. 1994. Kinship bias in *Cebus apella* (tufted capuchin monkey). Masters thesis, University of Georgia.

Martin, R. D. 1990. *Primate Origins and Evolution: A Phylogenetic Reconstruction*. London: Chapman and Hall.

Martin, R. D. and Harvey, P. H. 1985. Brain size allometry: ontogeny and phylogeny. In *Size and Scaling in Primate Biology*, ed. W. Jungers, pp. 147–173. New York: Plenum.

Martin, R. D., Chivers, D. J., MacLarnon, A. M. and Hladik, C. M. 1985. Gastrointestinal allometry in primates and other mammals. In *Size and Scaling in Primate Biology*, ed. W. L. Jungers, pp. 61–89. New York: Plenum.

Martinez, R. A., Moscarella, R. A., Aguilera, M. and Marquez, E. 2000. Update on the status of the Margarita Island capuchin, *Cebus apella margaritae*. *Neotropical Primates* **8**, 34–35.

Marzke, M. W. and Wullstein, K. L. 1996. Chimpanzee and human grips: a new classification with a focus on evolutionary morphology. *International Journal of Primatology* **17**, 117–139.

Masterson, T. J. 1995. Morphological relationships between the Ka'apor capuchin (*Cebus kaapori* Queiroz 1992) and other male *Cebus* crania: a preliminary report. *Neotropical Primates* **3**, 165–171.

1997. Sexual dimorphism and interspecific cranial form in two capuchin species: *Cebus albifrons* and *C. apella*. *American Journal of Physical Anthropology* **104**, 487–511.

Matheson, M. D., Johnson, J. S. and Feuerstein, J. 1996. Male reunion displays in tufted capuchin monkeys (*Cebus apella*). *American Journal of Primatology* **40**, 183–188.

Matsuzawa, T. 1996. Chimpanzee intelligence in nature and in captivity: isomorphism of symbol use and tool use. In *Great Ape Societies*, ed. W. C. McGrew, L. F. Marchant and T. Nishida, pp. 196–209. Cambridge: Cambridge University Press.

Matsuzawa, T. and Yamakoshi G. 1996. Comparison of chimpanzee material culture between Bossou and Nimba, West Africa. In *Reaching into Thought: The Mind of the Great Apes*, ed. A. E. Russon, K. A. Bard and S. T. Parker, pp. 211–232. Cambridge: Cambridge University Press.

Matsuzawa, T., Biro, D., Humle, T., Inoue- Nakamura, N., Tonooka, R. and Yamakoshi, G. 2001. Emergence of culture in wild chimpanzees: education by master-apprenticeship. In *Primate Origins of Human Cognition and Behavior*, ed. T. Matsuzawa, pp. 557–574. Tokyo: Springer-Verlag.

McCarthy, T. J. 1982. *Chironectes, Cyclopes, Cabassous* and probably *Cebus* in southern Belize. *Mammalia* **46**, 397–400.

McGonigle, B. and Chalmers, M. 2001. The growth of cognitive structure in monkeys and men. In *Animal Cognition and Sequential Behavior. Behavioral, Biological, and Computational Perspectives*. ed. S. Fountain *et al.*, pp. 287–332. New York: Kluwer Academic.

McGonigle, B., Chalmers, M. and Dickinson, T. 2003. Concurrent disjoint and reciprocal classification by *Cebus apella* in seriation tasks: evidence for hierarchical organization. *Animal Cognition* **6**, 185–197.

McGrew, W. C. 1992. *Chimpanzee Material Culture.* Cambridge: Cambridge University Press.

1998. Culture in nonhuman primates? *Annual Review Anthropology* **27**, 301–328.

McGrew, W. C. and Marchant, L. 1993. Primate ethology: a perspective on human and nonhuman handedness. In *Handbook of Psychological Anthropology*, ed. B. K. Bock, pp. 171–184.

McGrew, W. C. and Marchant, L. 1997. On the other hand: current issues and meta-analysis of the behavioral laterality of hand function in nonhuman primates. *Yearbook of Physical Anthropology* **40**, 201–232.

McKinney, M. L. 2000. Evolving behavioral complexity by extending development. In *Brains, Biology and Behavior*, ed. S. Parker, J. Langer and M. McKinney, pp. 25–40. Santa Fe: School of American Research.

McKinney, M. L. and McNamara, K. J. 1991. *Heterochrony: The Evolution of Ontogeny*. New York: Plenum.

Meldrum, D. and Lemelin, P. 1991. Axial skeleton of *Cebupithecia sarmientoi* (Pitheciinae, Platyrrhini) from the middle Miocene of La Venta, Colombia. *American Journal of Primatology* **25**, 69–89.

Mendes, F. D. C., Martins, L. B. R., Pereira, J. A. and Marquezan, R. F. 2000. Fishing with a bait: a note on behavioral flexibility in *Cebus apella*. *Folia Primatologica* **71**, 350–352.

Mendes Pontes, A. R. 1997. Habitat partitioning among primates in Maraca Island, Roraima, Northern Brazilian Amazonia. *International Journal of Primatology* **18**, 131–157.

Mendres, K. A. and de Waal, F. B. M. 2000. Capuchins do cooperate: the advantage of an intuitive task. *Animal Behaviour* **60**, 523–529.

Menzel, C. and Beck, B. 2000. Homing and detour behavior in golden lion tamarin social groups. In *On the Move: How and Why Animals Travel in Groups,* ed. S. Boinski and P. Garber, pp. 299–326. Chicago: Chicago University Press.

Miller, L. E. 1992a. The association between group size and foraging success in adult female wedge-capped capuchins (*Cebus olivaceus*). (Abstract) In *XIVth Congress of the International Primatological Society. Strasbourg, France.*

1992b. Socioecology of the wedge-capped capuchin monkey (*Cebus olivaceus*). Doctoral Dissertation, University of California, Davis.

1996. Behavioral ecology of wedge-capped capuchin monkeys (*Cebus olivaceus*). In *Adaptive Radiations of Neotropical Primates*, ed. M. A. Norconk, A. L. Rosenberger and P. A. Garber, pp. 271–288. New York: Plenum Press.

1997. Methods of assessing dietary intake: a case study from wedge-capped capuchins in Venezuela. *Neotropical Primates* **5**, 104–108.

1998a. Dietary choices in *Cebus olivaceus*: a comparison of data from Hato Pinero and Hato Masaguaral. *Primate Conservation* **18**, 42–50.

1998b. Fatal attack among wedge-capped capuchins. *Folia Primatologica* **69**, 89–92.

Milton, K. 1975. Urine rubbing behavior in the mantled howler monkey (*Alouatta palliata*). *Folia Primatologica* **23**, 105–112.

1984. The role of food processing factors in primate food choices. In *Adaptations for Foraging in Nonhuman Primates*, ed. P. S. Rodman and J. G. H. Cant, pp. 249–279. New York: Columbia University Press.

1985. Urine washing behavior in the woolly spider monkey (*Brachyteles arachnoides*). *Zeitschrift für Tierpsychologie* **67**, 154–160.

1993. Diet and primate evolution. *Scientific American* **269**, 70–77.

1996. Effects of botfly (*Alouattamyia baeri*) parasitism on a free-ranging howler monkey (*Alouatta palliata*) population in Panama. *Journal of the Zoological Society of London* **239**, 39–63.

Mineka, S. and Cook, M. 1993. Mechanisms involved in the observational conditioning of fear. *Journal of Experimental Psychology: General* **122**, 23–38.

Mineka, S., Davidson, M., Cook, M. and Keir, R. 1984. Observational conditioning of snake fear in rhesus monkeys. *Journal of Abnormal Psychology* **93**, 355–372.

Mitchell, B. J. 1989. Resources, group behavior and infant development in white-faced capuchin monkeys, *Cebus capucinus*. Doctoral Dissertation, University of California, Berkeley.

Mitchell, R. W. and Anderson, J. R. 1997. Pointing, withholding information, and deception in capuchin monkeys (*Cebus apella*). *Journal of Comparative Psychology* **111**, 351–361.

Mittermeier, R. A. 1977a. Distribution, synecology, and conservation of Surinam monkeys. Doctoral dissertation, Harvard University, Cambridge, MA.

1977b. *A Global Strategy for Primate Conservation*. Washington, DC: IUCN/SSC Primate Specialist Group.

1991. Hunting and its effect on wild primate populations in Suriname. In *Neotropical Wildlife Use and Conservation*, ed. J. G. Robinson and K. H. Redford, pp. 93–107. Chicago: The University of Chicago Press.

1996. New foundation dedicated to support for primate conservation. *Neotropical Primates* **4**, 65–66.

Mittermeier, R. A. and Cheney, D. L. 1987. Conservation of primates and their habitats. In *Primate Societies*, ed. B. B. Smuts, D. L. Cheney, R. M. Seyfarth, R. W. Wrangham and T. T. Struhsaker, pp. 477–490. Chicago: University of Chicago Press.

Mittermeier, R. A. and van Roosmalen, M. G. M. 1981. Preliminary observations on habitat utilization and diet in eight Surinam monkeys. *Folia Primatologica* **36**, 1–39.

Mittermeier, R. A., Kinzey, W. G. and Mast, R. B. 1993. Neotropical primate conservation. In *Primates of the Americas. Strategies for Conservation and Sustained Use in Biomedical Research*, ed. P. Arámbulo III, F. Encarnación, J. Estupiñán, H. Samamé, C. R. Watson and R. E. Weller, pp. 11–28. Columbus, OH: Battelle Press.

Mittermeier, R. A., Konstant, W. R. and Mast, R. B. 1994. Use of neotropical and malagasy primate species in biomedical research. *American Journal of Primatology* **34**, 73–80.

Mittermeier, R. A., Rylands, A. B. and Coimbra-Filho, A. F. 1988. Systematics: species and subspecies – an update. In *Ecology and Behavior of Neotropical Primates, Vol. 2*, ed. R. A. Mittermeier, A. B. Rylands, A. F. Coimbra-Filho and G. A. B. Fonseca, pp. 13–75. Washington, DC: World Wildlife Fund.

Mittermeier, R. A., Rylands, A. B. and Konstant, W. R. 1999. Primates of the world: an introduction. In *Walker's Primates of the World*, ed. R. M. Nowak, pp. 1–52. Baltimore, MD: Johns Hopkins University Press.

Møller, A. P. 1988. Ejaculate quality, testes size and sperm competition in primates. *Journal of Human Evolution* **17**, 479–488.

Moltedo, G. 1998. Il comportamento sessuale del cebo dai cornetti, *Cebus apella*, L., 1758. Un approccio sperimentale allo studio delle influenze sociali sulla scelta del partner. Tesi di Laurea, Università degli Studi di Roma "La Sapienza". Facoltà di Scienze Matematiche, Fisiche e Naturali.

Moscow, D. and Vaughn, C. 1987. Troop movement and food habits of white-faced monkeys in a tropical-dry forest. *Revista de Biologia Tropical* **35**, 287–297.

Moya, L., Encarnación, F., Aquino, R., Tapia, J., Ique, C. and Puertas, P. 1993. The status of the natural populations of primates and the benefits of sustained cropping. In *Primates of the Americas: Strategies for Conservation and Sustained Use in Biomedical Research*, ed. P. Arámbulo III, F. Encarnación, J. Estupiñán, H. Samamé, C. R. Watson and R. E. Weller, pp. 29–50. Columbus, OH: Batelle Press.

Moynihan. M. 1976. *The New World Primates*. Princeton: Princeton University Press.

Mudry de Pargament, M. D. and Slavutsky, I. R. 1987. Banding patterns of the chromosomes of *Cebus apella*: comparative studies between specimens from Paraguay and Argentina. *Primates* **28**, 111–117.

Myers, N., Mittermeier, R. A., Mittermeier, C. G., Fonseca, G. A. B. da and Kent, J. 2000. Biodiversity hotspots for conservation priorities. *Nature* **403**, 853–858.

Nadler, R. D., and Collins, D. C. 1991. Copulatory frequency, urinary pregnanediol, and fertility in great apes. *American Journal of Primatology* **24**, 167–179.

Nagle, C. A. and Denari, J. H. 1982. The reproductive biology of capuchin monkeys. In *International Zoo Yearbook*, ed. P. J. S. Olney, pp. 143–150. Dorchester: Dorset Press.

1983a. The Cebus monkey (*Cebus apella*). *Reproduction in New World Primates: New Models in Medical Science*, ed. J. Hearn, pp. 39–67. Lancaster: MTP Press.

1983b. The reproductive biology of capuchin monkeys. *International Zoo Yearbook* **22**, 144–150.

Nagle, C. A., Denari, J. H., Quiroga, S., Riarte, A., Merlo, A., Germino, N. I., Gomez-Argana, F. and Rosner, J. M. 1979. The plasma, a pattern of ovarian steroids during the menstrual cycle in capuchin monkeys (*Cebus apella*). *Biological Reproduction* **21**, 979–983.

Nagle, C. A., Digiano, L., Paul, N., Terlato, M., Quiroga, S. and Mendizabal, A. F. 1994. Interovarian communication for the control of follicular growth and corpus luteum function in the cebus monkey. *American Journal of Primatology* **34**, 19–28.

Nagle, C. A., Paul, N., Mazzoni, I., Quiroga, S., Torres, M., Mendizabal, A. F. and Farinati, Z. 1989. Interovarian relationships in the secretion of progesterone during the luteal phase of the capuchin monkey (*Cebus apella*). *Journal of Reproduction and Fertility* **85**, 389–396.

Nagle, C. A., Riarte, A., Quiroga, S., Azorero, R. M., Carril, M., Denari, J. H. and Rosner, J. M. 1980. Temporal relationship between follicular development, ovulation, and ovarian hormonal profile in the capuchin monkey (*Cebus apella*). *Biology of Reproduction* **23**, 629–635.

Napier, J. R. 1960. Studies of the hands of living primates. *Proceedings of the Zoological Society of London* **134**, 647–657.

1961. Prehensility and opposability in the hands of primates. *Symposia of the Zoological Society of London* **5**, 115–132.

1980. *Hands*. New York: Pantheon Books.

Napier, J. R. and Napier, P. 1967. *A Handbook of Living Primates*. London: Academic Press.

Natale, F. 1989. Patterns of object manipulation. In *Cognitive Structure and Development in Nonhuman Primates*, ed. F. Antinucci, pp. 145–161. Hillsdale, NJ: Lawrence Erlbaum Associates.

Naughton-Treves, L., Treves, A., Chapman, C. and Wrangham, R. 1998. Temporal patterns of crop-raiding by primates: linking food availability in croplands and adjacent forest. *Journal of Applied Ecology* **35**, 596–606.

Nepstad, D. C., Veríssimo, A., Alencar, A., Nobre, C., Lima, E., Lefebvre, P., Schelinger, P., Potter, C., Moutinho, P., Mendoza, E., Cochrane, M. and Brooks, V. 1999. Large-scale impoverishment of Amazonian forests by logging and fire. *Nature* **398**, 505–508.

Newcomer, M. W. and DeFarcy, D. D. 1985. White-faced capuchin (*Cebus capucinus*) predation on a nestling coati (*Nasua narica*). *Journal of Mammalogy* **66**, 185–186.

Niemitz, C. 1990. The evolution of primate skin structures in relation to gravity and locomotor patterns. In *Gravity, Posture, and Locomotion in Primates*, ed. F. Jouffroy, M. Stack and C. Niemitz, pp. 129–156. Florence: Editrice Il Sedecesimo.

Nishimura, A. 1988. Mating behavior of woolly monkeys, *Lagothrix lagotricha*, at La Macarena, Colombia. *Field studies of New World monkeys, La Macarena, Colombia* **1**, 19–27.

Nolte, A. 1958. Beobachtungen uber das Instinktverhalten von Kapuzineraffen (*Cebus apella*) in der Gegenshaft. *Behaviour* **12**, 183–207.

Norconk, M. A. 1990. Introductory remarks: ecological and behavioral correlates of polyspecific primate troops. *American Journal of Primatology* **21**, 81–85.

Norris, J. C. 1990a. The semantics of *Cebus olivaceus* alarm calls: object designation and attribution (Venezuela). Doctoral Dissertation, University of Florida.

1990b. The semantics of *Cebus olivaceus* alarm calls: object designation and attribution. *American Journal of Primatology* **20**, 216.

Nowak, R. M. 1999. *Walker's Primates of the World*. Baltimore: Johns Hopkins Press.

O'Brien, T. G. 1988. Parasitic nursing behavior in the wedge-capped capuchin monkey (*Cebus olivaceus*). *American Journal of Primatology* **16**, 341–344.

1991. Male–female social interactions in wedge-capped capuchin monkeys: benefits and costs of group living. *Animal Behaviour* **41**, 555–568.

1993a. Allogrooming behaviour among adult female wedge-capped capuchin monkeys. *Animal Behaviour* **46**, 499–510.

1993b. Asymmetries in grooming interactions between juvenile and adult female wedge-capped capuchin monkeys. *Animal Behaviour* **46**, 929–938.

O'Brien, T. G. and Robinson, J. G. 1991. Allomaternal care by female wedge-capped capuchin monkeys: effects of age, rank and relatedness. *Behaviour* **119**, 30–50.

1993. Stability of social relationships in female wedge-capped capuchin monkeys. In *Juvenile Primates: Life History, Development, and Behavior*, ed. M. E. Pereira and L. A. Fairbanks, pp. 197–210. New York: Oxford University Press.

O'Keefe, R. T. and Lifshitz, K. 1989. Nonhuman primates in neurotoxicity screening and neurobehavioral toxicity studies. *Journal of the American College of Toxicology* **8**, 127–140.

O'Malley, R. C. 2002. Variability in foraging and food processing techniques among white-faced capuchins (*Cebus capucinus*) in Santa Rosa National Park, Costa Rica. *Master's Thesis*, University of Alberta.

Oates, J. F. 1996. *African Primates: Status Survey and Conservation Action Plan*. Revised edition. Gland: The World Conservation Union (IUCN).

Oetting, J., Bolen, R., Evans, S., Garber, P. and Fragaszy, D. 1994. Individual differences in response to a novel environment in *Cebus apella*. *American Journal of Primatology* **33**, 231–232.

Oliver, W. L. R. and Santos, I. B. 1991. Threatened endemic mammals of the Atlantic forest region of south-east Brazil. *Wildlife Preservation Trust, Special Scientific Report* **4**, 1–126.

Olmos, F. 1990. Nest predation of *Plumbeous ibis* by capuchin monkeys and greater black hawk. *Wilson Bulletin* **102**, 169–170.

Oppenheimer, J. R. 1968. Behavior and ecology of the white-faced monkey, *Cebus capucinus*, on Barro Colorado Island, Canal Zone. Doctoral Dissertation, University of Illinois.

1973. Social and communicative behavior in the *Cebus* monkey. In *Behavioral Regulators of Behavior in Primates*, ed. C. R. Carpenter, pp. 251–271. Lewisburg, PA: Bucknell University Press.

1977. Forest structure and its relation to activity of the capuchin monkey (*Cebus*). In *Use of Non-human Primates in Biomedical Research*, ed. M. R. N. Prasad and T. C. Anand Kumar, pp. 74–84. New Delhi: Indian National Science Academy.

1982. *Cebus capucinus*: home range, population dynamics, and interspecific relationships. In *The Ecology of a Tropical Forest: Seasonal Rhythms and Long-Term Changes*, ed. E. G. Leigh, Jr., A. S. Rand, and D. M. Windsor, pp. 253–271. Washington, DC: Smithsonian Institution Press.

Oppenheimer, J. R. and Lang, G. E. 1969. *Cebus* monkeys: effect on branching of *Gustavia* trees. *Science* **165**, 187–188.

Ortiz, M. E., Gajardo, G., León, C. G., Herrera, E., Valdez, E. and Croxatto, H. B. 1995. Sperm migration through the female genital tracts of the New World monkey *Cebus apella*. *Biological Reproduction* **52**, 1121–1128.

Osorio, D. and Vorobyev, M. 1996. Color vision as an adaption to frugivory in primates. *Proceedings of the Royal Society London Series B* **263**, 593–599.

Ottoni, E. B. and Mannu, M. 2001 Semi-free ranging tufted capuchin monkeys (*Cebus apella*) spontaneously use tools to crack open nuts. *International Journal of Primatology* **22**, 347–358.

Overman, W. 1990. Performance on traditional match-to-sample, non-match-to-sample, and object discrimination tasks by 12–32 month-old children: A developmental progression. In *The Development and Neural Basis of Higher Cognitive Functions*, ed. A. Diamond, pp. 365–383. New York: New York Academy of Sciences.

Oxford, P. 2003. Cracking monkeys. *BBC Wildlife* **21**(2), 26–29.

Panger, M. 1998. Object-use in free-ranging white-faced capuchins (*Cebus capucinus*) in Costa Rica. *American Journal of Physical Anthropology* **106**, 311–321.

Panger, M., Perry, S., Rose L., Gros-Louis, J., Vogel, E., MacKinnon, K. and Baker, M. 2002. Cross-site differences in the foraging behavior of white-faced capuchins (*Cebus capucinus*). *American Journal of Physical Anthropology* **119**, 52–66.

Parker, C. E. 1974a. The antecedents of man the manipulator. *Journal of Human Evolution* **3**, 493–500.

1974b. Behavioral diversity in ten species of nonhuman primates. *Journal of Comparative and Physiological Psychology* **87**, 930–937.

Parker, S. T. and Gibson, K. R. 1977. Object manipulation, tool use and sensorimotor intelligence as feeding adaptations in cebus monkeys and great apes. *Journal of Human Evolution* **6**, 623–641.

1979. A developmental model for the evolution of language and intelligence in early hominids. *Behavioral and Brain Sciences* **2**, 367–408.

Parker, S. T. and Potì, P. 1990. The role of innate motor patterns in ontogenetic and experiential development of intelligent use of sticks in cebus monkeys. In *"Language" and Intelligence in Monkeys and Apes: Comparative Developmental Perspectives*, ed. S. T. Parker and K. R. Gibson, pp. 219–243. New York: Cambridge University Press.

Parr, L., Hopkins, W. and de Waal, F. 1997. Haptic discrimination in capuchin monkeys: evidence of manual specialization. *Neuropsychologia* **35**, 143–152.

Parr, L. A., Matheson, M. D., Bernstein, I. S. and de Waal, F. B. M. 1997. Grooming down the hierarchy: allogrooming in captive brown capuchin monkeys, *Cebus apella*. *Animal Behaviour* **54**, 361–367.

Passingham, R. E. 1981. Primate specialization in brain and intelligence. *Symposia of the Zoological Society of London* **46**, 361–387.

Pereira, L. H., de Resende, D. M., de Melo, A. L. and Mayrink, W. 1993. Primatas platirrinos e leishmanioses da região neotropical americana. In *A Primatologia no Brasil* (Vol. 4), ed. M. E. Yamamoto and M. B. Cordeiro de Souza, pp. 245–254. Brasília: Sociedade Brasileira de Primatologia.

Peres, C. A. 1994. Primate responses to phenological changes in an Amazonian Terra Firme Forest. *Biotropica* **26**, 98–112.

1999. Effects of subsistence hunting and forest types on the structure of Amazonian primate communities. In *Primate Communities*, ed. J. G. Fleagle, C. Janson and K. E. Reed, pp. 268–283. Cambridge: Cambridge University Press.

2000a. Evaluating the impact and sustainability of subsistence hunting at multiple amazonian forest sites.

In *Hunting for Sustainability in Tropical Forests*, ed. J. G. Robinson and E. L. Bennett, pp. 31–56. New York: Columbia University Press.

Peres, C. A. and Janson, C. H. 1999. Species co-existence, distribution and environmental determinants of neotropical primate richness: a community-level zoogeographic analysis. In *Primate Communities*, ed. J. G. Fleagle, C. Janson and K. E. Reed, pp. 55–74. Cambridge: Cambridge University Press.

Perry, S. 1995. Social relationships in wild white-faced capuchin monkeys, *Cebus capucinus*. Doctoral Dissertation, University of Michigan.

1996a. Female–female social relationships in wild white-faced capuchin monkeys, *Cebus capucinus*. *American Journal of Primatology* **40**, 167–182.

1996b. Intergroup encounters in wild white-faced capuchin monkeys, *Cebus capucinus*. *American Journal of Primatology* **17**, 309–330.

1997. Male–female social relationships in wild white-faced capuchins (*Cebus capucinus*). *Behaviour* **134**, 477–510.

1998a. A case report of a male rank reversal in a group of wild white-faced capuchins. (*Cebus capucinus*). *Primates* **39**, 51–70.

1998b. Male–male social relationships in wild white-faced capuchin monkeys, *Cebus capucinus*. *Behaviour* **135**, 139–172.

Perry, S., and Rose, L. 1994. Begging and transfer of coati meat by white-faced capuchin monkeys, *Cebus capucinus*. *Primates* **35**, 409–415.

Perry, S., Baker, M., Fedigan, L., Gros-Louis, J., Jack, K., MacKinnon, K., Manson, J., Panger, M., Pyle, K. and Rose, L. 2003a. Social conventions in wild white-faced capuchin monkeys: evidence for traditions in a neotropical primate. *Current Anthropology* **44**, 241–268.

Perry, S., Panger, M., Rose, L., Baker, M., Gros-Louis, J., Jack, K., MacKinnon, K., Manson, J., Fedigan, L. and Pyle, K. 2003b. Traditions in wild white-faced capuchin monkeys. In *The Biology of Traditions. Models and Evidence*, ed. D. Fragaszy and S. Perry, pp. 391–425. Cambridge: Cambridge University Press.

Pessoa, V., Tavares, M., Aguiar, L., Gomes, U. R. and Tomaz, C. 1997. Color vision discrimination in the capuchin monkey *Cebus apella*: evidence for trichromacy. *Behavioral Brain Research* **89**, 285–288.

Petras, J. 1979. Some efferent connections of the motor and somatosensory cortex of simian primates and felid, canid and procyonid carnivores. *Annals of the New York Academy of Science* **67**, 469–505.

Phillips, K. 1995. Resource patch size and flexible foraging in white-faced capuchins (*Cebus capucinus*). *International Journal of Primatology* **16**, 509–519.

1998. Tool use in wild capuchin monkeys *(Cebus albifrons trinitatis)*. *American Journal of Primatology* **46**, 259–261.

Phillips, K. A., Bernstein, I. S., Dettmer, E. L., Devermann, H. and Powers, M. 1994. Sexual behavior in brown capuchins (*Cebus apella*). *International Journal of Primatology* **15**, 907–917.

Piaget, J. 1952. The *Origins of Intelligence in Children*. New York: Harcourt Brace.

Pinto, L. P. de S., Coimbra-Filho, A. F. and Rylands, A. B. 1998. *Cebus apella xanthosternos* (Wied, 1820). In *Livro Vermelho das Espécies Ameaçadas de Extinção da Fauna de Minas Gerais*, ed. A. B. M. Machado, G. A. B. da Fonseca, R. B. Machado, L. M. de S. Aguiar, and L. V. Lins, pp. 86–89. Belo Horizonte: Fundação Biodiversitas.

Pinto, O. M. O. 1941. Da validez de *Cebus robustus* Kuhl e de suas relações com as formas mais afins. *Papéis Avulsos, Departamento de Zoologia, Secretaria da Agricultura, São Paulo* **1**, 111–120.

Podolsky, R. D. 1990. Effects of mixed-species associations on resource use by *Saimiri sciureus* and *Cebus apella*. *American Journal of Primatology* **21**, 147–158.

Portmann, A. 1990. A *Zoologist looks at Humankind*. New York: Columbia University Press.

Potì, P. 2000. Aspects of spatial cognition in capuchins (*Cebus apella*): frames of reference and scale of space. *Animal Cognition* **3**, 69–77.

Potì, P. and Antinucci, F. 1989. Logical operations. In *Cognitive Structure and Development in Nonhuman Primates*, ed. F. Antinucci, pp. 189–228. Hillsdale, NJ: Lawrence Erlbaum Associates.

Povinelli, D. J. 2000. *Folk Physics for Apes. The Chimpanzee's Theory of how the World Works*. Oxford: Oxford University Press.

Prance, G. T. 1980. A note on the probable pollination of *Combretum* by Cebus monkeys. *Biotropica* **12**, 239.

Provine, R. 2000. *Laughter*. New York: Penguin.

Queiroz, H. L. 1992. A new species of capuchin monkey, genus *Cebus* Erxleben 1977 (Cebidae, Primates) from eastern Brazilian Amazonia. *Goeldiana Zoologia* **15**, 1–3.

Radinsky, L. B. 1972. Endocasts and studies of primate brain information. In *The Functional and Evolutionary Biology of Primates*, ed. R. Tuttle, pp. 175–184. Chicago: Aldine Atherton.

Rafferty, K. L. and Ruff, C. B. 1994. Articular structure and function in *Hylobates, Colobus*, and *Papio*. *American Journal of Physical Anthropology* **94**, 395–408.

Rapaport, L. G. and Ruiz-Miranda, C. 2002. Tutoring in wild golden lion tamarins. *International Journal of Primatology* **23**, 1063–1070.

Reader, S. M. 2003. Relative brain size and the distribution of innovation and social learning across the nonhuman primates. In *Traditions in Nonhuman Animals: Models and Evidence*, ed. D. Fragaszy and S. Perry, pp. 56–93. Cambridge: Cambridge University Press.

Reaux, J. E. and Povinelli, D. J. 2000. The trap-tube problem. In *Folk Physics for Apes. The Chimpanzee's Theory of how the World Works*, ed. D. J. Povinelli, pp. 108–131. Oxford: Oxford University Press.

Recabarren, M. P., Vergara, M., Martínez, M. C., Gordon, K. and Serón-Ferrè, M. 2000. Impact of lactation upon fertility in the New World primate capuchin monkey (*Cebus apella*). *Journal of Medical Primatology* **29**, 350–360.

Reed, E. S. 1996. *Encountering the World. Toward an Ecological Psychology*. Oxford: Oxford University Press.

Reed, K. E. and Fleagle, J. G. 1995. Geographic and climatic control of primate diversity. *Proceedings of the National Academy of Science USA* **92**, 7874–7876.

Reid, F. A. 1997. *A Field Guide to the Mammals of Central America and Southeast Mexico*. Oxford: Oxford University Press.

Resende, B., Izar, P. and Ottoni, E. 2003. Interaction between social play and nutcracking behavior in semifree tufted capuchin monkeys (*Cebus apella*). *Revista de Etologia* **5**, 198–199.

Resende, M., Tavares, C. and Tomaz, C. 2002. Ontogenetic dissociation between habit learning and recognition memory in capuchin monkeys (*Cebus apella*). *Neurobiology of Learning and Memory* **79**, 19–24.

Rettig, N. L. 1978. Breeding behavior of the Harpy Eagle (*Harpia harpyja*). *Auk* **95**, 629–643.

Richard, A. F. 1985. *Primates in Nature*. San Francisco: W. H. Freeman and Co.

Richardson, K., Washburn, D., Hopkins, W., Savage-Rumbaugh, S. and Rumbaugh, D. 1990. The NASA/LRC computerized test system. *Behavior Research Methods, Instruments and Computers* **22**, 127–131.

Ridgely, R. S. 1976. *A Guide to the Birds of Panama*. Princeton: Princeton University Press.

Rilling, J. K. and Insel, T. 1999. The primate neocortex in comparative perspective using magnetic resonance imaging. *Journal of Human Evolution* **37**, 191–233.

Ritchie, B. G. and Fragaszy, D. M. 1988. Capuchin monkey (*Cebus apella*) grooms her infant's wound with tools. *American Journal of Primatology* **16**, 345–348.

Riviello, M. C. and Wirz, A. 2001. Haematology and blood chemistry for *Cebus apella* in relation to sex and age. *Journal of Medical Primatology* **30**, 308–312.

Riviello, M. C., Visalberghi, E. and Blasetti, A. 1993. Mirror responses in tufted capuchin monkeys (*Cebus apella*). *Hystrix* **4**, 35–44.

Robinson, J. G. 1979. Correlates of urine washing in the wedge-capped capuchin *Cebus nigrivittatus*. In *Vertebrate Ecology in the Northern Neotropics*, ed. J. F. Eisenberg, pp. 137–143. Washington, DC: Smithsonian Institution Press.

1981. Spatial structure in foraging groups of wedge-capped capuchin monkeys *Cebus nigrivittatus*. *Animal Behaviour* **29**, 1036–1056.

1982a. Vocal systems regulating within-group spacing. In *Primate Communication*, ed. C. R. Snowdon, C. H. Brown and M. R. Petersen, pp. 94–116. Cambridge: Cambridge University Press.

1982b. Intrasexual competition and mate choice in primates. *American Journal of Primatology* **Supplement 1**, 131–144.

1984a. Diurnal variation in foraging and diet in the wedge-capped capuchin *Cebus olivaceus*. *Folia Primatologca* **43**, 216–228.

1984b. Syntactic structures in the vocalizations of wedge-capped capuchin monkeys, *Cebus nigrivittatus*. *Behaviour* **90**, 46–79.

1986. Seasonal variation in use of time and space by the wedge-capped capuchin monkey *Cebus olivaceus*: implications for foraging theory. *Smithsonian Contributions to Zoology* **431**, iii-60.

1988a. Demography and group structure in wedge-capped capuchin monkeys, *Cebus olivaceus*. *Behaviour* **104**, 202–231.

1988b. Group size in wedge-capped capuchin monkeys (*Cebus olivaceus*) and the reproductive success of males and females. *Behavioral Ecology and Sociobiology* **23**, 187–197.

Robinson, J. G. and Janson, C. H. 1987. Capuchins, squirrel monkeys, and atelines: socioecological convergence with

Old World primates. In *Primate Societies*, ed. B. B. Smuts, D. L. Cheney, R. M. Seyfarth, R. W. Wrangham and T. T. Struhsaker, pp. 69–82. Chicago: University of Chicago Press.

Robinson J. G., Valderrama, X. and Melnick, D. J. 2000. Mechanisms of female dispersal in a female-bonded species. [Abstract] *American Journal of Physical Anthropology* **Suppl 30**, 264.

Rocha, J. V., dos Reis, N. R. and Sekiama, M. L. 1998. Uso de ferramentas por *Cebus apella* (Linnaeus) (Primates, Cebidae) para obtenção de larvas de Coleoptera que parasitam sementes de *Syagrus romanzoffianum* (Cham.) Glassm. (Arecaceae). *Revista Brasileira de Zoologia* **15**, 945–950.

Rodríguez-Luna, E., Cortés-Ortiz, L., Mittermeier, R. A. and Rylands, A. B. 1996. Plan de acción para los primates Mesoamericanos. Unpublished report, IUCN/SSC Primate Specialist Group, Gland.

Roeder, J. J. and Anderson, J. R. 1991. Urine washing in brown capuchin monkeys (*Cebus apella*): testing social and nonsocial hypotheses. *American Journal of Primatology* **24**, 55–60.

Rogers, S. A. 1996. Feeding behaviors of captive capuchin monkeys as a function of food type, abundance, and distribution. Doctoral Dissertation, University of Georgia.

Rolls, E. T. 2000. Memory systems in the brain. *Annual Review Psychology* **51**, 599–630.

Romanes, G. J. 1883/1977. *Significant contributions to the history of psychology 1750–1920*. Washington, DC: University Publications of America.

Rosa, M. G. P., Pinon, M. C., Gattass, R. and Sousa, A. P. B. 2000. "Third tier" ventral extrastriate cortex in the New World monkey, *Cebus apella*. *Experimental Brain Research* **132**, 287–305.

Rose, L. M. 1994a. Sex differences in diet and foraging behavior in white-faced capuchins (*Cebus capucinus*). *International Journal of Primatology* **15**, 95–114.

1994b. Benefits and costs of resident males to females in white-faced capuchins, *Cebus capucinus*. *American Journal of Primatology* **32**, 235–248.

1997. Vertebrate predation and food-sharing in *Cebus* and *Pan*. *International Journal of Primatology* **18**, 727–766.

1998. Behavioral ecology of white-faced capuchins (*Cebus capucinus*) in Costa Rica. Doctoral Dissertation, Washington University.

2001. Meat and the early human diet: insights from neotropical primate studies. In *Meat Eating and Human Evolution*, ed. C. B. Stanford and H. T. Bunn, pp. 141–159. Oxford: Oxford University Press.

Rose, L. M. and Fedigan, L. M. 1995. Vigilance in white-faced capuchins, *Cebus capucinus*, in Costa Rica. *Animal Behaviour* **49**, 63–70.

Rose, L., Perry, S., Panger, M., Jack, K., Manson, J., Gros-Louis, J., Mackinnon, K. and Vogel, E. 2003. Interspecific interactions between *Cebus capucinus* and other species: data from three Costa Rican sites. *International Journal of Primatology* **24**, 759–796.

Rose, M. D. 1992. Kinematics of the trapezium-1st metacarpal joint in extant anthropoids and Miocene hominids. *Journal of Human Evolution* **22**, 255–266.

Rosenberger, A. L. 1979. Phylogeny, evolution, and classification of New World monkeys (Platyrrhini, Primates). Doctoral Dissertation, City University of New York.

1981. Systematics: the higher taxa. In *Ecology and Behavior of Neotropical Primates, Vol. 1*, ed. A. F. Coimbra-Filho and R. Mittermeier, pp. 9–27. Rio de Janeiro: Academia Brasiliera de Ciencias.

1983. Tale of tails: parallelism and prehensility. *American Journal of Physical Anthropology* **60**, 103–107.

1992. Evolution of feeding niches in New World monkeys. *American Journal of Physical Anthropology* **88**, 525–562.

Rosenberger, A. L. and Strier, K. B. 1989. Adaptive radiation of the ateline primates. *Journal of Human Evolution* **18**, 717–750.

Rosengart, C. 2001. Techniques, demands, and success in structure construction in tufted capuchin monkeys (*Cebus apella*). Master's thesis, University of Georgia.

Ross, C. 1991. Life history patterns of New World monkeys. *International Journal of Primatology* **12**, 481–502.

Ross, R. A. and Giller, P. S. 1988. Observations on the activity patterns and social interactions of a captive group of black-capped or brown capuchin monkeys (*Cebus apella*). *Primates* **29**, 307–317.

Rowe, N. 1996. *The Pictorial Guide to the Living Primates*. East Hampton, NY: Pogonias Press.

Rowell, T. E. and Mitchell, B. J. 1991. Comparison of seed dispersal by guenons in Kenya and capuchins in Panama. *Journal of Tropical Ecology* **7**, 269–274.

Rubenstein, J., Martinez, S., Shimamura, K. and Puelles, L. 1994. The embryonic vertebrate forebrain: the prosomeric model. *Science* **266**, 578–579.

Rumbaugh, D. and Pate, J. 1984. The evolution of cognition in primates: a comparative perspective. In *Animal Cognition*, ed. L. Roitblat, T. Bever and H. Terrace, pp. 569–587. Hillsdale, NJ: Lawrence Erlbaum Associates.

Russon, A. 2003. Developmental perspectives on great ape traditions. In *The Biology of Traditions: Models and Evidence*, ed. D. Fragaszy and S. Perry, pp. 329–364. Cambridge: Cambridge University Press.

Rylands, A. B. 1999. The name of the weeper or wedge-capped capuchin in the Guianas. *Neotropical Primates* 7, 89–91.

2001. Two taxonomies of the New World primates – a comparison of Rylands *et al.* (2000) and Groves (2001). *Neotropical Primates* 9, 121–124.

Rylands, A. B. and Pinto, L. P. de S. 1994. Macaco-prego-do-peito- amarelo, *Cebus apella xanthosternos* (Wied, 1820). In *Livro Vermelho dos Mamíferos Brasileiros Ameaçados de Extinção*, ed. G. A. B. da Fonseca, A. B. Rylands, C. M. R. Costa, R. B. Machado and Y. L. R. Leite, pp. 219–26. Belo Horizonte: Fundação Biodiversitas.

Rylands, A. B., Fonseca, G. A. B. da, Leite, Y. L. R. and Mittermeier, R. A. 1996. Primates of the Atlantic forest: origin, endemism, distributions and communities. In *Adaptive Radiations of the Neotropical Primates*, ed. M. A. Norconk, A. L. Rosenberger and P. A. Garber, pp. 21–51. New York: Plenum Press.

Rylands, A. B., Mittermeier, R. A. and Rodriguez-Luna, E. 1995. A species list for the New World Primates (Platyrrhini): Distribution by country, endemism, and conservation status according to the Mace-Land System. *Neotropical Primates* 3, 113–160.

Rylands, A. B., Mittermeier, R. A. and Rodriguez-Luna, E. 1997. Conservation of neotropical primates: threatened species and an analysis of primate diversity by country and region. *Folia Primatologica* **68**, 134–160.

Rylands, A. B., Rodríguez-Luna, E. and Cortés-Ortiz, L. 1996/1997. Neotropical primate conservation – The species and the IUCN/SSC Primate Specialist Group network. *Primate Conservation* 17, 46–69.

Rylands, A., van Roosmalen, M. G. M. and Mittermeier, R. A. In press. *Monkeys of the Guianas*. Tropical Field Guide Series. Washington, DC: Conservation International.

Rylands, A. B., Schneider, H., Mittermeier, R. A., Groves, C. P. and Rodriguez-Luna, E. 2000. An assessment of the diversity of New World Primates. *Neotropical Primates* 8, 61–93.

Rylands, A. B., Spironelo, W. R., Tornisielo, V. L., Lemos de Sá, R. M., Kierulff, M. C. M. and Santos, I. B. 1988. Primates of the Rio Jequitinhonha valley, Minas Gerais, Brazil. *Primate Conservation* 9, 100–109.

Sabino, J. and Sazima, I. 1999. Association between fruit-eating fish and foraging monkeys in western Brazil. *Ichthyological Exploration of Freshwaters* **10**, 309–312.

Saito, A., Ueno, Y., Kawamura, S. and Hasegawa, T. 2001. Food search behavior of trichromatic and dichromatic capuchin monkeys (*Cebus apella*). *Advances in Ethology* **36**, 256–257.

Samonds, K. W. and Hegsted, D. M. 1973. Protein requirements of young Cebus monkeys (*Cebus albifrons* and *apella*). *American Journal of Clinical Nutrition* **26**, 30–40.

1978. Protein deficiency and energy restriction in young Cebus monkeys. *Proceeding of the National Academy of Sciences of the United States of America* **75**, 1600–1604.

Santa Cruz, A. M., Borda, J. T., de Rott, M. I. O. and Gomez, L. 2000. Endoparasitosis in captive *Cebus apella*. *Laboratory Primate Newsletter* **39**, 10–12.

Santa Cruz, A. M., Borda, J. T., Gomez, L. and de Rott, M. I. O. 1998. Pulmonary *Filariopsis arator* in capuchin monkeys (*Cebus apella*). *Laboratory Primate Newsletter* **37**, 15–16.

Santos, I. B. and Lernould, J.-M. 1993. 1st meeting of the International Committee for *Cebus apella xanthosternos* and *C. apella robustus*. *Neotropical Primates* **1**, 9–10.

Santos, I. B., Mittermeier, R. A., Rylands, A. B. and Valle, C. M. C. 1987. The distribution and conservation status of primates in southern Bahia, Brazil. *Primate Conservation* **8**, 126–142.

Sanz, V. and Márquez, L. 1994. Conservación del mono capuchino de Margarita (*Cebus apella margaritae*) en la Isla de Margarita, Venezuela. *Neotropical Primates* **2**, 5–8.

Sarnat, H. and Netsky, M. 1981. *Evolution of the Nervous System*. New York: Oxford University Press.

Schneider, H. and Rosenberger, A. L. 1996. Molecules, morphology, and platyrrhine systematics. In *Adaptive Radiations of Neotropical Primates*, ed. M. A. Norconk, A. L. Rosenberger and P. A. Garber, pp. 3–19, 533. New York: Plenum Press.

Schneider, H., Canavez, F. C., Sampaio, I., Moreira, M. A. M., Tagliaro, C. H. and Seuanez, H. N. 2001. Can molecular data place each neotropical monkey in its own branch? *Chromosoma* **109**, 515–523.

Schneider, H., Sampaio, I., Harada, M. L., Barroso, C. M. L., Schneider, M. P. C., Czelusniak, J. and Goodman, M.

1996. Molecular phylogeny of the New World monkeys (Platyrrhini, Primates) based on two unlinked nuclear genes: IRBP intron 1 and e-globin sequences. *American Journal of Physical Anthropology* **100**, 153–179.

Schneider, M. L. 1988. A rhesus monkey model of human infant individual differences. *Dissertation Abstracts International* **B48**, 2804.

Schneider, M. L. and Suomi, S. J. 1992. Neurobehavioral assessment in rhesus monkey (*Macaca mulatta*): developmental changes, behavioral stability, and early experience. *Infant Behavior and Development* **15**, 155–177.

Schoener, T. W. 1971. Theory of feeding strategies. *Annual Reviews of Ecological Systematics* **2**, 369–404.

Schusterman, R., Kastak, C. and Kastak, D. 2003. Equivalence classification as an approach to social knowledge: From sea lions to simians. In *Animal Social Complexity*, ed, F. de Waal and P. Tyack, pp. 179–206. Cambridge, MA: Harvard University Press.

Schwartz, G. G. and Rosenblum, L. A. 1985. Sneezing behavior in the squirrel monkey and its biological significance. *Handbook of Squirrel Monkey Research*, ed. L. A. Rosenblum and C. L. Coe, pp. 253–269. New York: Plenum Press.

Seuanez, H. N., Armada, J. L., Freitas, L., Rocha e Silva, R., Pissinatti, A. and Coimbra-Filho, A. 1986. Intraspecific chromosome variation in *Cebus apella* (Cebidae, Platyrrhini): the chromosomes of the yellow breasted capuchin *Cebus apella xanthosternos* Wied 1820. *American Journal of Primatology* **10**, 237–247.

Seyfarth, R. M. 1977. A model of social grooming among adult female monkeys. *Journal of Theoretical Biology* **65**, 671–698.

Shettleworth, S. 1998. *Cognition, Evolution, and Behavior*. Oxford: Oxford University Press.

Siegler, R. 1998. *Children's Thinking, Third edition*. Upper Saddle River, NJ: Prentice Hall.

Siemers, B. M. 2000. Seasonal variations in food resources and forest strata use by brown capuchin monkeys (*Cebus apella*) in a disturbed forest fragment. *Folia Primatologica* **71**, 181–184.

Silk, J. 1978. Patterns of food sharing among mother and infant chimpanzees at Gombe National Park, Tanzania. *Folia Primatologica* **29**, 129–141.

Silva Jr., J. de S. 2001. Especiação nos macacos-prego e caiararas, gênero *Cebus* Erxleben, 1777 (Primates, Cebidae). Doctoral thesis, Universidade Federal do Rio de Janeiro, Rio de Janeiro.

Silva Jr., J. de S. and Cerqueira, R. 1998. New data and historical sketch on the geographical distribution of the Ka'apor capuchin, *Cebus kaapori* Queiroz, 1992. *Neotropical Primates* **6**, 118–120.

Simmen, B. and Hladik, C. M. 1998. Sweet and bitter taste discrimination in primates: scaling effects across species. *Folia Primatologica* **69**, 129–138.

Simons, D. and Holtkotter, M. 1986. Cognitive process in cebus monkeys when solving problem-box tasks. *Folia Primatologica* **46**, 149–163.

Smith, R. J., and Jungers, W. L. 1997. Body mass in comparative primatology. *Journal of Human Evolution* **32**, 523–559.

Smitsman, A. 1997. The development of tool use: changing boundaries between organism and environment. In *Evolving Explanations of Development*, ed. C. Denton-Read and P. Zukow-Goldring, pp. 301–329. Washington, DC: American Psychological Association.

Sonntag, C. F. 1921. The comparative anatomy of the tongues of the mammalia. IV. Families 3 and 4. Cebidae and hapalidae. *Proceedings of the Zoological Society of London* **1921**, 497–524.

Sorenson, T. C. and Fedigan, L. M. 2000. Distribution of three monkey species along a gradient of regenerating tropical dry forest. *Biological Conservation* **92**, 227–240.

Spence, K. W. 1937 Experimental studies of learning and higher mental processes in infra-human primates. *Psychological Bulletin* **34**, 806–850.

Spinozzi, G. 1989. Early sensorimotor development in cebus (*Cebus apella*). In *Cognitive Structure and Development in Nonhuman Primates*, ed. F. Antinucci, pp. 55–66. Hillsdale, NJ: Lawrence Erlbaum Associates.

Spinozzi, G. and Cacchiarelli, B. 2000. Manual laterality in haptic and visual reaching tasks by tufted capuchin monkeys (*Cebus apella*). An association between hand preference and hand accuracy for food discrimination. *Neuropsychologia* **38**, 1685–1692.

Spinozzi, G. and Natale, F. 1989. Classification. In *Cognitive Structure and Development in Nonhuman Primates*, ed. F. Antinucci, pp. 163–187. Hillsdale, NJ: Lawrence Erlbaum Associates.

Spinozzi, G. and Truppa, V. 1999. Hand preference in different tasks by tufted capuchin monkeys *Cebus apella*. *International Journal of Primatology* **20**, 827–849.

2002. Problem-solving strategies and hand preferences for a multicomponent task by tufted capuchins (*Cebus apella*). *International Journal of Primatology* **23**, 621–638.

Spironelo, W. R. 1991. Importáncia de frutos de palmeiras (*Palmae*) na dieta de um grupo de *Cebus apella* (Cebidae, Primates) na Amazonia central. *A Primatologia no Brasil* **3**, 285–296.

Sponsel, L. E. 1997. The human niche in Amazonia. In *New World Primates. Ecology, Evolution and Behavior*, ed. W. G. Kinzey, pp. 143–165. New York: Aldine de Gruyter.

Stallings, J. D. 1985. Distribution and status of primates in Paraguay. *Primate Conservation* **6**, 51–57.

Stärk, A., Anke, T., Mocek, U. and Steglich, W. 1991. Omphalone, an antiobiotically active benzoquinone derivative from fermentations of Lentinellus omphalodes [1]. *Zeitschrift Naturforschung* **46c**, 989–992.

Staton, V. 1995. Goal directed object manipulation in tufted capuchins (*Cebus apella*). Master's Thesis, University of Georgia.

Stephan, H. 1972. Evolution of primate brains: a comparative anatomical investigation. In *The Functional and Evolutionary Biology of Primates*, ed. R. Tuttle, pp. 155–174. Chicago: Aldine-Atherton.

Stephan, H., Baron, G. and Frahm, H. D. 1988. Comparative size of brains and brain components. In *Comparative Primate Biology, Volume 4: Neurosciences*, ed. H. D. Steklis and J. Erwin, pp. 1–38. New York: Alan R. Liss, Inc.

Stephan, H., Bauchot, R. and Andy, O. J. 1970. Data on size of the brain and of various brain parts in insectivores and primates. *Advances in Primatology* **1**, 289–297.

Stephan, H., Frahm, H. and Baron, G. 1981. New and revised data on volumes of brain structures in insectivores and primates. *Folia Primatologica* **35**, 1–29.

Sterck, E., Watts, D. and van Schaik, C. 1997. The evolution of female social relationships in nonhuman primates. *Behavioral Ecology and Sociobiology* **41**, 291–309.

Steudel, K. 2000. The physiology and energetics of movement: effects on individuals and groups. In *On the Move: How and Why Animals Travel in Groups*, ed. S. Boinski and P. Garber, pp. 9–23. Chicago: University of Chicago Press.

Stevenson, P. R., Quinones, M. J. and Ahumada, J. A. 2000. Influence of fruit availability on ecological overlap among four neotropical primates at Tinigua National Park, Colombia. *Biotropica* **32**, 533–544.

Stewart, M. 1983. *Monkey Shines*. New York: Freundlich Books.

Stoffregen, T. A. and Bardy, B. G. 2001. On specification and the senses. *Behavioral and Brain Sciences* **24**, 195–261.

Stoner, K. E. 1996. Prevalence and intensity of intestinal parasites in mantled howling monkeys (*Alouatta palliata*) in northeastern Costa Rica: implications for conservation biology. *Conservation Biology* **10**, 539–546.

Stott, K. and Selsor, C. J. 1961. Association of trogons and monkeys on Barro Colorado. *The Condor* **63**, 508.

Straus, W. 1929. Studies on primate ilia. *American Journal of Anatomy* **43**, 403–460.

Strier, K. B. 1992. *Faces in the Forest*. New York: Oxford University Press.

1998. Menu for a monkey. In *The Primate Anthology*, ed. R. L. Ciochon and R. A. Nisbett, pp. 180–186. Upper Saddle River, NJ: Prentice Hall.

1999. Why is female kin bonding so rare? Comparative sociality of neotropical primates. In *Comparative Primate Socioecology*, ed. P. C. Lee, pp. 300–319. Cambridge: Cambridge University Press.

Strier, K. B. and Ziegler, T. E. 1997. Behavioral and endocrine characteristics of the reproductive cycle in wild muriqui monkeys, *Brachyteles arachnoides*. *American Journal of Primatology* **42**, 299–310.

Stuart, M. D. and Strier, K. B. 1995. Parasites and primates: a case for a multidisciplinary approach. *International Journal of Primatology* **6**, 577–592.

Stuart, M. D., Greenspan, L. L., Glander, K. E. and Clarke, M. R. 1990. A coprological survey of parasites of wild mantled howling monkeys, *Alouatta palliata palliata*. *Journal of Wildlife Disease* **26**, 547–549.

Stuart, M. D., Pendergast, V., Rumfelt, S., Pierberg, S., Greenspan, L., Glander, K. and Clarke, M. 1998. Parasites of wild howlers (*Alouatta spp*). *International Journal of Primatology* **19**, 493–512.

Stuart, M. D., Strier, K. B. and Pierberg, S. M. 1993. A coprological survey of parasites of wild muriquis, *Brachyteles arachnoides*, and brown howling monkeys, *Alouatta fusca*. *Journal of Helminthological Society of Washington* **60**, 111–115.

Suboski, M. D. 1990. Releaser-induced recognition learning. *Psychological Review* **97**, 271–284.

Sugiyama, Y. 1993. Local variation of tools and tool use among wild chimpanzee populations. In *The Use of Tools by Human and Non-Human Primates*, ed. A. Berthelet and J. Chavaillon, pp. 175–187. Oxford: Clarendon Press.

Susman, R. L. 1998. Hand function and tool behavior in early hominids. *Journal of Human Evolution* **35**, 23–46.

Sussman, R. W. and Phillips-Conroy, J. E. 1995. A survey of the distribution and density of primates of Guyana. *International Journal of Primatology* **16**, 761–791.

Sussman, R. W. and Raven, P. H. 1978. Pollination by lemurs and marsupials: an archaic coevolutionary system. *Science* **200**, 731–736.

Tavares, C. 2002. Biological rhythmicity and cognitive performance in human and nonhuman primates. Unpublished doctoral dissertation, University of São Paulo.

Tavares, M. C. H. and Tomaz, C. 2002. Working memory in capuchin monkeys (*Cebus apella*). *Behavioural Brain Research* **131**, 131–137.

Teaford, M. and Robinson, J. G. 1989. Seasonal or ecological differences in diet and molar microwear in *Cebus nigrivittatus*. *American Journal of Physical Anthropology* **80**, 391–401.

Tejedor, M. F. 1998. The evolutionary history of platyrrhines: old controversies and new interpretations. *Neotropical Primates* **6**, 77–82.

Terborgh, J. 1983. *Five New World Primates. A Study of Comparative Ecology*. Princeton, NJ: Princeton University Press.

Terpstra, A. H., Strucchi, A. F. and Nicolos, R. J. 1991. Estimation of HDL cholesteryl ester kinetic parameters in the Cebus monkey, an animal species with high plasma cholesteryl ester transfer activity. *Atherosclerosis* **88**, 243–248.

Thelen, E. and Smith, L. B. 1994. *A dynamic systems approach to the development of cognition and action*. Cambridge, MA: MIT Press.

Thierry, B., Wunderlich, D. and Gueth, C. 1989. Possession and transfer of objects in a group of brown capuchins (*Cebus apella*). *Behaviour* **110**, 294–305.

Thomas, R. 1986. Vertebrate intelligence: A review of the laboratory research. In *Animal Intelligence. Insights into the animal mind*, ed. R. Hoage and L. Goldman, pp. 37–56. Washington, DC: Smithsonian Institution Press.

Thomas, R. K. 1994. A critique of Nakagawa's "Relational rule learning in the rat". *Psychobiology* **22**, 347–348.

1996. Investigating cognitive abilities in animals: unrealized potential. *Cognitive Brain Research* **3**, 157–166.

Thomas, R. K. and Boyd, M. G. 1973. A comparison of *Cebus albifrons* and *Saimiri sciureus* on oddity performance. *Animal Learning and Behavior* **1**, 151–153.

Thorington Jr., R. W. 1967. Feeding and activity of *Cebus* and *Saimiri* in a Colombian forest. In *Neue Ergebnisse der Primatologie*, ed. D. Starck, R. Schneider and H.-J. Kuhn, pp. 180–184. Stuttgart, Gustav Fischer Verlag. [Reprinted in *Primate Ecology: Problem-Oriented Field Studies*, ed. R. W. Sussman. New York: John Wiley.]

Thorndike, E. L. 1901. The mental life of the monkeys. *Psychological Monographs* **3**, No. 5, pp. 57.

Thorpe, W. H. 1963. *Learning and Instinct in Animals, Second Edition* London: Methuen.

Thurm, D. A., Samonds, K. W. and Fleagle, J. G. 1975. *An Atlas for the Skeletal Maturation of the Cebus Monkey: The First Year*. Boston: Harvard School of Public Health.

Tokida, E., Tanaka, I., Takefushi, H. and Hagiwara, T. 1994. Tool-using in Japanese macaques: use of stones to obtain fruit from a pipe. *Animal Behaviour* **47**, 1023–1030.

Tomasello, M. 1999. *The Cultural Origins of Human Cognition*. Cambridge, MA: Harvard University Press.

Tomasello, M. and Call, J. 1997. *Primate Cognition*. Oxford: Oxford University Press.

Tomasello, M., Gust, D. and Frost G. T. 1989. A longitudinal investigation of gestural communication in young chimpanzees. *Primates* **30**, 35–50.

Tomblin, D. C. and Cranford, J. A. 1994. Ecological niche differences between *Alouatta palliata* and *Cebus capucinus* comparing feeding modes, branch use, and diet. *Primates* **35**, 265–274.

Tonooka, R. and Matsuzawa, T. 1995. Hand preferences of captive chimpanzees (*Pan troglodytes*) in simple reaching for food. *International Journal of Primatology* **16**, 17–35.

Torigoe, T. 1985. Comparison of object manipulation among 74 species of non-human primates. *Primates* **26**, 182–194.

Torres de Assumpção, C. 1981. *Cebus apella* and *Brachyteles arachnoides* as potential pollinators of *Mabea fistulifera* (Euphorbiaceae). *Journal of Mammalogy* **62**, 386–388.

Torres de Assumpção, C. 1983. *An ecological study of the primates of southeastern Brazil, with a reappraisal of* Cebus apella *races*. Doctoral Dissertation, University of Edinburgh.

Torres, C. 1988. Resultados preliminaries de reavaliacao das racas do macaco-prego *Cebus apella* (Primates: Cebidae). *Revista Nordestina Biologia* **6**, 15–28.

Toth, N., Schick, K. D., Savage-Rumbaugh, S., Sevcik, R. A. and Rumbaugh, D. M. 1993. *Pan* the tool-maker: investigations into the stone tool-making and tool-using capabilities of a bonobo (*Pan paniscus*). *Journal of Archaeological Science* **12**, 101–120.

Treves, A. 1998. Primate social systems: conspecific threat and coercion-defense hypotheses. *Folia Primatologica* **69**, 81–88.

Troise, A. 1991. Acquisizione e comprensione dell'uso di strumenti nei bambini: una comparazione con i Primati non umani. Master Thesis. Università di Roma, Facoltà di Magistero.

Tulving, E. 1995. Introduction (chapter VI: Memory). In *The Cognitive Neurosciences*, ed. M. Gazzaniga, pp. 751–753. Cambridge, MA: MIT Press.

Turvey, M. T. 1996. Active touch. *American Psychologist* **51**, 1134–1152.

Turvey, M. T., Shockley, K. and Carello, C. 1999. Affordance, proper function, and the physical basis of perceived heaviness. *Cognition* **73**, B17–B26.

Tutin, C. E. G. and McGinnis, P. R. 1981. Chimpanzee reproduction in the wild. In *Reproductive Biology of the Great Apes: Comparative and Biomedical Perspectives*, ed. C. E. Graham, pp. 239–264. New York: Academic Press.

Ueno, Y. 1994a. Responses to urine odor in the tufted capuchin (*Cebus apella*). *Journal of Ethology* **12**, 81–87.

1994b. Olfactory discrimination of eight food flavors in the capuchin monkey (*Cebus apella*): comparison between fruity and fishy odors. *Primates* **35**, 301–310.

1994c. Olfactory discrimination of urine odors from five species by tufted capuchins (*Cebus apella*). *Primates* **35**, 311–323.

2001. How do we eat? Hypothesis of foraging strategy from the viewpoint of gustation in primates. In *Primate Origins of Human Cognition and Behavior*, ed. T. Matsuzawa, pp. 104–111. Tokyo: Springer Verlag.

Urbani, B. 1998. An early report on tool use by neotropical primates. *Neotropical Primates* **6**, 123–124.

1999a. Nuevo mundo, nuevos monos: sobre primates neotropicales en los siglos XV y XVI. *Neotropical Primates* **7**, 121–125.

1999b. Spontaneous use of tools by wedge-capped capuchin monkeys (*Cebus olivaceus*). *Folia Primatologica* **70**, 172–174.

Valderrama, X., Robinson, J.G., Attygalle, A. B. and Eisner, T. 2000. Seasonal anointment with millipedes in a wild primate: a chemical defense against insect? *Journal of Chemical Ecology* **26**, 2781–2790.

Valderrama, X., Srikosamatara, S. and Robinson, J. G. 1990. Infanticide in wedge-capped capuchin monkeys, *Cebus olivaceus*. *Folia Primatologica* **54**, 171–176.

Valenzano, D. R., and Visalberghi, E. 2002. Facial expressions in tufted capuchins (*Cebus apella*). *Folia Primatologica* **73**, 298.

Valenzuela, N. 1992. Early development of three wild infant *Cebus apella* at La Macarena, Colombia. *Field Studies of New World Monkeys, La Macarena, Colombia* **6**, 15–23.

1993. Social contacts between infants and other group members in a wild *Cebus apella* group at La Macarena, Colombia. *Field Studies of New World Monkeys, La Macarena Colombia* **8**, 1–9.

van Schaik, C. P. 1983. Why are diurnal primates living in groups? *Behaviour* **87**, 120–144.

1989. The ecology of social relationships amongst female primates. In *Comparative Socioecology. The Behavioural Ecology of Humans and Other Mammals*, ed. V. Standen and G. Foley, pp. 195–218. Oxford: Blackwell Press.

2003. Local traditions in orangutans and chimpanzees: social learning and social tolerance. In *Traditions in Nonhuman Animals: Models and Evidence*, ed. D. Fragaszy and S. Perry, pp. 297–328. Cambridge: Cambridge University Press.

van Schaik, C. P. and Janson, C. H. 2000. The behavioral ecology of infanticide by males. In *Infanticide by Males and its Implications*, ed. C. P. van Schaik and C. H. Janson, pp. 469–494. Cambridge: Cambridge University Press.

van Schaik, C. P. and Kappeler, P. M. 1997. Infanticide risk and the evolution of male-female association in primates. *Proceedings of the Royal Society of London Series B* **264**, 1687–1694.

van Schaik, C. P. and van Noordwijk, M. A. 1989. The special role of male *Cebus* monkeys in predation avoidance and its effect on group composition. *Behavioral Ecology and Sociobiology* **24**, 265–276.

van Schaik, C. P., Ancrenaz, M., Borgen, G., Galdikas, B., Knott, C. D., Singleton, I., Suzuki, A., Utami, S. S. and Merrill, M. 2003. Orangutan cultures and the evolution of material culture. *Science* **299**, 102–105.

van Schaik, C. P., Fox, E. A. and Sitompul, A. F. 1996. Manufacture and use of tools in wild Sumatran orangutans. *Naturwissenschaften* **83**, 186–188.

Verbeek, P. and de Waal, F. 1997. Postconflict behavior of captive brown capuchins in the presence and absence of attractive food. *International Journal of Primatology* **18**, 703–726.

Vermeer, J. 2000. La vallée des singes – A new primate zoo in France. *International Zoo News* **47**, 297–300.

Vevers, G. and Weiner, J. 1963. Use of a tool by a captive capuchin monkey (*Cebus apella*). *Symposia of the Zoological Society of London* **10**, 115–118.

Vick, S. J. and Anderson, J. R. 2000. Learning and limits of use of eye gaze by capuchin monkeys (*Cebus apella*) in an object-choice task. *Journal of Comparative Psychology* **114**, 200–207.

Vickers, W. T. 1984. The faunal component of lowland South America hunting kills. *Interciencia* **9**, 366–376.

Vieira, C. da C. 1955. Lista remissiva dos mamíferos do Brasil. *Arquivos de Zoologia*, São Paulo **8**, 341–347.

Visalberghi, E. 1987. Acquisition of nut-cracking behavior by two capuchin monkeys (*Cebus apella*). *Folia Primatologica* **49**, 168–181.

1988. Responsiveness to objects in two social groups of tufted capuchin monkeys (*Cebus apella*). *American Journal of Primatology* **15**, 349–360.

1993a. Capuchin monkeys: A window into tool use activities by apes and humans. In *Tool, Language and Cognition in Human Evolution*, ed. K. Gibson and T. Ingold, pp.138–150. Cambridge: Cambridge University Press.

1993b. Tool use in a South American monkey species: an overview of the characteristics and limits of tool use in *Cebus apella*. In *The Use of Tools by Human and Non-Human Primates*, ed. A. Berthelet and J. Chavaillon, pp. 118–131. Oxford: Clarendon Press.

1994. Learning processes and feeding behavior in monkeys. In *Behavioral Aspects of Feeding. Basic and Applied Research on Mammals*, ed. B. G. Galef, M. Mainardi and P. Valsecchi, pp. 257–270. Chur, Switzerland: Harwood Academic Publishers.

1997. Success and understanding in cognitive tasks: a comparison between *Cebus apella* and *Pan troglodytes*. *International Journal of Primatology* **18**, 811–830.

2000. Tool use behaviour and the understanding of causality in primates. In *Comparer ou Prédire: Exemples de Recherches en Psychologie Comparative Aujourd'hui*, ed. E. Thommen and H. Kilcher, pp. 17–35. Fribourg (Switzerland): Les Editions Universitaires.

Visalberghi, E. and Addessi, E. 2000a. Seeing group members eating a familiar food enhances the acceptance of novel foods in capuchin monkeys. *Animal Behavior* **60**, 69–76.

2000b. Response to changes in food palatability in tufted capuchin monkeys (*Cebus apella*). *Animal Behaviour* **59**, 231–238.

2001. Acceptance of novel foods in *Cebus apella*: do specific social facilitation and visual stimulus enhancement play a role? *Animal Behaviour* **62**, 567–576.

2003. Food for thought: social learning about food in feeding capuchin monkeys. In *The Biology of Traditions*, ed. D. M. Fragaszy and S. Perry, pp. 187–212. Cambridge: Cambridge University Press.

In press. Are our assumptions rational? The impact of social influences on capuchin monkeys' feeding behaviour. In *Rationality in Animals*, ed. M. Nudds and S. Hurley. Oxford: Oxford University Press.

Visalberghi, E. and Anderson, J. R. 1993. Reasons and risks associated with manipulating primates' social environments. *Animal Welfare* **2**, 3–15.

1999. Capuchin monkeys. In *The Universities Federation for the Welfare of Animals Handbook on the Care and Management of Laboratory Animals, Seventh edition, Vol. 1*, ed. T. Poole, pp. 601–610. Oxford: Blackwell.

Visalberghi, E. and Fragaszy, D. M. 1990a. Food-washing behaviour in tufted capuchin monkeys (*Cebus apella*) and crabeating macaques (*Macaca fascicularis*). *Animal Behaviour* **40**, 829–836.

1990b. Do monkeys ape? In *"Language" and Intelligence in Monkeys and Apes*, ed. S. T. Parker and K. R. G. Gibson, pp. 247–273. Cambridge: Cambridge University Press.

1995. The behaviour of capuchin monkeys, *Cebus apella*, with novel foods: the role of social context. *Animal Behaviour* **49**, 1089–1095.

2002. Do monkeys ape? Ten years after. In *Imitation in Animals and Artifacts*, ed. K. Dautenhahn and C. L. Nehaniv, pp. 471–479. Cambridge, MA: MIT Press.

Visalberghi, E. and Guidi, C. 1998. Play behaviour in young tufted capuchin monkeys. *Folia Primatologica* **69**, 419–422.

Visalberghi, E. and Limongelli, L. 1992. Experiments on the comprehension of a tool-use task in tufted capuchin monkeys (*Cebus apella*). Video presented at *XIV Congress of the International Primatological Society*, August 16–21, 1992, Strasbourg, France.

1994. Lack of comprehension of cause-effect relations in tool-using capuchin monkeys (*Cebus apella*). *Journal of Comparative Psychology* **108**, 15–22.

1996. Action and understanding: tool use revisited through the mind of capuchin monkeys. In *Reaching into Thought. The Minds of the Great Apes*, ed. A. Russon, K. Bard and S. Parker, pp. 57–79. Cambridge: Cambridge University Press.

Visalberghi, E. and Moltedo, G. 2001. Sexual behaviour of tufted capuchin monkeys (*Cebus apella*): what affects the target? *Folia Primatologica* **72**, 134.

Visalberghi, E. and Néel, C. 2003. Tufted capuchins (*Cebus apella*) use of weight and sound to choose between full and empty nuts. *Ecological Psychology* **15**, 215–228.

Visalberghi, E. and Tomasello, M. 1998. Primate causal understanding in the physical and in the social domains. *Behavioural Processes* **42**, 189–203.

Visalberghi, E. and Trinca, L. 1989. Tool use in capuchin monkeys: distinguishing between performing and understanding. *Primates* **30**, 511–521.

Visalberghi, E. and Vitale, A. F. 1990. Coated nuts as an enrichment device to elicit tool use in tufted capuchins (*Cebus apella*). *Zoo Biology* **9**, 65–71.

Visalberghi, E. and Welker C. 1986. Sexual behavior in *Cebus apella*. *Anthropology Contemporanea* **9**, 164–165.

Visalberghi, E., Fragaszy, D. M. and Savage-Rumbaugh, S. 1995. Performance in a tool-using task by common chimpanzees (*Pan troglodytes*), bonobos (*Pan paniscus*), and orangutan (*Pongo pygmaeus*), and capuchin monkeys (*Cebus apella*). *Journal of Comparative Psychology* **109**, 52–60.

Visalberghi, E., Janson, C. H. and Agostini, I. 2003. Response towards novel foods and novel objects in wild tufted capuchins (*Cebus apella*). *International Journal of Primatology* **24**, 653–675.

Visalberghi, E., Quarantotti, B. P. and Tranchida, F. 2000. Solving a cooperation task without taking into account the partner's behavior: The case of capuchin monkeys (*Cebus apella*). *Journal of Comparative Psychology* **114**, 297–301.

Visalberghi, E., Sabbatini, G., Stammati, M. and Addessi, E. 2003. Preferences towards novel foods: the role of nutrients and social influences in *Cebus apella*. *Physiology and Behaviour* **80**, 341–349.

Visalberghi, E., Valente, M. and Fragaszy, D. M. 1998. Social context and consumption of unfamiliar foods by capuchin monkeys (*Cebus apella*) over repeated encounters. *American Journal of Primatology* **45**, 367–380.

Vitale, A. F., Visalberghi, E. and De Lillo, C. 1991. Responses to a snake model in captive crab-eating macaques (*Macaca fascicularis*) and captive tufted capuchins (*Cebus apella*). *International Journal of Primatology* **12**, 277–286.

Von Pusch, B. 1941. Die arten der gattung *Cebus*. *Zeitschrift fuer Säugetierkunde* **16**, 183–237.

Wallace, R. B., Painter, R. L. E., Rumiz, D. I. and Taber, A. B. 2000. Primate diversity, distribution and relative abundances in the Rios Blanco y Negro Wildlife Reserve, Santa Cruz Department, Bolivia. *Neotropical Primates* **8**, 24–30.

Wallis, J. 1997. From ancient expeditions to modern exhibitions: the evolution of primate conservation in the zoo community. In *Primate Conservation: the Role of Zoological Parks*, ed. J. Wallis, pp. 1–27. USA: The American Society of Primatologists.

Want, S. and Harris, P. L. 2001. Learning from other people's mistakes. *Child Development* **72**, 431–443.

Ward, J. and Hopkins, W. D. 1993. *Primate Laterality: Current Behavioral Evidence of Primate Asymmetries.* New York: Springer Verlag.

Warden, C., Koch, A. and Field, H. 1940. Instrumentation in cebus and rhesus monkeys. *Journal of Genetic Psychology* **40**, 297–310.

Warkentin, I. G. 1993. Presumptive foraging association between sharp-shinned hawks (*Accipiter striatus*) and white-faced capuchin monkeys (*Cebus capucinus*). (Letter) *Journal of Raptor Research* **27**, 46–47.

Washburn, D. and Rumbaugh, D. 1991. Rhesus monkeys' (*Macaca mulatta*) complex learning skills reassessed. *International Journal of Primatology* **12**, 377–388.

Watanabe, K. 1994. Precultural behavior of Japanese macaques: longitudinal studies of the Koshima troops. In *The Ethological Roots of Culture*, ed. R. A. Gardner *et al.*, pp. 81–94. New York: Kluwer Academic Publishers.

Watson, J. 1908. Imitation in monkeys. *Psychological Bulletin* **5**, 169–178.

Watts, E. 1990. Evolutionary trends in primate growth and development. In *Primate Life History and Evolution*, ed. DeRousseau, C. J., pp. 89–104. New York: Wiley Liss.

Weaver, A. C. 1999. The role of attachment in the development of reconciliation. Doctoral Dissertation, Emory University.

Weaver, A. C. and de Waal, F. B. M. 2000. The development of reconciliation in brown capuchins. In *Natural Conflict Resolution*, ed. F. Aureli and F. B. M. de Waal, pp. 216–218. Berkeley: University of California Press.

Weaver, A. C. and de Waal, F. B. M. 2002. An index of relationship quality based on attachment theory. *Journal of Comparative Psychology* **116**, 93–106.

Weigel, R. M. 1978. The facial expressions of the brown capuchin monkey (*Cebus apella*). *Behaviour* **68**, 250–276.

Weiskrantz, L. and Cowey, A. 1963. Striate cortex lesions and visual acuity of the rhesus monkey. *Journal of Comparative and Physiological Psychology* **56**, 225–231.

Welker, C. 1992. Long-term studies on the social behavior of the capuchin monkeys, *Cebus apella*. In *Perspectives in Primate Biology vol. 4*, ed. P. K. Seth and S. Seth, pp. 9–16. New Delhi: Today and Tomorrow's Publisher.

Welker, C., Hoehmann, H. and Schaefer-Witt, C. 1990. Significance of kin relations and individual preferences in the social behavior of *Cebus apella*. *Folia Primatologica* **54**, 166–170.

Welker, C., Hoehmann-Kroeger, H. and Doyle, G. A. 1992a. Social relations in groups of black-capped monkeys, (*Cebus apella*) in captivity: mother-juvenile relations from the second to the fifth year of life. *Zeitschrift für Saugetierkunde* **57**, 70–76.

Welker, C., Hoehmann-Kroeger, H. and Doyle, G. A. 1992b. Social relations in groups of black-capped capuchin monkeys (*Cebus apella*) in captivity: sibling relations from the second to the fifth year of life. *Zeitschrift für Saugetierkunde* **57**, 269–274.

Welker, C., Pippert, P. and Witt, C. 1983. Die kapuzinerkolonie der Universität Kassel. *Zeitschrift des Kölner Zoo* **26**, 115–126.

Welles, J. F. 1972. The anthropoid hand: A comparative study of prehension. Doctoral Dissertation, Tulane University.

1976. A comparative study of manual prehension in anthropoids. *Saeugetierkundliche Mitteilungen* **24**, 26–38.

Wesley, M. J., Fernandez-Carriba, S., Hostetter, A., Pilcher, D., Poss, S. and Hopkins, W. D. 2002. Factor analysis of multiple measures of hand use in captive chimpanzees: an alternative approach to the assessment of handedness in nonhuman primates. *International Journal of Primatology* **23**, 1155–1168.

Westergaard, G. C. 1995. The stone-tool technology of capuchin monkeys: possible implications for the evolution of symbolic communication in hominids. *World Archaeology* **27**, 1–9.

1999. Structural analysis of tool use by tufted capuchins (*Cebus apella*) and chimpanzees (*Pan troglodytes*). *Animal Cognition* **2**, 141–145.

Westergaard, G. C. and Fragaszy, D. M. 1985. Effects of manipulatable objects on the activity of captive capuchin monkeys (*Cebus apella*). *Zoo Biology* **4**, 317–327.

1987a. The manufacture and use of tools by capuchin monkeys (*Cebus apella*). *Journal of Comparative Psychology* **101**, 159–168.

1987b. Self-treatment of wounds by a capuchin monkey (*Cebus apella*). *Human Evolution* **2**, 557–562.

Westergaard, G. C. and Suomi, S. J. 1993. Use of a tool-set by capuchin monkeys (*Cebus apella*). *Primates* **34**, 459–462.

1994a. A simple stone-tool technology in monkeys. *Journal of Human Evolution* **27**, 399–404.

1994b. The use and modification of bone tools by capuchin monkeys. *Current Anthropology* **35**, 75–77.

1994c. Stone-tool bone-surface modification by monkeys. *Current Anthropology* **35**, 468–470.

1994d. Aimed throwing of stones by tufted capuchin monkeys (*Cebus apella*). *Human Evolution* **9**, 323–329.

1995a. The production and use of digging tools by monkeys: a nonhuman primate model of a hominid subsistence activity. *Journal of Anthropological Research* **51**, 1–8.

1995b. Stone-throwing by capuchins (*Cebus apella*): a model of throwing capabilities in *Homo habilis*. *Folia Primatologica* **65**, 234–238.

1995c. The manufacture and use of bamboo tools by monkeys: possible implications for the development of material culture among east Asian hominids. *Journal of Archeological Science* **22**, 677–681.

1997. Transfer of tools and food between groups of tufted capuchins (*Cebus apella*). *American Journal of Primatology* **43**, 33–41.

Westergaard, G. C., Chavanne, C. L. T. J. and Suomi, S. J. 1998a. Token-mediated tool-use by tufted capuchin monkey (*Cebus apella*). *Animal Cognition* **1**, 101–106.

Westergaard, G. C., Greene, J. A., Babitz, M. A. and Suomi, S. J. 1995. Pestle use and modification by tufted capuchins (*Cebus apella*). *International Journal of Primatology* **16**, 643–651.

Westergaard, G. C., Greene, J. A., Menuhin-Hauser, C. and Suomi, S. J. 1996. The emergence of naturally-occurring copper and iron tools by monkeys: Possible implications for the emergence of metal-tool technology in hominids. *Human Evolution* **11**, 17–25.

Westergaard, G. C., Kuhn, H. E. and Suomi, S. J. 1998a. Bipedal posture and hand preference in humans and other primates. *Journal of Comparative Psychology* **112**, 55–64.

1998b. Effects of upright posture on hand preference for reaching vs. the use of probing tools by tufted capuchins (*Cebus apella*). *American Journal of Primatology* **44**, 147–153.

1998c. Laterality of hand function in tufted capuchin monkeys (*Cebus apella*): comparison between tool use actions and spontaneous non-tool actions. *Ethology* **104**, 119–125.

Westergaard, G. C., Lundquist, A. L., Kuhn, H. E. and Suomi, S. J. 1997. Ant-gathering with tools by captive tufted capuchins (*Cebus apella*). *International Journal of Primatology* **18**, 95–104.

Whiten, A., Custance, D. M., Gomez, J. C., Teixidor, P. and Bard, K. A. 1996. Imitative learning of artificial fruit processing in children (*Homo sapiens*) and chimpanzees (*Pan troglodytes*). *Journal of Comparative Psychology* **110**, 3–14.

Whiten, A., Goodall, J., McGrew, W. C., Nishida, R., Reynolds, V., Sugiyama, Y., Tutin, C. E. G., Wrangham, R. W. and Boesch, C. 1999. Cultures in chimpanzees. *Nature* **399**, 682–685.

Whiten, A., Goodall, J., McGrew, W. C., Nishida, R., Reynolds, V., Sugiyama, Y., Tutin, C. E. G., Wrangham, 2001. Charting cultural variation in chimpanzees. *Behaviour* **138**, 1481–1516.

Wildt, D. E., Doyle, U., Stone, S. C. and Harrison, R. M. (1977). Correlation of perineal swelling with serum ovarian hormone levels, vaginal cytology and ovarian follicular development during the baboon reproductive cycle. *Primates* **18**, 216–270.

Wilen, R. and Naftolin, F. 1978. Pubertal age, weight, and weight gain in the individual female New World monkey (*Cebus albifrons*). *Primates* **19**, 766–773.

Willard, M. J., Dana, K., Stark, L., Owen, J., Zazula, J. and Corcoran, P. 1982. Training a capuchin (*Cebus apella*) to perform as an aide for a quadriplegic. *Primates* **23**, 520–532.

Willard, M. J., Levee, A. and Westbrook, L. 1985. The psychological impact of simian aides on quadriplegics. *Einstein Quarterly Journal of Biology and Medicine* **3**, 104–106.

Wilson, F. R. 1998. *The Hand*. New York: Pantheon Books.

Wolfheim, J. 1983. *Primates of the World. Distribution, Abundance and Conservation*. Seattle: Washington University Press.

Wrangham, R. W. 1980. An ecological model of female-bonded primate groups. *Behaviour* **75**, 262–299.

Wright, E. M. and Bush, D. E. 1977. The reproductive cycle of the capuchin (*Cebus apella*). *Laboratory Animal Science* **27**, 651–654.

Wright, R. V. S. 1972. Imitative learning of a flaked tool technology: the case of an orangutan. *Mankind* **8**, 296–306.

Yamada, E.S., Silveira, L. C. L. and Perry, V. H. 1996. Morphology, dendritic field size, somal size, density, and coverage of M and P retinal ganglion cells of dichromatic Cebus monkeys. *Visual Neuroscience* **13**, 1011–1029.

Yamakoshi, G. 1998. Dietary responses to fruit scarcity of wild chimpanzees at Bossou, Guinea: possible implications for ecological importance of tool use. *American Journal of Physical Anthropology* **106**, 283–295.

Yamakoshi, G. and Myowa-Yamakoshi, M. In press. New observations of anti-dipping techniques in wild chimpanzees at Bossou, Guinea. *Primates*.

Yerkes R. M. 1916. *The Mental Life of Monkeys and Apes: A Study of Ideational Behavior*. Cambridge, MA: H. Holt.

1927a. The mind of a gorilla. *Genetic Psychology Monographs* **2**, 1–193.

1927b. The mind of a gorilla: Part II. Mental development. *Genetic Psychology Monographs* **2**, 375–551.

Yost, J. A. and Kelley, P. M. 1983. Shotguns, blowguns, and spears: the analysis of technological efficiency. In *Adaptive Responses of Native Amazonians*, ed. R. B. Hames and W. T. Vickers, pp. 189–224. New York: Academic Press.

Youlatos, D. 1998. Positional behavior of two sympatric Guianan capuchin monkeys, the brown capuchin (*Cebus apella*) and the wedge-capped capuchin (*Cebus olivaceus*). *Mammalia* **62**, 351–365.

1999. Tail-use in capuchin monkeys. *Neotropical primates* **7**, 16–20.

Zahavi, A. 1977. The testing of a bond. *Animal Behaviour* **25**, 246–247.

Zhang, S. 1995a. Activity and ranging patterns in relation to fruit utilization by brown capuchins (*Cebus apella*) in French Guiana. *International Journal of Primatology* **16**, 489–507.

1995b. Sleeping habits of brown capuchin monkeys (*Cebus apella*) in French Guiana. *American Journal of Primatology* **36**, 327–335.

Zhang, S. and Wang, L.-X. 1995. Fruit consumption and seed dispersal of *Ziziphus cinnamomum* (*Rhamnaceae*) by two sympatric primates (*Cebus apella* and *Ateles paniscus*) in French Guiana. *Biotropica* **27**, 397–401.

Zhang, S.-Y. and Wang, L. 2000. Following of brown capuchin monkeys by white hawks in French Guiana. *The Condor* **102**, 198–201.

Zilles, K., Armstrong, E., Moser, K. H., Schleicher, A. and Stephan, H. 1989. Gyrification in the cerebral cortex of primates. *Brain Behavior and Evolution* **34**, 143–150.

Zuberbühler, K., Gygax, L., Harley, N. and Kummer, H. 1996. Stimulus enhancement and spread of a spontaneous tool use in a colony of long-tailed macaques. *Primates* **37**, 1–12.

Zunino, G. E. 1990. Reproduction and mortality of *Saimiri boliviensis* and *Cebus apella* in captivity. *Boletin Primatologico Latinamericano* **2**, 23–28.

Index

Printed in the United States
By Bookmasters